METHODS IN MOLECULAR BIOLOGY

For further volumes:
http://www.springer.com/series/7651

Electrophoretic Separation of Proteins

Methods and Protocols

Edited by

Biji T. Kurien and R. Hal Scofield

Department of Medicine, University of Oklahoma Health Sciences Center, Oklahoma City, OK, USA; Arthritis and Clinical Immunology, Oklahoma Medical Research Foundation, Oklahoma City, OK, USA; Department of Veterans Affairs Medical Center, Oklahoma City, OK, USA

Humana Press

Editors
Biji T. Kurien
Department of Medicine
University of Oklahoma Health Sciences Center
Oklahoma City, OK, USA

Arthritis and Clinical Immunology
Oklahoma Medical Research Foundation
Oklahoma City, OK, USA

Department of Veterans Affairs Medical Center
Oklahoma City, OK, USA

R. Hal Scofield
Department of Medicine
University of Oklahoma Health Sciences Center
Oklahoma City, OK, USA

Arthritis and Clinical Immunology
Oklahoma Medical Research Foundation
Oklahoma City, OK, USA

Department of Veterans Affairs Medical Center
Oklahoma City, OK, USA

ISSN 1064-3745 ISSN 1940-6029 (electronic)
Methods in Molecular Biology
ISBN 978-1-4939-8792-4 ISBN 978-1-4939-8793-1 (eBook)
https://doi.org/10.1007/978-1-4939-8793-1

Library of Congress Control Number: 2018960267

Cover Caption: Two-dimensional difference gel electrophoresis (see Chapter 20).

This Humana Press imprint is published by the registered company Springer Science+Business Media, LLC, part of Springer Nature.
The registered company address is: 233 Spring Street, New York, NY 10013, U.S.A.

Preface

This volume expands on the first edition of *Protein Electrophoresis* with more focus on electrophoresis methods. In the first edition we made an analogy between laboratory protocols and cooking recipes, and this analogy still seems appropriate to us. To make it a metaphor, research laboratory protocols are like recipes—neither is very explanatory. That is, both assume one already knows what to do. A recipe that directs one to fold one ingredient into another assumes the cook already knows how to fold. Likewise, when a protocol says "remove the supernatant," it rarely tells whether one should decant, aspirate with a pipette, or use a vacuum. In these chapters, each one outlining how to perform a specific electrophoresis technique, we have striven to tell the laboratory scientist exactly what to do. As in the previous volume, because of the specific step-by-step instructions given, we hope that readers will be able to open a chapter on the benchtop and perform the technique. If the instructions seem stupidly simple, then our idea is probably working. We thank everyone, including persons with little bench experience, who reviewed these chapters with this goal in mind. In the last volume we opened with a history of the development of gel electrophoresis by 2007 Nobel Laureate Oliver Smithies, one of the giants of biomedical research in the middle and late twentieth century. Not only did he develop gel electrophoresis in the 1950s but he also developed DNA recombination technology that are the underpinning of knock-out mice and recently potentiated gene therapy. His 1955 paper entitled "Zone electrophoresis in starch gels—Group variations in the serum proteins of normal humans" in volume 61 of *Biochemical Journal* has been cited 3186 times as of 15 November 2017. Since publication of *Protein Electrophoresis: Methods and Protocols* (2012), Doctor Smithies died at age 91. We are saddened by the death of such an important contributor to the biomedical research enterprise, who was still working at the bench until shortly before his death. In addition, our friend, colleague, and coauthor from the previous volume, Jim Fesmire, also died suddenly. He worked for more than four decades at the Oklahoma Medical Research Foundation, the last few years with us. We remembered him fondly while editing this present volume.

Oklahoma City, OK, USA

Biji T. Kurien
R. Hal Scofield

Contents

Contributors

RACHNA AGGARWAL • *Arthritis and Clinical Immunology, Oklahoma Medical Research Foundation, Oklahoma City, OK, USA*

SARAH E. ALTSCHULER • *Department of Biochemistry, UCB 596, University of Colorado Boulder, Boulder, CO, USA*

FABIOLA AMAIR-PINEDO • *Departament de Ciències Fisiològiques, IDIBELL, L'Hospitalet de Llobregat, Universitat de Barcelona, Barcelona, Spain*

CLAUDIA ARNDT • *Institute of Radiopharmaceutical Cancer Research, Department of Radio-/Tumorimmunology, Helmholtz-Zentrum Dresden Rossendorf (HZDR), Dresden, Germany*

MICHAEL BACHMANN • *Institute of Radiopharmaceutical Cancer Research, Department of Radio-/Tumorimmunology, Helmholtz-Zentrum Dresden Rossendorf (HZDR), Dresden, Germany*

RAMON BARTRONS • *Departament de Ciències Fisiològiques, IDIBELL, L'Hospitalet de Llobregat, Universitat de Barcelona, Barcelona, Spain*

RALF BERGMANN • *Institute of Radiopharmaceutical Cancer Research, Department of Radio-/Tumorimmunology, Helmholtz-Zentrum Dresden Rossendorf (HZDR), Dresden, Germany*

MARK D. BIGGIN • *Lawrence Berkeley National Laboratory (LBNL), Berkeley, CA, USA*

MALACHI BLUNDON • *Department of Biological Science, Carnegie Mellon University, Pittsburgh, PA, USA*

ENGELBERT BUXBAUM • *Kevelaer, NRW, Germany*

BRE'ANA BYRD • *Endocrinology Section, Department of Medicine, University of Oklahoma Health Sciences Center, Oklahoma City, OK, USA; Arthritis and Clinical Immunology, Oklahoma Medical Research Foundation, Oklahoma City, OK, USA; Department of Psychology, University of Central Oklahoma, Oklahoma City, OK, USA*

MUSTAFA ÇELEBIER • *Laboratory of Foodomics, CIAL, CSIC, Madrid, Spain; Department of Analytical Chemistry, Faculty of Pharmacy, Hacettepe University, Ankara, Turkey*

MEGAN CHOI • *Lawrence Berkeley National Laboratory (LBNL), Berkeley, CA, USA*

ALEJANDRO CIFUENTES • *Laboratory of Foodomics, CIAL, CSIC, Madrid, Spain*

MARÍA DÁMARIS CORTEZ-DÍAZ • *Chimie ParisTech PSL, Ecole Nationale Supérieure de Chimie, Unité de Technologies Chimiques et Biologiques pour la Santé, Paris, France; CNRS, Unité de Technologies Chimiques et Biologiques pour la Santé UMR 8258, Paris, France; Université Paris Descartes, Unité de Technologies Chimiques et Biologiques pour la Santé, Paris, France; INSERM, Unité de Technologies Chimiques et Biologiques pour la Santé (N° 1022), Paris, France; Departamento de Quimica, Universidad de Guanajuato, Guanajuato, Mexico*

LEAH COSGROVE • *CSIRO, Health and Biosecurity, Adelaide, SA, Australia*

GYORGY CSAKO • *Department of Laboratory Medicine, Clinical Center, National Institutes of Health, Bethesda, MD, USA*

MONICA CUBILLOS-ROJAS • *Departament de Ciències Fisiològiques, IDIBELL, L'Hospitalet de Llobregat, Universitat de Barcelona, Barcelona, Spain*

CHRIS CURSARO • *School of Biological Sciences, University of Adelaide, Adelaide, SA, Australia*

FANNY D'ORLYÉ • *Chimie ParisTech PSL, Ecole Nationale Supérieure de Chimie, Unité de Technologies Chimiques et Biologiques pour la Santé, Paris, France; CNRS, Unité de Technologies Chimiques et Biologiques pour la Santé UMR 8258, Paris, France; Université Paris Descartes, Unité de Technologies Chimiques et Biologiques pour la Santé, Paris, France; INSERM, Unité de Technologies Chimiques et Biologiques pour la Santé (N° 1022), Paris, France*

WILBERT A. DERBIGNY • *Department of Microbiology and Immunology, Indiana University School of Medicine, Indianapolis, IN, USA*

MING DONG • *Lawrence Berkeley National Laboratory (LBNL), Berkeley, CA, USA*

YASER DORRI • *Arthritis and Clinical Immunology, Oklahoma Medical Research Foundation, Oklahoma City, OK, USA; Department of Medicine Endocrinology, University of Oklahoma Health Sciences Center, Oklahoma City, OK, USA; University of Oklahoma, Norman, OK, USA*

GIULIANO ELIA • *Philochem AG, Otelfingen, Switzerland*

ANJA FELDMANN • *Institute of Radiopharmaceutical Cancer Research, Department of Radio-/Tumorimmunology, Helmholtz-Zentrum Dresden Rossendorf (HZDR), Dresden, Germany*

JAMES D. FESMIRE • *Arthritis and Clinical Immunology, Oklahoma Medical Research Foundation, Oklahoma City, OK, USA*

STEPHEN E. FLOWER • *Department of Chemistry, University of Bath, Bath, UK*

KIM Y. C. FUNG • *CSIRO, Health and Biosecurity, Adelaide, SA, Australia*

VINITHA GANESAN • *Department of Biological Science, Carnegie Mellon University, Pittsburgh, PA, USA*

GÜNTER GMEINER • *Doping Control Laboratory, Seibersdorf Labor GmbH, Seibersdorf, Austria*

J. P. DEAN GOLDRING • *Department of Biochemistry, University of KwaZulu-Natal, Scottsville, South Africa*

MARION L. GREASER • *Muscle Biology Laboratory, University of Wisconsin-Madison, Madison, WI, USA*

SHALINI GUPTA • *Protein-Peptide-Metabolite Research Laboratory, Research Resources Center, University of Illinois at Chicago, Chicago, IL, USA*

SYED R. HAIDER • *Department of Chemistry, Centre for Analytical Science, Loughborough University, Loughborough, UK*

HISAO HANIU • *Institute for Biomedical Sciences, Interdisciplinary Cluster for Cutting Edge Research, Shinshu University, Matsumoto, Nagano, Japan*

PETER HOFFMANN • *University of South Australia, Adelaide, SA, Australia*

JIN-SHENG HUANG • *Protein-Peptide-Metabolite Research Laboratory, Research Resources Center, University of Illinois at Chicago, Chicago, IL, USA*

CLARA IBÁÑEZ • *Laboratory of Foodomics, CIAL, CSIC, Madrid, Spain*

TONY D. JAMES • *Department of Chemistry, University of Bath, Bath, UK*

LASANTHI P. JAYATHILAKA • *Protein-Peptide-Metabolite Research Laboratory, Research Resources Center, University of Illinois at Chicago, Chicago, IL, USA*

JIAN JIN • *Lawrence Berkeley National Laboratory (LBNL), Berkeley, CA, USA*

OMAR KASSAAR • *Department of Biology and Biochemistry, University of Bath, Bath, UK*

KRISTI A. KOELSCH • *University of Oklahoma Health Sciences Center, Oklahoma City, OK, USA; Department of Veterans Affairs Medical Center, Oklahoma City, OK, USA; Oklahoma Medical Research Foundation, Oklahoma City, OK, USA*

NAOKA KOMORI • *Department of Biochemistry and Molecular Biology, University of Oklahoma Health Sciences Center, Oklahoma City, OK, USA*
STEFANIE KORISTKA • *Institute of Radiopharmaceutical Cancer Research, Department of Radio-/Tumorimmunology, Helmholtz-Zentrum Dresden Rossendorf (HZDR), Dresden, Germany*
ROBERT G. E. KRAUSE • *Biochemistry, University of KwaZulu-Natal, Scottsville, South Africa*
RAMESH KUMAR • *Department of Microbiology and Immunology, Indiana University School of Medicine, Indianapolis, IN, USA*
BIJI T. KURIEN • *Department of Medicine, University of Oklahoma Health Sciences Center, Oklahoma City, OK, USA; Arthritis and Clinical Immunology, Oklahoma Medical Research Foundation, Oklahoma City, OK, USA; Department of Veterans Affairs Medical Center, Oklahoma City, OK, USA*
BAO-SHIANG LEE • *Protein-Peptide-Metabolite Research Laboratory, Research Resources Center, University of Illinois at Chicago, Chicago, IL, USA*
HON-CHIU EASTWOOD LEUNG • *Department of Molecular and Cellular Biology, Baylor College of Medicine and Texas Children's Hospital, Houston, TX, USA*
TANYA LEWANOWITSCH • *GE Healthcare, Singapore, Singapore*
KAREN A. LEWIS • *Department of Chemistry and Biochemistry, Texas State University, San Marcos, TX, USA*
L. KEVIN LEWIS • *Department of Chemistry and Biochemistry, Texas State University, San Marcos, TX, USA*
SONGPING LIANG • *College of Life Sciences, Hunan Normal University, Changsha, Hunan, China*
DANIEL J. MACPHEE • *One Reproductive Health Research Group, Department of Veterinary Biomedical Sciences, Western College of Veterinary Medicine, University of Saskatchewan, Saskatoon, SK, Canada*
HIROYUKI MATSUMOTO • *Department of Biochemistry and Molecular Biology, University of Oklahoma Health Sciences Center, Oklahoma City, OK, USA; Clinical Proteomics and Gene Therapy Laboratory, Kurume University, Kurume City, Fukuoka, Japan*
BRIAN MCDONAGH • *Department of Physiology, School of Medicine, NUI Galway, Galway, Ireland*
JONATHAN S. MINDEN • *Department of Biological Science, Carnegie Mellon University, Pittsburgh, PA, USA*
EWA I. MISKIEWICZ • *One Reproductive Health Research Group, Department of Veterinary Biomedical Sciences, Western College of Veterinary Medicine, University of Saskatchewan, Saskatoon, SK, Canada*
MOHAMAD REZA MOHAMADI • *Laboratoire Physico Chimie Curie, Institut Curie, PSL Research University, CNRS UMR168, Paris cedex 05, France*
SWAPAN K. NATH • *Arthritis and Clinical Immunology Program, Oklahoma Medical Research Foundation, Oklahoma City, OK, USA*
MARKUS OTTO • *Department of Neurology, University of Ulm, Ulm, Germany*
MARTA P. PEREIRA MORAIS • *Department of Biology and Biochemistry, University of Bath, Bath, UK*
THIERRY RABILLOUD • *Chemistry and Biology of Metals, Univ Grenoble Alpes, CNRS UMR 5249, CEA-Grenoble, Grenoble, France*
JENNIFER A. REAM • *Department of Chemistry and Biochemistry, Texas State University, San Marcos, TX, USA*

BRENDAN REDLER • *Department of Biological Science, Carnegie Mellon University, Pittsburgh, PA, USA*
CHRISTIAN REICHEL • *Doping Control Laboratory, Seibersdorf Labor GmbH, Seibersdorf, Austria*
HELEN J. REID • *Department of Chemistry, Centre for Analytical Science, Loughborough University, Loughborough, UK*
PHILIPP REIHLEN • *Institute of Biochemistry, German Sport University Cologne, Cologne, Germany*
JOSE LUIS ROSA • *Departament de Ciències Fisiològiques, IDIBELL, L'Hospitalet de Llobregat, Universitat de Barcelona, Barcelona, Spain*
RONALD SARASWAT • *Department of Physiology, School of Medicine, NUI Galway, Galway, Ireland*
WILHELM SCHÄNZER • *Institute of Biochemistry, German Sport University Cologne, Cologne, Germany*
R. HAL SCOFIELD • *Department of Medicine, University of Oklahoma Health Sciences Center, Oklahoma City, OK, USA; Arthritis and Clinical Immunology, Oklahoma Medical Research Foundation, Oklahoma City, OK, USA; Department of Veterans Affairs Medical Center, Oklahoma City, OK, USA*
BARRY L. SHARP • *Department of Chemistry, Centre for Analytical Science, Loughborough University, Loughborough, UK*
CAROLINA SIMÓ • *Laboratory of Foodomics, CIAL, CSIC, Madrid, Spain*
BHUPINDER SINGH • *Oklahoma Medical Research Foundation (OMRF), Oklahoma City, OK, USA*
OLIVER SMITHIES • *Department of Pathology and Laboratory Medicine, University of North Carolina School of Medicine, Chapel Hill, NC, USA*
POTHUR R. SRINIVAS • *National Heart, Lung, and Blood Institute, National Institutes of Health, Bethesda, MD, USA*
NOBUAKI TAKEMORI • *Proteomics Core Laboratory, Proteo-Medicine Research Center, Ehime University, Toon City, Ehime, Japan*
IRA N. TARGOFF • *Arthritis and Clinical Immunology, Oklahoma Medical Research Foundation, Oklahoma City, OK, USA; Department of Medicine, University of Oklahoma Health Sciences Center, Oklahoma City, OK, USA; Veterans Affairs Medical Center of Oklahoma City, Oklahoma City, OK, USA*
IRANTZU TATO • *Departament de Ciències Fisiològiques, IDIBELL, L'Hospitalet de Llobregat, Universitat de Barcelona, Barcelona, Spain*
MYRIAM TAVERNA • *PNAS, Institut Galien de Paris-Sud, Faculté de Pharmacie, Université Paris-Sud, CNRS UMR8612, Chatenay Malabry, France*
MARIO THEVIS • *Institute of Biochemistry, German Sport University Cologne, Cologne, Germany*
ROHIT THOMAS • *Arthritis and Clinical Immunology Program, Oklahoma Medical Research Foundation, Oklahoma City, OK, USA; Oklahoma School of Science and Mathematics, Oklahoma City, OK, USA*
HUYEN TRAN • *Endocrinology Section, Department of Medicine, University of Oklahoma Health Sciences Center, Oklahoma City, OK, USA; Arthritis and Clinical Immunology, Oklahoma Medical Research Foundation, Oklahoma City, OK, USA; Department of Psychology, University of Central Oklahoma, Oklahoma City, OK, USA*
EDWARD P. TRIEU • *Arthritis and Clinical Immunology, Oklahoma Medical Research Foundation, Oklahoma City, OK, USA*

HENGHANG TSAI • *Department of Molecular and Cellular Biology, Baylor College of Medicine and Texas Children's Hospital, Houston, TX, USA*
PHU T. VAN • *Department of Biological Science, Carnegie Mellon University, Pittsburgh, PA, USA*
JEAN M. H. VAN DEN ELSEN • *Department of Biology and Biochemistry, University of Bath, Bath, UK*
ANNE VARENNE • *Chimie ParisTech PSL, Ecole Nationale Supérieure de Chimie, Unité de Technologies Chimiques et Biologiques pour la Santé, Paris, France; CNRS, Unité de Technologies Chimiques et Biologiques pour la Santé UMR 8258, Paris, France; Université Paris Descartes, Unité de Technologies Chimiques et Biologiques pour la Santé, Paris, France; INSERM, Unité de Technologies Chimiques et Biologiques pour la Santé (N° 1022), Paris, France*
FRANCESC VENTURA • *Departament de Ciències Fisiològiques, IDIBELL, L'Hospitalet de Llobregat, Universitat de Barcelona, Barcelona, Spain*
ROMAIN VERPILLOT • *PNAS, Institut Galien de Paris-Sud, Faculté de Pharmacie, Université Paris-Sud, CNRS UMR8612, Chatenay Malabry, France*
JEAN-LOUIS VIOVY • *Laboratoire Physico Chimie Curie, Institut Curie, PSL Research University, CNRS UMR168, Paris cedex 05, France*
XIANCHUN WANG • *College of Life Sciences, Hunan Normal University, Changsha, Hunan, China*
CHAD M. WARREN • *Department of Physiology and Biophysics and Center for Cardiovascular Research, University of Illinois at Chicago, Chicago, IL, USA*
ROBERT J. WILLIAMS • *Department of Biology and Biochemistry, University of Bath, Bath, UK*
DEBORAH S. WUTTKE • *Department of Biochemistry, UCB 596, University of Colorado Boulder, Boulder, CO, USA*

Chapter 1

How It All Began: A Personal History of Gel Electrophoresis

Oliver Smithies

Abstract

Arne Tiselius' moving boundary electrophoresis method was still in general use in 1951 when this personal history begins, although zonal electrophoresis with a variety of supporting media (e.g., filter paper or starch grains) was beginning to replace it. This chapter is an account of 10 years of experiments carried out by the author during which molecular sieving gel electrophoresis was developed and common genetic variants of two proteins, haptoglobin and transferrin, were discovered in normal individuals. Most of the figures are images of pages from the author's laboratory notebooks, which are still available, so that some of the excitement of the time and the humorous moments are perhaps apparent. Alkaline gels, acidic gels with and without denaturants, vertical gels, two-dimensional gels and gels with differences in starch concentration are presented. The subtle details that can be discerned in these various gels played an indispensable role in determining the nature of the change in the haptoglobin gene (*Hp*) that leads to the polymeric series characteristic of Hp^2/Hp^2 homozygotes. Where possible the names of scientific friends who made this saga of gel electrophoresis so memorable and enjoyable are gratefully included.

Key words Insulin, Filter paper, Starch gel, Serum proteins, Haptoglobin, Transferrin, Vertical gel, Two-dimensional gel, Denaturing gel

1 Introduction

Life with electrophoresis began for me almost 60 years ago as a post-doctoral fellow in the laboratory of Dr. Jack Williams, a physical chemist at the University of Wisconsin, Madison. His lab had a state-of-the-art Tiselius apparatus in which the electrophoretic migration of proteins was followed at the two boundaries formed by a U-tube of protein solution with buffer above the two vertical arms (Fig. 1, blue inset box). Arne Tiselius' development of the method [1], and his use of it to show four components in serum (albumin, and α-, β-, and γ-globulin) earned him a Nobel Prize in 1948. A particularly exciting example of the power of the method was its use by Linus Pauling and his colleagues [2] in defining sickle cell anemia as a molecular disease. The method, however, had two

Biji T. Kurien and R. Hal Scofield (eds.), *Electrophoretic Separation of Proteins: Methods and Protocols*, Methods in Molecular Biology, vol. 1855, https://doi.org/10.1007/978-1-4939-8793-1_1,

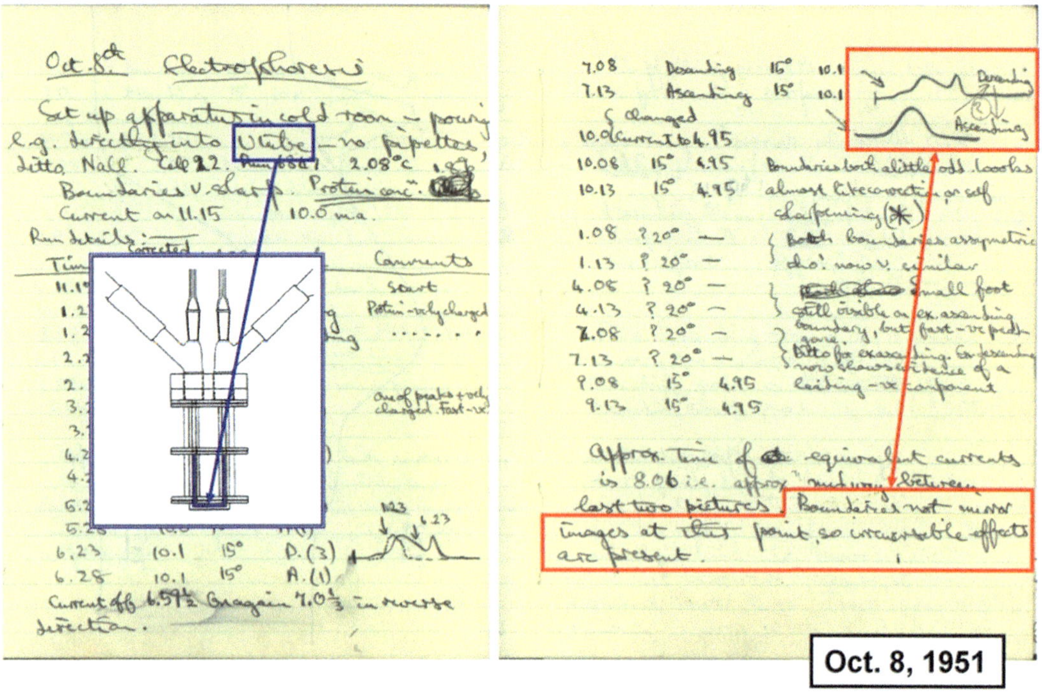

Fig. 1 Reproduced with permission from the Nobel Foundation

substantial disadvantages. First, because it was a boundary method, separations were overlapping rather than complete:

[A ···> [A + B ···> [A + B + C ··· U-tube ··· A + B + C] ···> B + C] ···> C].

Secondly, because the boundaries were detected by refractive index changes, the method required a cumbersome optical system and relatively large amounts of protein, of the order of half a gram or more.

The entry in my laboratory notebook (*see* Fig. 1), made with a fountain pen in 1951 before the invention of ballpoint pens, records that my first use of the Tiselius electrophoresis method was not very encouraging. I was trying to determine the purity of a recrystallized preparation of a protein, β-lactoglobulin, which I had made from cow's milk, but the ascending and descending boundaries were not mirror images, as they should be (Fig. 1 red boxes). I struggled with the boundary anomaly, off and on, for 2 years but never could get understandable results [3].

My next electrophoresis problem arose during the early days of my first job, which was with Dr. David A. Scott at the Connaught Laboratories, Toronto, Canada. Scott was the first person to crystallize insulin, as a zinc salt, and show its value as a slow release form of the hormone. Not surprisingly, he suggested that I should pick a research task related to insulin. After learning that insulin was

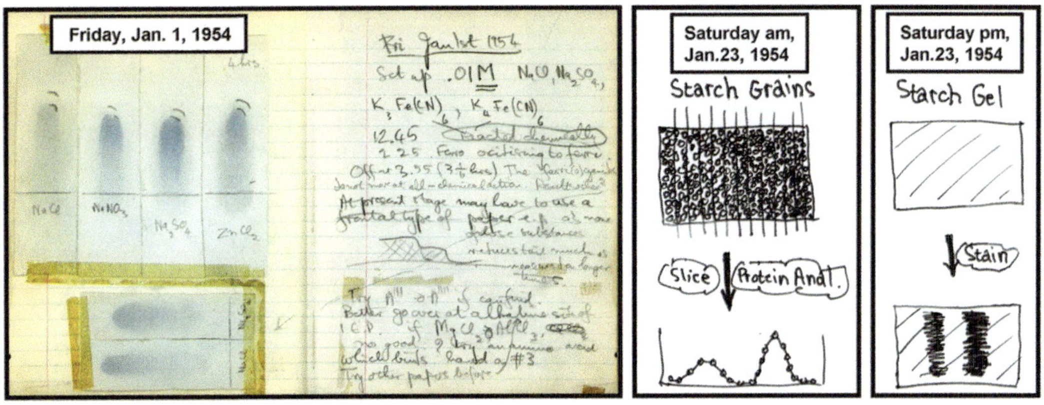

Fig. 2 Reproduced with permission from the Nobel Foundation

typically prepared by extracting pancreases with 70% alcohol containing 2% concentrated hydrochloric acid, I thought that insulin prepared with such a harsh procedure might be a breakdown product. I therefore chose, as my task, to look for a precursor.

In preparation for an expected need to distinguish the postulated precursor from insulin as normally prepared, I tried a new form of electrophoresis that had recently been championed by Kunkel and Tiselius [4]. The method separated proteins on buffer-soaked filter paper. It was zonal and therefore could separate proteins completely:

$$[A + B + C] \cdots> [A] \cdots> [B] \cdots> [C].$$

It also had the advantage of requiring very little protein (<1 mg). But when I used it, my New Year's Day experiment (Fig. 2, left panel) revealed one of its disadvantages – some proteins (including insulin) stick to filter paper and unroll like a carpet.

Kunkel was well aware of the absorption and trailing problem, and he and Slater [5] sought and found a supporting matrix that would be free from it. Their preferred matrix was a barely moist bed of potato starch grains. Proteins migrated in the buffer *surrounding* the impervious starch grains. With their method they could easily separate the five proteins of normal serum (by this time α-globulin had been subdivided into α_1 and α_2).

I heard that Dr. Andrew Sass-Kortsak, a scientific neighbor working at the Toronto Hospital for Sick Children, was using the method, and I visited his lab to see it in action on Saturday, January 23, 1954. The freedom from absorption was spectacular, but the labor required for each run was overfacing; Folin protein determinations of 40 segments cut from the moist starch-grain bed were needed to detect the migrating proteins and develop an electrophoretic pattern (Fig. 2 center panel). Impossible for me, I didn't even have someone to wash my lab glassware! But I remembered

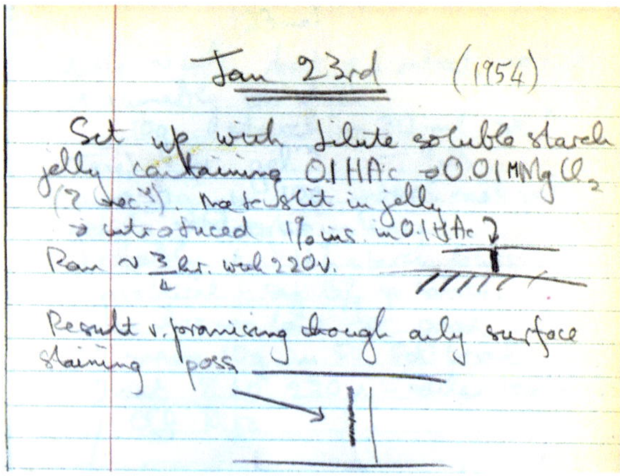

Saturday Jan.23rd, 1954

Fig. 3 Reproduced with permission from the Nobel Foundation

"helping" my mother do the laundry as a child, and noticing that the hot gooey liquid she made to starch the collars of my father's shirts set into a gel when left standing. Perhaps, if I boiled a starch suspension and allowed it to cool, proteins could migrate *through* the resulting jelly and could be detected by staining, in the way that was used to detect proteins in filter-paper electrophoresis (Fig. 2, right panel).

Back at Connaught that Saturday afternoon, I found a bottle of "Starch according to Lintner." It made a gel (jelly); insulin migrated through the gel without trailing; it could be detected by simply staining the gel, and no protein determinations were needed (Fig. 3). I had invented starch-gel electrophoresis!

Over the next few months, I worked at improving my experiments with insulin and helped my friend Gordon H. Dixon (a graduate student in the Department of Biochemistry, University of Toronto) use the starch-gel system to separate cabbage enzymes [6]. Then 1 day, "just for rough test," I tried electrophoresing serum (Fig. 4, left page). The results were promising, and after only a few more days of work, I was finding at least nine components in serum (Fig. 4, right page), although I was unsure how to label them (Fig. 4, red box and arrow, indicating an erasure).

This was new territory, and (with Dr. Scott's approval) I stopped working on insulin and settled down to making the method reproducible for separating serum proteins. Within a few months, I was ready to publish and had two pretty gels as examples with serum samples from Gordon (GD) and another graduate student friend, George Connell (Fig. 5, left panel). However, for no special reason, I ran another sample before having photographs made (not easy because I had no camera). The serum was from

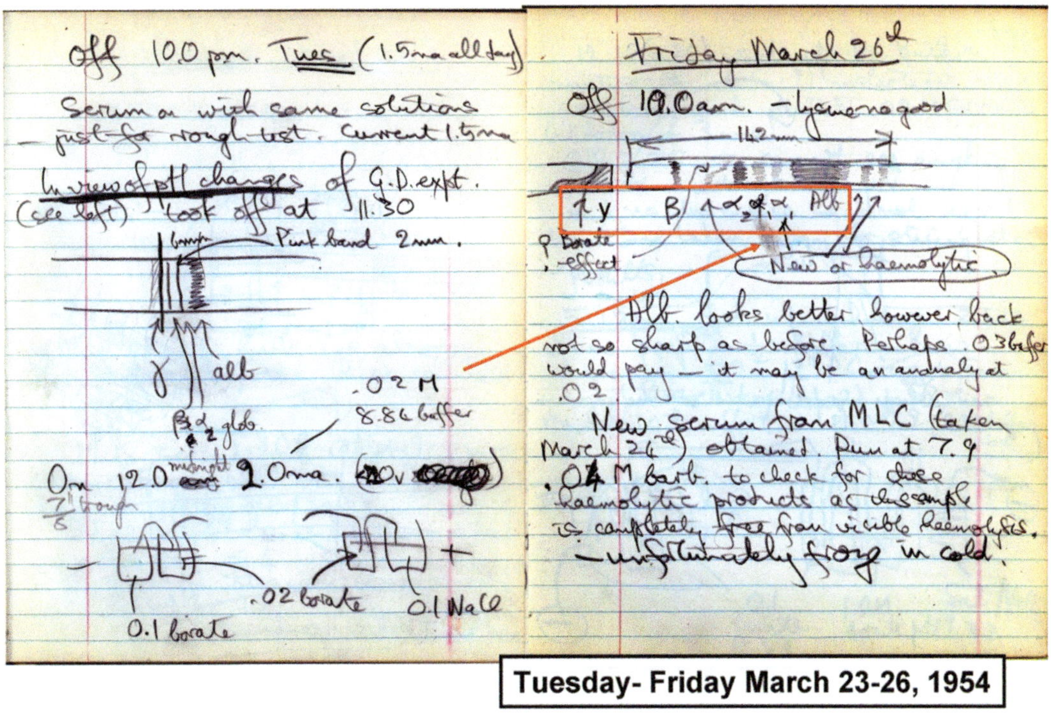

Fig. 4 Reproduced with permission from the Nobel Foundation

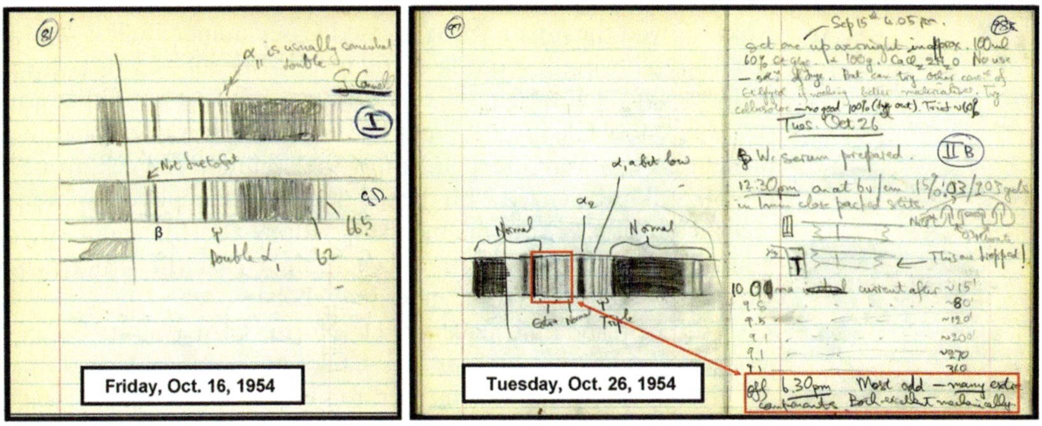

Fig. 5 Reproduced with permission from the Nobel Foundation

B. W. (Beth Wade), who was a young woman working as a technician in Charles Hanes' lab where Gordon and George were students. The results (Fig. 5, right panel) were "Most odd – many extra components." I thought I had found a male/female difference and a new way of telling men from women! Testing serum from another male and another female for each of the next few days appeared to confirm that there was a male pattern (M) and a female

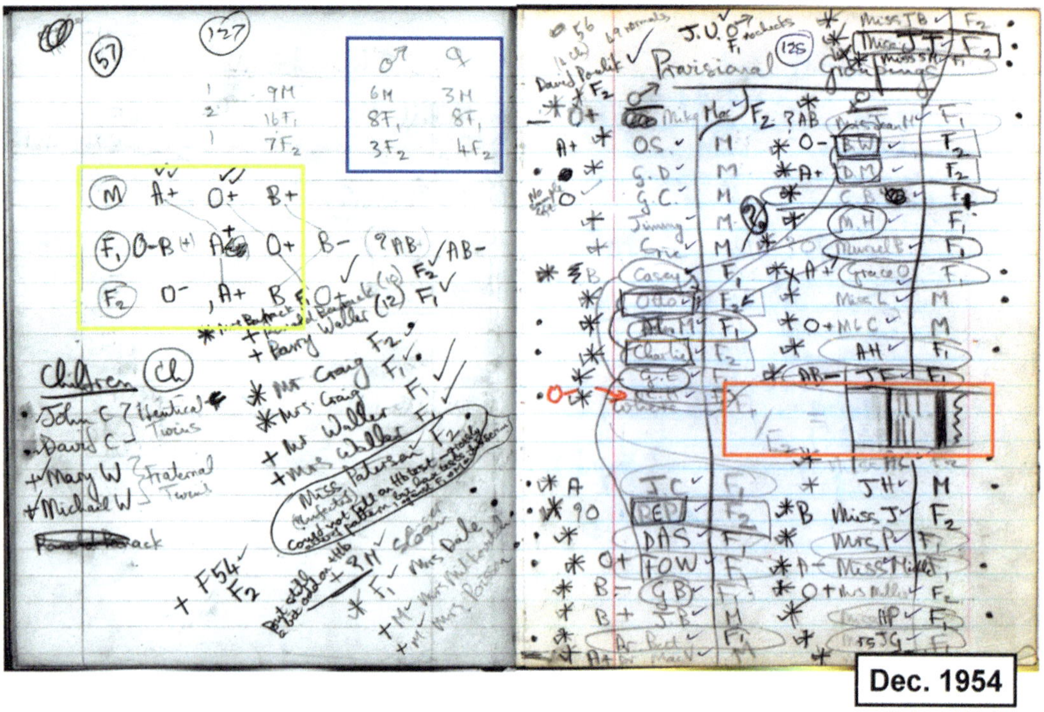

Fig. 6 Reproduced with permission from the Nobel Foundation

pattern (F). But very soon the serum from a male colleague (Casey Kok) showed the F pattern. We teased Casey unmercifully.

The following 2 months of intensive work showed that there really was no gender effect (Fig. 6 blue box), that the F pattern could be subdivided into F_1 and F_2 (Fig. 6 red box), and that the protein groups were independent of the common red cell groups (Fig. 6, yellow box). So, with the gender notation replaced by more realistic albeit less entertaining designations (I, IIA, and IIB), I published my method, with the comment that the differences would probably prove to be inherited, [7, 8].

With help and guidance from Dr. Norma Ford-Walker (Genetics Department, Hospital for Sick Children, Toronto, Canada), a delightful collaborator and teacher, this comment was quickly confirmed – the differences were indeed genetically determined [9]. Shortly thereafter, Uppsala Helge Laurell, University, pointed out to us that the protein with the inherited differences, which by then we knew was a hemoglobin-binding protein, was probably haptoglobin, a hemoglobin-binding protein already described by Jayle et al. [10]. Norma and I agreed, and we re-named the three serum groups Hp1-1, Hp2-1, and Hp2-2, with the genotypes being Hp^1/Hp^1, Hp^2/Hp^1 and Hp^2/Hp^2 [11].

Following this introduction to genetics, on the urging of another Connaught scientist, David M. Poulik, who subsequently

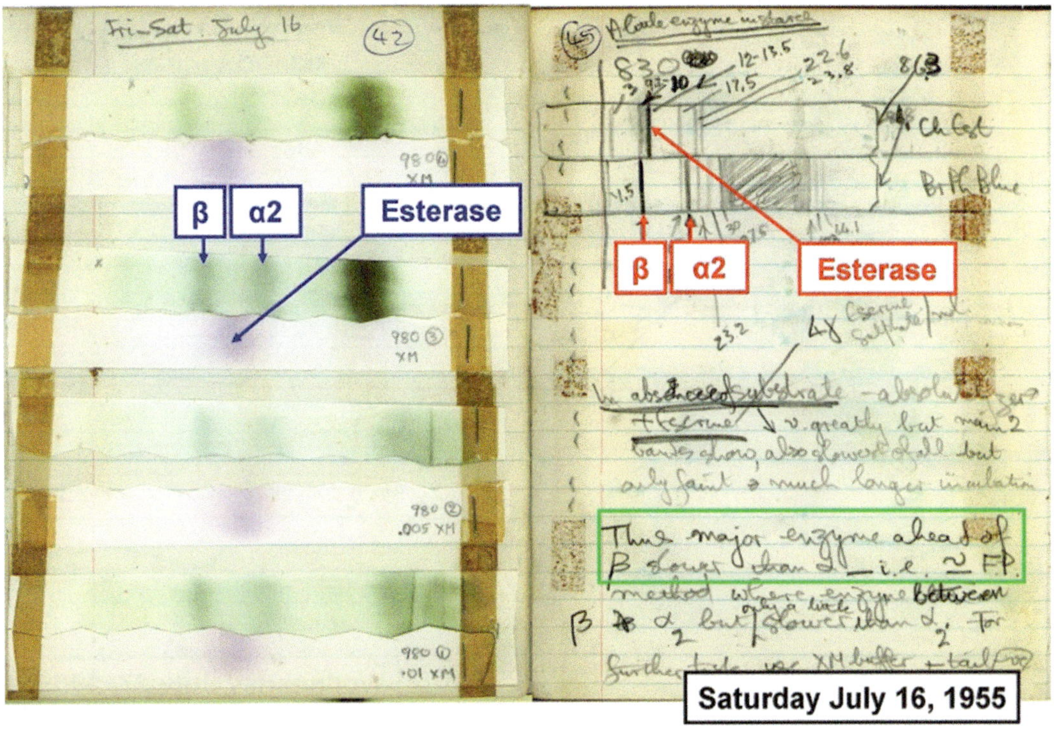

Fig. 7 Reproduced with permission from the Nobel Foundation

became a lifelong friend, I took up the task of correlating the serum components seen in starch gels with those seen on filter paper. Dave had been looking at serum samples from hospital patients using filter-paper electrophoresis and had begun switching to starch gel. As a result he was particularly keen to compare the two systems. He suggested that I should try a test with alpha naphthyl acetate, a chromogenic substrate which he had been using to detect esterases on filter paper. When I did, the results confirmed his finding that esterase activity was detectable on filter paper as a single zone between α_2- and β-globulin (Fig. 7, filter-paper nomenclature blue). In the starch gel, several enzyme bands were detectable, with the main one migrating between the bands in serum which I thought were α_2-, and β-globulins (Fig. 7, provisional starch-gel nomenclature red). I concluded that the two methods agreed (Fig. 7, green box).

But my next test quickly demonstrated that this conclusion was unlikely to be correct. For this test, I cut pieces expected to include α_1-, α_2- and β-globulin from two filter-paper electrophoresis strips that had been run with Dave's serum, inserted each piece into a starch gel alongside a second piece of filter paper soaked in serum, and ran the gel electrophoresis. My July 19, 1955, notebook entry indicates my surprise – "β (fp) – lin[es] up with α_2". This was the beginning of a realization that gel separations differ qualitatively

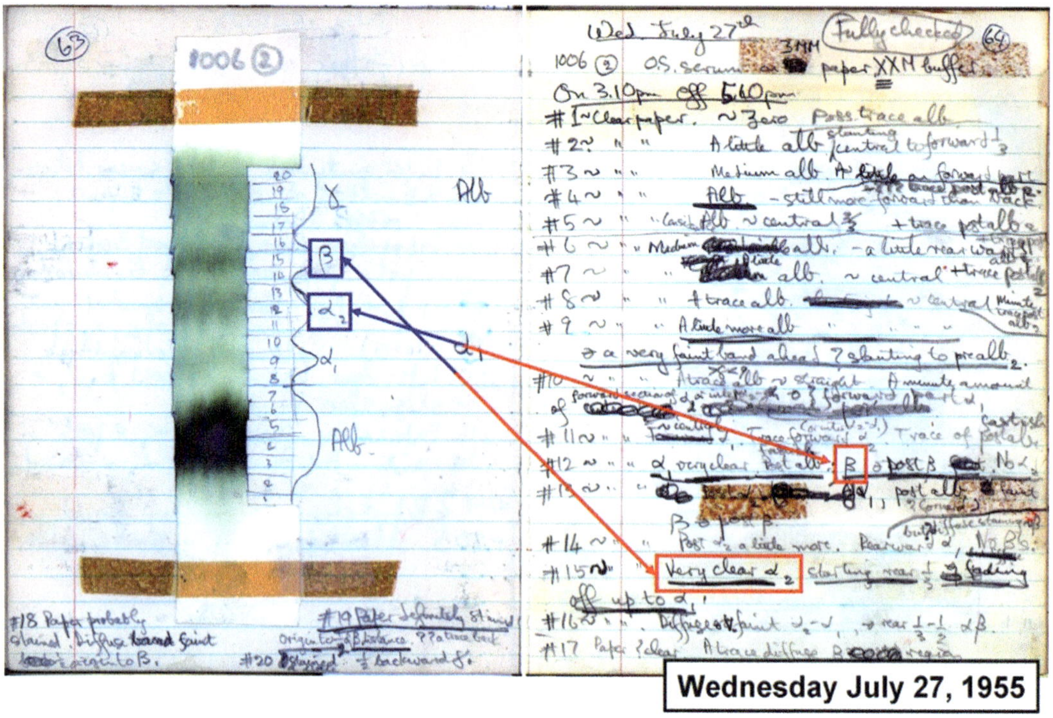

Fig. 8 Reproduced with permission from the Nobel Foundation

from separations in free solution or in buffer surrounding an impervious matrix.

Within a few days, I expanded the comparison experiments to cover all parts of the filter-paper electrophoresis strip. This involved cutting out 20 consecutive transverse pieces from the filter-paper electrophoresis strip (Fig. 8, left page) and running 20 gels to compare the proteins in each piece with whole serum inserted alongside in the same slot in the gel using a second piece of filter paper soaked in serum (Fig. 8, right page). The result confirmed that the gel band which I had provisionally labeled α_2 was in fact a β-globulin and showed that the gel band which I had provisionally labeled β was an α_2-globulin. Dave and I repeated these laborious multi-gel experiments many times. All led to the same conclusion—the two electrophoresis systems were qualitatively different. Filter-paper mobilities approximate those in free solution; the high concentration of starch used in the gels markedly enhances the retarding effects on mobility of increasing molecular size.

However, we were still having a problem in identifying the filter-paper equivalents of the starch-gel pre-albumins. In attempting to overcome this problem, I cut the filter paper *longitudinally* instead of transversely, inserted the forward part of the filter-paper strip into a wide starch gel, and ran the second electrophoresis *at right angles* to the first (Fig. 9, lower part of left page and upper part

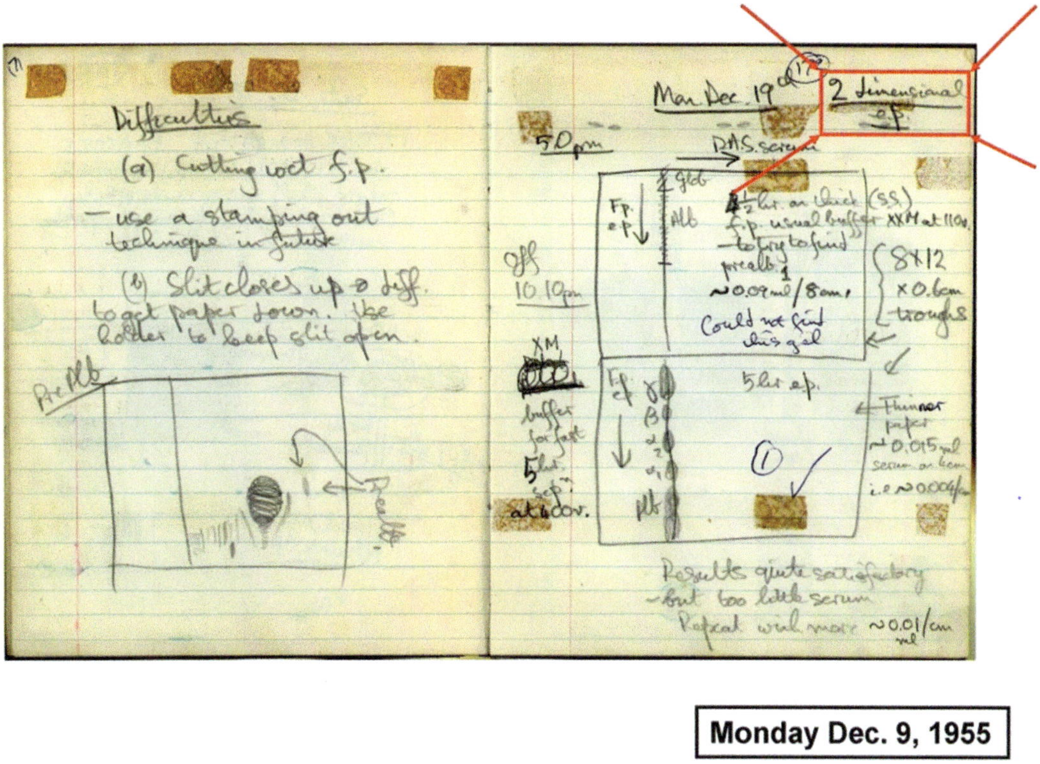

Fig. 9 Reproduced with permission from the Nobel Foundation

of right page). This 90° turn was, I realized, all that was required to have a two-dimensional gel electrophoresis system (Fig. 9, red box), and I tried it immediately with a complete longitudinal strip cut from a still wet filter-paper run (Fig. 9, lower part of right page). [Later, I used a narrow strip of filter paper for the first electrophoresis to avoid one of the difficulties that I noted – "(a) cutting wet f.p."]

The moment when I thought of the 90° turn is one that remains vivid in my memory; I can still remember where I was standing! The result was that, instead of 5 proteins in serum, Dave and I could find 20 [12, 13], at that time a very pleasing number, although it pales in comparison to the several hundred protein spots that are now detected in serum with the two-dimensional O'Farrell-type system [14] which combines iso-electric focusing with Laemmli SDS electrophoresis [15].

The filter-paper/starch-gel two-dimensional system proved very useful for studying an unusual variant first found by Ralph Engle and Jim Pert (New York Hospital, New York) in serum from an African-American which they sent to me in 1956 for characterization. My November 1, 1956, two-dimensional gel (Fig. 10), with the first electrophoresis now on a narrow filter paper strip, gave a "V[ery] good result. Shows that the extra zone is a slightly slow β".

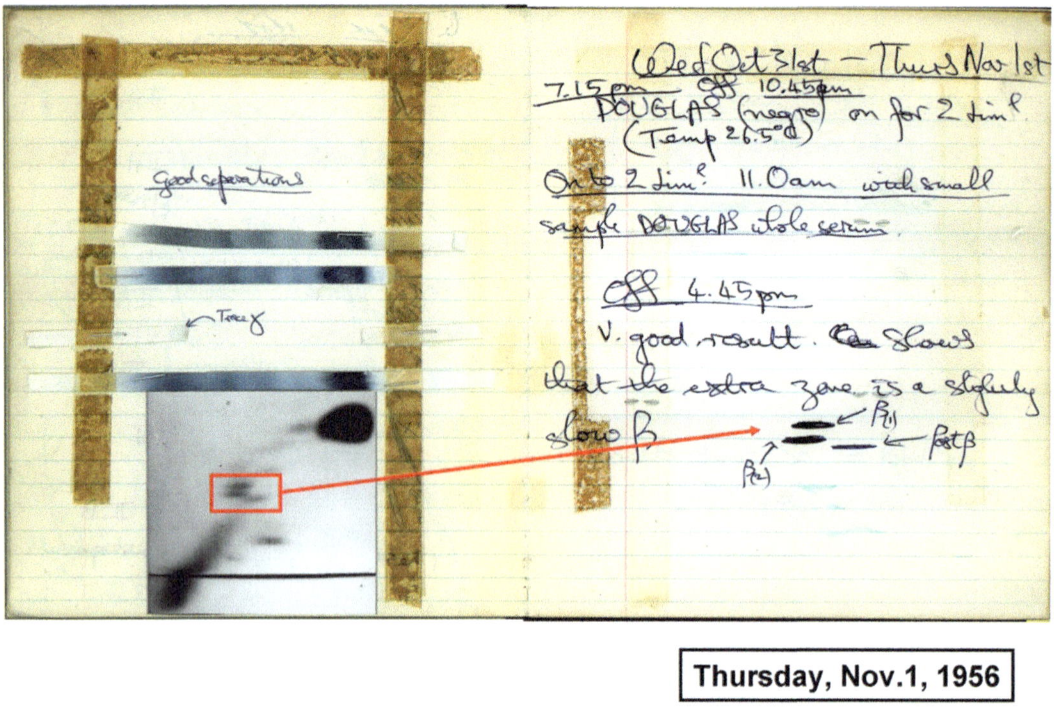

Fig. 10 Reproduced with permission from the Nobel Foundation

Elo Giblett, Charlie Hickman, and I subsequently identified the variable β-globulin as transferrin [16]. Elo, who ran the King County Central Blood Bank, Seattle, was an ardent blood grouper and became one of the most successful discoverers of new haptoglobin and transferrin variants. Charlie was as keen when it came to cattle [17].

The β-globulin variant first seen in an African-American turned up again in some samples from Australian aborigines that were sent to me by Bill Horsfall, Commonwealth Health Laboratories, Queensland, Australia [18]. However, the full blown two-dimensional method was cumbersome for the family studies needed to establish the mode of inheritance. So, in its place, my technician (I now had one, Otto Hiller) and I used a cutdown version of the two-dimensional system. Only the β-globulin-containing region of the narrow filter-paper electrophoresis strip was used, and it could be inserted into a normal-sized one-dimensional starch gel for the second electrophoresis at right angles. With this method we could test 16 samples a day (Fig. 11, white boxes indicate β-globulin variants), and we used it to study the many samples that Bill collected from his aborigine friends. They were, he explained, very interested and willing to take part in family studies but only occasionally came "out of the bush." Together we proved that the β-globulin variants were inherited [19].

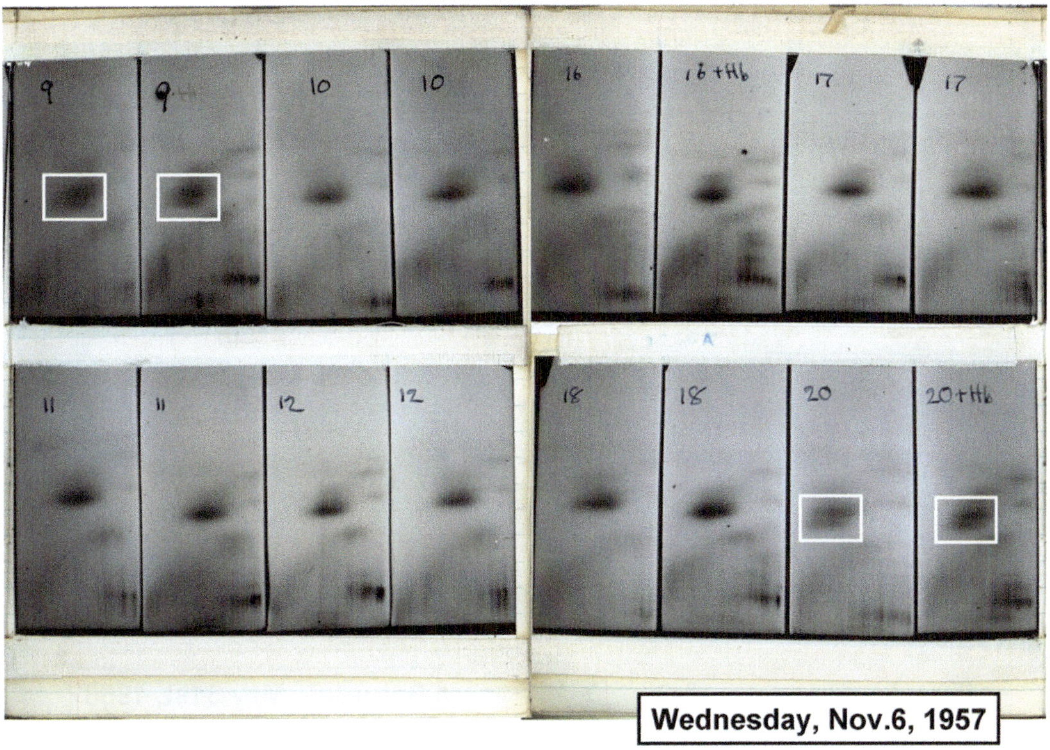

Fig. 11 Reproduced with permission from the Nobel Foundation

Meanwhile, I was unhappy with any of the several methods that I had tried for inserting samples into the gels, which were all at that time being run horizontally. A piece of filter paper soaked in the sample was adequate for many purposes, but absorption was often a problem. A suspension of starch grains was good but cumbersome. A simple opening in the gel was plagued with edge effects, and electrodecantation occurred as the proteins migrated across the opening in the gel. Eventually I realized the now obvious solution—run the gel *vertically* with open slots [20]. The problems of absorption and electrodecantation completely disappeared, and the full resolving power of the gels was achieved. I illustrate the beauty of these early vertical starch gels with two runs. The first (Fig. 12, left page) is a VAC (vertical air cooled) gel, run downward on May 31, 1958. This gel compares serum samples having the previously described β-globulin variants with a serum sample having a new faster migrating variant that Otto had found after running over 400 repetitions of the cutdown version of the two-dimensional system [21, 22]. The white box in the figure outlines the β-globulin variants. The resolving power of the vertical starch gel is well illustrated by the easily seen polymers of haptoglobin apparent in this air-cooled gel (Fig. 12, blue box). [I also tried water cooling, but the improvement was not worth the added complexity for pursuing the studies that most interested me.]

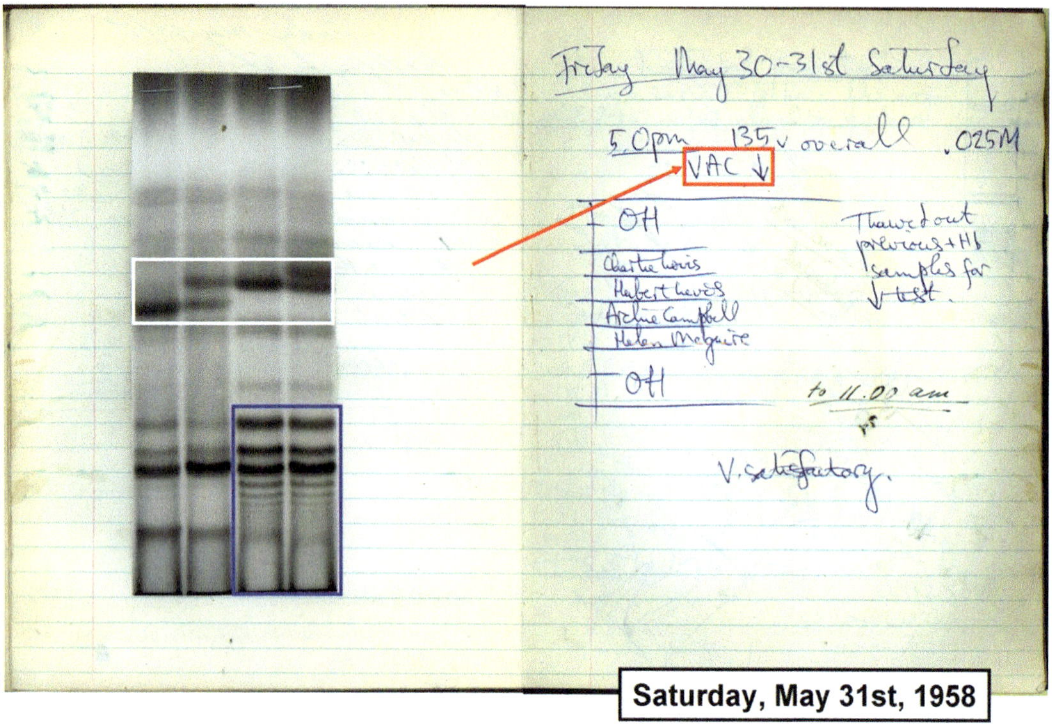

Fig. 12 Reproduced with permission from the Nobel Foundation

The second gel (Fig. 13), run a few days later on June 4, 1958, compares George Connell's purified samples of the four inherited types of haptoglobin known at that time [23]. At least ten increasingly more polymeric forms of haptoglobin are resolved in this vertical starch gel. The right page shows my attempts to find a relationship between their mobilities and the band number (*n*).

The molecular basis underlying the differences between the haptoglobin variants was of considerable interest, because a single protein species in the Hp^1/Hp^1 homozygotes was replaced by a polymeric series in the Hp^2/Hp^2 homozygotes, and the heterozygotes had haptoglobins that were not the same as a mixture of the haptoglobins from the homozygotes. Despite this complexity, the problem seemed likely to be approachable using the methods that Vernon Ingram had developed in working out the difference between normal and sickle cell hemoglobin—a glutamic acid residue in the normal β-globin polypeptide was replaced by valine in the sickle cell β-globin [24]. My graduate student friends, George E. Connell and Gordon H. Dixon, were now back at the University of Toronto as assistant professors, and together we decided to tackle the haptoglobin problem. My task was to try to find some procedure that would convert the complex polymeric haptoglobins into simpler products that would better mirror the simplicity of their inheritance [25]. George's tasks included scaling up

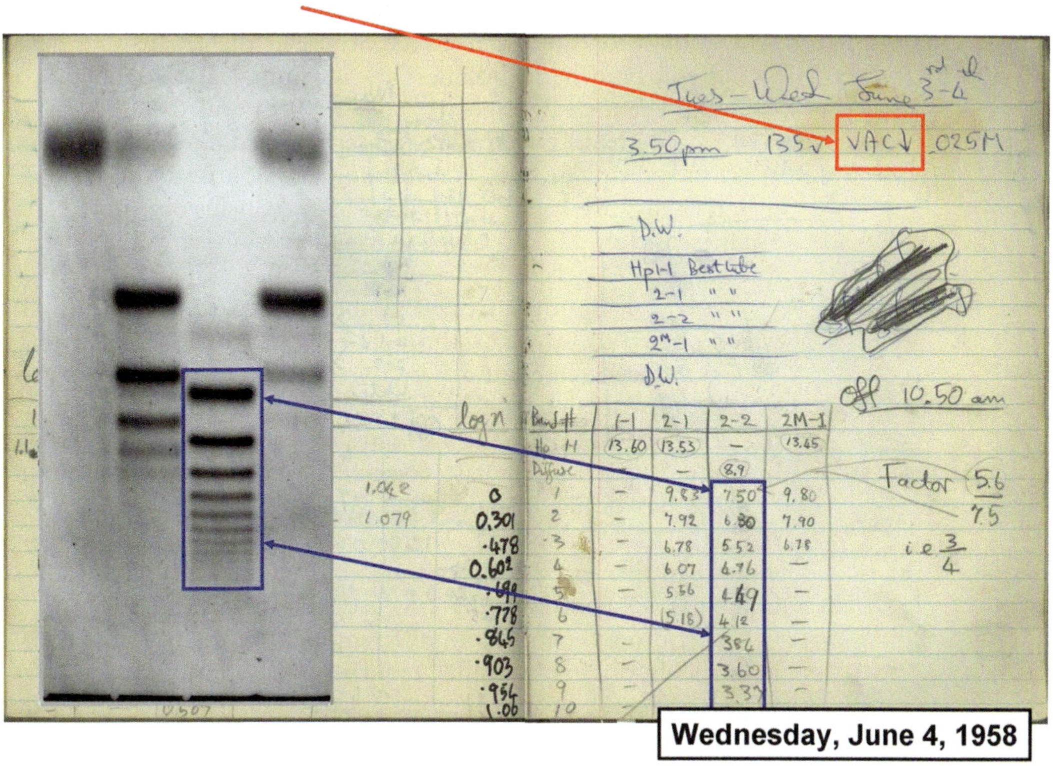

Fig. 13 Reproduced with permission from the Nobel Foundation

production of the purified proteins, developing an assay, and investigating the general physical properties of the products that I was hoping to make [26]. Gordon's task was to apply Ingram's methods to the hoped-for simpler products [27].

Starch forms gels under a wide range of conditions. This versatility proved of enormous help in carrying out my task, which I approached by testing the electrophoretic properties of purified haptoglobins in gels of many types. For example, I found that the patterns were not affected by running the proteins in a gel made with 0.05 M HCl (pH $<$ 2); so the polymers were unlikely to be due to simple ionic bands. The proteins of all three types did not survive 0.05 M NaOH (pH $>$ 12) in any form that migrated in gels, not very informative. The usual patterns were seen in the presence of 0.05 M EDTA; so the polymers were unlikely to be due to calcium or similar cross-links. Neither alkaline nor acidic 6 M or 8 M urea disassociated the polymers; nor did non-denaturing reduction with thioglycollate. On the other hand, alkaline 8 M urea plus thioglycollate caused depolymerization, and I could see the beginnings of a simpler pattern, although I could not completely understand it (my Wednesday, April 15, 1959, notebook comment was "Splitting has occurred!") [28].

Over the next few months, I tried numerous variations of the urea/thiol procedure and tested whether the treated proteins would still bind hemoglobin. They did not. I also found that a stringy precipitate developed when the urea and the thiol (now β-mercaptoethanol) were dialyzed away. Adding alkali did not affect the precipitate, nor did acid. But adding urea to the acidified material dissolved the precipitate. My conclusion, noted on Wednesday, July 22, 1959, was that "aggregation is trouble. Therefore run in 4 M gels and try acid gels."

Next day, I ran the treated samples overnight at an acid pH, which proved to be a breakthrough – the products obtained after the urea/thiol treatment at last bore a clear relationship to the genetics. The treated haptoglobins gave markedly different migrating components that were distinguishable in a pH 3.15 formate buffer gel. "Hp1-1 has fast component " (Fig. 14, blue box). "Hp2-2 has diffuse component ~2/3-1/2 mobility." (Fig. 14, red box) "2-1 has both." There was also material that did not migrate in the acid gel without urea but migrated to some extent in the pH 3.7 gel containing 4 M urea (Fig. 14, green boxes).

After experimenting with different concentrations of urea, different acids, different disulfide reducing agents, and different ways of protecting the thiol groups generated by the reductant, I finally got a procedure which gave stable reproducible products that

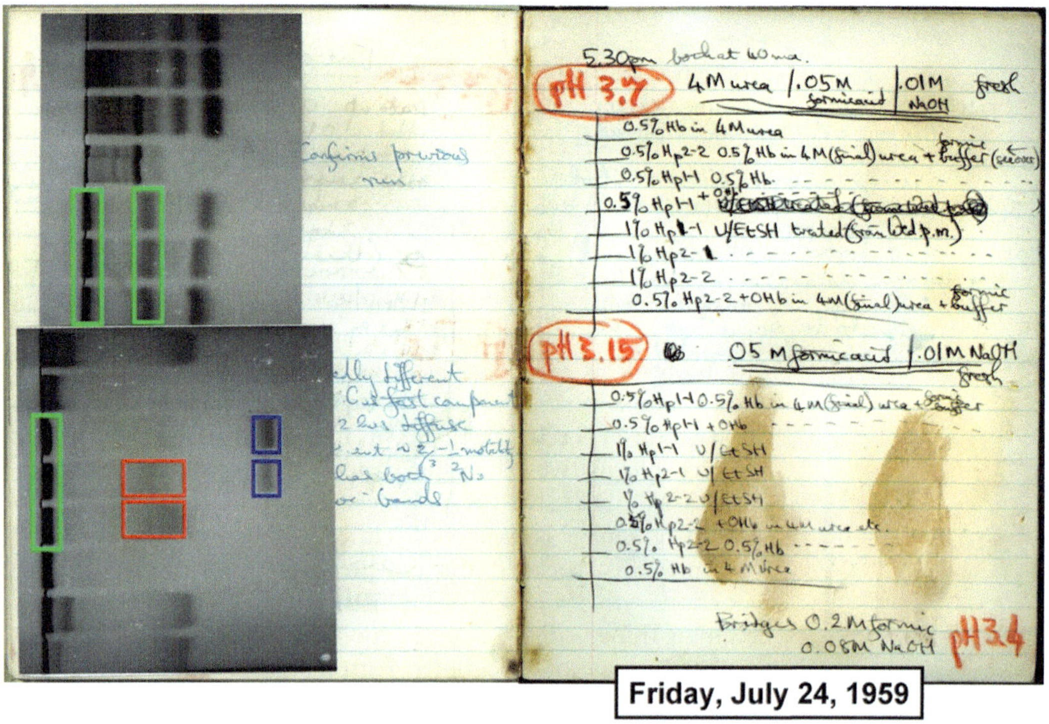

Fig. 14 Reproduced with permission from the Nobel Foundation

migrated in 8 M urea acid gels and mirrored the genetics. The procedure was to reduce the disulfide cross-links with mercaptoethanol at an alkaline pH in the presence of 8 M urea, alkylate the exposed thiol groups with iodoacetamide, and electrophorese the products in 8 M urea gels containing 20% neutralized formic acid. This procedure was very powerful and was a critical element in solving the haptoglobin problem. It showed that the haptoglobins had two polypeptide chains, an invariant slow-migrating β chain and a genetically variable faster-migrating α chain. And it proved equally powerful in Edelman's use of it to show that urinary Bence Jones proteins were the light chains of the unique immunoglobulins present in the plasma of patients with multiple myeloma [29, 30].

Despite the power of the procedure, the results obtained when we used it to study the haptoglobins were at first somewhat confusing. A gel run on February 18, 1960, to compare the merits of alkylating the urea-/thiol-treated products with iodoacetamide or iodoacetic acid illustrates one problem.

We could not understand why the fast component of reduced and alkylated Hp1-1 gave *two* fast migrating bands in the 8 M urea/formate gels (Fig. 15, red box), while Hp2-2 gave only a *single*

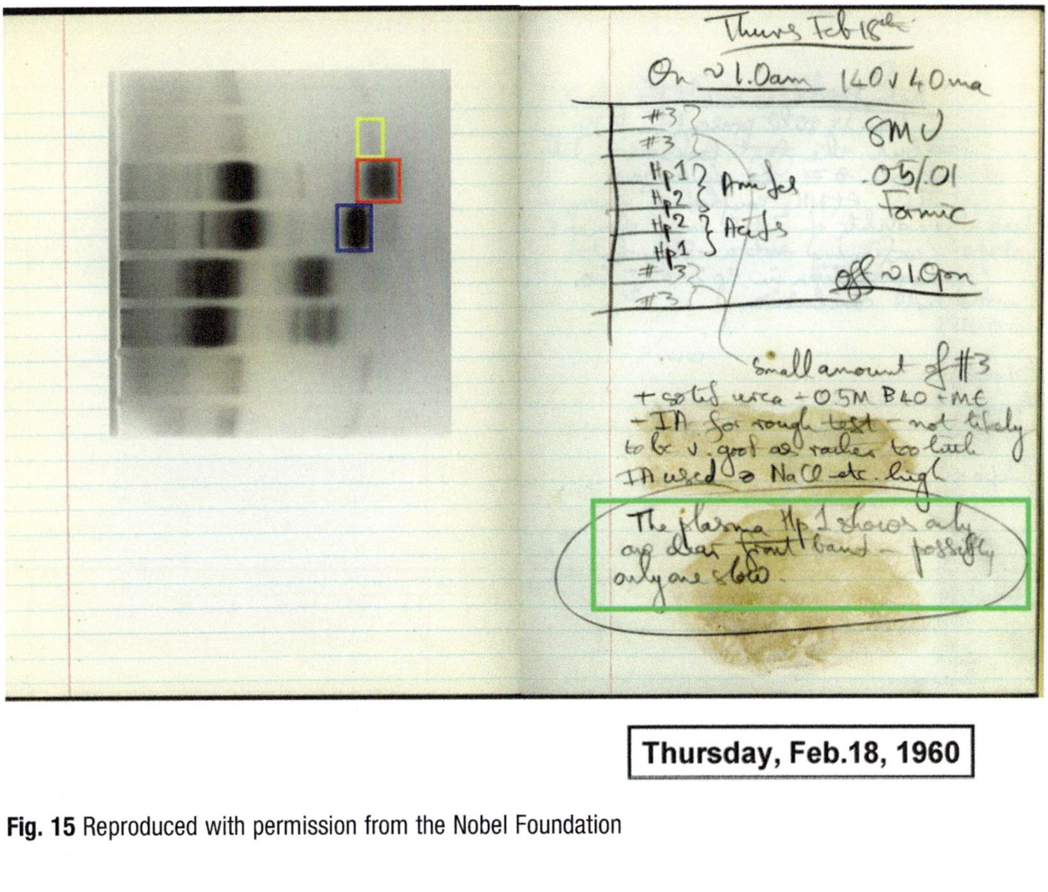

Thursday, Feb.18, 1960

Fig. 15 Reproduced with permission from the Nobel Foundation

band (Fig. 15, blue box). One possibility was that the Hp1-1 product was susceptible to proteolysis during its preparation and this gel included a small amount of a haptoglobin sample (#3) designed to test this idea—the sample was prepared from plasma that had been treated with a protease inhibitor. It gave a single Hp1α band (Fig. 15, yellow box) and so appeared to support the proteolysis idea (Fig. 15, green box). But the answer turned out to be different—George Connell discovered that the common Hp1-1 type could be divided into three subtypes, as we reported in 1962 [31], and that there were consequently two forms of the Hp^1 gene [32]. The red box sample was from an Hp^{1F}/Hp^{1S} heterozygote; the yellow box sample was from an Hp^{1S}/Hp^{1S} homozygote!

The 8 M urea/formate gel also revealed a second puzzle which proved to be more complicated. The problem is easily seen in a urea/formate gel (Fig. 16, left image on page 52) run on November 22, 1961, to compare the products obtained after reducing and alkylating purified preparations of the six common haptoglobins, Hp1F-1F, Hp1F-1S, Hp1S-1S, Hp2-1F, Hp2-1S, and Hp2-2. In the Hp1F-1S heterozygote, the amounts of Hp1Fα and Hp1Sα (Fig. 16, yellow box) are equal, but in the Hp2-1F and Hp2-1S heterozygotes, the amount of Hp2α (Fig. 16, red box) is

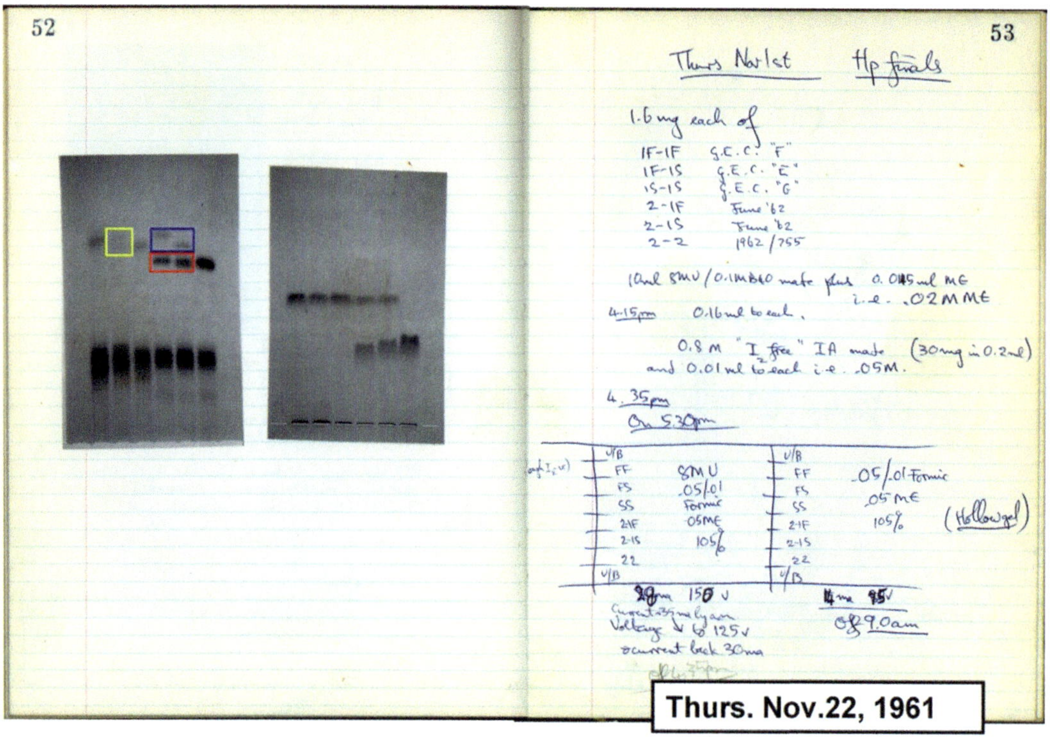

Fig. 16 Reproduced with permission from the Nobel Foundation

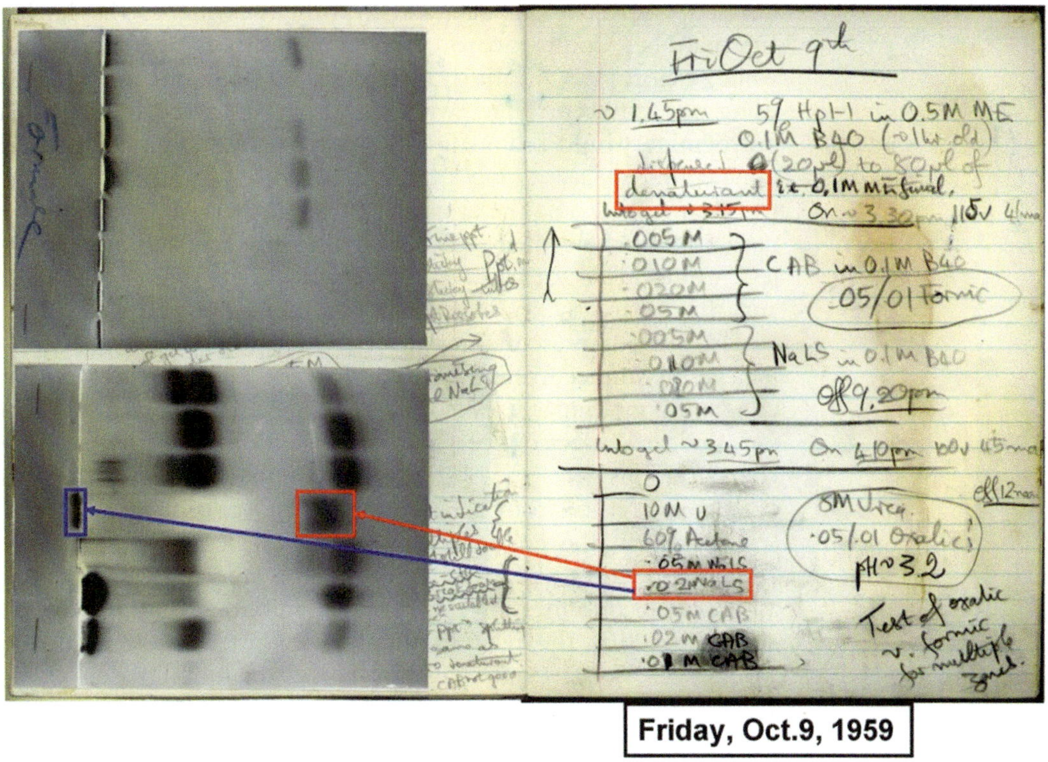

Fig. 17 Reproduced with permission from the Nobel Foundation

considerably more than the amount of Hp1Fα and Hp1Sα (Fig. 16, blue box). This problem was only solved when we had the results from Gordon's peptide studies of purified α polypeptides.

A method of purifying the α polypeptides was discovered in the course of testing different denaturants, including cetyltrimethyl ammonium bromide (CAB), sodium lauryl sulfate (SDS), and acetone, for use in reducing the haptoglobins. A test gel run on October 9, 1959, shows that the β chain was precipitated by low concentrations of SDS (Fig. 17, blue box), but the α chains were not (Fig. 17, red box). We used this feature to prepare the variable polypeptides for the next task—Gordon's analysis of the peptides resulting from digesting the purified α chains with various proteolytic enzymes.

Gordon's results, combined with the gel data, led us to the conclusion that almost the whole of the amino-acid sequence of Hp1Fα (ABCDE**F**GHI) together with most of Hp1Sα (ABCDE**S**GHI) occurs in a single molecule of Hp2α (ABCDE**F**GCDE**S**GHI), making it nearly twice as long. We hypothesized that the Hp^2 gene must have been formed by a unique non-homologous crossing-over event between the Hp^{1F} and Hp^{1S} genes, and we decided to present our data and hypothesis at the 1961 Second International Conference of Human Genetics in Rome. Meanwhile, before the meeting, George would test the hypothesis using the ultracentrifuge, with the expectation that the Hp2α polypeptide would sediment faster than the Hp1Fα and Hp1Sα polypeptides.

We met in Rome the evening before the meeting, and George gave us the news – their sedimentations were identical! Nevertheless, we decided that we would present our data, our hypothesis, and the apparently contradicting ultracentrifuge result!! But we would end our presentation with the statement that we did not believe the ultracentrifuge result and would invent a new way of comparing molecular sizes!!! This led to the final adventure in this saga of gel electrophoresis—the usefulness of varying gel concentrations.

On my return from Rome, I varied the concentration of the starch used in the gels, predicting (Fig. 18, left page) that the *relative* mobilities of Hp1Fα and Hp1Sα would remain constant as the starch concentration was increased, since they only differed in charge but that the relative mobility of the larger-sized Hp2α would decrease. The right page shows that the mobilities changed

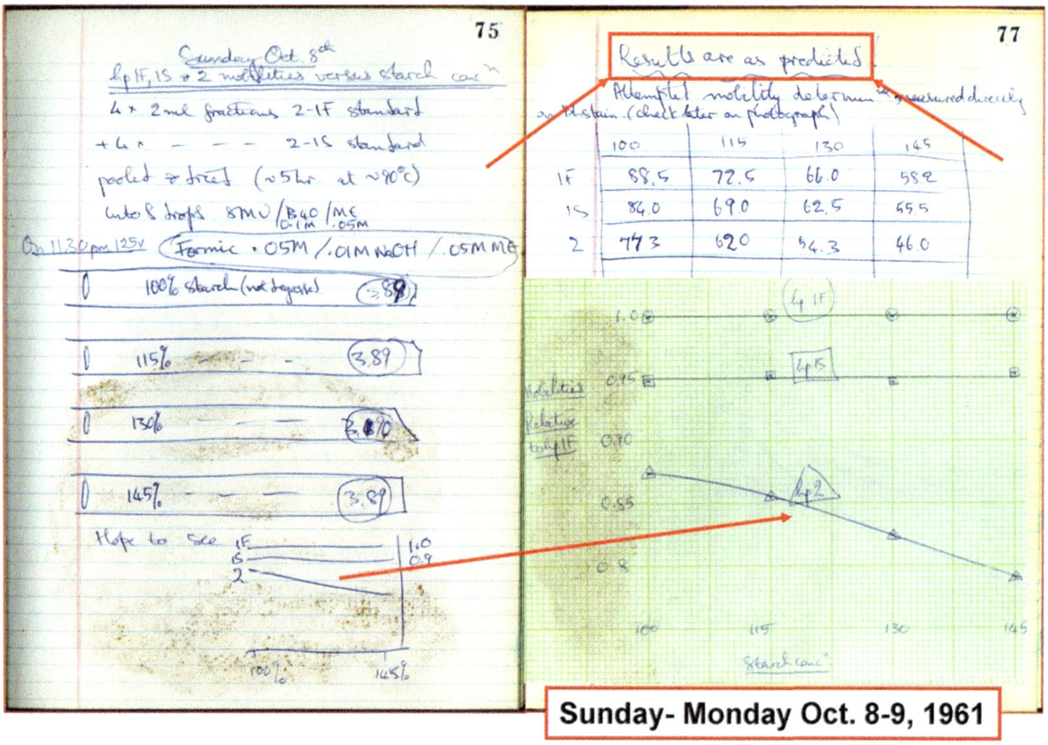

Fig. 18 Reproduced with permission from the Nobel Foundation

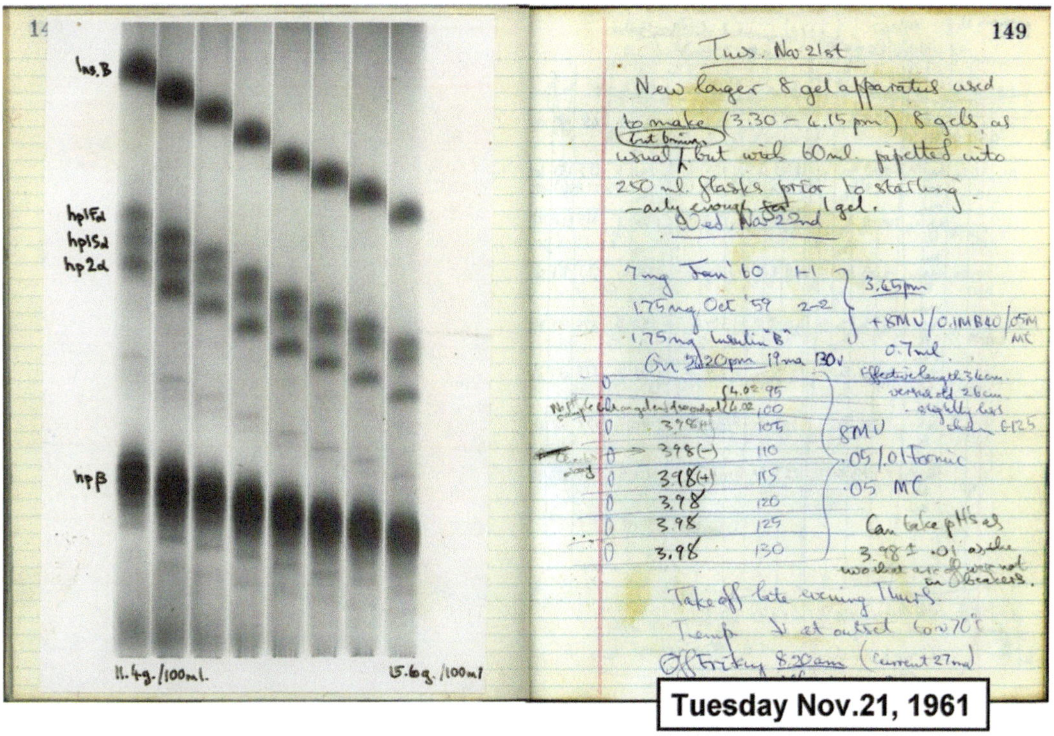

Fig. 19 Reproduced with permission from the Nobel Foundation

as predicted and therefore that Hp2α was indeed larger than Hp1Fα and Hp1Sα, which were equal in size. "Results are as predicted" (Fig. 18, red box).

George subsequently showed that, in the presence of urea, the ultracentrifuge gave the same result [26]. So together the three of us published our hypothesis [33]. And I published my new method (Fig. 19) under the title "Molecular Size and Starch-Gel Electrophoresis" [34]. The paper was part of a tributary volume celebrating Tiselius' 60th birthday. It felt particularly appropriate, because in 1953 Tiselius had written that "there is a great need for zone separations according to differences in molecular size."

Gradient gels are the modern variants of this principle, and I use them (the only gels that I buy!) enjoyably. The gel in Fig. 20 is a 4–20% polyacrylamide gradient gel (run, like many other of my gels, on a Saturday). It tells me that the gold nanoparticles which I have been making to test a hypothesis related to kidney function [35] have beautiful colors related to their sizes.

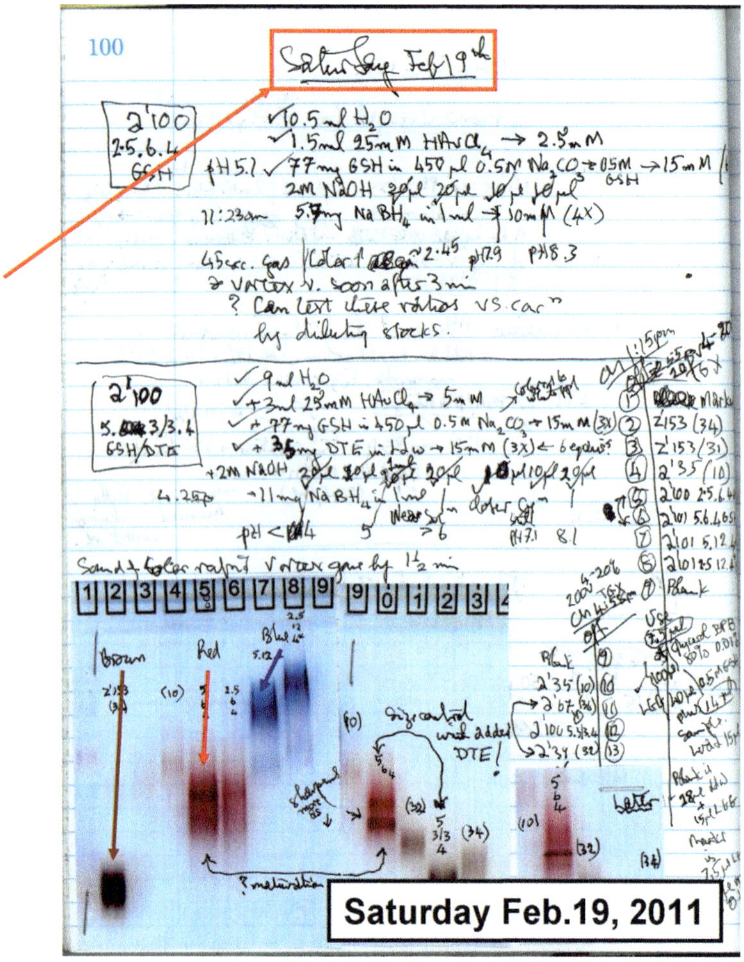

Fig. 20 Reproduced with permission from the Nobel Foundation

References

1. Tiselius A (1937) Electrophoresis of serum globulin. I. Biochem J 31:313–317
2. Pauling L, Itano HA, Singer J et al (1949) Sickle cell anemia a molecular disease. Science 110:543–548
3. Smithies O (1954) The application of four methods for assessing protein homogeneity to crystalline beta-lactoglobulin: an anomaly in phase rule solubility tests. Biochem J 58:31–38
4. Kunkel HG, Tiselius A (1951) Electrophoresis of proteins on filter paper. J Gen Physiol 35:89–118
5. Kunkel HG, Slater RJ (1952) Zone electrophoresis in a starch supporting medium. Proc Soc Exp Biol Med 80:42–44
6. Dixon GH, Smithies O (1957) Zone electrophoresis of cabbage enzymes in starch gels. Biochim Biophys Acta 23:198–199
7. Smithies O (1955a) Grouped variations in the occurrence of new protein components in normal human serum. Nature 175:307–308
8. Smithies O (1955b) Zone electrophoresis in starch gels: group variations in the serum proteins of normal human adults. Biochem J 61:629–641
9. Smithies O, Walker NF (1955) Genetic control of some serum proteins in normal humans. Nature 176:1265–1266
10. Jayle MF, Boussier G, Badin J (1952) Electrophorese de l'haptoglobine et de son complexe

hemoglobinique. Bull Soc Chim Biol (Paris) 34:1063–1069
11. Smithies O, Walker NF (1956) Notation for serum-protein groups and the genes controlling their inheritance. Nature 178:694–695
12. Smithies O, Poulik MD (1956) Two-dimensional electrophoresis of serum proteins. Nature 177:1033
13. Poulik MD, Smithies O (1958) Comparison and combination of the starch-gel and filter-paper electrophoretic methods applied to human sera: two-dimensional electrophoresis. Biochem J 68:636–643
14. O'Farrell PH (1975) High resolution two-dimensional electrophoresis of proteins. J Biol Chem 250:4007–4021
15. Laemmli UK (1970) Cleavage of structural proteins during the assembly of the head of bacteriophage T4. Nature 227:680–685
16. Giblett ER, Hickman CG, Smithies O (1959) Serum transferrins. Nature 183:1589–1590
17. Smithies O, Hickman CG (1958) Inherited variations in the serum proteins of cattle. Genetics 43:374–385
18. Smithies O (1957) Variations in human serum beta-globulins. Nature 180:1482–1483
19. Horsfall WR, Smithies O (1958) Genetic control of some human serum beta-globulins. Science 128:35
20. Smithies O (1959) An improved procedure for starch-gel electrophoresis: further variations in the serum proteins of normal individuals. Biochem J 71:585–587
21. Smithies O (1958) Third allele at the serum beta-globulin locus in humans. Nature 181:1203–1204
22. Smithies O, Hiller O (1959) The genetic control of transferrins in humans. Biochem J 72:121–126
23. Connell GE, Smithies O (1959) Human haptoglobins: estimation and purification. Biochem J 72:115–121
24. Ingram VM (1957) Gene mutations in human haemoglobin: the chemical difference between normal and sickle cell haemoglobin. Nature 180:326–328
25. Smithies O, Connell GE, Dixon GH (1966) Gene action in the human haptoglobins. I. Dissociation into constituent polypeptide chains. J Mol Biol 21:213–224
26. Connell GE, Smithies O, Dixon GH (1966) Gene action in the human haptoglobins. II. Isolation and physical characterization of alpha polypeptide chains. J Mol Biol 21:225–229
27. Black JA, Dixon GH (1970) Gene action in the human haptoglobins. IV. Amino acid sequence studies on the haptoglobin alpha chains. Can J Biochem 48:133–146
28. Smithies O, Connell GE (1959) Biochemical aspects of the inherited variations in human serum haptoglobins and transferrins. In: Wostenholme GEW (ed) Ciba foundation symposium, biochemistry of human genetics. Little, Brown and Company, Boston p. 178–193.
29. Edelman GM, Poulik MD (1961) Studies on structural units of the gamma-globulins. J Exp Med 113:861–884
30. Edelman GM, Gally JA (1962) The nature of Bence-Jones proteins. Chemical similarities to polypetide chains of myeloma globulins and normal gamma-globulins. J Exp Med 116:207–227
31. Connell GE, Dixon GH, Smithies O (1962) Subdivision of the three common haptoglobin types based on 'hidden' differences. Nature 193:505–506
32. Smithies O, Connell GE, Dixon GH (1962a) Inheritance of haptoglobin subtypes. Am J Hum Genet 14:14–21
33. Smithies O, Connell GE, Dixon GH (1962b) Chromosomal rearrangements and the evolution of haptoglobin genes. Nature 196:232–236
34. Smithies O (1962) Molecular size and starch-gel electrophoresis. Arch Biochem Biophys Suppl 1:125–131
35. Smithies O (2003) Why the kidney glomerulus does not clog: a gel permeation/diffusion hypothesis of renal function. Proc Natl Acad Sci U S A 100:4108–4113

Chapter 2

Introduction to Protein Electrophoresis

Pothur R. Srinivas

Abstract

This chapter briefly discusses the developments in electrophoresis of proteins from Tiselius' moving-boundary electrophoresis to the modern-day two-dimensional polyacrylamide gel electrophoresis. It also touches upon the staining methods used to visualize total proteins post electrophoresis.

Key words Electrophoresis, Proteins, SDS-PAGE, Isoelectric focusing, Capillary electrophoresis

1 Introduction

Since proteins are the functional units of the cellular machinery they provide significant information regarding the molecular basis of health and disease. Hence techniques to separate and isolate the various proteins are critical to studying and understanding their functional characteristics. One of the widely used techniques for this purpose is electrophoresis which predates many of the modern-day proteomic technologies. Protein electrophoresis involves the migration of charged protein molecules in the context of an applied electric field. The electrophoretic mobility of an individual protein depends on a variety of factors including the aggregate charge on its surface, size, shape and strength of the applied electric field. Further, the pH of the electrophoresis buffer also influences protein migration since it affects the net charge on the protein surface.

A typical electrophoretic run today involves separation of samples on a gel support that is immersed in buffer with the gel sandwich spanning both electrodes and the proteins migrating as anions toward the cathode. Once separation is achieved, the gel is (a) stained by a variety of methods to visualize the separated proteins, (b) further processed to detect incorporated radioactive or fluorescent tags, (c) processed for immunoblotting (d) characterized for separated polypeptides, or (e) processed for other procedures.

Biji T. Kurien and R. Hal Scofield (eds.), *Electrophoretic Separation of Proteins: Methods and Protocols*, Methods in Molecular Biology, vol. 1855, https://doi.org/10.1007/978-1-4939-8793-1_2,

Electrophoretic techniques have progressed from separation in free solution originally described in the moving-boundary method by Tiselius [1] to separation on support-based systems that include paper and gel electrophoresis. Tiselius initially used electrophoresis to separate proteins in human plasma in free solution without the use of any support medium. One drawback of this "moving-boundary" method was that the heat generated by the current caused the protein bands to become diffuse with only the heads and tails being separated as clean zones [2]. This drawback led to the introduction and use of solid support mediums such as paper, starch and polyacrylamide that helped prevent the separated proteins from sedimenting. The first of the support mediums to be used was paper.

2 Electrophoresis on a Solid Support

2.1 Paper Electrophoresis

Paper was used as a support medium initially for the separation of snake venom [3]. Cellulose-based filter paper was one of the first support media to be used in electrophoresis for the separation of small-molecular-weight molecules such as amino acids [4]. The biggest improvement that this brought about was that all of the sample could now be loaded at the same starting point or line rather than being equally distributed in one part of the electrophoresis apparatus. This allowed for much better separation of the proteins in the sample. Further, the migration of the proteins was not prone to heat-generated convection currents prevalent in free solution electrophoresis. One of the features of apparatus used in paper electrophoresis was the physical separation of the electrode chambers from the electrophoresis compartment to help prevent contamination by products of oxidation and reduction formed at the electrodes [2]. Further, a nonconducting liquid lighter than water was used to dissipate heat generated during the run as well as to keep the filter paper moist. It is interesting to note that paper electrophoresis coupled with chromatography was instrumental in discovering that sickle cell anemia was caused by a single amino acid substitution in the hemoglobin protein [5].

2.2 Gel Electrophoresis

2.2.1 Polyacrylamide Gel Electrophoresis (PAGE)

The demonstration and use of starch gels for the separation of serum haptoglobins was one of the biggest advances in gel-based electrophoresis [6] that also brought with it the concept of sieving. Another significant advance of the starch-gel electrophoresis system was the use of discontinuous buffers [7] that were used to great advantage in the PAGE system. The success of the starch gel was instrumental in the development of PAGE. PAGE has been one of the most widely used techniques for protein separation and characterization since the 1970s.Two significant factors contributing to the separation of proteins in PAGE are molecular weight and surface charge. The polyacrylamide support matrix was initially described in 1959 [8] and its use in electrophoresis using a vertical slab gel platform and the disc gel system was demonstrated in 1964 [9, 10]. The polyacrylamide gel matrix is formed through copolymerization of acrylamide and a bifunctional cross-linking agent [11]. One of the most commonly used cross-linking agents is *N*, *N'*-methylenebisacrylamide (Bis) [11, 12]. The cross-linking reaction is catalyzed in the presence of *N*,*N*,*N*,*N*-tetramethylethylenediamine (TEMED) and ammonium persulfate [13]. One big advantage of the polymerization process is that the pore size of the gel can be varied by changing the ratios of acrylamide and the cross-linker, thus enabling the separation of proteins with a wide range of molecular weights. Generally, the pore size is one of the factors that in addition to the electrical field, size, shape and surface charge determine the distance that a protein migrates and is a function of the molecular sieving capacity of the gel matrix [14]. The pore size of a polyacrylamide gel is inversely proportional to the concentration of acrylamide. Higher percentages of acrylamide generally result in better separation of low-molecular-weight proteins. Hence, the concentrations of acrylamide can be tailored for the desired polyacrylamide pore size based on the proteins that need to be separated. While the introduction of porosity gradients for resolving complex protein samples into sharp separated bands enhanced the versatility of the PAGE platform [15, 16], the use of the anionic detergent sodium dodecyl sulfate (SDS) was a significant step in improving the resolution of the separation [17]. SDS was initially used in electrophoresis for the separation and identification of polypeptide chains and subsequently for molecular weight determination which was facilitated by minimizing the effects of the charge differences and ensuring that all the proteins in the sample migrate as anions [17]. SDS helps denature and unfold most protein structures thus eliminating differences in secondary and tertiary structures while rendering the proteins as polypeptides.

2.2.2 Buffers

PAGE can be performed at any pH ranging from 2.0 to 10.0 without adversely affecting the proteins being separated [13]. In general, buffers with low ionic strengths facilitate faster rates of protein migration while buffers with higher ionic strengths reduce migration rates but allow for clearer separation. Concentrated buffers with higher ionic strengths usually result in sharper protein separations since diffusion coefficients are generally lower at higher ionic strengths [13]. However, systems should be in place to dissipate the excess heat that is generated under conditions of high ionic strengths. The discontinuous gel and buffer system first demonstrated by Ornstein [18] and Davis [19] is more widely used today than the uniform gel and buffer phase system since it helps increase the resolution of protein separation. The Laemmli SDS-PAGE system [20] is the most popular version of the discontinuous system. The discontinuous system employs different pH and voltage gradients in different gel compartments. In this system, a gel is polymerized as an upper stacking gel and a lower resolving gel. The stacking gel generally has a lower pH and is of a larger pore size (typically 4% acrylamide) than the resolving gel. The pH and ionic strengths of the buffer used in the stacking gel, the large pore size and the sample preparation buffer all help concentrate the proteins into a sharp and thin starting zone before they enter the resolving gel [13]. The sample preparation buffer usually contains SDS and beta-mercaptoethanol to solubilize, denature and break up the disulfide bonds in proteins. In addition, they include a dye such as bromophenol blue to help visualize migration. The Tris/HCl buffers used in the stacking gel have a lower pH (at least 2 pH units) than the Tris/glycine buffer used for the electrophoresis run and the protein molecules are entrapped as sharp bands between these leading chloride ions and the trailing glycine ions. The resolving gel has a higher percentage of acrylamide (typically 10% acrylamide) and hence a smaller pore size and a higher pH and salt concentration. Once the protein bands enter the resolving gel, the glycine is ionized, the voltage gradient disappears and the individual proteins are separated based on their sizes [13].

Once the proteins have been separated by SDS-PAGE electrophoresis, they can be visualized through non-covalent chemical staining or metal deposition methods through a series of steps that include (a) fixing to remove substances that would interfere with the stain and to limit background staining, (b) staining with appropriate reagents and (c) destaining to stop the staining process, remove the excess stain and prepare the gel for further processing [21]. Staining with Coomassie Brilliant Blue is one of the most widely used methods for visualizing total proteins in the microgram range. Coomassie dyes are disulfonated triphenylmethane dyes which stain protein bands blue through interaction with aminoacids [22]. Silver staining provides better and enhanced sensitivity than Coomassie dyes for detecting proteins usually at the nanogram range [23]. Stains such as Nile red, a phenoxazone dye; SYPRO Ruby, an organometallic ruthenium ion-based stain; and epicocconone, a fungal derivative are a few of the fluorescent agents used for ultrasensitive detection of proteins [24–26]. In addition, specialized staining products such as Pro-Q Diamond phosphoprotein stains and Phos-tag phosphoprotein stains are available to view posttranslationally modified proteins [27, 28].

2.2.3 Two-Dimensional Gel Electrophoresis

Two-dimensional gel electrophoresis (2-DE) was the main electrophoretic technology used for separating proteins before high-throughput mass spectrometry methods matured and were more widely adopted. Initially described in the mid-1970s [29], this method involves the separation of proteins in two dimensions. Separation in the first dimension is based on the respective isoelectric points and the separation in the second dimension is based on molecular weight. In addition to separating proteins from blood and other fluids, two-dimensional gel electrophoresis has been used to evaluate whole-cell as well as tissue proteins. Using narrow, immobilized pH gradients in the first dimension increases the resolving power which in turn can help detect low-abundant proteins. A number of techniques that include radioactive or fluorescent labeling and silver staining allow for visualization of the separated proteins. Comparisons between samples can be achieved by quantifying the ratios of spot intensities, and a number of computer-assisted image processing tools and software packages have been developed recently to help automate this process and also facilitate data analysis. Spots can be excised and further processed through mass spectrometry platforms to infer qualitative information about the individual proteins. One of the important uses of 2-DE is in the study of posttranslational modifications and hence it remains an important technique for the study of proteins even in the face of recent high-throughput proteomics technologies.

2.2.4 Capillary Electrophoresis

Capillary electrophoresis (CE) has been gaining popularity as it is less cumbersome than 2-DE. Initially developed in 1981 [30], CE has widely adapted due to its ability for microfluidic applications and seamless upfront integration with sensitive detection systems including mass spectrometry platforms [31]. By facilitating better upstream separation of proteins, CE has helped improve both qualitative and quantitative detection of proteins by mass spectrometers, laser-induced fluorescence and ultraviolet instruments [32]. While initial use of CE was predominantly for DNA its adaptability has made it a versatile tool for protein analysis. Significant contributing factors for the wide adoption of CE for protein separation include high throughput, low reagent use, short run and analysis times, and the ease of automation [33]. Proteins are separated in CE by virtue of their size-to-charge ratio in the background of a high electric field. One of the major limitations of CE is the adsorption of proteins to the inner walls of the separating capillary tubes. This is predominantly due to electrostatic or hydrophobic interactions of proteins with the silica capillary tubes [34, 35]. However, recent advances involving dynamic coating approaches [36, 37] and use of high or low pH electrolytes [38] have helped alleviate some of the adsorption concerns. Another limitation of CE is its low concentration sensitivity [39, 40]; however, pre-concentration methods are helping to address this to a great extent [41, 42].

Two variations of CE include capillary gel electrophoresis (CGE) and free-flow electrophoresis (FFE). Faster separation times, ability to recover samples, and ease of on-line detection are some of the advantages of CGE relative to conventional SDS-PAGE [43, 44]. CGE appears to be an effective tool for analyzing protein-protein interactions [45]. In FFE, proteins are continuously separated in an electric field in the background of an orthogonal hydrodynamic electrolyte flow [46, 47]. While FFE was initially used for protein purification, it has now been adapted for separation of candidate proteins from complex samples [48–50].

As protein electrophoresis technologies continue to evolve and adapt to emerging needs, they will serve as important facilitators of protein analysis and proteomics research.

References

1. Tiselius A (1938) Elektrophoretische Messungen am Eiweiss. Kolloid-Ztschr 85:129
2. Righetti PG (2005) Electrophoresis: the march of pennies, the march of dimes. J Chromatogr AJ Chromatogr A 1079:24–40
3. Von Klobusitzky P, Konig P (1939) Biochemische Studien uber die Gifte der Schlan gengattung. Arch Exp Path Pharm 192:271–275
4. Evered DF (1959) Ionophoresis of acidic and basic amino acids on filter paper using low voltages. Biochim Biophys Acta 36:14–19
5. Ingram VM (1958) The comparison of normal human and sickle-cell haemoglobins by fingerprinting. Biochim Biophys Acta 28:539–545
6. Smithies O (1955) Zone electrophoresis in starch gels: group variations in serum proteins of normal human adults. Biochem J 61:629–641
7. Poulik MD (1957) Starch gel electrophoresis in a discontinuous system of buffers. Nature 180:1477–1479
8. Raymond S, Weintraub L (1959) Acrylamide gel as a supporting medium for zone electrophoresis. Science 130:711–712
9. Raymond S (1964) Acrylamide gel electrophoresis. Ann N Y Acad Sci 121:35–65
10. Ornstein L (1964) Disc electrophoresis.I. Background and theory. Ann N Y Acad Sci 28:32–49
11. Gelfi C, Righetti PG (1981) Polymerization kinetics of polyacrylamide gels I. Effect of different cross-linkers. Electrophoresis 2:213 219
12. Righetti PG, Gelfi C, Bosisio AB (1981) Polymerization kinetics of polyacrylamide gels III. Effect of catalysis. Electrophoresis 2:291–295
13. Andrews AT (1986) Polyacrylamide gel electrophoresis (PAGE):homogenous gel and buffer systems. In: Electrophoresis:theory, techniques, and biochemical and clinical applications. Oxford University Press, New York
14. Maurer HR (1971) Disc electrophoresis and related techniques of polyacrylamide gel electrophoresis, 2nd edn. Walter de Gruyter, Berlin, New York
15. Margolis J, Kenrick KG (1967) Polyacrylamide gel-electrophoresis across a molecular sieve gradient. Nature 214:1334–1336
16. Margolis J, Kenrick KG (1967) Electrophoresis in polyacrylamide concentration gradient. Biochem Biophys Res Commun 27:68–73
17. Shapiro AL, Vinuela E, Maizel JV (1967) Molecular weight estimation of polypeptide chains by electrophoresis in SDS-polyacrylamide gels. Biochem Biophys Res Commun 28:815–820
18. Ornstein L (1964) Disc electrophoresis. I. Background and theory. Ann N Y Acad Sci 121:321–349
19. Davis BJ (1964) Disc electrophoresis. II. Method and application to human serum proteins. Ann N Y Acad Sci 121:404–427
20. Laemmli UK (1970) Cleavage of structural proteins during the assembly of the head of bacteriophage T4. Nature 227:680–685
21. Steinberg TH (2009) Protein gel staining methods: an introduction and overview. Methods Enzymol 463:541–563
22. NeuhoffV, Stamm R, Hansjorg E (1985) Clear background and highly sensitive protein staining with Coomassie blue dyes in polyacrylamide gels: a systematic analysis. Electrophoresis 6:427–488
23. Switzer RC, Merril CR, Shifrin S (1979) A highly sensitive silver stain for detecting proteins and peptides in polyacrylamide gels. Anal Biochem 98:231–237
24. Daban J-R, Bartolome S, Samso M (1991) Use of the hydrophobic probe Nile red for the fluorescent staining of protein bands in sodium dodecyl sulfate-polyacrylamide gels. Anal Biochem 199:169–174

25. Berggren K, Chernokalskaya E, Steinberg TH et al (2000) Background-free, high sensitivity staining of proteins in one- and two-dimensional sodium dodecyl sulfate-polyacrylamide gels using a luminescent ruthenium complex. Electrophoresis 21:2509–2521
26. Mackintosh JA, Choi H-Y, Bae S-H et al (2003) A fluorescent natural product for ultrasensitive detection of proteins in one-dimensional and two-dimensional gel electrophoresis. Proteomics 3:2273–2288
27. Steinberg TH, Agnew BJ, Gee KR et al (2003) Global quantitative phosphoprotein analysis using multiplexed proteomics technology. Proteomics 3:1128–1144
28. Kinoshita E, Kinoshita-Kikuta E, Takiyama K, Koike T (2006) Phosphate binding tag, a new tool to visualize phosphorylated proteins. Mol Cell Proteomics 5:749–757
29. O'Farrell PH (1975) High resolution two-dimensional electrophoresis of proteins. J Biol Chem 250:4007–4021
30. Jorgenson JW, Lukacs KD (1981) Free-zone electrophoresis in glass capillaries. Anal Chem 53:1298–1302
31. Sisu E, Flangea C, Serb A, Rizzi A, Zamfir AD (2011) High-performance separation techniques hyphenated to mass spectrometry for ganglioside analysis. Electrophoresis 32:1591–1609
32. Stepanova S, Kasicka V (2016) Recent applications of capillary electromigration methods to separation and analysis of proteins. Anal Chim Acta 933:23–42
33. Zhu Y, Li Z, Wang P, Shen L, Zhang D, Yamaguchi Y (2018) Factors affecting the separation performance of proteins in capillary electrophoresis. J Chromatogr B 1083:63–67
34. Gilges M, Kleemiss MH, Schomburg G (1994) Capillary zone electrophoresis separations of basic and acidic proteins using poly(vinyl alcohol) coatings in fused silica capillaries. Anal Chem 66:2038–2046
35. Novotny MV, Cobb KA, Liu JP (1990) Recent advances in capillary electrophoresis of proteins, peptides and amino acids. Electrophoresis 11:735–749
36. Huhn C, Ramautar R, Wuhrer M, Somsen GW (2010) Relevance and use of capillary coatings in capillary electrophoresis-mass spectrometry. Anal Bioanal Chem 396:297–314
37. Horvath J, Dolnik V (2001) Polymer wall coatings for capillary electrophoresis. Electrophoresis 22:644–655
38. Dawod LM, Arvin NE, Kennedy RT (2017) Recent advances in protein analysis by capillary and microchip electrophoresis. Analyst 142:1847–1866
39. Breadmore MC, Shallan AI, Rabanes HR, Gstoettenmayr D, Keyon AS, Gaspar A, Dawod M, Quirino JP (2013) Recent advances in enhancing the sensitivity of electrophoresis and electrochromatography in capillaries and microchips (2010–2012). Electrophoresis 34:29–54
40. Breadmore MC, Thabano JRE, Dawod M, Kazarian AA, Quirino JP, Guijt RM (2009) Recent advances in enhancing the sensitivity of electrophoresis and electrochromatography in capillaries and microchips (2006–2008). Electrophoresis 30:230–248
41. Dawod M, Chung DS (2011) High-sensitivity capillary and microchip electrophoresis using electrokinetic supercharging. J Sep Sci 34:2790–2799
42. Polikarpov N, Potolytsyna V, Bessonova E, Tripp S, Appelhans D, Voit B, Kartsova L (2015) Dendritic glycopolymers as dynamic and covalent coating in capillary electrophoresis: view on protein separation processes and detection of nanogram-scaled albumin in biological samples. J Chromatogr A 1378:65–73
43. Zhu Z, Lu JJ, Liu S (2012) Protein separation by capillary gel electrophoresis: a review. Anal Chim Acta 709:21–31
44. Wu D, Regnier FE (1992) Sodium dodecyl sulfate-capillary gel electrophoresis of proteins using non-cross-linked polyacrylamide. J Chromatogr 608:349–356
45. Ouimet CM, Shao H, Rauch JN, Dawod M, Nordhues B, Dickey CA, Gestwicki JE, Kennedy RT (2016) Protein cross-linking capillary electrophoresis (PXCE) for protein-protein interaction analysis. Anal Chem 88:8272–8278
46. Agostino FJ, Krylov SN (2015) Advances in steady-state continuous-flow purification by small-scale free-flow electrophoresis. TrAC Trends Anal Chem 72:68–79
47. Krivankova L, Boceck P (1998) Continuous free-flow electrophoresis. Electrophoresis 19:1064–1074
48. Hannig K (1978) Continuous free-flow electrophoresis as an analytical and preparative method in biology. J Chromatogr 1978 (159):183–191
49. Hannig K (1982) New aspects in preparative and analytical continuous and free-flow electrophoresis (a review). Electrophoresis 3:235–243
50. Shen QY, Guo CJ, Yan J, Zhang Q, Xie HY, Jahan S, Fan LY, Xiao H, Cao CX (2015) Target protein separation and preparation by free-flow electrophoresiscoupledwith charge-to-mass ratio analysis. J Chromatogr A 1397:73–80

Chapter 3

Measuring Protein Concentration with Absorbance, Lowry, Bradford Coomassie Blue, or the Smith Bicinchoninic Acid Assay Before Electrophoresis

J. P. Dean Goldring

Abstract

Measuring the concentration of proteins is an essential part of enzyme analysis or serves to monitor protein yields and losses during protein isolation procedures. Decisions on the usefulness of any protein isolation procedure depend on knowing the concentration of proteins before and after a procedure. Protein concentration in solution is generally measured with spectrophotometry in the UV range or in the presence of dyes or copper interacting with the protein. This review describes absorbance at 280 nm, the Lowry, Bradford (Coomassie Blue), and Smith (bicinchoninic acid) assays for measuring protein and includes suggestions for optimizing each method.

Key words Protein staining, Coomassie, Bicinchoninic acid, Lowry, Absorbance, Bradford, Protein concentration

1 Introduction

During protein isolation procedures from plant, animal, insect, yeast, or bacterial materials it is important to determine the concentration of protein at each step in the purification procedure. Measuring protein concentration is essential for enzyme activity assays and is important in clinical measurements. Protein quantification methods detect particular amino acids based on absorbance by aromatic ring structures or the interaction of dyes or copper with charged amino acid residues or peptide bonds. The choice of method depends on the amount of protein available, the detection limit of the method, the ease of use, the time required to complete the method, the presence of substances that interfere with a method, and whether one can afford to lose some protein for the

Biji T. Kurien and R. Hal Scofield (eds.), *Electrophoretic Separation of Proteins: Methods and Protocols*,
Methods in Molecular Biology, vol. 1855, https://doi.org/10.1007/978-1-4939-8793-1_3,

measurement. To ensure accurate and reproducible results it is imperative to exercise accuracy and precision with every step and addition of reagents. The methods described here are limited to those that detect 10–50 μg protein per sample.

1.1 Detecting Proteins with Absorbance at 280 nm

Proteins comprising aromatic rings in their primary sequence absorb light at 280 nm. The absorbance at 280 nm is primarily due to the presence of the amino acids tryptophan (λ_{max} 279.8 nm) and tyrosine (λ_{max} 274.6 nm) which have extinction coefficients, ε, of 5.6 and 1.42 M^{-1} cm^{-1} respectively. Phenylalanine (λ_{max} 257 nm, ε 0.197 M^{-1} cm^{-1}) makes a minor contribution [1]. Since the method is based on absorbance by tryptophan and tyrosine residues in a protein, the absorbance values of each protein will differ in relation to the number of tryptophan and tyrosine residues present (*see* **Note 1**). Should an individual protein contain a high or low proportion of tryptophan residues in its primary amino-acid sequence, the corresponding estimation of protein concentration at 280 nm will be an overestimation or underestimation of protein concentration. The absorbance measured is directly proportional to the concentration of the protein solution and the path length, that is, follows the Beer–Lambert law, and enables the measurement of the "extinction coefficient" for a particular protein at a defined wavelength. An alternative is to plot a standard curve for a protein of known concentration and determine the comparative concentration of another protein or protein mixture deduced from its absorbance. The method can be used to detect protein in the 20–3000 μg range. The method is particularly useful for the detection of proteins eluting from gel filtration, ion-exchange, affinity, hydrophobic, and chromatofocusing chromatography columns as there is no loss of protein and protein is measured in real time (*see* **Note 2**). Nucleic acids are often present in protein solutions and contribute to absorbance values at 280 nm. A compensation for the presence of nucleic acids should be made (*see* **Note 3**). For general comments on reagents and assays *see* **Notes 10–21**.

1.2 Detecting Proteins with the Lowry Method

The Biuret reaction detects a cupric complex formed between copper and four nitrogen atoms of peptides. Lowry, in 1951 modified the biuret reaction producing a more sensitive method of protein detection. Lowry [2] added phosphomolybdic/phosphotungstic acid (Folin–Ciocalteau reagent) which stains the protein with a blue green color after interaction with the cuprous ions and the side chains of tryptophan, tyrosine, and cysteine residues. The blue green color can be detected between 650 and 750 nm [3]. The protein detection range is 2 to 100 μg. For some 25 years the Lowry assay was the dominant method to quantify proteins until being replaced with the Bradford (Coomassie G-250) assay and the introduction of bicinchoninic acid to the biuret reaction by Smith

[4, 5]. The Lowry assay has a number of steps and many common laboratory reagents interfere with the method (*see* **Notes 1**, **4**, **5**, and **10–21**).

1.3 Detecting Proteins with the Smith Bicinchoninic Acid Method

The biuret reaction, the first part of the Lowry assay above, is also the basis of the bicinchoninic acid assay (BCA) developed in 1985 by Smith [5]. The cuprous complex from the biuret reaction reacts with the sodium salt of bicinchoninic acid to produce a deep blue color with an absorbance detected at 562 nm. The method detects proteins in the range of 0.2 to 50 μg [5]. The BCA reagent reacts with cysteine, tryptophan, tyrosine, and peptide bonds [5].The BCA reagent is stable under alkali conditions so can be included in the biuret alkaline copper solution. The assay can be run at room temperature or sensitivity can be increased at 60 °C. The assay is compatible with detergents, giving it an advantage over both the Lowry and Coomassie dye assays (*see* **Notes 1**, **6**, 7, and **10–21**).

1.4 Detecting Proteins with the Bradford Coomassie Blue G-250 Method

The Coomassie Blue G-250 dye is reddish/brown with an absorbance maximum of 465 nm and the dye interacts with proteins in the Bradford method producing a change in color [4]. The blue color produced under acidic conditions is the result of the dye reacting with arginine and to a lesser extent with lysine, histidine, tyrosine, tryptophan, and phenylalanine residues in proteins. Hydrophobic interactions are also taking place. The absorbance maximum is 595 nm (the absorption range is between 575 and 615 nm) [4, 6] and 0.2 to 20 μg of protein can be detected. The interaction between dye and protein at different concentrations of protein, when plotted on a graph produces a curve. An alternate form of the assay is to take readings at 595 and 450 nm and plot the ratio of the two readings (A_{595}/A_{450}) against protein concentration to obtain a linear relationship [7]. The method is the easiest and fastest of the protein determination methods currently in use (*see* **Notes 1** and **8–21**).

2 Materials

2.1 Detecting Proteins by Measuring Absorbance at 280 nm or Using a Standard Curve of Absorbance Against Protein Concentration at 280 nm

1. 1 mL quartz cuvettes are recommended. Some plastic cuvettes are now available where the plastic does not absorb light at 280 nm.
2. A UV spectrophotometer.

2.2 Detecting Proteins with the Lowry Method

1. Spectrophotometer absorbing light in the visible range and 1 mL plastic or glass cuvettes.
2. Copper reagent: Dissolve 20 g of Na_2CO_3 in 260 mL water, 0.4 g $CuSO_4{\cdot}5H_2O$ in 20 mL water, and 0.2 g sodium potassium tartrate in 20 mL water and mix.
3. SDS solution: 10 g sodium dodecyl sulfate in 200 mL water (1% w/v).
4. NaOH solution: Dissolve 4 g NaOH in 100 mL water to make a 1 M solution.
5. 2× Lowry working reagent: Mix 3 parts of copper reagent with 1 part SDS solution and 1 part NaOH solution. This reagent is stable for 3 weeks.
6. 0.2 N Folin reagent: Mix 10 mL Folin reagent with 90 mL water. The reagent is stable for several months if stored in an amber bottle.

2.3 Detecting Proteins with the Smith Bicinchoninic Acid Method

1. Spectrophotometer absorbing in the visible range and 1 mL plastic or glass cuvettes.
2. Reagent A: 1 g sodium bicinchoninate, 2 g Na_2CO_3, 0.16 g sodium tartrate, 0.4 g NaOH, and 0.95 g $NaHCO_3$ made up to 100 mL with water. Adjust the pH to 11.25 with NaOH.
3. Reagent B: 0.4 g $CuSO_4{\cdot}5H_2O$ in 10 mL water. Reagents A and B are stable.
4. Mix 100 volumes of reagent A with 2 volumes reagent B for a working solution. The solution is stable for a week at room temperature.

2.4 Detecting Proteins with the Bradford Coomassie Blue G-250 Method

1. Spectrophotometer absorbing in the visible range and 1 mL plastic or glass cuvettes.
2. Dissolve 100 mg Coomassie Brilliant Blue G250 in 100 mL of 85% (v/v) phosphoric acid and 50 mL 95% (v/v) ethanol. After the dye is dissolved adjust the volume to 1 L with water. The reagent is stable for several months. Should a precipitate form, filter through filter paper and determine the standard curve again.

3 Methods

3.1 Detecting Proteins by Measuring Absorbance at 280 nm

1. Set the UV spectrophotometer to read at 280 nm allowing 15 min for the instrument to equilibrate.
2. Set the absorbance reading to zero with a solution of the buffer and all components except the protein present. Alternately take a reading of solution without protein and subtract the value from each reading of solution containing protein.

3. Place the protein solution in the 1 mL cuvette and determine the absorbance. This step should be repeated with a new sample to obtain duplicate readings. If a reading is obtained with an absorbance value greater than 2, then dilute the protein sample with the parent buffer and determine the absorbance value. An initial 1 in 10 dilution is suggested. The dilution and readings of samples should be performed in duplicate. It is advisable to use the same cuvette or matched cuvettes for samples.
4. If the extinction coefficient of the protein is known, then the following equation can be employed. Absorbance = Extinction coefficient x concentration of protein × path length (1 cm) to determine the concentration of the protein.

3.2 Detecting Proteins Using a Standard Curve of Protein Against Absorbance at 280 nm

Follow **steps 1–3** as set out in Subheading 3.1 above.

5. Use a range of duplicate concentrations (20 to 2000 μg) of the known protein (often bovine serum albumin—*see* **Note 1**). Measure the absorbance of each concentration of the known protein. Plot the absorbance readings against the protein concentration to produce a standard curve.
6. Place the protein solution of unknown concentration in a 1 mL cuvette (as above) and measure the absorbance of the unknown at 280 nm. The concentration of the unknown can be deduced from the absorbance reading obtained and using the protein standard curve generated in **step 4**.

3.3 Detecting Proteins with the Lowry Method

1. Set the spectrophotometer to read at 750 nm allowing 15 min for the instrument to equilibrate.
2. Prepare a bovine serum albumin (BSA) protein standard at 1 mg/mL concentration in duplicate. Dilute the protein standard in a volume of 0.4 mL to give 5 concentrations over the range of 10 to 100 μg protein. Perform the dilution and readings for both standard solutions to give four readings at each protein concentration (*see* **Note 1**).
3. To 0.4 mL protein sample add 0.4 mL of the Lowry working reagent and incubate for 10 min at room temperature.
4. Add 0.2 mL of the 0.2 N Folin reagent, vortex and leave for 30 min and read the absorbance values at 750 nm in the spectrophotometer. Readings can be done between 650–750 nm, but are more sensitive at 750 nm.
5. A graph of absorbance readings at 750 nm against protein concentration is plotted. The protein concentration of the test sample is derived using the absorbance value of the sample and the graph.

3.4 Detecting Proteins with the Smith Bicinchoninic Acid Method

1. Set the spectrophotometer to read at 562 nm, allowing 15 min for the instrument to equilibrate.
2. Prepare a BSA protein standard at 1 mg/mL concentration in duplicate. Dilute the protein standard in a volume of 20 μL to give 5 concentrations over a range of 10 to 50 μg protein. Each dilution of each of the two standards should be done in duplicate to produce 4 readings per protein concentration.
3. Add 20 μL of protein sample to 1 mL of working solution and incubate for 30 min at 37 °C or 60 °C for increased sensitivity. Cool the sample and place in a 1 mL cuvette and read the absorbance.
4. A graph of absorbance readings at 562 nm against protein concentration is plotted. The protein concentration of the unknown is derived using the absorbance value of the sample and the graph.

3.5 Detecting Proteins with the Bradford Coomassie Blue G-250 Method

1. Set the spectrophotometer to read at 595 nm, allowing 15 min for the instrument to equilibrate.
2. Prepare a BSA protein standard at 1 mg/mL concentration in duplicate. Dilute the protein standard in a volume of 20 μL to give 5 concentrations over a range of 10 to 50 μg protein. Each dilution of each of the two standards should be done in duplicate to produce 4 readings per protein concentration.
3. Add 20 μL of protein solution to 1 mL of dye reagent, mix, incubate for 2 min at room temperature and measure the absorbance in a cuvette.
4. A graph of absorbance readings at 595 nm against protein concentration is plotted. The protein concentration of the unknown is derived using the absorbance value of the sample and the graph.

4 Notes

1. The recommended calibration protein for protein determination is BSA due to its availability and common use. One should be aware that each individual protein contains unique amounts of each amino acid. For example BSA (66.3 kDa) in serum contains 2 tryptophan, 20 tyrosine, 23 arginine, and 60 lysine; mature ovalbumin (44.3 kDa) contains 3 tryptophan, 10 tyrosine, 15 arginine, and 20 lysine amino acid residues. The two proteins will thus provide different standard curves in all the assays described above.
2. Advantages of using absorbance at 280 nm: the method is (a) fast, (b) easily automated, (c) and reasonably sensitive;

(d) does not destroy protein; and (e) can be automated; and (f) most proteins contain tryptophan and tyrosine residues.

3. Disadvantages of using the absorbance at 280 nm to detect proteins: (a) it should be noted that DNA and RNA have absorbance maxima at 260 nm, but still absorb at 280 nm and have tenfold higher absorbance values at 280 nm compared to the equivalent concentration of protein. Groves et al. developed an elegant method to correct for the presence of nucleic acid in a protein solution [8]. The formula 1.55 A_{280}–0.76 A_{260} can be used. A pure protein solution has a ratio of absorbance values (A_{280}/A_{260}) of greater than 1.7 and pure nucleic acid of less than 0.5. (b) Cuvettes should be handled carefully as fingerprints on the cuvette will distort readings.
4. Advantages of Lowry method: (a) the method is 100-fold more sensitive than the original biuret reagent, and (b) the method is more sensitive than determining absorbance at 280 nm.
5. Disadvantages of Lowry method: (a) many common substances (K^+, Mg^{2+}, NH_4^+, EDTA, Tris, carbohydrates, and reducing agents), interfere with the method; (b) lipids and detergents may lead to precipitation of components; (c) the Folin reagent is reactive for only a short period of time after addition; (d) the method is complicated and requires more steps and reagents than the BCA or Bradford assays, and (e) the method is destructive to proteins, that is, once the protein sample has reacted with the dye, the protein cannot be used for other assays.
6. Advantages of Smith bicinchoninic acid assay: (a) the method is less complicated than the Lowry assay; (b) the method is more sensitive than the Lowry assay; (c) the method has less protein–protein variability than the Bradford assay; (d) the volume of reagents can be reduced and a microassay can be performed in 96-well plates; (e) detergents are compatible with the assay (up to 5%), and (f) the method can be automated.
7. Disadvantages of Smith bicinchoninic acid assay: (a) reducing reagents (e.g., ascorbic acid, dithiothreitol, 2-mercaptoethanol) that reduce copper interfere with the assay; (b) chelating agents (e.g., EDTA) that chelate copper interfere with the assay; (c) the method is destructive to proteins, that is, once the protein sample has reacted with the copper, the protein cannot be used for other assays.
8. Advantages of Bradford Coomassie Blue G-250 assay: (a) The method is fast. (b) The method is simple. (c) The reagents are stable for long periods of time. (d) The volume of reagents can be reduced and the assay performed in a 96-well microtiter

place. (e) The method is sensitive. (d) The method is compatible with reducing reagents, and (f) can be automated.

9. Disadvantages of Bradford Coomassie Blue G-250 assay: (a) the dye stains cuvettes. Cuvettes can be cleaned by washing with a dilute sodium dodecyl sulfate solution; (b) dye binding depends on the basic amino acid content that can vary between proteins; (c) concentrated protein solutions can form a precipitate upon contact with the dye reagent. If this is observed, the protein solution should be diluted to determine protein concentration, and (d) the method is destructive to proteins, that is, once the protein sample has reacted with the dye, the protein cannot be used for other assays.
10. It is advisable to prepare the protein standard in duplicate. If a protein standard is weighed or solubilized incorrectly in a single preparation, then all protein determinations will be inaccurate.
11. All reagents, unless otherwise stated, should be at room temperature. Cold reagents in a cuvette result in condensation on the outside of the vessel, distorting readings.
12. Should readings of a particular sample "waver," that is, increase and decrease constantly. This is often caused by the presence of particulate matter. Samples can be filtered through a 0.2 μm filter (that does not bind protein) or centrifuged briefly for example in a microfuge at 12,000 × g for 5 s and a sample taken from the filtrate or the supernatant.
13. When working with small volumes, it is critical to pipet carefully. Dipping a pipette tip deep into a liquid picks up unwanted excess liquid outside the outside surface of the tip, which can lead to inconsistencies in the quantities of liquid and hence protein delivered to a cuvette.
14. A plot of the raw data (rather than the average), that is, 2 per point in duplicate = 4 readings when plotting the standard curve enables one to see how precisely each measurement was conducted. If any individual data point has a wide range of absorbance values, that measurement should be repeated. As the determination of many proteins may be compared to the standard curve, it is important that the standard curve is as accurate as possible.
15. As a general rule, if absorbance values of an unknown protein are greater than 2.0, the sample should be diluted and then added to the protein determination reagents and the new absorbance value determined.
16. Two different dilutions of protein solutions of unknown concentration in duplicate for determining protein concentration are recommended to ensure accurate determination of protein concentration.

17. All the assays described should be conducted at the wavelength of the absorbance maximum for each assay. Absorbance readings at absorbance maxima have the largest change in absorbance reading for each change in protein concentration.
18. Protein quantification with A_{280}, Lowry method, BCA, or Coomassie Blue reagents should always be considered as one representing a relative concentration of protein. It is an accepted practice to express the relative concentrations of proteins during protein purification steps, provided that all protein concentrations are determined with the same method.
19. Lowry, copper (BCA), and Coomassie Blue protein determination methods all produce a nonlinear curve. Readings should be taken on the ascending linear portion of the curve. When using graphical software note that if the software plots a straight line by linear regression, only a poor approximation of protein concentration is obtained from the line. Rather treat the data with curvilinear regression to produce a polynomial trend line, and the line with the best fit is considered optimal. Absorbance at 280 nm produces a straight line as protein concentration increases.
20. Lowry, copper (BCA), and Coomassie Blue protein curves all proceed from the origin of the graph, provided that the correct "blank" or "control" has been used.
21. A standard curve needs to be constructed each time new stocks of color reagents are made up. The Bradford reagent obtained from different suppliers can produce different standard curves [9].

References

1. Fasman GD (1990) Practical handbook of biochemistry and molecular biology. CRC Press, Boston
2. Lowry OH, Rosebrough NJ, Farr AL, RJ R (1951) Protein measurement with the F olin phenol reagent. J Biol Chem 193:265–275
3. Stoscheck CM (1990) Quantitation of protein. Methods Enzymol 182:50–68
4. Bradford MM (1976) Rapid and sensitive method for quantitation of microgram quantities of protein utilizing principle of protein-dye binding. Anal Biochem 72:248–254
5. Smith PK, Krohn RI, Hermanson GT, Mallia AK, Gartner FH, Provenzano MD, Fujimoto EK, Goeke NM, Olson BJ, Klenk DC (1985) Measurement of protein using bicinchoninic acid. Anal Biochem 150:76–85
6. Compton SJ, Jones CG (1985) Mechanism of dye response and interference in the Bradford protein assay. Anal Biochem 151:369–374
7. Zor T, Selinger Z (1996) Linearization of the Bradford protein assay increases its sensitivity: theoretical and experimental studies. Anal Biochem 236:302–308
8. Groves WE, Davis FC, Sells BH (1968) Spectrophotometric determination of microgram quantities of protein without nucleic acid interference. Anal Biochem 22:195–210
9. Noble JE, Knight AE, Reason AJ, Di Matola A, Bailey MJ (2007) A comparison of protein quantitation assays for biopharmaceutical applications. Mol Biotechnol 37:99–111

Chapter 4

Concentrating Proteins by Salt, Polyethylene Glycol, Solvent, SDS Precipitation, Three-Phase Partitioning, Dialysis, Centrifugation, Ultrafiltration, Lyophilization, Affinity Chromatography, Immunoprecipitation or Increased Temperature for Protein Isolation, Drug Interaction, and Proteomic and Peptidomic Evaluation

J. P. Dean Goldring

Abstract

In protein isolation, drug interaction studies, and proteomic or peptidomic procedures, protein solutions are often concentrated to remove solvents and undesirable molecules, to separate protein fractions, or to increase protein concentrations. Proteins can be concentrated by precipitation from solution with ammonium sulfate, polyethylene glycol, organic solvents, trichloroacetic acid, potassium chloride/sodium dodecyl sulfate thermal denaturation, and three-phase partitioning. Solvents can be removed by passage through a semipermeable barrier where protein solutions are forced against the barrier in a centrifuge tube or with increased pressure, concentrating proteins in the remaining solution. The semipermeable barrier can be surrounded by a hygroscopic reagent to draw the solvent across the membrane. Proteins can be concentrated by freeze-drying (lyophilization). Unique ligand interactions with proteins can be used to select for proteins by affinity purification or immunoprecipitation. All these methods to concentrate proteins are discussed.

Key words Protein, Proteomic, Peptidomic, Concentration, Dialysis, Precipitation, Ultrafiltration, Vacuum centrifugation, SDS/KCl, Potassium dodecyl sulfate, KDS, Trichloroacetic acid, Three-phase partitioning, Absorption, Freeze-drying, Lyophilization, Affinity purification, Immunoprecipitation, Thermal denaturation

1 Introduction

The study of proteins, enzymes, and peptides at the macro- and microlevel or in proteomics frequently requires concentrating the protein solution. The recombinant expression of proteins from animal, plant, insect, yeast, or bacterial cells or the isolation of proteins from animal, plant, insect, yeast, or bacterial tissues

Biji T. Kurien and R. Hal Scofield (eds.), *Electrophoretic Separation of Proteins: Methods and Protocols*, Methods in Molecular Biology, vol. 1855, https://doi.org/10.1007/978-1-4939-8793-1_4,

involves several biochemical steps. Many of the steps employed in protein isolation procedures dilute the protein of interest. Buffers are added to maintain pH and ionic concentration, salt-induced precipitates increase in volume upon dialysis, and samples are often diluted during chromatography such as molecular exclusion chromatography (gel filtration). Decreases in the presence of cellular components and dilute solutions lead to reduced enzyme activity and can complicate the purification of a target protein [1–3]. Proteomic studies to detect biomarkers for disease or metabolic perturbations require protocols to enhance (and concentrate) proteins or peptides. The method employed to concentrate a protein solution depends on the protein(s) involved, the composition of the solvent, and what an investigator wishes to do with the protein after concentration.

Methods to concentrate dilute protein solutions can either target the selection of a particular protein, thus removing the target protein from the general milieu, or the removal of the surrounding solvent. Proteins can be "selected" by their affinity for particular ligands, their behavior on chromatographic matrices, or their precipitation. Affinity methods using either combinatorial peptide ligand library (CPLL) or affinity chromatography are very powerful techniques to concentrate proteins. The CPLL method has been reported to detect as little as 1 ng/mL of protein [4], while a 1 mL peptide affinity matrix can isolate 2.4 mg of anti-peptide antibody from 1082 mg of protein solution (Goldring personal observation). Chromatographic methods include size exclusion, reverse phase, or ion-exchange chromatography exploiting the molecular mass, hydrophobicity, or surface charge on proteins. Due to the broad range of chromatographic techniques, only affinity chromatography will be described in detail here. Antibodies raised against a protein or a peptide region of the protein can be covalently coupled to an affinity matrix to purify the protein or the antibody bound to the antigen fished out of solution using an immunoprecipitation protocol. Drug binding to proteins can lead to increased thermal stability of the protein, which can be exploited to isolate the drug protein partners from tissues, cells, or cell lysates.

Proteins can be precipitated from solution with salts and solvents that compete for water of solvation or neutralize protein-protein charge repulsion forces. Removal of the solvent can be achieved in two ways. The first involves a semipermeable barrier (membrane) of defined pore dimensions to permit solvent passage (ultrafiltration) but retain proteins as they are too big to pass through the barrier. The passage of solvent is promoted by increased pressure or centrifugation. The second way to remove solvent is using reduced pressure (vacuum) on the sample. For a frozen protein solution, the solvent sublimates under vacuum, thereby concentrating the protein. A number of common methods employing selection or solvent removal to concentrate proteins are discussed here.

1.1 Concentrating Proteins by Precipitation with Salts and Polyethylene Glycol

Ammonium sulfate is the most popular salt for protein precipitation as it is cheap, is readily available, is "gentle" on proteins, stabilizes some proteins, and its solubility changes little between 5 and 30 °C [1]. Ammonium sulfate competes for water of hydration surrounding proteins in solution and neutralizes charges on proteins in solution leading to the precipitation of protein. The salt has to be removed (by gel filtration or dialysis) for downstream analysis, e.g., LS-MS due to potential interface contamination [5]. Ammonium sulfate can be removed by washing the pellet in 90% (v/v) ice-cold acetone, air-dried and suspended in sample loading buffer for isoelectric focusing or SDS-PAGE for proteomic analysis [6]. Polyethylene glycol M_r 6000 is a hydrophilic polymer which takes up water upon addition to a protein solution leading to precipitation of the protein [7]. Polyethylene glycol was first introduced by Polson [8] and is particularly effective for the isolation of antibodies from chicken egg yolk [9, 10]. The ideal concentration of ammonium sulfate or polyethylene glycol to precipitate and concentrate a particular protein or a pool of proteins from a protein solution has to be determined by trial and error.

1.2 Concentrating Proteins by Precipitation with Organic Solvents and Trichloroacetic Acid (TCA)

Organic solvents, when added to aqueous solutions, reduce the polarity of the solution and decrease the dielectric constant of the solution leading to the precipitation of proteins. The organic solvent may precipitate the required protein at a particular concentration which can be determined empirically. To precipitate and thus concentrate most of the proteins in solution, acetone is added to 50% (v/v) and ethanol to 60% (v/v). Improved yields of small amounts of protein can be obtained with a chloroform/methanol/water mix as described by Wessel and Flugge in 1984 [11]. Acetonitrile appears to be the most efficient organic solvent to precipitate and concentrate serum proteins or to selectively deplete high abundance proteins prior to mass spectrometry proteomic analysis [5, 12]. Trichloroacetic acid disrupts hydrogen bonding with water and forms precipitates with cations of proteins. The concentration of the protein in solution needs to be greater than 5 μg/mL, and TCA denatures proteins [13]. TCA precipitation is used to precipitate proteins when measuring the incorporation of radioactive precursors in metabolic studies and can be used to prepare samples for liquid chromatography tandem mass spectrometry [5, 14]. Organic solvents and TCA are particularly useful for the analysis of samples by SDS-PAGE followed by Western blotting as the methods can denature proteins for evaluation on the SDS-polyacrylamide gel.

1.3 Concentrating Proteins for SDS-PAGE with SDS/KCl Precipitation

When potassium chloride is added to sodium dodecyl sulfate (SDS), the potassium dodecyl sulfate (KDS) precipitates [15]. This interaction is exploited in the removal of proteins during the isolation of nucleic acids. Proteins in SDS solution bind to the SDS, and in the presence of potassium chloride, the KDS/protein complexes precipitate [16]. SDS is effective for protein extraction and denaturation but in proteomic and peptidomic studies needs to be removed before LC-MS-based analysis. The SDS solubilizes protein samples and can be diluted for enzymatic digestion, and then the SDS is removed with the SDS/KCl method leaving the peptide fraction [17]. Alternately for peptide analysis, the SDS/KCl method can be used to remove proteins leaving peptides. The SDS/KCl method has been used for small volumes of a solution containing low concentrations of proteins. Samples precipitated by the SDS/KCl method can be resuspended in sample buffer for direct loading onto a SDS polyacrylamide electrophoresis gel for direct analysis. Protein concentration can be determined with the Goldring and Ravaioli method that is not affected by the presence of KCl or SDS [18].

1.4 Concentrating Proteins with Three-Phase Partitioning

Three-phase partitioning (TPP) was developed by Dennison and Lovrien in 1997 [19]. The technique involves mixing t-butanol with the protein solution and then adding ammonium sulfate. The t-butanol becomes insoluble in aqueous solutions with concentrations of ammonium sulfate above 0.6 and up to 2.6 M. Centrifuging the t-butanol/protein/ammonium sulfate sample generates three phases with t-butanol on top, proteins precipitated by ammonium sulfate in the middle and a bottom aqueous layer containing ammonium sulfate. An advantage of the technique is that the upper t-butanol layer dissolves lipids that may be present, and therefore lipids are also removed. The precipitated protein sample, when recovered from the middle layer, does not require dialysis, e.g., for ion-exchange chromatography, as salt concentrations around the protein are minimal [20]. When preparing samples for isoelectric focusing, SDS-PAGE, or 2-D electrophoresis, trace amounts of ammonium sulfate can be removed by dialysis or washing the precipitate with 90% ice-cold acetone, air-dried and suspended in the appropriate sample buffer. TPP works well with dilute protein solutions. TPP has been used on crude extracts, in whole *E.coli* cells, and in the isolation of cathepsin D and green fluorescent protein (*see* Dennison [20] for a review [21–23]).

1.5 Concentrating Protein by the Absorption of Solvent

There is a range of dialysis membranes available with different pore sizes restricting the passage of molecules above a particular size. Solvents can pass through the membranes, and larger molecules remain trapped by the membranes. For many proteins a 12 kDa cutoff dialysis membrane is suitable. The protein solution is placed inside the dialysis tubing and the tube sealed. The tube containing the protein solution is surrounded by a material that has a high capacity to absorb water, for example, polyethylene glycol (PEG) M_r 20000. The solvent is drawn out of the dialysis tube and absorbed by the PEG 20000. Sucrose or polyvinylpyrrolidone can be used in place of PEG but tend to enter the dialysis tube and contaminate the protein solution. PEG 20000 is too big to pass through the pores and cross the dialysis membrane, so does not contaminate the concentrated protein solution. The PEG can be dried and reused without a decrease in performance. Concentrating the protein takes 1 h for 10 mL, 4 h for 30 mL, and up to 6 h for 50 mL of sample. One needs to be careful, or all the solvent can be removed from the protein solution. Sephadex G-100 and Sephadex G-200 can be used, but this increases the cost. Some suppliers have water-absorbing products as alternatives to PEG 20000. For small (1.5 mL) volumes, Sephadex G-25 can be added to the protein solution. The gel swells four to six times its volume absorbing buffer. The process can be carried out in a 1.5 mL microfuge tube with a small puncture hole in the base of the tube. Most proteins are too big to enter the pores in the Sephadex beads. The concentrated protein is obtained by centrifugation though the hole in the bottom of the microfuge tube into a second tube.

1.6 Centrifugation to Concentrate Proteins

Porous membranes with pores of differing dimensions are employed in a number of methods to concentrate proteins. The membranes are made of materials which are compatible with and do not "stick" most proteins.

One approach is to use the forces generated by centrifugation and can be described as centriconcentration. Specially made centrifuge tubes with differing designs to enhance solvent passage are available. Volumes range from 0.5 to 70 mL, tubes are constructed of material that is compatible with pH ranges from pH 1 to 9, and tubes are designed to retain a minimal volume of liquid to prevent samples from drying out. There are a range of tubes with different pore sizes available to retain proteins of differing sizes. The tubes can operate over a wide range of temperatures. The centrifugal field, generated during centrifugation, forces the protein solution against the membrane. The solvent passes though the membrane, and proteins with a molecular mass larger than the cutoff value of the membrane are trapped. The tubes are designed to ensure a small volume of liquid is retained. This is for those of us who forget to keep a watchful eye on the experiment! For smaller volumes of sample, the 1.5 mL centrifugation tube contains a small insert for concentrating protein solutions. After concentration, the insert can be inverted in a clean 1.5 mL centrifuge tube, and the sample is again centrifuged to capture the concentrated protein sample. Care needs to be taken as some of the tubes used to concentrate proteins require centrifugation in a fixed angle rotor and others in a swing out rotor. The pore size of the membrane can be chosen to both concentrate the protein of interest and to separate it from smaller components, for example, a 100 kDa "cutoff" membrane will retain immunoglobulins (156 kDa) but not serum albumin (bovine 67 kDa). The technique is used to separate low-molecular-weight from high-molecular-weight protein fractions in proteomic samples [24].

1.7 Concentrating Proteins with Increased Pressure

In this technique the permeable membrane is placed in a sealed apparatus, referred to as a "stirred cell." The protein solution is added to a vessel above the membrane which includes a suspended magnetic stir bar and the unit sealed and placed on a magnetic stirrer. Pressure is applied to the top of the stirred protein solution. The solvent passes through the membrane, and the protein solution is retained. Unlike the centriconcentration method, there is not a design at present that retains some solution for the unwary operator and protein solutions can be "dried out." The gas introduced to increase pressure should be inert (nitrogen or argon) to prevent the oxidation of the components of the solution. Like the centriconcentration method above, protein samples can be separated by molecular mass and concentrated at the same time. Peptides can be separated from high-molecular-weight proteins for proteomic studies [24].

1.8 Concentrating Protein Solutions Using a Vacuum

When the pressure above a liquid is reduced, the liquid boils. Water is thus removed from the solution leaving behind buffers, dissolved chemicals, and protein. This can be achieved in the laboratory with a centrifuge encompassing a vacuum. The centrifugal force helps drive out the gas bubbles and reduces the foaming of protein solutions. This method works well for small volumes of samples where the surface to volume ratio is high leading to more efficient removal of the liquid.

An adaptation of the dialysis method is to place the protein solution in a dialysis tube. The tube is sealed and placed in a side-arm vessel attached to a Venturi pump on the water tap in the laboratory. The vacuum is applied, and the solvent drips off the outside of the membrane. Vacuum dialysis takes longer than the techniques mentioned above due to the lower pressures involved.

1.9 Concentrating Protein Solutions Using a Vacuum and Freezing: Freeze–Drying (Lyophilization)

A popular method to concentrate protein solutions which also aids in preserving the protein is to remove water exploiting the effects of decreased pressure on solutions. The process is known as freeze-drying (lyophilization) and has been used in the food industry for many years, for example, in the production of dried soups. Three steps are involved. The protein solution is frozen and then placed in a vacuum. Initially water is removed by sublimation followed by desorption leaving concentrated protein [25]. The method leaves less than 5% water in the sample and is used for the commercial preparation of many of the antibodies employed in Western blotting. Care must be taken with the components of the solution surrounding the protein as they will also be concentrated and may influence downstream processing. Components of the solution may reduce the efficiency of the method. Consideration should be given for the inclusion of cryoprotectants and lyoprotectants [26].

1.10 Concentrating Proteins Using Affinity Chromatography

Proteins and enzymes can have unique protein/ligand interactions which can be exploited to isolate a particular protein from solution. Protein egg-white avidin binds to biotin, and the interaction has been exploited to isolate biotin-containing enzymes like acetyl-CoA carboxylase using an avidin affinity matrix. Other enzymes, like glutathione-S-transferase, can be isolated using the enzyme's substrate, in this case glutathione, on an affinity matrix. Recombinant proteins are often expressed with a fusion partner, for example, glutathione-S-transferase or maltose-binding protein, or are expressed with additional histidine residues in their amino acid structure. His-tagged proteins are purified using metal chelate chromatography [27]. Passage of the crude *E. coli* protein lysate over the affinity matrix isolates the recombinant protein from other unwanted *E. coli* proteins and concentrates the protein. A His-tagged protein as an example of affinity purification is described.

1.11 Concentrating Proteins with Antibodies in Immunoprecipitation Assays

Antibodies can be raised in animals against proteins or peptides with amino acid sequences differing from the amino acid sequence of the animal's own proteins. Antibodies against proteins from an infectious agent can be isolated and purified from the serum of animals during or after an infection. Single-chain variable fragment antibodies can be obtained from phage display libraries by panning with an antigen or peptide. Antibodies against a protein or antibodies against a peptide domain in the protein's amino acid sequence can be used to obtain the protein partner from a cell lysate or a pool of proteins and in the process both isolate and/or concentrate the partner protein. Antibodies, in the technique, interact with the native protein structure, unlike in a Western blot. A technique that is used to do this is known as immunoprecipitation. Immunoprecipitation is similar to "pull-down" assays which use a bait ligand in place of an antibody to interact with the target protein.

The solution containing the protein is incubated with the antibody raised/produced against the protein. The binding of antibody to antigen, taking place in this step, is often improved if the experiment is performed at 4 °C. To separate the antibody-protein complex, protein A (or other forms of the protein known as protein G or H) covalently attached to Sephadex or Sepharose beads or a magnetic particle is added to the solution. Protein A (or G, etc.) has a specific affinity for the Fc region of antibodies and thus forms a protein A/antibody/target protein complex. The complex at this step is still surrounded by all other proteins in the starting protein mix. A layer of 1 M sucrose is introduced under the protein solution as a one-step gradient. The entire mix is centrifuged. The Sepharose protein A/antibody/target protein complex passes through the sucrose bed due to the weight of the Sephadex bead when centrifuged, while the sucrose bed keeps the remainder of the protein solution separate. The protein that was dispersed in a mixture of proteins is separated and concentrated. An alternate method is to use protein A attached to a magnetic bead, and after incubation with the antibody and protein solution, the magnetic bead protein A/antibody/target protein is obtained with a magnet placed against the side of the tube and the unwanted protein solution removed by aspiration. Proteins A, G, and H have different affinities for antibodies and isotypes of antibodies from different species of animals and need to be chosen appropriately. There are a number of variations on the treatment of source material, steps using primary and secondary antibodies, bead composition, and antibody-bead conjugates [28].

The pelleted complex is heated in a SDS-PAGE sample buffer and evaluated on a SDS-PAGE gel. The protein can either be visualized by staining or electrophoretically transferred to nitrocellulose and detected with the same antibody used in the isolation procedure. An alternate method of detection is to metabolically label the proteins of interest in a cell sample with radioactive S^{35}-methionine, and the immunoprecipitated protein can be detected in the SDS- PAGE gel by autoradiography [14]. The method concentrates the proteins enabling the study of antibodies or their generation during an infection or immunization protocol [28].

1.12 Concentrating Proteins Using Thermal Denaturation

Proteins in solution denature with an increase in temperature. Increases in temperature have been used to denature and remove undesirable proteins, for example, when plasma is heated to 56 °C for an hour to "heat inactivate" complement proteins to reduce the complement lysis of cells in culture. Proteins that are thermally stable can be isolated by heat denaturation of other proteins. Studying the thermal stability of a protein has required the use of circular dichroism or differential scanning calorimetry and has been made easier with the use of a thermocycler and a Thermofluor. As a protein is heated in the thermal cycler, its structural integrity decreases exposing hydrophobic regions that can interact with a thermofluor like SYPRO orange and produce fluorescence. The excitation and emission wavelengths of the dye are suitable for use with a *q*PCR machine. The term thermal shift assay has been coined to describe the increased stability some protein drug partnerships have compared to the protein alone. The thermal shift assay has been adapted to identify drug partners in whole cells or cell lysates. A drug is incubated with cells or a cell lysate and then the sample heated to denature unwanted proteins which are removed by centrifugation leaving the drug partner protein(s), a technique known as a cellular thermal shift assay (CETSA) [29, 30].

1.13 Concentrating Protein Solutions for Proteomic and Peptidomic Studies

Biomarkers of disease and peptides and proteins playing metabolic regulatory roles are often present in low abundance and in low concentration in cells, body fluids, and tissue extracts. To enhance the levels of these molecules for proteomic and peptidomic analysis, sample complexity is reduced with the depletion of highly abundant proteins, and then components present in low concentration are selected and analyzed (reviewed by Finoulst 24). Samples for proteomic and peptidomic analysis can be concentrated using ammonium sulfate, polyethylene glycol, organic solvent (particularly TCA and acetonitrile, 5, 6) precipitation, centriconcentration, three-phase partitioning, dialysis, or lyophilization as described in this chapter. Depletion of abundant proteins can be achieved with monoclonal or polyclonal antibodies raised against the individual or a pool of proteins. Chicken antibodies are attractive for depletion studies due to the evolutionary distance of chickens from mammals, and thus chickens have a propensity to generate high-titer antibodies against mammalian proteins [31]. A recently described method employs Amphipol A8–35, an amphipathic surfactant, to precipitate and concentrate proteins which were analyzed by mass spectroscopy [32]. An attractive technique is the use of a combinatorial peptide ligand library to select for and enrich proteins. Abundant proteins in the solution will saturate the appropriate ligand, while the concentration of low-abundance proteins can be enhanced by repeatedly exposing the library to the target protein solution [4]. The combinatorial peptide ligand library approach can detect 1 ng/mL of casein in a liter of sample [4]. The particular combination of techniques to select a protein or remove a solvent for proteomic and peptidomic studies will depend on the starting material and the desired fraction to be analyzed.

2 Materials

2.1 Concentrating Proteins by Precipitation with Ammonium Sulfate and Polyethylene Glycol

1. Ammonium sulfate, polyethylene glycol M_r 6000.
2. 10–50 mL glass beakers, 15 mL centrifuge tubes, magnetic stir bars, and magnetic stirrer.
3. Centrifuge for 15 mL test tubes (10,000 × *g*).

2.2 Concentrating Proteins by Precipitation with Organic Solvents and Trichloroacetic Acid (TCA)

1. Analar grade acetone, ethanol, acetonitrile, trichloroacetic acid.
2. 10–50 mL glass beakers and 15 mL centrifuge tubes (compatible with acetone if using acetone), magnetic stir bars, and magnetic stirrer.
3. A dry-ice bath (or ice-salt bath).
4. Centrifuge for 15 mL test tubes (12,000 × *g*).

2.3 Concentrating Proteins for SDS–PAGE with SDS/KCl Precipitation

1. Sodium dodecyl sulfate (5% w/v), 1 mL of 3 M KCl.
2. 1.5 mL microfuge tube.
3. Microfuge for 1.5 mL microfuge tubes.

2.4 Concentrating Proteins with Three-Phase Partitioning

1. Tertiary butanol (note the solution solidifies <25 °C), ammonium sulfate.
2. 15 mL centrifuge tube.
3. Centrifuge for 15 mL test tubes, swing out rotor (12,000 × *g*).

2.5 Concentrating Protein by the Absorption of Solvent

1. Polyethylene glycol M_r 20,000.
2. Dialysis membrane (Mwt cutoff M_r 12,000). Clamps (to seal the dialysis tube).

2.6 Concentrating Proteins with Centrifugation

1. Centrifuges for 1.5, 15, or 50 mL centrifuge tubes.
2. Vivaspin® (Sartorius), Centricon® (Millipore), Protein Concentrators (Fisher Scientific) or Nanosep® (Pall), or other supplier of centrifuge concentrating tubes. Tubes come in small (1.5 mL or less) up to 70 mL ranges (depending on the supplier). Choose the most appropriate molecular weight cutoff limit of the tube to suit the protein being concentrated.
3. Centrifuge with a fixed angle or swing out rotor. The trend is a fixed angle rotor for small volumes and swing out rotor for larger volumes. Check which rotor type with appropriate speed (*g*) capability is recommended for each tube type and size with the manufacturer.

2.7 Concentrating Proteins with Increased Pressure

1. Pressure chamber, e.g., Amicon® stirred cell, Sartorius Vivacell, Sterlitech (small-volume disposable unstirred and large volume—liters), and helium gas cylinder.
2. Filter membrane. Choose the membrane pore limit (depending on proteins being concentrated) and chemical compatibility.

2.8 Concentrating Protein Solutions Using Decreased Pressure: Vacuum

1. Centrifuge with attached vacuum pump (Vacufuge®, SpeedVac®, CentriVap®).
2. Water tap, Venturi pump, side-arm flask, dialysis membrane (Mwt cutoff M_r 12,000). Clamps (to seal the dialysis tube).

2.9 Concentrating Protein Solutions Using Freeze-Drying (Lyophilization)

1. Freeze-dryer, −80 °C freezer.

2.10 Concentrating Proteins Using Affinity Chromatography

1. Nickel-nitriloacetic acid (or cobalt carboxymethylaspartate) affinity matrix and chromatography column.
2. Protein solution containing polyhistidine-tagged recombinant protein (5–10 mg/mL resin) in loading buffer (300 mM NaCl, 50 mM NaH_2PO_4, pH to 8.0 with NaOH) maintained at 4 °C.
3. Column wash buffer—Loading buffer containing 10 mM (or a higher concentration) imidazole.

4. Elution buffer made up of column wash buffer which contains either a gradient of imidazole or a specific concentration of imidazole determined by trial and error.
5. Eppendorf tubes or test tubes for sample collection.
6. Spectrophotometer absorbing light in the UV range.

2.11 Concentrating Proteins with Antibodies in Immunoprecipitation

1. Centrifuge to take 1.5 mL centrifuge tubes.
2. 1.5 mL centrifuge tubes.
3. Antibody against the target protein. Protein A (or protein G or L) conjugated to Sepharose or a magnetic bead.
4. 1 M sucrose, phosphate-buffered saline (pH 7.4).

3 Methods

3.1 Concentrating Proteins by Precipitation with Ammonium Sulfate and Polyethylene Glycol

1. Add chosen concentrations of ammonium sulfate or polyethylene glycol M_r 6000 to the protein sample (*see* **Notes 1–3**).
2. Stir the sample continuously and allow 15–30 min for the sample to equilibrate.
3. Place the sample in a centrifuge tube and centrifuge for a minimum of 10 min at 10,000 × *g*. Times can be reduced but need to be determined experimentally.
4. Carefully remove the supernatant.
5. Resuspend the pellet in a minimal volume of appropriate buffer for downstream experimentation (*see* **Notes 4** and **5**).

3.2 Concentrating Proteins by Precipitation with Organic Solvents and Trichloroacetic Acid

1. Trichloroacetic acid. Dissolve 100 g TCA in 85 mL deionized water, and after dissolving the acid, water is added to make 100% (w/v).
2. Add, dropwise, with stirring to reduce local heating and denaturation of protein, ice-cold acetone (to 50% v/v), acetonitrile (1.5 volumes), ethanol (to 60% v/v), and 10% (v/v) TCA to cold protein solution (sample in an ice-salt bath) (*see* **Notes 1–3, 6**).
3. Mix samples thoroughly and keep on ice for 30 min (*see* **Note 7**). If the desired result is the optimum precipitation of protein, then it may be necessary to incubate samples overnight.
4. Centrifuge at 12,000 × *g* for 10 min.
5. Remove the supernatant and resuspend the pellet in an appropriate buffer (*see* **Notes 8–11**).

3.3 Concentrating Proteins for SDS-PAGE with SDS/KCl Precipitation

1. To a 100 μL protein solution, add 10 μL of SDS (5% w/v) in a 1.5 mL centrifuge tube.
2. Mix by inverting the tube.

3. Add 10 μL of 3 M KCl and mix again.
4. Centrifuge at 12,000 × *g* for 5 min.
5. Resuspend the pellet in SDS-PAGE sample loading buffer (*see* **Notes 12** and **13**).

3.4 Concentrating Proteins with Three-Phase Partitioning

1. Add 30% (v/v) tertiary butanol to the sample.
2. Add the desired concentration of ammonium sulfate and allow 15–30 min for equilibration of the salt (*see* **Notes 1–3**).
3. Centrifuge the sample at 12,000 × *g* for 10 min.
4. Carefully (with a Pasteur pipette) remove the upper t-butanol layer.
5. Collect the suspended protein sample—plastic spatula (not metal).
6. Add a minimal volume of appropriate buffer to the protein sample to resuspend the protein (*see* **Notes 14** and **15**).

3.5 Concentrating Protein by the Absorption of Solvent

1. Soak the dialysis membrane (tube) in buffer.
2. Seal (clamp or tie) one end of dialysis membrane.
3. Pipette protein solution into the open end of the dialysis tube which is held upright, and seal (clamp or tie) the top of the tube. The larger the surface area of the tube, the faster the removal of solvent, i.e., a long thin tube is more efficient than a short fat tube.
4. Cover the sealed dialysis tube thoroughly with polyethylene glycol M_r 20,000. Periodically monitor the volume of the liquid in the dialysis tube to ensure that the sample does not dry out. The solvent tends to come out more rapidly as the volume reduces.
5. Once the desired volume is reached, the sample is removed from the dialysis tubing (*see* **Notes 1, 2, 16, 17**).

3.6 Concentrating Proteins with Centrifugation

1. Place the protein solution in the chosen centrifuge concentrating tube.
2. Place the concentrating tube in the rotor. Some tubes require a fixed angle rotor and others a swing out rotor—check with the manufacturer's instructions.
3. Centrifuge at the *g* (rpm) recommended for the chosen tube by the manufacturer. The time required will depend on the volume of solute to be removed.
4. When the desired volume or the trapped volume of liquid is all that remains, decant the liquid (*see* **Notes 1, 2, 18, 19**).

3.7 Concentrating Proteins with Increased Pressure

1. Place the protein solution in the "stirred cell" apparatus. Seal the apparatus and place on a magnetic stirrer.
2. Attach compressed gas supply (helium or nitrogen) by a stiff plastic hose to the apparatus, and place a container to collect solvent under the exit pipe from the chamber. Turn on the magnetic stirrer and the gas supply.
3. Monitor the experiment to ensure liquid remains in the sample chamber. When the desired volume is attained, disconnect the gas supply, and pour out the concentrated protein solution (*see* **Notes 1, 2, 20, 21**).

3.8 Concentrating Proteins Using Decreased Pressure: Vacuum

1. Centrifugation: place sample in the vacuum concentrator (vacuum centrifuge). Centrifuge the sample under vacuum. The time required to achieve the desired volume needs to be tested empirically.
2. Dialysis tube: Place the dialysis membrane in buffer to soak. After soaking, seal (clamp or tie a knot) one end of the dialysis membrane, and add the protein solution. Seal the second end of the dialysis tube.
3. Suspend (by a sewing thread tied around the upper clamp or knot) the dialysis tube in a side-arm flask attached by a hose to a Venturi pump on the water tap. Turn on the water flow. Some laboratories have modified this method to enable the addition of more protein solution to the tube as the volume decreases.
4. Monitor the experiment until the volume has been reduced to the required level, and carefully remove the upper seal (clamp or tie), and remove the remaining protein solution (*see* **Notes 1, 2, 22, 23**).

3.9 Concentrating Proteins Using a Vacuum and Freezing: Freeze–Drying (Lyophilization)

1. Freeze the protein sample. Rapid freezing is recommended.
2. Place sample in vacuum chamber.
3. Turn on the vacuum. Times for the experiment depend on the volume and surface area of the frozen protein solution (*see* **Notes 1, 2, 24, 25**).

3.10 Concentrating Proteins Using Affinity Chromatography

1. The recombinant protein sample is either cycled over 1.0 mL Ni-NTA affinity matrix (50% w/v slurry) in a column or incubated and mixed end over end with the Ni-NTA slurry for 1 h at 4 °C.
2. The sample containing the affinity matrix slurry and the protein mixture is placed into a column. Whether starting as a slurry or already in a column, the column is washed with 20 column volumes of wash buffer. Samples of the unbound protein solution should be retained for analysis to see if the desired protein did indeed bind to the column.

3. Bound protein is eluted from the affinity column by either an imidazole gradient or a one-step increase in imidazole concentration (previously worked out by trial and error with different concentrations of imidazole) and collected in 1 mL fractions. Samples are monitored at A_{280} nm and analyzed by SDS-PAGE and Western blotting to identify fractions containing the desired protein. If unwanted proteins adhere to the column and are eluted with low concentrations of imidazole, that concentration of imidazole can be added to the wash buffer to prevent nonspecific binding (*see* **Notes 1**, **2**, **26**, **27**).

3.11 Concentrating Proteins with Antibodies in Immunoprecipitation/Pull-Down Assays

1. Add 1–5 μL of immune serum or 1 μg antibody solution to 1.0 mL of the protein solution, and incubate with end-over-end mixing (gentle) at 4 °C for 1–2 h (overnight may be necessary with some antibody-protein partners). The control should contain 1–5 μL pre-immune serum or 1 μg of a species and isotype-matched antibody (corresponding to the antibody used in the experiment). All samples and the controls should be treated as follows.
2. Add 10 μL of the protein A agarose to the sample, and incubate with end-over-end mixing (gentle) at 4 °C for 1–2 h (or longer if necessary).
3. Carefully underlay 100 μL of a 1.0 M sucrose bed, and centrifuge the sample (12,000 × *g* for 10 min).
4. Carefully aspirate the supernatant fluid, being careful not to disturb the pellet.
5. Wash the pellet in 200 μL of cold (4 °C) PBS by resuspending the pellet and centrifugation. Repeat this step at least two more times.
6. The final pellet is ready for the addition of SDS-PAGE sample buffer and analysis by SDS polyacrylamide gel electrophoresis or can be analyzed by other downstream processing methods (*see* **Notes 1**, **2**, **28**, **29**).

4 Notes

1. Protein concentration and/or enzyme activity should be determined in solutions before the use of any concentration method and should be determined after using the method to determine the efficiency of the method and potential loss of protein or enzyme.
2. Most of these techniques are conducted at or below 4 °C to preserve protein structure (*1*). This is less of a concern if proteins are being prepared for SDS-PAGE.

3. The increase of volume generated by the addition of chemicals needs to be taken into consideration when calculating the final concentration (in percent) of the additive. For organic solvents it is common practice to indicate that "assuming additive volumes" is used so that others can repeat the technique.
4. Advantages of precipitation with ammonium sulfate and polyethylene glycol: (a) the techniques are "gentle" on proteins; (b) reagents are readily available and; (c) easy to use; (d) proteins are not denatured; and (e) polyethylene glycol can be left in the final purification product.
5. Disadvantages of precipitation with ammonium sulfate and polyethylene glycol: (a) concentrations to precipitate particular proteins need to be determined by trial and error; (b) ammonium sulfate needs to be removed by dialysis or gel filtration before undertaking further procedures except hydrophobic chromatography; (c) salt takes longer at 4 °C than at 25 °C to equilibrate; and (d) specific concentrations of polyethylene glycol may alter the performance of proteins on some size exclusion chromatography columns and can be removed by ultrafiltration (Subheadings 2.6 and 2.7).
6. When protein concentrations are less than 1 μg/mL, 0.15% deoxycholate and TCA 10% v/v can be added to the protein solution to precipitate proteins [13].
7. Some samples, e.g., those in acetonitrile, can be sonicated.
8. For small volumes of acetonitrile, the supernatant can be evaporated in a centrifugal evaporator to concentrate proteins not precipitated by the solvent.
9. Glass vessels should be used when protein solutions need to be kept cold as plastic is a poor conductor of heat.
10. Advantages of concentrating proteins by precipitation with organic solvents and trichloroacetic acid: (a) reagents are readily available and the methods easy to use; (b) samples can be added to sample buffer for running on SDS-PAGE gels; and (c) organic solvents can be removed by vacuum centrifugation.
11. Disadvantages of concentrating proteins by precipitation with organic solvents and trichloroacetic acid: (a) local heating effects when adding organic solvent leads to denaturation of proteins; (b) TCA denatures most proteins; (c) compatible test tubes are needed for acetone; and (d) samples need to be kept in an ice-salt bath.
12. Advantages of concentrating proteins for SDS-PAGE with SDS/KCl precipitation: (a) small volumes of protein sample can be precipitated rapidly, and (b) samples need no further treatment before the addition of SDS-PAGE sample buffer.
13. Disadvantages of concentrating proteins for SDS-PAGE with SDS/KCl precipitation: (a) samples are denatured; (b) small

sample volumes are required; (c) SDS interferes with protein determination methods; and (d) the method cannot be used for zymograms.

14. Advantages of concentrating proteins with three-phase partitioning (TPP): (a) TPP is simple and rapid; (b) TPP is very effective for concentrating a wide range of protein solutions; (c) TPP is preferred to precipitation with ammonium sulfate as relatively low concentrations of ammonium sulfate remain with the proteins; (d) some enzymes are stabilized with t-butanol; (e) some enzyme activities are inhibited with t-butanol; and (f) TPP denatures proteins with quaternary structure (hemoglobin), which is an advantage when isolating blood proteins.
15. Disadvantages of concentrating proteins with three-phase partitioning (TPP): (a) trace quantities of ammonium sulfate may need to be removed; (b) some enzymes may lose activity; (c) protein structure can be altered by t-butanol; (d) TPP is not suitable for the isolation of IgG antibodies or serum albumin; and (e) TPP denatures proteins with quaternary structure (hemoglobin).
16. Advantages of concentrating proteins by the absorption of solvent. PEG 20000 can be reused without a decrease in efficiency, and the method is simple to use.
17. Disadvantages of concentrating proteins by the absorption of solvent. The experiment has to be monitored, and the sample can dry out.
18. Advantages of concentrating proteins with centrifugation: (a) a range of tubes with different capacity are available; (b) the method is simple to execute; (c) membranes with a range of solvent compatibilities are available; and (d) membranes can be chosen to both separate proteins of differing sizes while concentrating the solution.
19. Disadvantages of concentrating proteins with centrifugation: (a) samples may dry out if devices without a minimum retention volume are used; (b) there can be some losses of protein due to sticking to the filtration membrane; (c) there can be some clogging of the filter membrane with protein preventing concentration; (d) samples can take a long time to concentrate; and (e) the tubes are not reusable which increases costs.
20. Advantages of concentrating proteins with increased pressure: (a) devices with a range of volumes are available; (b) membranes with differing compatibilities are available; (c) the technique can be used to separate proteins based on size during concentration; and (d) the technique is preferred to concentration by centrifugation as the proteins solutions are stirred which aids solvent flow through the semipermeable membrane.

21. Disadvantages of concentrating proteins with increased pressure: (a) unsuitable for small volumes; (b) experiment has to be monitored to ensure samples do not dry out; (c) some proteins may stick to the membrane; and (d) some protein solutions can take a long time to concentrate.
22. Advantages of concentrating proteins with decreased pressure—vacuum. Useful for small volumes. The vacuum centrifuge has other uses, e.g., to remove solvents from DNA precipitates.
23. Disadvantages of concentrating proteins with decreased pressure. Concentrating proteins can take a long time, and only small volumes can be used.
24. Advantages of concentrating proteins with freeze-drying (lyophilization): (a) freeze-dried proteins are stable for long periods; (b) the bulk of samples is decreased which reduces space requirements for storage or costs for transport; and (c) samples can be solubilized in desired solvents.
25. Disadvantages of concentrating proteins with freeze-drying (lyophilization): (a) the addition of cryoprotectants and lyoprotectants may be required which could interfere with downstream protein work; (b) salts and other ingredients are concentrated with the protein sample; (c) in the addition of moisture, some proteins undergo disulfide interchange; (d) drying tends to promote a decrease in alpha-helix and random structure and an increase in in beta-sheet structure; and (e) it is difficult to control moisture content of the end product (this is likely to change with the introduction of new technology).
26. Advantages of concentrating proteins with affinity purification: (a) small amounts of protein can be concentrated from a large volume of liquid; (b) in one step, a recombinant protein can be serparated from all the *E. coli* proteins in an *E. coli* lysate; (c) most contaminating proteins can be removed by the judicious use of different concentrations of imidazole in the wash buffer; (d) the purified protein can be obtained in a few hours; and (e) remaining imidazole has little effect on many downstream techniques.
27. Disadvantages of concentrating proteins with affinity purification: (a) an additional step may be necessary to remove imidazole; (b) some *E. coli* proteins may bind to the matrix; and (c) several trials may need to be done to reduce background and to obtain a single concentration of imidazole to use.
28. Advantages of concentrating proteins with immunoprecipitation: (a) the protocol can be completed within a few hours; (b) individual proteins can be isolated from a complex mixture of proteins; (c) small quantities of antibody are required;

(d) small quantities of initial protein solution are required; (e) far less initial sample is required than routine protein isolation techniques; (f) the target protein structure remains intact after concentration; (g) antibodies interact with the native structure of the protein; and (h) samples are ready for downstream analysis or can be heated with SDS-PAGE sample buffer and analyzed directly on a SDS polyacrylamide electrophoresis gel. The heating step in sample buffer separates protein and antibody.

29. Disadvantages of concentrating protein with immunoprecipitation: (a) antibodies are required, and this may necessitate extra experimentation to obtain the antibodies; (b) some antibodies, e.g., IgY, from chickens do not bind to protein A, and a second step can be introduced where a secondary anti-IgY antibody, like mouse or rabbit antibody that binds protein A or G, is introduced after the primary antibody incubation; (c) some antibodies may cross-react with and nonspecifically bind other proteins; (d) protein A or protein G may nonspecifically bind proteins, so controls with protein A alone need to be included in the experiment; (e) an extra step is required if proteins are to be separated from the antibody/antigen complex; and (f) proteins that are the same size as the heavy or light chain of the antibody used cannot be differentiated on a stained SDS-PAGE gel and require additional steps to identify the isolated protein.

References

1. Scopes RK (1988) Protein purification: principles and practice. Springer-Verlag, New York
2. Dennison C (2003) A guide to protein isolation, 2nd edn. Kluwer Academic Publishers, Boston
3. Pohl T (1990) Concentrations of proteins and removal of solutes. In: Deutsher MP (ed) Methods in enzymology, vol 182. Academic Press, London, pp 68–83
4. Righetti PG, Fasoli E, Boschetti E (2011) Combinatorial peptide ligand libraries: the conquest of the "hidden proteome" advances at great strides. Electrophoresis 32:960–966
5. Polson C, Sarkar P, Incledon B, Raguvaran V, Grant R (2003) Optimization of protein precipitation upon effectiveness of protein removal and ionization effect in liquid chromatography-tandem mass spectrometry. J Chromatogr B 785:263–275
6. Jaing L, He L, Fountoulakis M (2004) Comparison of protein precipitation methods for sample preparation prior to proteomic analysis. J Chromatogr A 1023:3170320
7. Ingham KC (1990) Precipitation of proteins with polyethylene glycol. In: Deutscher MP (ed) Guide to protein purification, Methods in enzymology, vol 120. Elsevier Inc., New York, NY, pp 301–308
8. Polson A, Potgieter JF, Larier JF, Mears GEF, Joubert FJ (1964) The fractionation of protein mixtures by linear polymers of high molecular weight. Biochim Biophys Acta 82:463–475
9. Polson A, Coetzer T, Kruger J, von Maltzahn E, van der Merwe KJ (1985) Improvements in the isolation of IgY from the yolks of eggs laid by immunized hens. Immunol Investig 14:323–327
10. Goldring JPD, Coetzer THT (2003) Isolation of chicken immunoglobulins (IgY) from egg yolk. Biochem Mol Biol Ed 31:185–187
11. Wessel D, Flugge UI (1984) A method for the qualitative recovery of protein in dilute solution in the presence of detergents and lipids. Anal Biochem 138:141–143
12. Kay R, Barton C, Ratcliffe L, Matharoo-Ball B, Brown P, Roberts J, Teale P, Creaser C (2008)

Enrichment of low molecular weight serum proteins using acetonitrile precipitation for mass spectrometry based proteomic analysis. Rapid Commun Mass Spectrom 22:3255–3260

13. Koontz L (2014) TCA precipitation. Methods Enzymol 541:4–10
14. Goldring JPD, Brake DA, Cavacini LA, Long CA, Weidanz W (1989) Cloned T cells provide help for malaria specific polyclonal antibody responses. Eur J Immunol 19:559–561
15. Zhitkovich A, Costa M (1992) A simple, sensitive assay to detect DNA-protein crosslinks in intact cells and *in vivo*. Carcinogenesis 13:1485–1489
16. Liu LF, Rowe TC, Liu Y, Tewey KM, Chen GL (1983) Cleavage of DNA by mammalian DNA topoisomerase II. J Biol Chem 258:15365–15370
17. Zhou J-Y, Dann GP, Shi T, Wang L, Gao X, Su D, Nicora CD, Shukla AK, Moore DJ, Liu T, Camp DG II, Smith RD, Qian W-J (2012) A simple sodium dodecyl sulfate-assisted sample preparation method for LC-MS-based proteomic applications. Anal Chem 84:2862–2867
18. Goldring JPD, Ravaioli L (1996) Solubilization of protein dye complexes on nitrocellulose to quantify proteins spectrophotometrically. Anal Biochem 242:197–201
19. Dennison C, Lovrien R (1977) Three phase partitioning: concentration and purification of proteins. Protein Expr Purif 11:149–161
20. Dennison C (2012) Three-phase partitioning. In: Tschesche H (ed) Methods in protein biochemistry. De Gruyter, Berlin
21. Ward W (2012) The art of protein purification. In: Ahmed R (ed) Protein purification. IntechOpen Limited, London
22. Pillay D, Boulange AF, Coetzer THT (2010) Expression, purification and characterization of two variant cysteine peptidases from *Trypanosoma congolense* with active site substitutions. Protein Expr Purif 74:264–271
23. Pike R, Dennison C (1989) Protein fractionation by three-phase partitioning (TPP) in aqueous/t-butanol/water/ammonium sulfate. Anal Biochem 180:169–171
24. Finoulst I, Pinkse M, Van Dongen W, Verheart P (2011) Sample preparation techniques for the untargeted LS-MS-based discovery of peptides in complex biological matrices. J Biomed Biotechnol 2011:245291
25. Jennings A (1999) Lyophilization: introduction and basic principles. Taylor & Francis, Boca Raton, FL
26. Roy I, Gupta MN (2004) Freeze-drying of proteins: some emerging concerns. Biotechnol Appl Biochem 39:165–177
27. Bonhorst JA, Falke JJ (2000) Purification of proteins using polyhistidine affinity tags. Methods Enzymol 326:245–254
28. Bonifacio JS, Dell'Angelica TC, Springer TA (2001) Immunoprecipitation. Curr Protoc Immunol 41:8.3.1–8.3.28
29. Martinez Molina D, Nordlund P (2016) The cellular thermal shift assay: a novel biophysical assay for in situ drug target engagement and mechanistic biomarker studies. Annu Rev Pharmacol Toxicol 56:141–161
30. Alshareef A, Zhang HF, Huang YH, Wu C, Zhang JD, Wang P, El-Sehemy A, Fares M, Lai R (2016) The use of cellular thermal shift assay (CETSA) to study Crizotinib resistance in ALK-expressing human cancers. Sci Rep 6:33710
31. Schade R, Behn I, Erhard M, Hlinak A, Staak C (2001) Chicken egg yolk antibodies, production and application. Springer Lab Manual, Berlin
32. Ning Z, Hawley B, Seebun D, Figeys D (2014) APols-aided protein precipitation: a rapid method for concentrating proteins for proteomic analysis. J Membr Biol 247:941–947

Chapter 5

Lysis Buffer Choices Are Key Considerations to Ensure Effective Sample Solubilization for Protein Electrophoresis

Ewa I. Miskiewicz and Daniel J. MacPhee

Abstract

The efficient extraction of proteins of interest from cells and tissues can be challenging. Here we demonstrate the differences in extraction of the focal adhesion protein Kindlin-2 and the transcriptional repressor Snail from choriocarcinoma cells using NP-40 and RIPA lysis buffer. We also show the use of a more denaturing urea/thiourea lysis buffer for solubilization, by comparing its effectiveness with the often utilized RIPA lysis buffer for solubilization of heat shock proteins (HSP) B1 and B5 and the cytoplasmic adapter protein integrin-linked kinase (ILK) from smooth muscle. Overall, the results demonstrate the importance of optimizing lysis buffers for specific protein solubilization prior to finalizing the experimental workflow.

Key words NP-40 lysis buffer, RIPA lysis buffer, Urea/thiourea lysis buffer, Protein extraction, Protein solubilization, HSPB1, HSPB5, Kindlin-2, Integrin-linked kinase, Snail

1 Introduction

The appropriate extraction of proteins from cells and tissues can be fraught with challenges. For example, intracellular proteins may represent only a tiny fraction of total cellular protein, and therefore it would be difficult to extract these proteins in reasonable quantities without proper lysis and extraction protocols [1]. The hydrophobicity of membrane proteins can also be a significant stumbling block, particularly to 2D gel electrophoresis [2]. The advent of proteomics accelerated the development and modification of strategies for cell lysis as well as protein extraction at both small- and large-scale levels. The result is that in addition to more traditionally known cell and tissue lysis buffers [3, 4], there are now quite a number of commercially available reagents for use in such procedures [1]. Nonetheless, one must consider if a particular lysis buffer is optimally designed to extract the protein(s) of interest,

Biji T. Kurien and R. Hal Scofield (eds.), *Electrophoretic Separation of Proteins: Methods and Protocols*, Methods in Molecular Biology, vol. 1855, https://doi.org/10.1007/978-1-4939-8793-1_5,

particularly the use of the appropriate denaturants and detergent(s) for protein extraction and solubilization.

Several basic criteria are considered for the production of an effective lysis buffer for protein extraction: (1) pH, (2) ionic strength, (3) detergents and denaturants, and (4) constituents to improve protein stability or prevent proteolysis. Specific details on these parameters are described in detail elsewhere (*see* **Note 1**) [1, 5–8]. Although Gromov et al. [9] have reported the development of a lysis buffer that could be used for several different proteomic strategies, it has so far proven difficult to design a lysis buffer effective for isolation of a broad array of proteins from many subcellular compartments. This has necessitated the tailoring of buffers and protocols for individual use.

The small heat shock protein B (HSPB) family is comprised of 11 small molecular weight proteins (HSPB1–HSPB11; 15–40 kDa) that are key for cellular homeostasis and are induced by physiological stressors such as temperature, oxidative stress, and biophysical forces [10, 11]. These proteins have the ability to form low and high molecular mass (>300 kDa) complexes with other heat shock proteins and possess a dynamic quaternary structure [12–14]. In periods of stress, HSPB family members can bind denatured proteins and prevent their irreversible aggregation [15], thus aiding in the assembly, disassembly, stabilization, and internal transport of intracellular proteins [16]. Some HSPB members, such as HSPB1, can be found in high quantities (e.g., 2 mg per g of tissue protein) within specific tissues such as different types of muscle [15]. HSPB members like HSPB1 and HSPB5 also appear to interact with integrins and the actin cytoskeleton at focal adhesions and to be present in endosomes [10]. These characteristics and interactions present challenges to the effective extraction and solubilization of such proteins for SDS-PAGE and immunoblot analysis. Clearly, the determination of optimal lysis buffer conditions would be especially important if relative abundance of a specific protein(s) was going to be calculated between control and experimental conditions with downstream immunoblot analysis. Efficient immunoprecipitation of proteins could also be affected by suboptimal lysis conditions.

The use of more standard tissue lysis buffers such as NP-40 and RIPA lysis buffers is quite common for extraction of proteins from tissues, such as muscle, or established mammalian cell lines [17–22]. Using the focal adhesion protein Kindlin-2 and the transcriptional repressor Snail as examples, differences in extraction of these proteins from BeWo choriocarcinoma cells using NP-40 and RIPA lysis buffers were demonstrated. Tyson and colleagues [23] showed very effective extraction of small stress proteins from uterine smooth muscle with a urea/thiourea lysis buffer (*see* **Note 2**) and subsequent SDS-PAGE and immunoblot analysis. Thus, using stress proteins such as HSPB1 and HSPB5 and the cytoplasmic adapter protein integrin-linked kinase (ILK) as examples, the ability of RIPA lysis buffer and urea/thiourea lysis buffer to solubilize such proteins from uterine smooth muscle was assessed. Overall, the data shown here demonstrate that lysis buffer choice should be evaluated prior to the establishment of an experimental workflow.

2 Materials

All aqueous solutions utilized ultrapure water (~18 MΩ/cm^2), and all reagents used were electrophoresis or analytical grade.

1. Phosphate-buffered saline: Dissolve 4 g NaCl, 0.1 g KCL, 0.72 g Na_2HPO_4, and 0.12 g KH_2PO_4 in 500 mL of ultrapure water, and adjust the pH to 7.4 with HCl. Filter sterilize and store at room temperature.
2. Modified RIPA lysis buffer: 50 mM Tris–HCl (pH 7.5), 150 mM NaCl, 1% (v/v) Triton X-100, 1% (w/v) sodium deoxycholate, and 0.1% (w/v) SDS. Dissolve 3.02 g Tris base, 4.38 g NaCl, 5 g deoxycholic acid, and 0.5 g SDS in 400 mL of ultrapure water. Add 5 mL Triton X-100; mix and bring up the volume to 500 mL following adjustment of pH to 7.5 with HCl. Filter sterilize and store at 4 °C (*see* **Note 3**).
3. NP-40 lysis buffer: 50 mM Tris (pH 8.0), 150 mM NaCl, 1% Nonidet P-40 [3]. Dissolve 0.88 g Tris base and 0.61 g NaCl in 80 mL of ultrapure water. Add 10 mL NP-40 detergent; mix and bring up the volume to 100 mL following adjustment of pH to 8.0 with HCl (*see* **Note 3**).
4. Urea/thiourea lysis buffer: 7 M urea, 2 M thiourea, and 4% (w/v) CHAPS, in 30 mM Tris–HCl solution (pH 8.5) [23]. Combine 4.2 g urea, 1.52 g thiourea, and 0.4 g CHAPS in 10 mL of 30 mM Tris–HCl (pH 8.5). Dissolve one tablet each of Mini EDTA-free protease and PhosSTOP phosphatase inhibitors (Roche Applied Science, Indianapolis, IN, USA), and store the buffer in 1 mL aliquots at −80 °C (*see* **Note 4**).
5. 1.5 M Tris–HCl resolving gel buffer: Dissolve 18.2 g of Tris base in ultrapure water to a final volume of 100 mL following adjustment of pH to 8.8 with HCl. Store at 4 °C.
6. 0.5 M Tris–HCl stacking gel buffer: Dissolve 6.1 g Tris base in ultrapure water to a final volume of 100 mL following adjustment of pH to 6.8 using HCl. Store at 4 °C.
7. SDS-PAGE 12% resolving gel composition: Combine 6 mL 30% acrylamide mix (29:1), 3.8 mL 1.5 M Tris-Cl (pH 8.8), 0.15 mL 10% SDS, 0.15 mL freshly made 10% ammonium

persulfate (w/v), and 0.006 mL TEMED in ultrapure water to a final volume of 15 mL.

8. SDS-PAGE 4% stacking gel composition: Combine 1 mL 30% acrylamide mix (29:1), 2 mL 0.5 M Tris–HCl (pH 6.8), 0.08 mL 10% SDS, 0.04 mL freshly prepared 10% ammonium persulfate, and 0.008 mL TEMED in ultrapure water to a final volume of 8 mL.
9. SDS-PAGE running buffer (5×): Dissolve 15.1 g Tris base and 94 g glycine in ultrapure water. Add 50 mL 10% SDS and adjust volume to 1 L in ultrapure water. Refrigerate until use. Dilute to 1× with ultrapure water when required.
10. SDS-PAGE loading dye (2×): Mix 2 mL 0.5 M Tris–HCl (pH 6.8), 4 mL 10% SDS, 2 mL glycerol, 1 mL 2-mercaptoethanol, and 0.02 g bromophenol blue. Adjust volume to 10 mL with ultrapure water. Aliquot and store at −20 °C.
11. Transfer membranes: 0.22 μm nitrocellulose membranes.
12. Gel transfer buffer: Dissolve 2.9 g glycine, 5.8 g Tris base, and 0.37 g SDS in 800 mL of ultrapure water, and then add 200 mL of methanol for a final volume of 1 L.
13. TBST: Dissolve 2.42 g Tris base and 8.0 g NaCl in ultrapure water, and adjust the pH to 7.6 with HCl. Add 1.0 mL of Tween-20 and bring the volume to 1 L. Buffer can be stored in the refrigerator.
14. Immunoblot blocking buffer: Dissolve 5 g fat-free skim milk powder in TBST buffer, and mix by vigorous shaking. For antisera requiring bovine serum albumin (BSA) for blocking, dissolve 5 g BSA (Fraction V) in 100 mL TBST buffer and mix vigorously.
15. Colloidal blue protein stain: Colloidal Blue Staining Kit.
16. Electrophoresis and transfer system: Mini-PROTEAN 3 system.
17. Whatman paper: 3 mm chromatography paper.
18. Molecular weight protein markers: Precision Plus All Blue Protein Standards.
19. Pierce Reversible Protein Stain Kit for nitrocellulose membranes.
20. Antisera: Rabbit polyclonal anti-Kindlin-2 (Sigma-Aldrich Chemical Co., St. Louis, MO, USA), anti-GAPDH (Abcam, Inc., Cambridge, MA, USA), anti-ILK (Cell Signaling Technology, Inc., Danvers, MA, USA), anti-HSPB1 (EMD-Millipore, Etobicoke, ON, Canada), anti-HSPB5 (Enzo Life Sciences, Farmingdale, NY, USA), or mouse monoclonal anti-Snail (Cell Signaling Technology).

21. Tissue: Rat uterine smooth muscle tissue from d23 of pregnancy.
22. Cells: BeWo choriocarcinoma cells (American Type Culture Collection, Manassas, VA, USA).

3 Methods

3.1 Tissue Collection

Tissue samples should be isolated quickly and placed in ice-cooled PBS for rapid washing. Subsequently, place tissues in polypropylene vials, and freeze in liquid nitrogen (*see* **Note 5**).

3.2 Protein Solubilization

1. For BeWo cells, wash cells with PBS, then add 0.2 mL of lysis buffer to the cells, and collect with a plastic cell scraper. Place the mixture in a pre-chilled 0.5 mL lysis tube (Precellys CK14 lysis kit) containing ceramic beads and homogenize cells with a Minilys Bead Mill using a 10 s burst. Centrifuge the sample at full speed for 15 min in a microcentrifuge, and collect supernatants for protein analysis.
2. For uterine smooth muscle samples, chip off pieces of frozen tissues, weigh the fragments (~100–250 mg) in pre-cooled weigh boats, and place in a pre-cooled mortar on dry ice. Grind the samples into a fine powder with a pestle under liquid nitrogen. The use of a fume hood is suggested to avoid inhalation of vapors from the liquid nitrogen. Transfer the powdered samples to 1 mL urea/thiourea lysis buffer (*see* **Notes 2** and **4**) or 1 mL RIPA lysis buffer in 15 mL polypropylene tubes and homogenize up to 1 min on ice with a Polytron PT10–35 homogenizer (*see* **Note 6**).
3. For samples homogenized in urea/thiourea lysis buffer, allow the lysates to settle at room temperature for 30 min while RIPA tissue lysates are kept on ice. Subsequently, transfer all sample lysates to appropriately labeled microcentrifuge tubes, centrifuge at full speed for 15 min in a microcentrifuge, and collect supernatants for protein analysis.
4. Determine sample protein concentrations using the Bradford assay [24].

3.3 SDS–PAGE and Electroblotting

1. Prepare a polyacrylamide gel casting module according to the instructions provided by the appropriate manufacturer (*see* **Note** 7).
2. Immediately after the addition of TEMED to the resolving gel mixture, add the mixture to the prepared gel cassette with a Pasteur pipet (*see* **Notes 8** and **9**). Add isopropanol over the top of the resolving gel to ensure that gel polymerization is not inhibited. After 45 min, remove the overlaid isopropanol by

tipping the gel molds to pour off the solvent, and soak up residual isopropanol with Kimwipes. Pour the stacking gel in the same manner as the resolving gel. Insert the appropriate gel comb into the stacking gels ensuring that no air bubbles are trapped under the teeth of the comb.

3. Once gel polymerization is complete and the gel is assembled in the electrophoresis tank, incubate protein samples (e.g., volumes equivalent to 20 μg protein) with equal volumes of 2× SDS-PAGE loading dye at 95 °C for 5 min prior to gel loading (*see* **Note 10**). Run the gel at 50 V until samples and the prestained molecular mass standards enter the resolving gel; then separate proteins at 100 V until the dye front reaches the bottom of the gel (*see* **Note 11**).
4. Following electrophoresis, gently pry open the gel plates with a plastic wedge to recover the gel.
5. To help assess the effective solubilization of sample proteins with the different lysis buffers, stain the polyacrylamide gel with a Colloidal Blue Staining Kit according to the manufacturer's instructions. Photograph the gel with a gel documentation system.
6. For immunoblot analysis skip **step 5** and place the gel in transfer buffer. Cut a nitrocellulose membrane to the same size as the gel, and also place it in transfer buffer. Assemble the gel for electroblotting as has been described in detail elsewhere [25, 26] (*see* **Note 12**), and conduct electroblotting for 1 h at 300 mA in transfer buffer with constant buffer stirring at 4 °C.
7. To help assess effective solubilization and transfer of sample proteins in an alternative way to **step 5** above, reversibly stain the immunoblot using a Pierce Reversible Protein Stain Kit according to the manufacturer's instructions, and photograph the blot with a digital immunoblot imaging system. Following erasure of the protein staining, proceed to Subheading 3.4, **step 1**.

3.4 Immunoblot Analysis

Unless otherwise stated, all incubations should be conducted at room temperature and with constant agitation.

1. Rinse the membrane with TBST for 5 min.
2. Block the blot in 5% skim milk powder/TBST or 5% BSA/TBST (depending on antisera) for 1 h (*see* **Note 13**).
3. Incubate the membrane in appropriate antisera, diluted in blocking solution, for 1 h.
4. Rinse the blot 1 × 15 min in TBST, followed by 2 × 5 min in TBST.

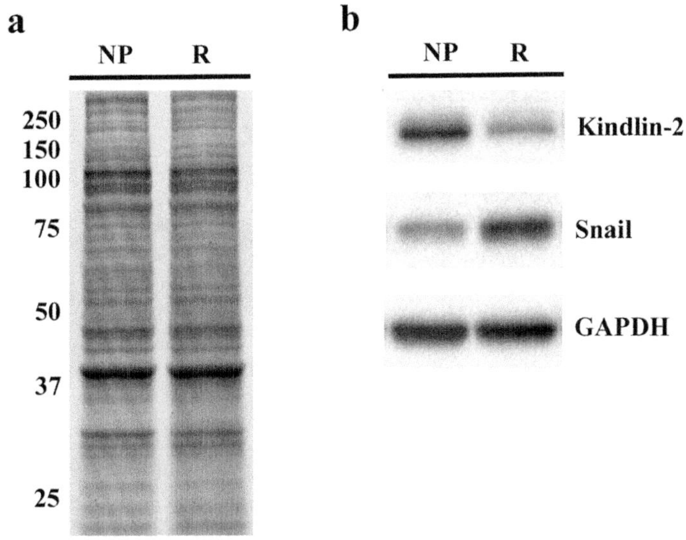

Fig. 1 SDS-PAGE and immunoblot analysis of Kindlin-2 and Snail expression using NP-40 and RIPA lysis buffers. Proteins were extracted from BeWo choriocarcinoma cells with NP-40 (NP) or RIPA (R) lysis buffer followed by SDS-PAGE and immunoblot analysis. (**a**) A Mini-PROTEAN TGX stain-free precast acrylamide gel (Bio-Rad) was utilized for SDS-PAGE. Thus, total protein was activated and photographed with a ChemiDoc MP imaging system. (**b**) Immunoblot analysis demonstrated that Kindin-2 was solubilized to a greater degree in NP-40 lysis buffer, while Snail was solubilized to a greater degree in RIPA lysis buffer even though the overall extent of total protein extraction was comparable. These results indicate the care that must be taken to predetermine the optimal lysis buffer prior to an experimental workflow as proteins of interest could remain in the insoluble fraction. Representative immunoblots are shown ($n = 4$)

5. Incubate membranes for 1 h in HRP-conjugated goat anti-rabbit IgG (H + L) or HRP-conjugated goat anti-mouse IgG (H + L) antisera (1:10,000 dilution) diluted in blocking solution.
6. Wash the blot 1 × 15 min in TBST and then 4 × 5 min in TBST.
7. Detect proteins on the immunoblot using the Pierce Super-Signal West Pico Chemiluminescent Substrate detection system (Figs. 1 and 2). Generate multiple images by capturing a timecourse of chemiluminescence signal with a digital imaging documentation system.

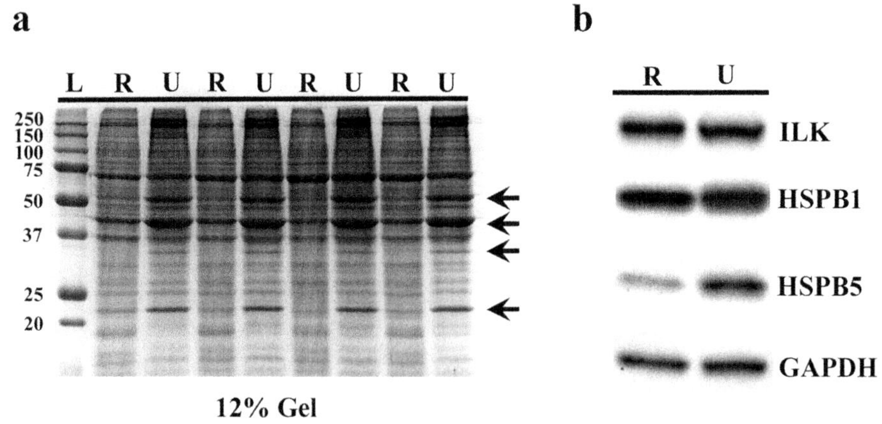

Fig. 2 SDS-PAGE and immunoblot analysis using RIPA and urea/thiourea lysis buffer. (**a**) Proteins were extracted from pregnant rat uterine smooth muscle with RIPA (R) or urea/thiourea (U) lysis buffer and loaded on a 12% polyacrylamide gel for SDS-PAGE. The gel was stained with Colloidal Blue and destained. Arrows indicate examples of protein species that appeared to solubilize to different extents in the two lysis buffers. L, protein molecular mass ladder listed in kDa. Paired lanes represent tissue lysates obtained from different animals ($n = 4$). (**b**) Representative immunoblot analyses of ILK, HSPB1, HSPB5, and GAPDH. Proteins extracted were separated by SDS-PAGE and electroblotted. Protein-specific primary antisera were then used for immunoblot analysis of the proteins obtained from the different lysis conditions. Both ILK and HSPB1 were comparably detected from smooth muscle tissue lysates prepared in both lysis buffers. In contrast, HSPB5 was more readily detected from lysates prepared in the urea/thiourea lysis buffer indicating it may be a more optimal lysis buffer for solubilization of this protein of interest

4 Notes

1. Some researchers like to modify or develop their own lysis buffers, and Harlow and Lane [3] recommended considering the following range of variables to optimize the lysis buffer for downstream western blot analyses: salt concentrations 0–1 M, nonionic detergents 0.1–2%, ionic detergents 0.01–0.5%, divalent cation concentrations 0–10 mM, EDTA concentrations 0–5 mM, and pH 6–9.
2. The urea/thiourea lysis buffer appears to have been originally used for extraction of skeletal muscle-specific proteins in 1983 by Yates and Greaser [27, 28]. Their utilization of thiourea was based on the report of Pace and Marshall [29] indicating thiourea was a potent protein denaturant. The use of this type of buffer, particularly in 2D gel electrophoresis, has been described in detail elsewhere [30, 31]. The combined use of urea and thiourea increases protein solubility since urea is effective at disrupting hydrogen bonds aiding protein unfolding and denaturation, while thiourea is much better at reducing hydrophobic interactions between proteins [32]. The volumes

of lysis buffers utilized are also very important for lysis efficiency. Gorg and colleagues [31] have previously reported the use of 1 mL of this lysis buffer with ~50–100 mg of mammalian tissue. In our hands, it has proven reasonable to use up to a maximum of 250 mg of tissue with a 1 mL volume. It is recommended that the appropriate volume be determined by the investigator on a case-by-case basis.

3. When required for lysis, take 10 mL aliquots, and completely dissolve one tablet each of Mini EDTA-free protease and PhosSTOP phosphatase inhibitors prior to use. Any unused buffer can be frozen at −80 °C for future use. As with just about every lysis buffer, care should be taken to ensure that all components are in solution prior to use as cold storage can lead to precipitation of some constituents (e.g., SDS).
4. It is recommended that the urea/thiourea lysis buffer be prepared fresh whenever possible, but it can also be aliquoted (1 mL) and stored at −80 °C for up to several months. It has also been reported that once the buffer is thawed, it should not be re-frozen [30, 31].
5. Cells and tissues should be frozen rapidly with liquid nitrogen to avoid protease degradation of proteins in the sample or collected and lysed quickly, preferably while chilled. Since proteases as well as phosphatases can be released during lysis and act on your target protein(s), protease and phosphatase inhibitors should be included in the lysis buffers. Many of these are produced as cocktails in tablet form for easy purchase, and their use is as simple as dissolving the tablet in the lysis buffer prior to utilization.
6. Methods utilized for tissue disruption clearly depend on the tissue origin. There are a large number of other means to lyse cells/tissues, and readers are directed to Simpson [33] for specific details and discussion of these protocols. When using mechanical homogenization, it is important to avoid the production of excessive amounts of foam as this could decrease your recovery volume (i.e., becomes difficult to recover from the homogenizer). Short bursts of mechanical homogenization, while the sample(s) is cooled with ice, are usually best.
7. Place two to three folded Kimwipes under the thermoplastic rubber gaskets of the casting module and a strip of Parafilm on top of each gasket. This prevents leaks by increasing the thickness of the rubber gaskets upon which the glass plates are held against with a spring-loaded lever. Leakage can be a problem as the gaskets age and lose their flexibility and overall thickness from constant use.

8. It is critical to use high-quality SDS from a single source and polyacrylamide that is free of contaminating metal ions. Sambrook and Russell [25] have reported that the migration pattern of polypeptides can change significantly when SDS from different manufacturers are interchanged. Purchase of pre-made acrylamide from reputable companies is becoming the normal procedure. Acrylamide solutions with a 1:29 bisacrylamide/acrylamide ratio are usually employed as they are capable of resolving polypeptides differing in size as little as 3% [25], but solutions can be purchased with different ratios if necessary to vary the pore size of the gel [26]. In addition, Tris base should always be used for the preparation of gel buffers to avoid production of diffuse protein bands and even improper polypeptide migration [26].
9. Leave approximately 1 cm of space below the eventual bottom of the combs for the later addition of the stacking gel mixture.
10. It has been noted that heating of samples containing urea for 2D gel electrophoresis can result in some decomposition of urea and release of isocyanate leading to protein carbamylation and charge heterogeneities of the samples [34]. However, in this instance there is no need to worry about protein carbamylation during heating of the samples in 2× SDS loading dye at 95 C as the samples are not being used for isoelectric focusing. This heating step is necessary to produce SDS-polypeptide complexes for subsequent SDS-polyacrylamide electrophoresis.
11. The system used here is a discontinuous buffer system. As a result, the SDS-polypeptide complexes in the 4% stacking gel become deposited and concentrated on the surface of the resolving gel. The SDS-polypeptide complexes are then separated in the resolving gel according to size by molecular sieving in a zone of uniform voltage and pH. Greater details on the mechanism of polyacrylamide gel electrophoresis are found elsewhere [25].
12. There are now many options for transfer of polypeptides to membranes, and the reader is directed to a review of these procedures [26]. It is also imperative that no air bubbles be trapped between the nitrocellulose membrane and the polyacrylamide gel as this will result in the lack of polypeptide transfer to the membrane in these regions. Use a blot roller to remove any bubbles between the gel and membrane.
13. The researcher should consider the blocking buffer that is most appropriate for the specific antiserum (e.g., skim milk powder vs BSA). Blocking a blot serves two important purposes. The first is well known in that it can help mask any potential non-specific binding sites on the membrane itself. The second

purpose, being less known and even less understood, is that blocking a membrane can promote renaturation of antigenic sites [35]. However, it has been reported that prolonged blocking times (>24 h) can actually remove antigens [36].

Acknowledgment

This work was supported by a Natural Sciences and Engineering Research Council Discovery Grant (#250218), an Establishment Grant from the Saskatchewan Health Research Foundation (SHRF; #2695), and a regional partnership program grant from SHRF (#2776) and the Canadian Institutes of Health Research (#ROP-101051) to DJM.

References

1. Grabski AC (2009) Advances in preparation of biological extracts for protein purification. Methods Enzymol 463:285–305
2. Cordwell SJ (2008) Sequential extraction of proteins by chemical reagents. In: Posch A (ed) 2D PAGE: sample preparation and fractionation, Methods in molecular biology, vol 424. Humana Press, New Jersey, pp 139–146
3. Harlow E, Lane E (1988) Immunoprecipitation. In: Harlow E, Lane D (eds) Antibodies: a laboratory manual. Cold Spring Harbor Laboratory, Cold Spring Harbor, NY, p 231
4. Rosenberg IM (2005) Protein analysis and purification, 2nd edn. Birkhauser, Boston, p 37
5. Helenius A, Simons K (1975) Solubilization of membranes by detergents. Biochim Biophys Acta 415:29–79
6. Helenius A, McCaslin DR, Fries E, Tanford C (1979) Properties of detergents. Methods Enzymol 56:734–749
7. Dawson RMC, Elliot DC, Elliot WH, Jones KM (1986) pH, buffers, and physiological media. In: Data for biochemical research. Oxford University Press, New York, pp 417–448
8. Linke D (2009) Detergents: an overview. Methods Enzymol 463:603–617
9. Gromov P, Celis JE, Gromova I, Rank F, Timmermans-Wielenga V, Moreira JMA (2008) A single lysis solution for the analysis of tissue samples by different proteomic technologies. Mol Oncol 2:368–379
10. MacPhee DJ, Miskiewicz EI (2017) The potential functions of small heat shock proteins in the uterine musculature during pregnancy. Adv Anat Embryol Cell Biol 222:95–116. https://doi.org/10.1007/978-3-319-51409-3
11. Kampinga HH, Garrido C (2012) HSPBs: small proteins with big implications in human disease. Int J Biochem Cell Biol 44:1706–1710
12. Kato K, Goto S, Inaguma Y, Hasegawa K, Morishita R, Asano T (1994) Purification and characterization of a 20 kDa protein that is highly homologous to alpha B crystallin. J Biol Chem 269:15302–15309
13. Pipkin W, Johnson JA, Creazzo TL, Burch J, Komalavilas P, Brophy CM (2003) Localization, macromolecular associations, and function of the small heat shock related protein HSP20 in rat heart. Circulation 107:469–476
14. Sun X, Fontaine J-M, Rest JS, Sheldon EA, Welsh MJ, Benndorf R (2004) Interaction of human hsp22 (HSPB8) with other small heat shock proteins. J Biol Chem 279:2394–2402
15. Gusev NB, Bogatcheva NV, Marston SB (2002) Structure and properties of small heat shock proteins and their interaction with cytoskeleton proteins. Biochem Mosc 67:511–519
16. Drieza CM, Komalavilas P, Furnish EJ, Flynn CR, Sheller MR, Smoke CC, Lopes LB, Brophy CM (2010) The small heat shock protein, HSPB6, in muscle function and disease. Cell Stress Chaperones 15:1–11
17. Williams SJ, Shynlova O, Lye SJ, MacPhee DJ (2009) Spatiotemporal expression of α1, α3, and β1 integrin subunits is altered in rat myometrium during pregnancy and labour. Reprod Fertil Dev 22:718–732
18. Palliser HK, Zakar T, Symonds IM, Hirst JJ (2010) Progesterone receptor isoform expression in the Guinea pig myometrium from

normal and growth restricted pregnancies. Reprod Sci 17:776–782
19. Shynlova O, Dorogin A, Lye SJ (2010) Stretch-induced uterine myocyte differentiation during rat pregnancy: involvement of caspase activation. Biol Reprod 82:1248–1255
20. Huo P, Zhao L, Li Y, Luo F, Wang S, Song J, Bai J (2014) Comparative expression of thioredoxin-1 in uterine leiomyomas and myometrium. Mol Hum Reprod 20:148–154
21. Elustondo PA, Hannigan GE, Caniggia I, MacPhee DJ (2006) Integrin-linked kinase (ILK) is highly expressed in first trimester human chorionic villi and regulates migration of a human cytotrophoblast-derived cell line. Biol Reprod 74:959–968
22. Butler TM, Elustondo PA, Hannigan GE, MacPhee DJ (2009) Integrin-linked kinase can facilitate syncytialization and hormonal differentiation of the human trophoblast-derived BeWo cell line. Reprod Biol Endocrinol 7:51
23. Tyson EK, MacIntyre DA, Smith R, Chan E-C, Read M (2008) Evidence that a protein kinase a substrate, small heat shock protein 20, modulates myometrial relaxation in human pregnancy. Endocrinology 149:6157–6165
24. Bradford MM (1976) A rapid and sensitive method for the quantification of microgram quantities of protein utilizing the principle of protein-dye binding. Anal Biochem 72:248–254
25. Sambrook J, Russell D (2001) Molecular cloning: a laboratory manual. Cold Spring Harbor Laboratory Press, Cold Spring Harbor, NY, p 2344
26. MacPhee DJ (2010) Methodological considerations for improving western blot analysis. J Pharmacol Toxicol Meth 61:171–177
27. Yates LD, Greaser ML (1983) Quantitative determination of myosin and actin in rabbit skeletal muscle. J Mol Biol 168:123–141
28. Yates LD, Greaser ML (1983) Troponin subunit stoichiometry and content in rabbit skeletal muscle and myofibrils. J Biol Chem 258:5770–5774
29. Pace CN, Marshall HF Jr (1980) A comparison of the effectiveness of protein denaturants for β-lactoglobulin and ribonuclease. Arch Biochem Biophys 199:270–276
30. Weiss W, Gorg A (2008) Sample solubilization buffers for two-dimensional electrophoresis. In: Posch A (ed) 2D PAGE: sample preparation and fractionation, Methods in molecular biology, vol 424. Humana Press, New Jersey, pp 35–42
31. Gorg A, Drews O, Weiss W (2006) Extraction and solubilization of total protein from mammalian tissue samples. Cold Spring Harb Protoc. https://doi.org/10.1101/pdb.prot4226 pp 3
32. Rabilloud T (1998) Use of thiourea to increase the solubility of membrane proteins in two dimensional electrophoresis. Electrophoresis 19:755–760
33. Simpson RJ (2009) Preparation of cellular and subcellular extracts. In: Simpson RJ, Adams PD, Golemis EA (eds) Basic methods in protein purification and analysis. Cold Spring Harbor Laboratory Press, Cold Spring Harbor, NY, pp 39–78
34. O'Farrell PJ (1975) High resolution two-dimensional electrophoresis of proteins. J Biol Chem 250:4007–4021
35. Towbin H, Gordon J (1984) Immunoblotting and dot immunobinding—current status and outlook. J Immunol Methods 72:313–340
36. DenHollander N, Befus D (1989) Loss of antigens from immunoblotting membranes. J Immunol Methods 122:129–135

Chapter 6

The Cydex Blue Assay: A One-Step Protein Assay for Samples Prior to SDS Electrophoresis

Thierry Rabilloud

Abstract

Determination of protein concentrations prior to (sodium dodecyl sulfate) SDS electrophoresis is made difficult by the simultaneous presence of SDS and reducers in the buffers used for protein extraction. Reducers interfere with the copper-based assays, while SDS interferes with the dye-binding assays. The combined use of cyclodextrins with a commercial Bradford reagent concentrate, described in this chapter, allows to determine protein concentrations in a Laemmli-type buffer, containing both SDS and reducers, in a single step (without any precipitation) with a simple spectrophotometric assay. The use of various cyclodextrins brings compatibility not only with SDS but also with other nonionic and ionic detergents such as sodium deoxycholate or detergents of the Triton type.

Key words Cyclodextrin, Detergents, Dye-binding, Protein assay, RIPA buffer, SDS electrophoresis, SDS buffer, Bradford assay

1 Introduction

In-gel detection methods are generally used for two purposes. The first one is just protein visualization prior to band excision, often prior to a proteomic analysis of the excised band(s) [1]. The second and more important use is a comparative quantitative analysis for sample comparisons. A typical case is the one of 2D gel-based proteomics, in which the comparisons between samples are based on the intensity of the signals detected on the gels for the separated proteins [2]. In the frame of SDS electrophoresis, comparative quantification between different gel lanes is used either for not-too-complex samples or as a normalization method prior to specific detection, e.g., by Western blotting (e.g., [3, 4]).

However, gel detection has its own limits, and optimal performances require similar amounts of proteins to be loaded on the different 2D gels making a gel series, or on the different lanes making a SDS gel or a series of gels. This is due to threshold effects,

Biji T. Kurien and R. Hal Scofield (eds.), *Electrophoretic Separation of Proteins: Methods and Protocols*, Methods in Molecular Biology, vol. 1855, https://doi.org/10.1007/978-1-4939-8793-1_6,

e.g., lower limits of detection on the one side and saturation on the other side, and these limitations occur for any detection method.

This situation means in turn that the protein concentration must be reasonably determined in the samples that will be used for the electrophoretic separations. For various reasons, e.g., avoidance of protein degradation during sample preparation and storage, samples are generally directly solubilized in strongly denaturing buffers, i.e., containing high concentrations of urea for 2D electrophoresis or high concentrations of SDS and reducers for SDS electrophoresis. Alternatively, lysis buffers that do not destroy the nuclei and release most of the cytoplasmic proteins can be used prior to dilution in the final electrophoresis buffer. A classic example is the so-called RIPA (Radio Immuno Precipitation Assay), which contains a cocktail of nonionic and anionic detergents.

Sensitive protein assays are based on two major principles. The first and most ancient one is the reducing effect of the peptide bond on the copper ion at high pH (the so-called biuret reaction). The Cu(I) ion thus generated is then used to produce a color signal, either using its reducing properties (in the Lowry assay [5]) or using a specific chelator (in the BCA assay [6]).

The second principle used in protein assay is metachromasia, i.e., the ability of a dye to change its color (i.e., its absorption wavelength) depending on the environment, here the binding to proteins. The most popular method is the Bradford assay using Coomassie Blue G250 [7], but other dyes or metal-dye complexes have been used to build protein assays (e.g., [8]).

These principles also determine in turn the interfering chemicals that will be able to bias the protein assays. In the copper-based assays, reducers present in the samples will induce copper reduction without the presence of any protein. In the dye-based assays, chemicals that will be able to bind to the dye and induce the metachromatic shift will be interfering, and this is the case of almost all detergents used for biochemical purposes.

This means in turn that buffers used for SDS electrophoresis, which contain both high concentrations of reducers and SDS, are not easily amenable to protein assays as the buffer strongly interferes with both types of assay. The classical solution is then to precipitate the sample first, in order to remove the interfering substances, and then to measure the protein concentration on the redissolved protein precipitate (e.g., in [9, 10]). Although efficient, such protocols are more cumbersome than direct assays and more prone to errors because of an increased number of steps.

A protocol has been however recently described for a single-step assay of protein concentrations in SDS reducers containing samples. This protocol is based on the Bradford assay and uses cyclodextrins to complex the detergent present in the sample and thus to remove the detergent interference [11]. A simplified version of this assay is presented here.

2 Materials

All solutions are prepared with ultrapure, deionized water and stored at +4 °C without the addition of any preservative.

1. Semi-diluted Bradford reagent: This reagent is prepared from a commercial concentrate (Bio Rad). 20 mL of this reagent are added to 30 mL of ultrapure water to prepare the working semi-diluted reagent (*see* **Note 1**).
2. BSA standard: Bovine serum albumin is dissolved at 10 mg/mL in ultrapure water.
3. Cyclodextrin reagent: Alpha-cyclodextrin and beta-cyclodextrin are dissolved at 5 mg/mL in ultrapure water. The dissolution is facilitated if the water is warmed prior to its addition to the cyclodextrin powder.
4. SDS sample buffer: A classical composition is used. The sample buffer contains 2% (w/v) SDS, 5% (v/v) beta mercaptoethanol, 10% (v/v) glycerol, and 62.5 mM Tris–HCl buffer pH 6.8. Low concentrations of bromophenol blue (up to 0,004%) can be tolerated (*see* **Note 2**). For practical reasons we use a 2× concentrated buffer (4% SDS, 10% mercaptoethanol, 20% glycerol, 125 mM Tris buffer pH 6.8).
5. RIPA buffer: There are several variants of this buffer described in the literature. The RIPA buffer used here contains 1% (v/v) IGEPAL 630, 0.5% sodium deoxycholate, 0.1% SDS, 150 mM sodium chloride, and 50 mM Tris–HCl buffer, pH 8.0. For practical reasons we use a 2× concentrated buffer (2% (v/v) IGEPAL 630, 1% sodium deoxycholate, 0.2% SDS, 300 mM sodium chloride, and 100 mM Tris–HCl buffer, pH 8.0).
6. Semimicro spectrophotometric cuvettes (maximum volume 3 mL).
7. A spectrophotometer able to operate at 595 nm. As the assay proceeds in the visible region of the spectrum, UV transparency is not needed, offering more flexibility regarding the material of the cuvettes.

3 Methods

3.1 BSA Standard Preparation

1. Using the 2× concentrated buffers (either RIPA or SDS sample buffer), water, and the 10 mg/mL BSA solution, prepare serial dilutions of BSA in the working buffer (RIPA 1× or SDS sample buffer 1×). The principle is that different amounts of BSA will be added in the same sample volume to

build the standard range. We use a 0–2 mg/mL concentration range by steps of 0.2 mg/mL, i.e., 0, 0.2, 0.4, 0.6, etc. mg/mL up to 2 mg/mL. Wider steps (e.g., 0.25 mg/mL) can be used but will result in a slightly lower precision. The BSA standard tubes at their final dilution can be stored frozen at −20 °C for several weeks.

3.2 Sample Preparation

1. Prepare samples either in the RIPA buffer or in the SDS sample buffer. Depending on the expected concentration, serial dilutions (e.g., twofold, fivefold, and tenfold) may be required (*see* **Note 3**). Prepare these dilutions in the same buffer as the one used for the sample preparation and for the standard curve.

3.3 Working Reagent for SDS Samples

1. Mix one volume of 5 mg/mL alpha-cyclodextrin solution and one volume of semi-diluted Bradford reagent. Mix by inversion of the tube.
2. Each point (blank, standard curve, and samples) requires 1 mL of this reagent.
3. The working reagent is stable at least 1 week at room temperature.

3.4 Working Reagent for RIPA Samples

1. Mix one volume of 5 mg/mL beta-cyclodextrin solution and one volume of semi-diluted Bradford reagent. Mix by inversion of the tube.
2. Each point (blank, standard curve, and samples) requires 1 mL of this reagent.
3. The working reagent is stable at least 1 week at room temperature.

3.5 Assay Procedure

1. Prepare the required numbers of 1.5 mL or 2 mL microcentrifuge tubes (*see* **Note 4**).
2. In each tube, pipet 1 mL of working reagent (*see* **Note 5**).
3. Add 10 μL of sample per tube, either buffer alone for the blank or BSA standards for building the standard curve or samples to determine their concentration.
4. Mix by several inversions of the tube. The color takes 5 min to develop and is stable for 1 h.
5. Read the absorbance at 595 nm. The simplest procedure is to use the blank tube (buffer + color reagent) to zero the spectrophotometer. Alternatively, use water to zero the spectrophotometer, and read the absorbance of all tubes including the blank(s).

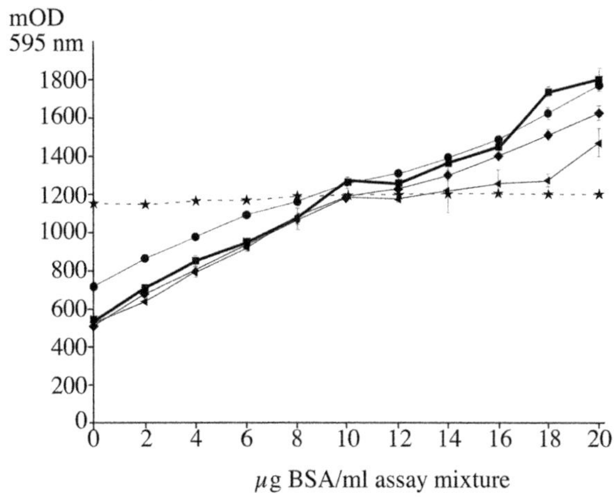

Fig. 1 Standard curves for BSA in various conditions. Standard curves were constructed using BSA in various conditions. Squares and thick line: sample in water, standard Bradford reagent. Stars and dotted line: sample in 10 μL SDS buffer, standard Bradford reagent. Circles and gray line: sample in 10 μL RIPA buffer, Bradford reagent with beta-cyclodextrin. Triangles: sample in water, Bradford reagent with alpha-cyclodextrin. Diamonds: sample in SDS buffer, Bradford reagent with alpha-cyclodextrin. Determinations made in duplicate using water as a blank. Non-visible error bars mean that they are smaller than the symbols. Note that the SDS buffer results in such an interference that no assay is feasible with the classical Bradford reagent. The linearity of the assay is optimal below 10 μg of protein per mL of assay. Average slopes for the 0–10 μg concentration range. Standard Bradford: 73 mOD/μg protein in 1 mL assay. Bradford with alpha-cyclodextrin, sample in water: 66 mOD/μg protein in 1 mL assay. Bradford with alpha-cyclodextrin, sample in SDS buffer: 67 mOD/μg protein in 1 mL assay. Bradford with beta-cyclodextrin, sample in RIPA buffer: 54 mOD/μg protein in 1 mL assay. Note also the higher blank value with beta–cyclodextrin

6. Use the BSA dilution series to determine the standard curve (*see* **Note 6**); then determine the protein concentration of your samples. Typical standard curves are shown in Fig. 1.
7. Use your samples for SDS separations (*see* **Note 7**), as shown in Fig. 2.

4 Notes

1. The commercial concentrated Bradford reagent is a rather viscous liquid and is uneasy to pipet reproducibly in series, as

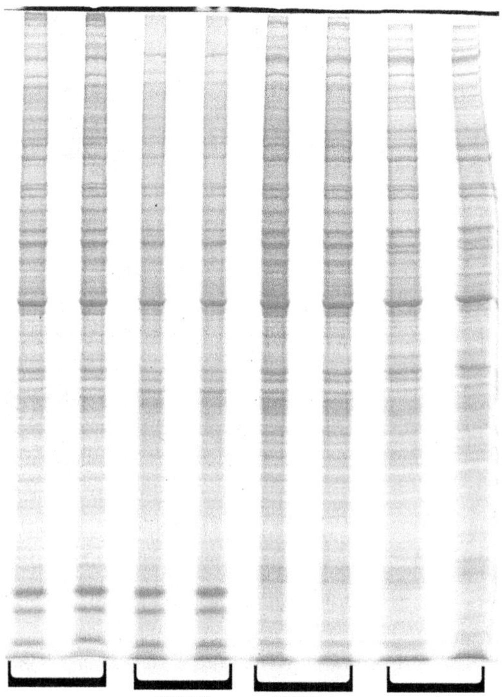

Fig. 2 Example of SDS separation after protein determination. RAW264.7 cells were harvested and distributed in several aliquots, on which different protein extractions were performed. Native: cell pellets were treated with 6 volumes of native lysis buffer (Hepes NaOH 10 mM pH 7.5, KCl 50 mM, MgCl2 2 mM, 3-[tetradecyldimethylammonio]-propane-1 sulfonate (SB 3–14) 0.1% for 30 min at 0 °C. The lysate was cleared by centrifugation (2000 × *g*, 10 min, 4 °C). The supernatant was recovered and the protein concentration measured by a classical Bradford assay. Urea: cell pellets were resuspended in an equal volume of PBS, and the suspension was extracted by four times its volume of urea lysis buffer (9 M urea, 1.25 M thiourea, 5% CHAPS, 12.5 mM TCEP, 25 mM spermine base, 25 mM spermine tetrahydrochloride), and the proteins were extracted for 1 h at room temperature. The lysate was cleared by centrifugation (15,000 × *g*, 15 min 20 °C). The supernatant was recovered and the protein concentration measured by a classical Bradford assay. RIPA: cell pellets were treated with 8 volumes of 1× RIPA buffer for 30 min at 0 °C. The lysate was cleared by centrifugation (2000 × *g*, 10 min, 4 °C). The supernatant was recovered and the protein concentration measured by the modified Bradford assay using the beta-cyclodextrin working reagent. SDS: cell pellets were treated with 8 volumes of 1× SDS sample buffer for 30 min 70 °C. The lysate was cooled down at room temperature and the DNA sheared by passing through a syringe fitted with a 0.8 mm needle. The supernatant was recovered and the protein concentration measured by the modified Bradford assay using the alpha-cyclodextrin working reagent. Protein determinations for RIPA and SDS samples were made on 10× dilutions in the corresponding buffer (10 μL of the final dilution in 1 mL of final assay mixture). Forty micrograms of proteins, as determined from the various assays, were diluted in SDS sample buffer with

needed for a protein assay (standard curve + samples). The use of an intermediate dilution makes the pipetting much easier.

2. Bromophenol blue turns yellow in the very acidic environment of the Bradford assay, and this yellow form does not interfere with the reading at 595 nm.
3. The assay operates between 1 and 20 μg of protein in a 1 mL assay, and the best linearity is achieved between 2 and 10 μg of protein (*see also* Fig. 1). As the sample volume needs to be kept constant, this implies to use serial dilutions of the sample to get at least one measurement in the linear part of the standard curve. As the color is obtained after 5 min, additional sample dilutions can be performed and measured during the 1-h stability period if needed.
4. It is advised not to rely on a single measurement, neither for the standard curve nor for the samples. Triplicate measurements result in a very high confidence determination of the protein concentration, but duplicate measurements are often sufficient.
5. Different working reagents are used for samples in RIPA buffer and for samples in SDS buffer. This means in turn that samples of each type must be read against a standard curve built in the same buffer type and with the adequate working reagent.
6. Different batches of the commercial concentrate and/or aging of the reagent result in slightly different response curves. Maximum precision requires the standard curve to be determined for each series of assays, but an often sufficient precision is obtained when the standard curve is run every week and at every change of a reagent, whichever comes first.
7. If RIPA samples are to be used in SDS electrophoresis, they will be diluted in SDS buffer. This means in turn that the protein concentration can be determined directly in the RIPA buffer or after dilution in the SDS buffer. It is preferable to determine the protein concentration directly in the RIPA buffer, using the beta-cyclodextrin reagent. The various concentrations of IGEPAL and sodium deoxycholate present in the sample after dilution in SDS induce variable interferences with the alpha-cyclodextrin reagent.

Fig. 2 (continued) bromophenol blue (40 μL final volume). Native and RIPA samples, after dilution in SDS, were heated at 95 °C for 5 min. The samples were loaded on the gel, migrated, and stained with colloidal Coomassie Blue [12]. Each lane corresponds to a different sample, starting from a different culture aliquot. This means that the sample preparations and protein concentration determination processes were carried out independently for each lane. The comparable staining intensities in the various lanes show that the protein determination was accurate

References

1. Schirle M, Heurtier MA, Kuster B (2003) Profiling core proteomes of human cell lines by one-dimensional PAGE and liquid chromatography-tandem mass spectrometry. Mol Cell Proteomics 2:1297–1305
2. Rabilloud T (2012) The whereabouts of 2D gels in quantitative proteomics. Methods Mol Biol 893:25–35
3. Welinder C, Ekblad L (2011) Coomassie staining as loading control in western blot analysis. J Proteome Res 10:1416–1419
4. Eaton SL, Roche SL, Llavero Hurtado M, Oldknow KJ et al (2013) Total protein analysis as a reliable loading control for quantitative fluorescent western blotting. PLoS One 8:e72457
5. Lowry OH, Rosebrough NJ, Farr AL, Randall RJ (1951) Protein measurement with the Folin phenol reagent. J Biol Chem 193:265–275
6. Smith PK, Krohn RI, Hermanson GT, Mallia AK et al (1985) Measurement of protein using bicinchoninic acid. Anal Biochem 150:76–85
7. Bradford MM (1976) A rapid and sensitive method for the quantitation of microgram quantities of protein utilizing the principle of protein-dye binding. Anal Biochem 72:248–254
8. Fujita Y, Mori I, Kitano S (1984) Determination of proteins by using the color reaction with pyrocatechol violet-molybdenum(VI) complex. Chem Pharm Bull (Tokyo) 32:4161–4164
9. Bensadoun A, Weinstein D (1976) Assay of proteins in the presence of interfering materials. Anal Biochem 70:241–250
10. Peterson GL (1977) A simplification of the protein assay method of Lowry et al. which is more generally applicable. Anal Biochem 83:346–356
11. Rabilloud T (2016) A single step protein assay that is both detergent and reducer compatible: the cydex blue assay. Electrophoresis 37:2595–2601
12. Neuhoff V, Arold N, Taube D, Ehrhardt W (1988) Improved staining of proteins in polyacrylamide gels including isoelectric focusing gels with clear background at nanogram sensitivity using Coomassie brilliant blue G-250 and R-250. Electrophoresis 9:255–262

Chapter 7

Cellulose Acetate Electrophoresis of Hemoglobin

Ramesh Kumar and Wilbert A. Derbigny

Abstract

The electrophoresis on cellulose acetate membrane is most widely used because of its simplicity, and is without the use of any sophisticated instrument other than electrophoresis apparatus and the cellulose acetate strip. Here we describe a modified version of cellulose acetate membrane electrophoresis for hemoglobin separation from blood sample. Sharp, clear bands without tailing effects can be obtained with this method. The method and apparatus described here would be appropriate to separate protein fractions under 1 h at voltages up to 60 V/cm measured between the electrodes.

Key words Electrophoresis, Amino acid, Protein, Membrane, Samples

1 Introduction

Among the electrophoretic supporting media, cellulose acetate membrane is the most widely used medium in the electrophoretic process. Kohn first introduced it in strip form [1]. The cellulose acetate electrophoresis (also called zone electrophoresis) is simple, rapid, reliable, and suitable for processing several samples at the same time and at low cost of analysis. This technique does not require any sophisticated instruments other than electrophoresis apparatus that is most commonly found in the laboratories, and the cellulose acetate membrane strip. This method has advantages over both agar and filter paper [2]. The cellulose acetate membrane has better resolving power with less absorption of proteins. It gives sharp protein bands without tailing effects. Under identical conditions, moist cellulose acetate membrane is less conductive than filter paper; hence, generation of heat by the Joule effect is not a critical issue, and electrophoresis can be done at higher working voltages. However, evaporation is a serious problem, which could be avoided by using vapor pressure depressants (glycol or glycerol at 5–10%) with the running buffer [3, 4]. The separation of proteins is primarily by charge. This method is being used in both

Biji T. Kurien and R. Hal Scofield (eds.), *Electrophoretic Separation of Proteins: Methods and Protocols*, Methods in Molecular Biology, vol. 1855, https://doi.org/10.1007/978-1-4939-8793-1_7,

clinical diagnosis and basic research. The cellulose acetate membrane electrophoresis may be used for both qualitative and quantitative analysis. Several original papers and book chapters describe this technique for the separation or quantitative analysis of amino acids [5], glycosaminoglycans (GAGs) [6], hemoglobin (Hb) and Hb variants [7–11], and urinary proteins [12].

Here we describe a modified version of cellulose acetate membrane electrophoresis for hemoglobin separation from blood sample. The authors of this study showed sharp, clear bands. Subsequently, some other investigators have also shown the similar results (*see* ref. 4). The method and apparatus described here would be appropriate to separate protein fractions under 1 h at voltages up to 60 V/cm measured between the electrodes.

2 Materials

Prepare all solutions using Milli-Q water (Millipore, sensitivity of 18 MΩ-cm at 25 °C) and analytical grade reagents.

2.1 Electrophoresis

1. Any horizontal electrophoresis tank that allows a bridge gap of 7 cm (alternatively, electrophoresis apparatus from Sebia, Issy-les-Moulineaux, France).
2. A power pack capable of delivering supply 350 V at 50 mA.
3. Cellulose acetate membrane sheets (78 × 150 mm from Shandon Scientific Co. Ltd. or Schleicher and Schuell).
4. A microsyringe for sample application (e.g., Hamilton).
5. Wicks of filter or chromatography paper grade no. 3.
6. Glass or plastic supporting platform.

2.2 Reagents

1. Electrophoresis buffer (Tris-EDTA Boric Acid (TEB) buffer, pH 8.4). Weigh 10.2 g Tris (hydroxymethyl-aminomethane), 0.6 g EDTA, disodium salt (ethylene diamine tetraacetic acid), 3.2 g boric acid, and 10% glycerol (v/v). Dissolve in about 700 mL water and adjust to 1000 mL with water (*see* **Notes 1** and **2**).
2. Fixative/staining solution: Weigh 5 g Ponceau S and 7.5 g trichloroacetic acid (TCA). Dissolve in about 700 mL water and then adjust to 1000 mL with water.
3. Destaining solution: 3% (v/v) acetic acid in methanol.
4. Phenol red, saturated aqueous solution.
5. Bromophenol blue.

3 Methods

Perform all procedures at room temperature unless otherwise specified.

3.1 Cellulose Acetate Membrane Electrophoresis

1. Prepare the electrophoresis apparatus by adding equal amounts of electrophoresis buffer (TEB buffer) in each of the buffer compartments (*see* **Note 3**).
2. Cut two wicks from grade no. 3 chromatography paper, wet, and place one along each bridge ensuring that they make good contact with buffer (*see* **Note 4**).
3. Cut cellulose acetate membrane into strips in 40 × 100 mm each, and soak (shiny side down) in electrophoresis (TEB) buffer for at least 10 min with gentle agitation (*see* **Note 5**).
4. Blot the strips in filter paper to remove excess of liquid. Do not dry. Immediately install strips carefully in electrophoresis tank (*see* **Notes 6** and 7).
5. Turn on the voltage current at 250 V for 5 min to equilibrate the cellulose acetate strips with the buffer.
6. Turn off the voltage current; then from cathode (−) end leaving 1 cm, mark 5 mm traits with a phenol red indicating the places for sample loading. Let the liquid dry.
7. Load diluted samples or standards (e.g., hemolysate 1 μg/μL) by 1–3 μL in bromophenol blue using microsyringe. Let the liquid dry every time (Fig. 1) (*see* **Notes 8–10**).

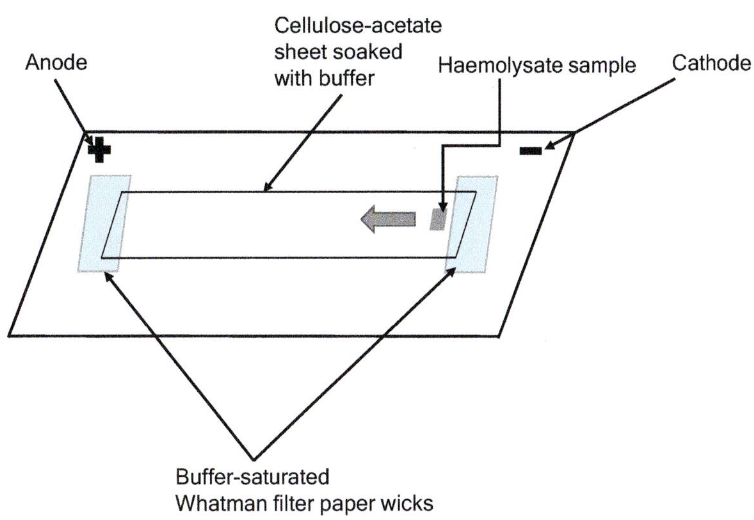

Fig. 1 Schematic presentation of cellulose acetate electrophoresis. Cellulose acetate strip soaked with buffer lying in the electrophoresis tank, hemolysate applied onto marked area of strip on cathode (−) end, and arrow indicates migration of sample toward anode (+) end

8. Then set the voltage at 35–60 V/cm working at constant current of 2 mA for each strip.
9. Run the electrophoresis for approximately 30–45 min until there is a clear area between the bands.

3.2 Staining

1. After the electrophoresis, immediately transfer the cellulose acetate strips to Ponceau S in a reservoir, fix, and stain for 5 min with gentle agitation.
2. Remove excess stain by washing strips in the acetic acid reservoir for 5 min.
3. Repeat washing step two more times in the acetic acid reservoir for 10 min each.
4. Blot once, using clean blotting filter paper, and deposit strips on glass plate, and dry at 80 °C for 10 min.
5. After cooling at room temperature, label the strips; unstuck from the glass plate with a spatula or scalpel.
6. Scan the strips with a scanner and store in a protective plastic envelope.

4 Notes

1. Using magnetic stirrer bar helps to dissolve Tris relatively easily. Tris can be dissolved faster if the water is warmed to about 37 °C. However, care should be taken to bring the solution to room temperature before adjusting pH.
2. To adjust pH of TEB buffer, concentrated HCl (12 N) can be used at first to bring closer to the required pH. From then on it would be better to use lower ionic strength HCl (6 N) to avoid a sudden drop in pH below the required pH.
3. Keep the electrophoresis (TEB) buffer at 4 °C before electrophoresis. Cold TEB buffer would help in heat management during electrophoresis.
4. Any horizontal electrophoresis apparatus may be easily adapted to support cellulose acetate strips. The strips are lying in the gel platform and are connected with the electrode chambers (buffer tanks) by wicks. The wicks are cut from filter paper grade no. 3. Do not submerge the strips in the buffer.
5. Glycerol is very good anti-vapor pressure agent; adding 5–10% glycerol (v/v) to TEB buffer would suppress evaporation during electrophoresis.
6. Do not dry the strips in the electrophoresis tank when the edges sink in the buffer.

7. If air bubbles are caught in the cellulose mesh, keep strips in between two clean filter papers, and remove air bubbles carefully by pressing.
8. 0.1% saponin helps red blood cells in lysis. Blood anticoagulant EDTA is suitable with the samples.
9. It is stressed that for good result use less sample for loading onto the cellulose acetate strips.
10. The samples to be separated are mixed by addition of bromophenol blue to make deposition easier.

Acknowledgment

This work was supported by NIH grant 5RO1AI104944-04 to W.A.D.

References

1. Kohn J (1957) A cellulose acetate supporting medium for zone electrophoresis. Clin Chim Acta 2(4):297–303
2. Desai S, Colah R, Gupte S, Mohanty D (1997) Is cellulose acetate electrophoresis a suitable technique for detection of Hb Bart's at birth? Hum Hered 47(4):181–184
3. Svensson H (1956) Ciba foundation symposium on paper electrophoresis. J. and A. Churhill, London, p 99
4. Afonso E (1961) On the electrophoresis of proteins on cellulose acetate membrane. Clin Chim Acta 6:883–885
5. Scherr GH (1962) Use of cellulose acetate strips for electrophoresis of amino acids. Anal Chem 34(7):777–777
6. Wegrowski Y, Maquart FX (2001) Cellulose acetate electrophoresis of glycosaminoglycans. In: Iozzo RV (ed) Proteoglycan protocols, Methods in molecular biology, vol 171. Humana Press Inc., Totowa, NJ, pp 175–179
7. Evans DIK (1971) Haemoglobin electrophoresis on cellulose acetate using whole blood samples. J Clin Pathol 24(9):877–878
8. Galanello R, Melis MA, Muroni P, Cao A (1977) Quantitation of Hb A2 with DE-52 microchromatography in whole blood as screening test for beta-thalassemia heterozygotes. Acta Haematol 57(1):32–36
9. Mario N, Baudin B, Aussel C, Giboudeau J (1997) Capillary isoelectric focusing and high performance cation-exchange chromatography compared for the qualitative and quantitative analysis of haemoglobin variants. Clin Chem 43:2137–2142
10. Cotton F, Lin C, Fontaine B, Gulbis B, Janssens J, Varteongen F (1999) Evaluation of a capillary electrophoresis method for routine determination of haemoglobins A_2 and F. Clin Chem 45:237–243
11. Barbara JW, Barbara JB (2017) Investigation of variant haemoglobins and thalassaemias. In: Dacie and lewis practical haematology 12th ed.
12. Kubota R, Machii R, Hiratsuka N, Hotta O, Itoh Y, Kobayashi S, Shiba K (2003) Cellulose acetate membrane electrophoresis in the analysis of urinary proteins in patients with tubulointerstitial nephritis. J Clin Lab Analysis 17:44–51

Chapter 8

Native Polyacrylamide Gels

Claudia Arndt, Stefanie Koristka, Anja Feldmann, and Michael Bachmann

Abstract

Proteins can easily be separated by polyacrylamide gel electrophoresis (PAGE) in the presence of a detergent and under (heat-) denaturing and (non- or) reducing conditions. The most commonly used detergent is sodium dodecyl sulfate (SDS). The major function of SDS is to shield the respective charge of the proteins present in the mixture to be analyzed and to provide all proteins with a negative charge. As a consequence, the proteins will be separated according to their molecular weight. Electrophoresis of proteins can also be performed in the absence of SDS. Using such "native" conditions, the charge of each of the proteins, which will depend on the primary amino acid sequence of the protein (isoelectric point) and the pH during electrophoresis, will mainly influence the mobility of the respective protein during electrophoresis. Here we describe a starting protocol for "native" PAGE.

Key words "Native" polyacrylamide gels, "Blue native" polyacrylamide gels, Proteins

1 Introduction

A first hallmark in protein electrophoresis was the introduction of sodium dodecyl sulfate (SDS) (e.g., [1, 2]). In the presence of SDS, the individual charge of a protein is no more relevant for its mobility in a polyacrylamide gel. The detergent forms micelles around the protein molecule. The surface of the micelle is negatively charged. The size of the protein determines the size of the micelle and thereby the mobility. Similar to gel filtration, the separation of a sample during electrophoresis depends on the starting volume. During electrophoresis this initial protein "band" gets broader. Thus, the "sharper" the protein band at the start, the sharper the band after electrophoresis. Therefore, a second hallmark in protein electrophoresis was the introduction of discontinuous gels. A discontinuous gel consists of a stacking and a separation gel. The function of the stacking gel is to concentrate the protein sample at the beginning of the electrophoresis. For this purpose, discontinuous electrophoresis systems use two different buffer systems. At the beginning of the electrophoresis, the pH in the stacking gel is

Biji T. Kurien and R. Hal Scofield (eds.), *Electrophoretic Separation of Proteins: Methods and Protocols*, Methods in Molecular Biology, vol. 1855, https://doi.org/10.1007/978-1-4939-8793-1_8,

adjusted in a way that it results in an acceleration of all the proteins present in the sample. This is mainly due to the fact that the proteins are the only charged ions in the sample at the starting pH in the stacking gel. This low number of charged molecules results in a strong electric field which finally leads to a strong acceleration of the proteins present in the electrophoresis sample. During further electrophoresis, however, this pH in the stacking gel changes usually from an acidic pH around 6 to the pH of the electrophoresis buffer and the separation gel (usually around pH 8). At the starting pH, amino acids (aa, such as glycine) present in the electrophoresis buffer are not charged. However, when the protein sample reaches the separating gel, these aa will now become charged. As a result the electric field breaks down, and during further electrophoresis the proteins are now separated according to their individual mobility. Commonly, the proteins will be heat denatured prior to electrophoresis to separate non-covalent complexes. In parallel, thiol group containing chemicals such as ß-mercaptoethanol or dithiothreitol (DTT) can be added resulting in reduction of disulfide bridges if present either within the protein or between protein subunits. In the latter case, subunits of complexes will be separated. All these developments and modifications dramatically improved the resolution and separation of proteins during electrophoresis. The major obvious disadvantage of SDS-PAGE is that the separated proteins will be heat denatured and complexed by the detergent SDS.

Thus, under certain circumstances electrophoresis using non-denaturing conditions may be of interest, for example, it is possible to recover proteins in their native state after the separation. However, such "native" gel electrophoresis attempts usually are limited due to the lack of all the above described advantages of SDS-PAGE. Especially protein isoforms with only slightly different isoelectric points will not efficiently be separated. Posttranslational modifications such as a varying phosphorylation pattern, methylation, acetylation, and glycosylations will more or less alter the mobility of the protein and influence its separation. In contrast to simple and robust SDS-PAGE protocols, the use of native gel electrophoresis usually requires an optimization of the separation conditions for the respective sample. Although discontinuous native gel electrophoresis system has also been described for native gel electrophoresis (e.g., [3]) and the use of gradient gels may also be helpful to improve the separation, the easiest way to establish a native gel electrophoresis approach is to start by using a continuous polyacrylamide gel electrophoresis protocol consisting of a separation gel which can be prepared by just leaving the SDS in the gel, the sample solution, and the gel electrophoresis buffer. Such a system can already result in a good separation (*see* Fig. 1). As the stacking gel is missing, the starting volume of the sample should be as small as possible and the protein concentration as high as possible. Aside such simple native gel electrophoresis protocols, a

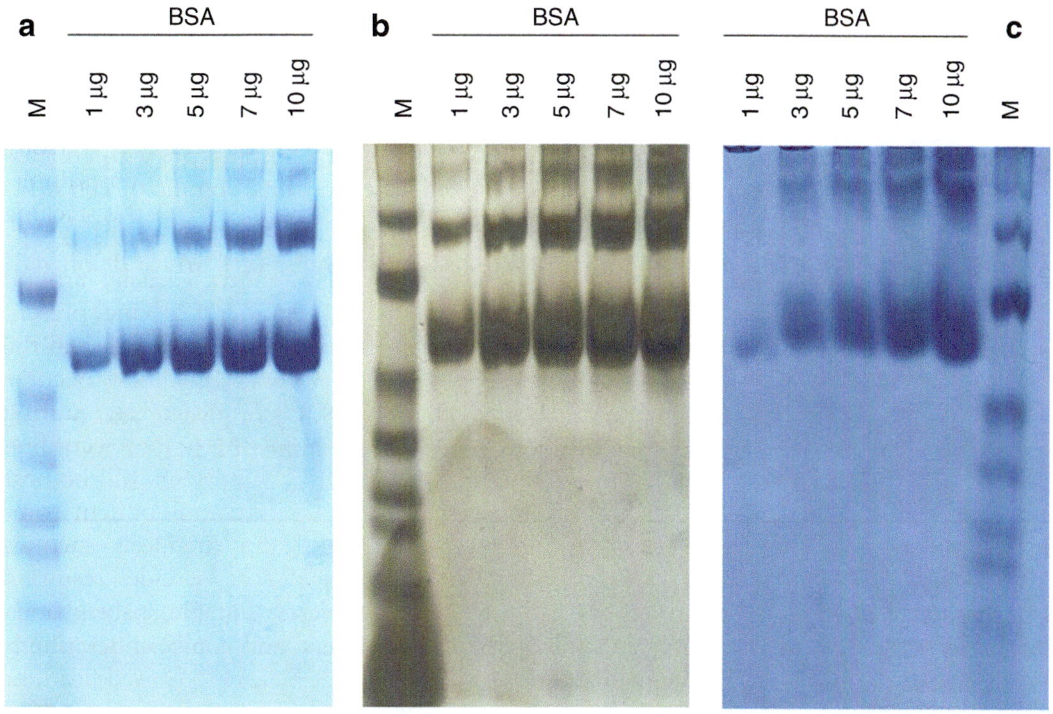

Fig. 1 Examples of a one-dimensional either native polyacrylamide gel (**a**, **b**) or blue native polyacrylamide gel (**c**). The native gel was either stained with Coomassie Brilliant Blue (**a**) or with silver (**b**)

protocol termed as "blue native" polyacrylamide gel electrophoresis has also been described. Blue native electrophoresis is based on the ability of Coomassie Brilliant Blue to form stable complexes with proteins. The resulting complexes are negatively charged. Thus, in case of blue native gel electrophoresis, Coomassie Brilliant Blue plays a similar role as SDS in case of SDS-PAGE. Although Coomassie Brilliant Blue seems to be less critical compared to SDS, protein complexes may also be disrupted by Coomassie Brilliant Blue. Here we describe simple protocols for a native and a blue native polyacrylamide gel electrophoresis which can be used as an experimental starting protocol for further improvements according to the respective requirements. It should be mentioned that all downstream processing steps as staining of the gels with Coomassie Brilliant Blue, silver, transfer by electroblotting, etc. can be performed as described for SDS-PAGE.

2 Materials

1. Coomassie Brilliant Blue G-250 solution: Dissolve 2.5 mg of Coomassie Brilliant Blue in 0.45 mL methanol. Add 0.45 mL of deionized water (*see* **Note 1**). Add 0.1 mL of glacial acetic acid.

2. Acrylamide solution (40%): Dissolve 20 g of acrylamide and 5.3 g of bis-acrylamide in deionized water (*see* **Note 1**) to a final volume of 50 mL.
3. Separating gel buffer (4×): Dissolve 18.15 g of Tris in about 75 mL of deionized water (*see* **Note 1**). Adjust pH with HCl to pH 8.8, and add deionized water to a final volume of 100 mL.
4. Ammonium persulfate solution (10%): Dissolve 0.1 g of ammonium persulfate in a final volume of 1 mL of deionized water (*see* **Note 2**).
5. Sample solution (2×): 0.187 M Tris–HCl (pH 6.8), 30% glycerol, and 80 μg/mL Bromophenol blue.
6. Electrophoresis buffer: Dissolve 28.8 g of glycine and 6 g of Tris in deionized water to a final volume of 2 L.

3 Methods

3.1 Preparation of a Continuous (10%) Polyacrylamide Gel for Native and Blue Native Gel Electrophoresis

1. Thoroughly clean and dry glass plates, suitable spacers, and comb. Assemble glass plates, spacers, and comb as described by the manufacturer.
2. Mix 2.5 mL of acrylamide solution (40%), 2.5 mL of separating gel buffer (4×), and 5 mL of deionized water (*see* **Note 3**).
3. Degas the mixture to avoid air bubbles in the gel after polymerization and to remove oxygen in the gel solution which otherwise accelerates the polymerization process.
4. Add 50 μL of ammonium persulfate solution (10%) and 10 μL of TEMED which will start the polymerization process.
5. Pour the mixed solution between the glass plates and add the comb.
6. Polymerize the acrylamide for 1 h.
7. Remove the comb carefully. The gel is ready to use.

3.2 Sample Preparation

3.2.1 Sample Preparation for Native Polyacrylamide Gel Electrophoresis

1. For native gel electrophoresis, mix 10 μL of protein solution with 10 μL sample solution (2×) containing glycerol and the dye bromophenol blue.

3.2.2 Sample Preparation for Blue Native Polyacrylamide Gel Electrophoresis

1. For blue native gel electrophoresis, mix 10 μL of protein solution in sample solution (2×) containing glycerol and the dye bromophenol blue with 10 μL of Coomassie Brilliant Blue solution.

3.2.3 Marker Sample

1. To follow the electrophoresis, you can run in parallel prestained marker proteins as commonly used for SDS-PAGE.

3.3 (Blue) Native Polyacrylamide Gel Electrophoresis Conditions

1. Fill apparatus with gel electrophoresis buffer.
2. Start electrophoresis immediately. (For a gel of 1 mm thickness and 15 cm length, apply about 150 volts (constant voltage) which will result in about 20 mA of current (*see* **Notes 4** and **5**).)
3. Remove the gel from between the glass plates.
4. After native gel electrophoresis, stain the gel with either Coomassie Brilliant Blue (Fig. 1a) or with silver (Fig. 1b).
5. Destain blue native gel with deionized water (the results of blue native gel electrophoresis are shown in Fig. 1c) (*see* **Note 6**).

4 Notes

1. Unless stated otherwise, all solutions should be prepared in water that has a resistance greater than 18 M ohm and total organic content of less than 5 parts per billion. This standard is referred to as "water" in this text.
2. Prepare this solution freshly before preparation of the gel.
3. If necessary, adjust the volume of the gel solution according to your gel equipment.
4. Current will fall during electrophoresis (when constant voltage is applied).
5. The bromophenol blue dye front takes about 3 h to reach the bottom of the gel. Greater voltage speeds up electrophoresis but generates more heat in the gel.
6. Native bovine serum albumin has a heart-like structure that is stabilized by a series of disulfide bridges. BSA can partially unfold, and the different forms can be separated by native polyacrylamide gel electrophoresis [3, 4].

References

1. Laemmli UK (1970) Cleavage of structural proteins during the assembly of the head of bacteriophage T4. Nature 227:680–685
2. Studier FW (1973) Analysis of bacteriophage T7 early RNAs and proteins on slab gels. J Mol Biol 79:237–248
3. Niepmann M, Zheng J (2006) Discontinuous native protein gel electrophoresis. Electrophoresis 27:3949–3951
4. He XM, Carter DC (1992) Atomic structure and chemistry of human serum albumin. Nature 358:209–215

Chapter 9

Isoelectric Focusing on Non-Denaturing Flatbed Gels

Biji T. Kurien and R. Hal Scofield

Abstract

Isoelectric focusing (IEF) serves as a very useful procedure for cell protein separation and characterization. We have used this method to study antibody clonotype changes. Here we discuss the use of a sensitive native flatbed isoelectric focusing method to analyze specific antibody clonotype changes in a patient with systemic lupus erythematosus, who developed autoantibodies to the Ro 60 autoantigen under observation. Patient sera samples collected over several years were used for analysis using flatbed IEF. Following electrofocusing, the gel is analyzed by affinity immunoblotting utilizing Ro 60-coated nitrocellulose membrane to determine oligoclonality of the anti-Ro 60 containing sera.

Key words Affinity immunoblotting, Clonotype distribution, Systemic lupus erythematosus, Ro 60 autoantigen, Flatbed IEF

1 Introduction

Non-denaturing isoelectric focusing (IEF) electrophoresis is a valuable method for investigating the heterogeneity of antibody and immunoglobulin (Ig) clonotypes [1]. Patterns of antigen-specific antibody clonotype can convey whether fluctuations in cell population occur during ongoing immune responses in response to regulatory influences. It can also show whether changes in hybridoma cell lines can occur with time [2]. Previous studies investigated these changes by immobilizing the separated antibody clonotypes following IEF and incubating them with radioactive antigen. Radiolabeled hapten was allowed to diffuse into a gel, in one method, prior to precipitation of Ig with sodium sulfate followed by detection of hapten-specific clonotype distribution by autoradiography [3]. Ig was precipitated in the gel with sodium sulfate, in another study, immediately following completion of the focusing run. It was then cross-linked with glutaraldehyde followed by the addition of labeled antigen or anti-Ig [4]. Ensuing studies showed that fixation with glutaraldehyde could reduce the antigen-binding capability of certain Ig [5]. In addition, it was shown that the earlier

Biji T. Kurien and R. Hal Scofield (eds.), *Electrophoretic Separation of Proteins: Methods and Protocols*, Methods in Molecular Biology, vol. 1855, https://doi.org/10.1007/978-1-4939-8793-1_9,

study was unable to define optimal cross-linker (glutaraldehyde or suberimidate) concentration, since certain antibodies could not be fixed at cross-linker concentrations that substantially inactivated others. The excess time needed to diffuse antigen into the gel and for rinsing the unbound antigen out of the gel is another drawback of these methods, which can take several days especially when using radioactive probes.

A technique for immobilizing focused antibodies involved the use of nitrocellulose membranes. Focused antibodies are transferred electrophoretically or non-electrophoretically to nitrocellulose membrane, and labeled antigen was used to detect antigen-specific clonotypes [6]. Yet another procedure involved laying the gel with the focused antibodies with agarose containing antigen-coated sheep erythrocytes [7]. In this method, antibodies diffuse into the RBC-containing gel, bind the antigen-coated cells, and lyse the cells after complement addition.

Here, we describe a technique in which a 60,000-molecular weight Ro autoantigen was first passively immobilized on nitrocellulose membrane and placed in contact with an IEF gel-containing autoantibodies (derived from a patient with systemic lupus erythematosus who developed antibodies to the Ro 60 autoantigen over time) focused according to its isoelectric point. After diffusion-mediated transfer to membrane, the non-specific antibody clonotypes are removed by washing, and the antigen-specific antibody clonotypes are detected using alkaline phosphatase-conjugated anti-Ig [8].

SLE (systemic lupus erythematosus) is a complex, chronic autoimmune disease in which autoantibodies target self-antigens, including the Ro (or SS-A) ribonucleoprotein complex. SS-A or Ro 60 autoantigen is a 60,000-molecular weight protein. Anti-Ro 60 antibodies occur in up to 40% of patients with SLE [8]. The epitopes of the Ro 60 autoantigen targeted by SLE patients have been characterized previously [9, 10]. Even though anti-Ro 60 sera commonly bound short Ro 60 peptides, it did not bind the denatured Ro 60 antigen well. In addition, the antibodies that bound Ro 60 octapeptides were also found to bind the native protein [10].

The literature provides examples of some SLE autoantibodies appearing and disappearing at times in association with specific disease manifestations, therapy, or generalized clinical disease activity which have been reported. For example, antibodies to native DNA are associated with renal disease, and the detection of this autoantibody may be an indication of disease exacerbation [11]. Antibodies to the P autoantigen (ribosomal P antigens) can appear with an increase of neurologic or renal disease. Autoantibodies like anti-Ro 60, on the other hand, occur in some normal subjects, as well as in SLE patients prior to onset of disease [8], and develop only rarely during the course of SLE.

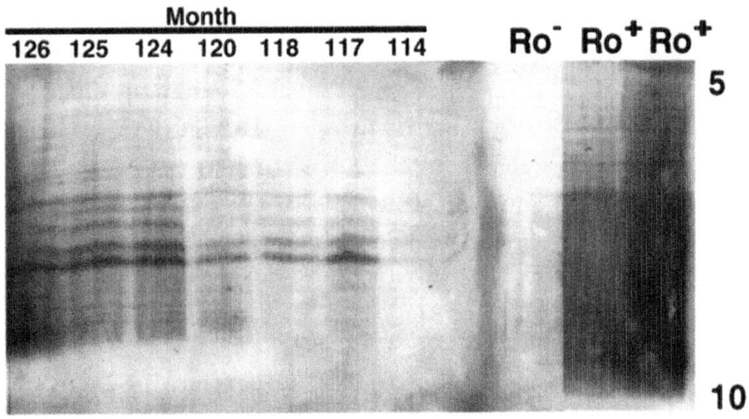

Fig. 1 Affinity immunoblotting of the patient's sera obtained at different time points following first observation, showing anti-Ro 60-specific IgG clonotypes. Sera from an anti-Ro 60-negative SLE patient (Ro$^-$) and from two typical anti-Ro 60-positive patients (Ro$^+$) are shown for comparison. The pH range of the IEF gel is shown on the right

This study was carried out after the identification of a patient with SLE who developed antibodies to the Ro 60 autoantigen after about 10 years of illness. Figure 1 shows that anti-Ro 60 clonality increases in complexity, and affinity to the Ro 60 antigen also increases as the response developed.

2 Materials

Make all solutions with ultrapure water (prepared by purifying deionized water, to attain a sensitivity of 18 M Ω cm at 25 °C) and analytical grade reagents. Prepare and store all reagents at room temperature (unless indicated otherwise). Diligently follow all waste disposal regulations when disposing waste materials. We do not add sodium azide to buffers or reagents.

1. Twenty-five percent glycerol (v/v): Add 25 mL glycerol to 75 mL of distilled water. Mix well.
2. 5× acrylamide (26.5% T, 3% C): Add about 25 mL water to a 100 mL graduated cylinder or a glass beaker. Weigh 12.84 g acrylamide and 0.4098 g bis-acrylamide, and transfer to the cylinder (*see* **Note 1**). Add a spatula of AG 501-X8 (D) mixed resin beads (Bio-Rad, Hercules, CA, USA), and stir using a magnetic stir bar on a magnetic plate for about 30 min. Make up to 50 mL (after removing the stir bar) with water, and filter through a 0.45 μm Corning filter (*see* **Note 2**). Store at 4 °C, with bottle wrapped with aluminum foil (*see* **Note 3**).

3. Ten percent Tween-20: Add 90 mL of distilled water into a glass beaker. Add 10 mL Tween-20 and mix.
4. Two percent ammonium persulfate: Weigh 0.02 g ammonium persulfate, and dissolve in 1 mL of distilled water (*see* **Note 4**).
5. N.N.N.N′-Tetramethylethylenediamine. Store at 4 °C (*see* **Note 5**).
6. Alkaline phosphatase buffer: Weigh 6.1 g of Tris, 2.9 g sodium chloride, and 0.51 g magnesium chloride-6H_2O, and make it to 500 mL with water after adjusting pH to 9.3 with HCl (*see* **Note 1**). Store at 4 °C.
7. Nitro blue tetrazolium (NBT)/5-bromo-4-chloro-3-indolyl phosphate (BCIP): Dissolve 1 g NBT in 20 mL of 70% dimethylformamide (DMF). Dissolve 1 g BCIP in 20 mL of 100% DMF. Add 33 μL of BCIP and 66 μL of NBT to 10 mL of alkaline phosphatase buffer just before adding to membrane.
8. Nitrocellulose membrane.
9. 0.5 M sodium bicarbonate solution, pH 9.5.
10. Phosphate-buffered saline (PBS), pH 7.4.
11. PBS containing 0.05% Tween-20 (PBST).
12. Ampholytes: pH 3–10 and pH 8–10.5.
13. Ro 60 autoantigen (Immunovision, Springdale, AK, USA).
14. Glass plates: Two 5″ by 4″ glass plates.
15. Medium binder clips (1 ¼ inch).
16. Small binder clips (¾ inch).
17. Gasket with three edges, about 3 mm wide, to serve as spacer between the plates.
18. LKB-2117 Multiphor apparatus for IEF.
19. Model 3000/300 power supply (any powerful power supply would suffice).
20. pH 3 and pH 10 solutions (Serva Electrophoresis GmbH, Heidelberg

    Germany).
21. Helium gas.
22. Sample applicator strip.
23. Paper wicks.

3 Methods

All procedures are performed at room temperature unless otherwise specified.

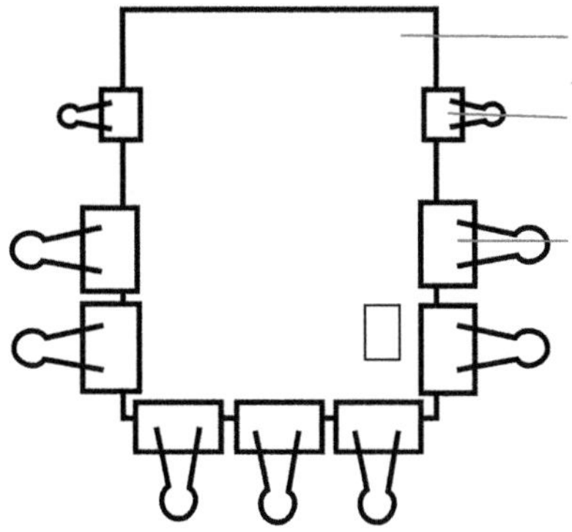

Fig. 2 Gel assembly for isoelectric focusing

1. Pipet 5.6 mL of distilled water into a conical flask. Add 2 mL of 25% glycerol followed by 2.1 mL of the 5× acrylamide solution. Then add 300 μL of pH 3–10 ampholytes followed by 100 μL of pH 8–10.5 ampholytes.
2. Degas this solution by bubbling helium through it for 15 min. Rinse the metal end of degassing tube first with water, and wipe dry with Kimwipes.
3. While the solution is degassing, set up the gel apparatus. Soak the gasket in water for few min. Mop dry with Kimwipes.
4. Take one glass plate, and lay the gasket on top of the glass plate around the edges so that it will seal the bottom and two sides of the plates. Lay the other glass plate on top of the gasket. Clamp the clips around the edges of the plates (the bottom, the left side, and the right side (*see* Fig. 2)). Stand the gel upright on using the base of the clips (*see* Fig. 2) to pour the gel. Prepare a 2% ammonium persulfate solution fresh.
5. After degassing is complete, the metal end of the degassing tube is cleaned with water. To the degassed solution, add 100 μL of 10% Tween-20 and mix gently. Then add 100 μL of 2% APS. Have a Pasteur pipette ready for pouring the gel. Add 10 μL of TEMED and mix gently. Pipet the gel mixture into the Pasteur pipette, and transfer into the gel apparatus quickly. Attempts should be made to avoid bubbles. Fill up the gel apparatus to the top. Polymerization should begin within minutes. However, let the assembly stand for 2 h without disturbance.
6. Turn cooling unit on and set it on 4 °C in preparation for focusing.
7. After 2 h carefully remove one of the glass plates and gasket. The gel will remain on one of the glass plates.
8. Lay the glass plate on top of the IEF unit, with the gel side facing up (wipe off water on top of the unit beforehand). Place

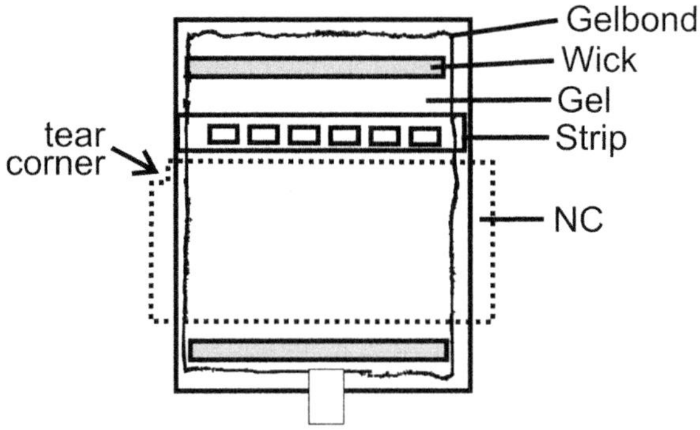

Fig. 3 The membrane-gel assembly following flatbed IEF. The gel bond is shown in this figure. However, we did not use the gel bond to support the gel. The gel was directly in contact with the glass plate

the smaller cover in place, and press down slightly so as to make imprints for the wicks. Cut two wicks to the size of the gel (be as close as possible). Soak the top wick in Serva pH 3 solution and the bottom wick in Serva pH 10 solution. Dab off excess solution, and place where imprints were made by cover (*see* Fig. 3).

9. Put smaller cover back on, making sure connection is made with both wicks. Connect red and black wires. Put on larger cover and make connections to power supply (red = $+^{ve}$; black = $-^{ve}$).
10. Prefocus by setting constant voltage 200 V for 20 min; then increase voltage to 400 V for another 20 min. Prepare samples for application.
11. Turn of power supply, disconnect wires, and remove covers. Take applicator strip, and lay on top of gel 1–2 cm below top wick. Make sure strip is stuck to the gel well. Strip can hang over gel a little (*see* Fig. 3; *see* **Note 6**).
12. Apply the samples, being careful not to spill over into other wells. Replace the covers and make the connections. Turn the power supply on to 12 W constant power. Focus for approximately 1–2 h. When focusing, the voltage will rise, and the current will drop. The rate at which these two parameters change is much faster in the beginning than the end. The run is complete when the voltage is between 1800 and 2000 V and the current 3–5 A. When the change appears to be very slow or not at all, turn off the unit (*see* **Note 7**).
13. After the run is complete, transfer the focused protein from the gel to membrane. Take gel off the flatbed and remove applicator strip.
14. Stain gel with appropriate stain or carry out transfer to membrane [12–14].

4 Notes

1. Having water at the bottom of the cylinder helps to dissolve the Tris relatively easily, allowing the magnetic stir bar to go to work immediately. If using a glass beaker, the Tris can be dissolved faster if the water is warmed to about 37 °C. However, the downside is that care should be taken to bring the solution to room temperature before adjusting pH.
2. Wear a mask when weighing acrylamide. To avoid exposing acrylamide to co-workers, cover the weigh boat containing the weighed acrylamide with another weigh boat (similar size to the original weigh boat containing the weighed acrylamide) when transporting it to the fume hood. Transfer the weighed acrylamide to the cylinder inside the fume hood, and mix on a stirrer placed inside the hood. Unpolymerized acrylamide is a neurotoxin, and care should be exercised to avoid skin contact. Mixed resin AG 501 –X8 (D) (anion- and cation-exchange resin) is used when acrylamide solution is made, since it removes charged ions (e.g., free radicals) and allows longer storage. Some investigators store the prepared acrylamide along with this resin in the refrigerator. However, we filter them out before storage. The used mixed resin should be disposed of as hazardous waste. Manufacturer's warning states that this resin is explosive when mixed with oxidizing substances. The resin contains a dye that changes from blue-green to gold when the exchange capacity is exhausted.
3. The acrylamide solution can be stored at 4 °C for 1 month. Acrylamide hydrolyzes to acrylic acid and ammonia. The acrylamide mixture, buffer, and water can be prepared in large batches, frozen in aliquots (for greater day-to-day reproducibility), and used indefinitely (*see* ref. 12). Remove the required amount, bring to room temperature, and add the other ingredients for polymerization. However, in our laboratory we make the acrylamide solution fresh about every month when we cast our own gels.
4. We find it is best to prepare this fresh each time.
5. We find that storing at 4 °C reduces its pungent smell.
6. Large well, 10 all; medium wells, 5 all; small well, 1 all. Only lay one size of wells onto the gel. Cut if necessary. Strips may be used again.
7. During the end of the run, the gel must be watched carefully in case a fire starts. Many times the gel will burn near the applicator strip. If this happens, turn off the unit. The gel can still be used if it had been focused for a long time. The bands are usually below the strip.

References

1. Awed ZL, Williamson AR, Askonas BA (1968) Isoelectric focusing in polyacrylamide gel and its application to immunoglobulins. Nature 219:66–67
2. Knisley KA, Rodkey LS (1986) Affinity immunoblotting. High resolution isoelectric focusing analysis of antibody clonotype distribution. J Immunol Methods 95:79–87
3. Askonas BA, Williamson AR, Wright BE (1970) Selection of a single antibody-forming cell clone and its propagation in syngeneic mice. Proc Natl Acad Sci U S A 67:1398–1403
4. Keck K, Grossberg AL, Pressmann D (1973) Specific characterization of isoelectrofocused immunoglobulins in polyacrylamide gel by reaction with 125 I-labeled protein antigens or antibodies. Eur J Immunol 3:99–102
5. Nicolotti RA, Briles DE, Schroer JA, Davie JM (1980) Isoelectric focusing of immunoglobulins: improved methodology. J Immunol Methods 33:101–115
6. Friedenson B, Soong CJ (1984) A simple general method of exactly comparing different binding activities and antigenic properties of antibodies by obtaining duplicate copies from a single isoelectric focusing gel. J Immunol Methods 67:235–242
7. Phillips JM, Dresser DW (1973) Antibody isoelectric spectra visualized by antigen-coated erythrocytes. Eur J Immunol 3:738–740
8. Scofield RH, Zhang F, Kurien BT et al (1996) Development of the anti-Ro autoantibody response in a patient with systemic lupus erythematosus. Arthritis Rheum 39:1664–1668
9. Scofield RH, Harley JB (1991) Autoantigenicity of Ro/SSA antigen is related to a nucleocapsid protein of vesicular stomatitis virus. Proc Natl Acad Sci U S A 88:3343–3347
10. Huang S-C, Yu H, Scofield RH, Harley JB (1995) Human anti-Ro autoantibodies bind peptides accessible to the surface of the native Ro autoantigen. Scand J Immunol 41:220–228
11. Lloyd W, Schur PH (1981) Immune complexes, complement, and anti-DNA in exacerbations of systemic lupus erythematosus (SLE). Medicine (Baltimore) 60:208–217
12. Harlow E, Lane D (1988) Electrophoresis. Appendix I. In: Harlow E, Lane D (eds) Antibodies. A laboratory manual. Cold Spring Harbor Laboratory, New York, p 638
13. Wu M, Stockley PG, Martin WJ 2nd. (2002) An improved western blotting technique effectively reduces background. Electrophoresis 23:2373–2376
14. Kurien BT, Scofield RH (2012) Native flatbed isoelectric focusing for determining antibody clonotype distribution. Methods Mol Biol 869:259–266

Chapter 10

Determination of Protein Molecular Weights on SDS-PAGE

Hiroyuki Matsumoto, Hisao Haniu, and Naoka Komori

Abstract

An apparent molecular weight (MW) of a protein can be determined from the migration distance of a protein complexed with a strong cationic detergent sodium dodecyl sulfate (SDS) separated on sodium dodecyl sulfate-polyacrylamide gel electrophoresis (SDS-PAGE). This method was established around 1969 and has been utilized substantially even today because of its simplicity. During the following half a century, although it has been reported that many proteins show some deviation in MW when determined on SDS-PAGE especially when their peptide chains are posttranslationally modified, this versatile method is still being used very often in current biochemical works. In this protocol, a simple method to estimate MW by running SDS-PAGE of standard proteins is explained by an example in which proteins extracted from mouse retina were analyzed by two-dimensional isoelectric focusing (2-D IEF) SDS-PAGE followed by protein identification by peptide mass fingerprinting.

Key words Protein molecular weight (MW), SDS-PAGE, Protein identification by mass spectrometry, 2-D gel

1 Introduction

Proteins, when solubilized in one of the strong cationic detergents sodium dodecyl sulfate (SDS, aka sodium lauryl sulfate SLS) and electrophoresed on an SDS polyacrylamide gel, the migration distance of each protein depends on the molecular weight (MW) [1–4]. The method is simple, rapid, and inexpensive and therefore has been utilized even now. Since the electrophoresis protocol uses SDS as a solubilizing agent, it is generally called "sodium dodecyl sulfate-polyacrylamide gel electrophoresis" or simply SDS-PAGE. It has been well accepted that MW determination (or estimation) of proteins on SDS-PAGE is accurate generally within ±10% with an exception that larger deviation is rarely observed when a protein exhibits abnormality in migration caused by posttranslational modification [5] such as phosphorylation and

Biji T. Kurien and R. Hal Scofield (eds.), *Electrophoretic Separation of Proteins: Methods and Protocols*, Methods in Molecular Biology, vol. 1855, https://doi.org/10.1007/978-1-4939-8793-1_10,

Table 1
Proteins identified by mass spectrometry and their estimated molecular weight (MW) on SDS-PAGE. Each spot # corresponds to that shown in Fig. 1

Spot #	Protein identity	MW (theoretical)	pI	MW (estimated)	Error (%)
1	Aconitase 2, mitochondrial	85,464	8.1	95,000	+10.0
2	Heat shock 70 kD protein 8	70,838	5.4	78,000	+9.2
3	Transketolase	60,584	6.5	71,000	+14.7
4	Pyruvate kinase 3	57,845	7.2	62,000	+6.7
5	ATP synthase, H^+ transporting, mitochondrial F1 complex, α subunit, isoform 1	59,753	9.2	53,000	-12.7
6	ATP synthase, H^+ transporting mitochondrial F1 complex, β subunit	56,301	5.2	51,000	-10.4
7	Enolase 2, γ neuronal	47,297	5.0	43,000	-10.0
8	α-enolase	47,141	6.4	43,000	-9.6
9	α-enolase	47,141	6.4	47,000	-0.3
10	Glutamine synthetase	42,120	6.6	43,000	+2.0
11	Phosphoglycerate kinase 1	44,537	7.5	43,000	-3.6
12	trancated β-actin (aa 27-375) γ-actin	39,186 41,019	5.8 5.6	42,000 42,000	+6.7 +2.3
13	Aldolase 1, A isoform	39,356	8.3	41,000	+4.0
14	Aldolase 3, C isoform	39,395	6.7	40,000	+1.5
	Cytosolic aspartate aminotransferase	46,232	6.7	40,000	-15.6
15	Glyceraldehyde-3-phosphate dehydrogenase	35,810	8.4	36,000	+0.5
16	Guanine nucleotide-binding protein, β-1 subunit	37,377	5.6	34,000	-9.9
17	Lactate dehydrogenase 1, A chain	36,499	7.6	33,000	-10.6
18	Phosphoglycerate mutase 1	28,832	6.7	28,000	-3.0
19	Triose-phosphate isomerase (EC 5.3.1.1)	26,696	6.9	26,000	-2.7
20	Crystallin, α A	22,490	6.4	19,000	-18.4

glycosylation. In this article, we describe a generally accepted method to estimate protein MW on SDS-PAGE and show an example how to apply it to estimate proteins separated and displayed on a 2-D IEF (isoelectric focusing) SDS-PAGE.

In Table 1, 20 protein spots on the 2-D SDS-PAGE shown in Fig. 1 were identified by peptide mass fingerprinting [6], and the theoretical MW and the pI of each protein spot are illustrated. It should be noted that the MWs of 20 proteins estimated in Table 1

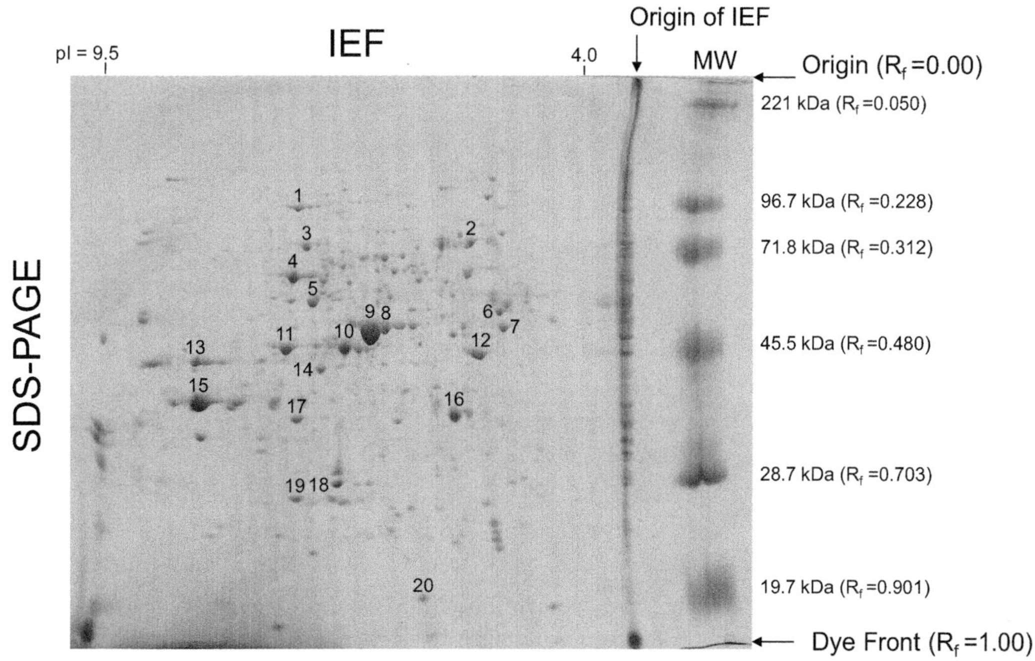

Fig. 1 2-D IEF SDS-PAGE gel of mouse retina. Two-dimensional gel profiles of C57/BL6 mouse retinal proteins were separated on 2-D gel stained with Coomassie Brilliant Blue [8]. MW marker proteins were run in a small well built at the edge of the SDS-PAGE slab. The SDS-PAGE was run until a trace amount of tracking dye, bromophenol blue, reached the bottom of the slab gel. R_f is a standardized migration distance of each MW marker protein as shown in the figure

are their theoretical MW values and experimentally determined MW values based on the measured R_f value. Such typical example of MW analysis illustrated in Table 1 shows that experimental errors are mostly within ±10%. Many factors are conceivable regarding the origins of such errors; among all, it should be noted that any posttranslational processing of proteins [5] that might have happened and the modification of side chains by posttranslational modification [5] are ignored in the theoretical MW values in Table 1. Even considering these uncertainties, the MW estimates rendered by this protocol usually give a practical reliability.

2 Materials

1. Standard Proteins for Molecular Weight Determination (MW Standards): In general, two groups of MW Standards are commercially available, each of which covers a smaller molecular weight region (lower than ~20 K Daltons) or a larger molecular weight region (higher than ~30 K Daltons) (*see* **Note 1**).
2. Dissolve protein samples in a standard SDS-PAGE lysis buffer, and run on an SDS-PAGE (*see* **Notes 2** and **3**).

3. Stain SDS-PAGE slab gel with Coomassie Brilliant Blue to visualize the MW standards and also relevant protein bands or spots. In this article, a typical case of molecular weight estimation of proteins separated on a two-dimensional (2-D IEF SDS-PAGE) gel electrophoresis will be explained.

3 Methods

1. Run a 2-D gel according to the protocol described previously [7] and stain (Fig. 1).
2. Determine migration distance R_f, a relative value standardized to the dye front, of each MW protein as shown to the right of the MW lane in Fig. 1.
3. Plot the MW of protein standards on a base 10 logarithmic scale against R_f as shown in Fig. 2. The data points define the MW calibration curve for the SDS-PAGE system in use.
4. In order to determine the MW of an unknown protein, first, measure the R_f value of the protein spot (or band).
5. Then, calculate, by using the calibration curve established in the same gel system, the estimated MW of an unknown protein.

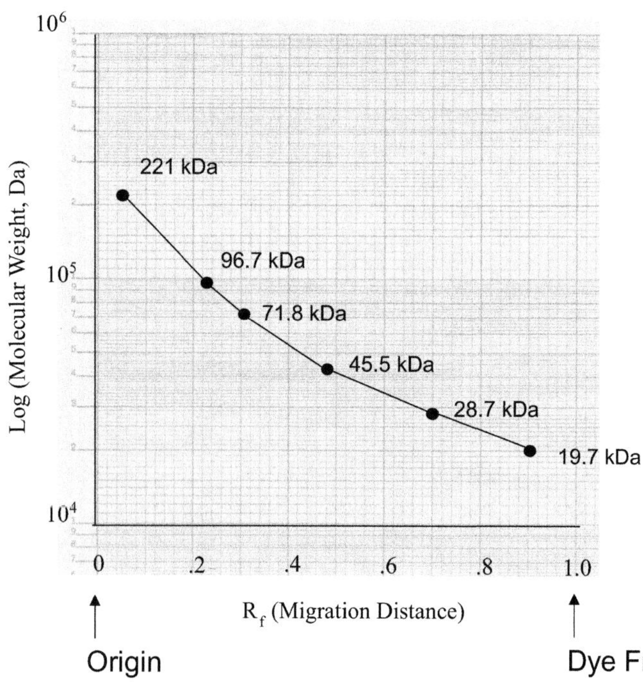

Fig. 2 Molecular weight (MW) calibration curve made based on the 2-D gel separation shown in Fig. 1

4 Notes

1. Several companies including Sigma-Aldrich, Bio-Rad, Clontech, and Thermo Fisher Scientific sell protein markers that can serve as molecular weight standards for this protocol.
2. For SDS-PAGE any commercially available precast gel and electrophoresis apparatus can be used.
3. A typical protocol for SDS-PAGE can be found elsewhere, for example, see articles published in this series, Protein Electrophoresis [8].

Acknowledgment

The projects described in this article were partially supported by Oklahoma Center for Advancement of Science and Technology (OCAST) Health Research Program Grant HR10-120 and NIH R21 EY017888.

References

1. Shapiro AL, Vinuela E, Maizel JV Jr (1967) Molecular weight estimation of polypeptide chains by electrophoresis in SDS-polyacrylamide gels. Biochem Biophys Res Commun 28(5):815–820
2. Shapiro AL, Maizel JV Jr (1969) Molecular weight estimation of polypeptides by SDS-polyacrylamide gel electrophoresis: further data concerning resolving power and general considerations. Anal Biochem 29(3):505–514
3. Dunker AK, Rueckert RR (1969) Observations on molecular weight determinations on polyacrylamide gel. J Biol Chem 244 (18):5074–5080
4. Weber K, Osborn M (1969) The reliability of molecular weight determinations by dodecyl sulfate-polyacrylamide gel electrophoresis. J Biol Chem 244(16):4406–4412
5. Wold F (1981) In vivo chemical modification of proteins (post-translational modification). Annu Rev Biochem 50:783–814
6. Matsumoto H, Komori N (2000) Ocular proteomics: cataloging photoreceptor proteins by two-dimensional gel electrophoresis and mass spectrometry. Methods Enzymol 316:492–511
7. Matsumoto H et al (2012) Two-dimensional gel electrophoresis: glass tube-based IEF followed by SDS-PAGE. Methods Mol Biol 869:267–273
8. Kurien BT, Scofield RH (2012) Protein electrophoresis: methods and protocols, methods in molecular biology. Methods Mol Biol 869

Chapter 11

Two-Dimensional Gel Electrophoresis by Glass Tube-Based IEF and SDS-PAGE

Hiroyuki Matsumoto, Hisao Haniu, Biji T. Kurien, and Naoka Komori

Abstract

The genome information combined with data derived from modern mass spectrometry enables us to determine the identity of a protein once it is isolated from a complex mixture. Two-dimensional gel electrophoresis established more than four decades ago serves as a powerful protocol to isolate many proteins at once for such protein analysis. In the first two decades, the original procedure to use a glass tube-based IEF had been commonly used. Since an IEF in glass tubes is rather difficult to maneuver, a new method to use an IEF on a thin agarose slab backed by a plastic film (IPG Dry Strip) had been invented and is now widely used. In this chapter, we describe a protocol that uses a glass tube-based IEF because the capacity of protein loading and resolving power of this type of classic two-dimensional gel is still indispensable for many applications, not only for protein identification but also for protocols that are benefited by larger amounts of materials, i.e., analysis of posttranslational modification of proteins such as phosphorylation, methylation, glycosylation, and others.

Key words 2D-gel, Glass tube-IEF

1 Introduction

Identification of isolated proteins by peptide mass fingerprinting [1, 2] became simplified because of the availability of genome information. This makes isolation of proteins of interest from a complex mixture an important protocol in biomedical sciences. Among various protocols of two-dimensional (2D) gel electrophoresis [3–7] published more than 35 years ago, those described by O'Farrell [5] and Ames and Nikaido [7] have been used frequently. Although these 2D gel protocols are powerful, their disadvantages are that the entire procedures are rather tedious and require skills. Especially somewhat difficult steps involve (1) preparing reliable isoelectric focusing (IEF) gel in a glass tube and (2) focusing proteins in the disc IEF gels in the first dimension. Because of technical difficulty in these steps, a simpler and user-friendly protocol for the IEF step using a thin layer of agarose IEF made on a plastic backing

Biji T. Kurien and R. Hal Scofield (eds.), *Electrophoretic Separation of Proteins: Methods and Protocols*, Methods in Molecular Biology, vol. 1855, https://doi.org/10.1007/978-1-4939-8793-1_11,

(i.e., IPG Dry Strip) has been developed recently. One clear difference between original glass tube IEF and the IPG Dry Strip IEF is the loading capacity of protein mixture; several hundred micrograms of proteins can be loaded on an IEF gel made in a glass tube, whereas an IPG Dry Strip IEF gel can take only ~150 micrograms of proteins. There appear to be some differences in focusing patterns of proteins on the final 2D gels on a slab SDS-PAGE. However, since the IEF gradients made on a glass tube and on an IPG Dry Strip are usually different, it is impossible to draw any scientific conclusion on the difference of 2D gel patterns between these two systems without doing detailed experiments. In this chapter, we will describe a 2D gel electrophoresis based on a glass tube IEF gel that our group has been using for more than three decades. The protocol described in this paper is based on O'Farrell [5] and its modification by Miyazaki et al. [8–10].

2 Materials

1. An isoelectric focusing (IEF) apparatus with a jacketed beaker (10.5-cm i.d., 12.5-cm inner height) used as a lower-buffer chamber (Buchler Instrument) or a similar tube gel apparatus such as one from Bio-Rad (Hercules, CA).
2. An SDS-PAGE slab gel apparatus such as a vertical slab gel apparatus Model V16 (GIBCO-BRL, Gaithersburg, MD) or equivalent.
3. Borosilicate glass tubes (14–15 cm in length with 3.0-mm id and 5.0-mm od).
4. Power supply appropriate for IEF focusing and SDS-PAGE.

2.1 Stock Solutions for IEF Gel

1. Solution A: Acrylamide (30%, w/v)-*bis*-Acrylamide (1.5%, w/v).
2. Solution B: Riboflavin (0.004%, w/v)-N,N,N',N'- tetramethylethylenediamine (TEMED; 0.45%, v/v).
3. Solution C: Triton X-100 or NP-40 (20%, w/v).
4. Solution D: Ammonium persulfate (1.5%, w/v).
5. Ampholyte: Bio-Lyte 3/10 (Bio-Rad) or equivalent.

2.2 IEF Lysis Buffer

1. Urea (9.5 M), Triton X-100 or NP-40 (2.0%, w/v), Bio-Lyte 3/10 (2.0%, w/v), 2-mercaptoethanol (5%, v/v), high-quality water.
2. To make this solution, weigh urea in a small gradient cylinder. Add other ingredients and dissolve the urea and adjust the final volume.

2.3 IEF Overlay Buffer

1. Urea, 5.0 M, Triton X-100 or NP-40 (2.0%, w/v), Bio-Lyte 3/10 (1.0%, w/v).
2. Use the same procedure as described for IEF lysis buffer. IEF lysis and IEF overlay buffers can be stored in aliquots at –20°C.

2.4 Electrode Solutions

1. Top chamber (positive polarity): 20 mM H_3PO_4.
2. Lower chamber (negative polarity): 1 M NaOH.
3. These solutions are designed to run IEF from the acidic side to the basic side [8, 9]. To run an IEF gel from the basic side, one should refer to O'Farrell [5] for the concentration of buffers. We reuse 1 M NaOH solution in the lower chamber several times. Store the solutions at room temperature.

2.5 Equilibration Buffer

1. Tris–HCl (pH 6.8), 62.5 mM SDS (2%, w/v), 2-mercaptoethanol (5%, v/v).
2. It is convenient to make 1 L of this solution.
3. Store the solution at room temperature.

2.6 Agarose Stock Solution

1. Prepare 1% agarose (w/v) in equilibration buffer. Microwave to dissolve the agarose, make 2- to 3-mL aliquots in 15-mL tubes, and store at room temperature. One should liquefy agarose by boiling prior to use.
2. Use this agarose solution for overlaying the first-dimension IEF gel on top of the second-dimension SDS-polyacrylamide slab gel.

3 Methods

3.1 Sample Preparation

Prepare IEF samples from either a tissue block, culture cells, or a tissue/cell homogenate solution (*see* **Note 1**).

1. Homogenize a tissue block weighing 1–5 mg in 90 mL of IEF lysis buffer. In this case, the dilution of the lysis buffer by the tissue fluid will not affect the performance of IEF significantly.
2. After homogenization, the sample is homogenized further by sonication in a sonication bath for at least 20 min to break down chromosomal DNA.
3. The sample is then centrifuged at 15,000 × *g* for 5 min at 25 °C to precipitate insoluble debris or aggregated DNAs.
4. The recovered supernatant (usually 50–100 μL) is loaded on top of the IEF gel.

3.2 Isoelectric Focusing Gel Electrophoresis (First Dimension)

It appears that preparation of high-quality IEF gels is crucial for a successful IEF. By "high quality" we mean homogeneously polymerized IEF gel in clean glass tubes. The method developed by Miyazaki et al. [8] utilizes a combination of photoinduced polymerization by riboflavin and light-independent polymerization by ammonium persulfate of acrylamide. Such a combination appears to make high-quality IEF gels routinely. IEF tubes must be cleaned thoroughly by soaking overnight in chromic-sulfuric acid cleaning solution (Fisher, Pittsburgh, PA) and by rinsing with hot water followed by deionized water. The tubes are dried in a vacuum oven at 80 °C for 1 h. Chromic-sulfuric acid solution can be reused many times, until its brown color becomes greenish. Special care needs to be taken for the disposal of the used chromic acid solution, which makes the use of chromic acid inconvenient. We have observed that, as long as IEF tubes are kept wet, that is, not allowed to dry after each electrophoresis, cleaning them with a common laboratory detergent solution is sufficient. When tubes are clean and dry, mark them 1.5–2.0 cm from the top, seal the bottom end of each with Parafilm, and set up the tubes on a tube stand.

1. IEF gel solution [8, 9] (made in a 125-mL filtering flask): Urea (8.5 M), acrylamide (4.0%, w/v), bisacrylamide (0.2%, w/v), riboflavin (0.0005%, w/v) TEMED (0.056%, v/v), Triton X-100 or NP-40 (2.0%, w/v), Bio-Lyte 3/10 (2.0%, w/v), ammonium persulfate (0.01%, w/v). To make IEF gel solution, the following steps should be performed. This recipe makes 12 mL of IEF gel mixture.

3.2.1 Steps to Make IEF Gels in Glass Tubes

1. Weigh 6.13 g of urea in a 125-mL filtering flask.
2. Add the following ingredients in order: 1.6 mL of stock solution A; 1.5 mL of stock solution B; 1.2 mL of stock solution C; 0.6 mL of Bio-Lyte 3/10 (40%, w/v); and 2.5 mL of high-quality water.
3. Dissolve the urea completely and degas the mixture for 5 min. During the process, we usually cover the flask with aluminum foil in order to keep riboflavin from light. It is also recommended that the ambient light be kept minimal by turning off some of the room lights until the polymerization is started at **step** 7.
4. Add 0.08 mL of stock solution D and mix quickly by swirling the flask.
5. By using a 1-mL syringe with appropriate tubing attached to the needle, fill the glass tubes with the IEF gel mixture up to the 1.5–2.0-cm mark.
6. Overlay water to about 5 mm on top of the IEF gel solution.

7. Start polymerization by illuminating the tubes with a fluorescent lamp.
8. Fifteen minutes after the start of polymerization, seal the top of each tube with Parafilm.
9. Continue illumination for at least 4 h or longer. Alternatively, an overnight illumination under normal room light will also polymerize IEF gels efficiently.

3.2.2 Loading the IEF Sample Solution

1. Set up the tubes on a tube gel apparatus, load samples, and slowly overlay IEF buffer on top of each sample (*see* **Note 2**).
2. Electrophoresis is performed first at 100 V for 1 h and then at 300 V for 15–18 h. At the end of the run, we often increase the voltage to 500 V and run the apparatus for 1–2 h. Consequently, the total voltage-hour is somewhere between 5000 and 7000. It is necessary to perform preliminary studies to determine the optimum voltage-hours for each system.
3. After a run, push the IEF gel out of the tube, using a syringe filled with water, and shake the gel in 20 mL of equilibration buffer for 15 min in order to equilibrate the gel in the SDS-PAGE environment.
4. Change the equilibration buffer three times before proceeding to SDS-PAGE. If necessary, freeze the gels in the same buffer at −20 °C until use; however, it should be kept in mind that some low molecular mass proteins (smaller than15 kDa) may diffuse out by freezing and thawing. When we freeze IEF gels, we usually perform equilibration only twice.

3.3 Sodium Dodecyl Sulfate-Polyacrylamide Gel Electrophoresis (Second Dimension)

1. Prepare separating and stacking gels according to a conventional SDS-polyacrylamide gel recipe [11].
2. The length of our stacking gel is somewhere between 1.5 and 2.0 cm.
3. To make a stacking gel, seal its top by means of a glass rod that is slightly shorter than the opening width of the glass plate, and pour stacking gel solution until it touches the glass rod. The glass rod will prevent acrylamide from being exposed to air and allow the gel to polymerize quickly and evenly. After the gel is polymerized, remove the glass rod, and load the IEF tube gel equilibrated with equilibration buffer. Seal the space between the tube gel and the stacking gel with 1% (w/v) agarose. At this time, a well for loading molecular weight markers can be made by placing a piece of spacer (10 × 5 × 0.8 mm) over a stacking gel. After the agarose solidifies, remove the spacer, and load prestained molecular weight markers (GIBCO-BRL). SDS-PAGE is carried out according to Laemmli [11]. The running front can be tracked by dropping 0.1% (w/v) bromophenol blue into the upper chamber or by using equilibration

buffer containing bromophenol blue (0.01%, w/v). The resulting 2D gels can be stained by Coomassie Brilliant Blue and dried.

4 Notes

It is recommended that the 2D gel protocol described in this chapter should be modified to each project in order to optimize the results. We have applied the protocol to mouse retinas, human vitreous samples, and rat dorsal root ganglion samples [12–14]. It is noted that the 2D gel electrophoresis by glass tube-based IEF in the first dimension gives a merit over a Dry Strip IEF because of its capacity to load and separate substantially higher amounts of proteins. It not only serves as a tool for protein identification, but also it provides materials for deeper analysis of chemical structures of the moieties that are altered by posttranslational modification such as phosphorylation, desmethyonylation, methylation, glycosylation [15–18] and potentially applicable to others also.

1. For cultured cells the procedures described by Steinberg and Coffino [19] are used with some modifications. Briefly, ~1.5×10^6 cells are spun down, and the cell pellet is dissolved in IEF lysis buffer to make the final volume less than 90 μL. In the case of bacterial cells, a 0.5 $A_{590\text{-}600}$ equivalent is usually adequate to start with. Again, in these two cases, the dilution of lysis buffer can be disregarded. If tissue homogenates are the starting material, mix homogenate (equivalent to 100–500 μg of total protein) with IEF lysis buffer to make the final volume less than 90 μL. One hundred micrograms of total protein is usually enough to see many protein spots by silver staining, but may not be enough for Coomassie Brilliant Blue staining if the protein of interest is in minor abundance. With our system, we can observe a linear increase in the size and intensity of protein spots up to 500 μg of total protein without detectable loss of small protein spots. In preparing IEF samples from culture cells or tissue homogenates, do not dilute the IEF lysis buffer too much; we have observed that a dilution of up to 30% (at least) of lysis buffer, caused by an aqueous sample, does not affect the IEF. Thus, the volume of IEF buffer should be decided on the basis of the sample volume.
2. IEF gels can be kept at 25 °C for at least 1 week or as long as the bottom and top ends of the gels do not dry out. Before loading IEF gels onto a tube gel apparatus, remove the water layer over the gels, using a Kimwipes, and also remove the Parafilm from the bottom ends. (Although we have never had an IEF gel

come out of a tube during electrophoresis, it could happen; if this appears to be the case, dialysis membrane can be used to seal the bottom end in order to prevent the gel from slipping out.)

References

1. Matsumoto H, Komori N (2000) Ocular proteomics: cataloging photoreceptor proteins by two-dimensional gel electrophoresis and mass spectrometry. Methods Enzymol 316:492–511
2. Matsumoto H, Kurono S, Matsumoto M, Komori N (2005) Mass spectrometry of biomolecules in proteomics. In: Meyers RA (ed) Encyclopedia of molecular and cell biology and molecular medicine. Wiley-VCH Verlag GmbH, Germany, pp 557–585
3. Kenrick KG, Margolis J (1970) Isoelectric focusing and gradient gel electrophoresis: a two-dimensional technique. Anal Biochem 33:204–207
4. Klose J (1975) Protein mapping by combined isoelectric focusing and electrophoresis of mouse tissues. A novel approach to testing for induced point mutations in mammals. Humangenetik *26*:231–243
5. O'Farrell PH (1975) High resolution two-dimensional electrophoresis of proteins. J Biol Chem 250:4007–4021
6. Scheele GA (1975) Two-dimensional gel analysis of soluble proteins. Characterization of guinea pig exocrine pancreatic proteins. J Biol Chem 250:5375–5385
7. Ames GF, Nikaido K (1976) Two-dimensional gel electrophoresis of membrane proteins. Biochemistry 15:616–623
8. Miyazaki K, Hagiwara H, Yokota M, Kakuno T, Horio T (1978) In: Ui N, Horio T (eds) Isoelectric focusing and isotachophoresis. Kyoritsu Shuppan, Tokyo, p. 183. [In Japanese]
9. Matsumoto H, Pak WL (1984) Light-induced phosphorylation of retina-specific polypeptides of *Drosophila in vivo*. Science 223:184–186
10. Matsumoto H, Kurien B, Takagi Y, Kahn ES, Kinumi T, Komori N, Yamada T, Hayashi F, Isono K, Pak WL, Jackson KW, Tobin SL (1994) Phosrestin I undergoes the earliest light-induced phosphorylation by a calcium/calmodulin-dependent protein kinase in *Drosophila* photoreceptors. Neuron 12:997–1010
11. Laemmli UK (1970) Cleavage of structural proteins during the assembly of the head of bacteriophage T4. Nature 227:680–685
12. Haniu H, Komori N, Takemori N, Singh A, Ash JD, Matsumoto H (2006) Proteomic trajectory mapping of biological transformation: Application to developmental mouse retina. Proteomics 6:3251–3261
13. Komori N, Takemori N, Kim HK, Kurono S, Singh A, Hwang S-H, Foreman RD, Chung K, Chung JM, Matsumoto H (2007) Proteomics study of neuropathic and non-neuropathic dorsal root ganglia: Altered protein regulation following segmental spinal nerve ligation injury. Physiol Genomics 29:215–230
14. Shitama T, Hayashi H, Noge S, Uchio E, Oshima K, Haniu H, Takemori N, Komori N, Matsumoto H (2008) Proteome profiling of vitreoretinal diseases by cluster analysis. Proteomics Clin Appl. 2:1265–1280
15. Kinumi T, Jackson KW, Ohashi M, Tobin SL, Matsumoto H (1997) The phosphorylation site and desmethionyl N-terminus of *Drosophila* phosrestin I *in vivo* determined by mass spectrometric analysis of proteins separated on two-dimensional gel electrophoresis. Eur Mass Spectrom 3:367–378
16. Takemori N, Komori N, Matsumoto H (2006) Highly sensitive multistage mass spectrometry enables small scale analysis of protein glycosylation from two-dimensional polyacrylamide gel. Electrophoresis 27:1394–1406
17. Takemori N, Komori N, Thompson J Jr, Yamamoto M-T, Matsumoto H (2007) Novel eye-specific calmodulin methylation characterized by protein mapping in *Drosophila melanogaster*. Proteomics 7:2651–2658
18. Takemori N, Komori N, Matsumoto H (2009) Chapter 150: MS analysis of protein glycosylation. In: Walker JM (ed) The protein protocols handbook, 3rd edn. Humana Press, New York, pp 1381–1388
19. Steinberg RA, Coffino P (1979) Two-dimensional gel analysis of cyclic AMP effects in cultured S49 mouse lymphoma cells: protein modifications, inductions and repressions. Cell 18:719–733

Chapter 12

Cationic Electrophoresis

Engelbert Buxbaum

Abstract

Denaturing, discontinuous electrophoresis in the presence of SDS has become a standard method for the protein scientist. However, there are situations where this method produces suboptimal results. In these cases electrophoresis in the presence of positively charged detergents like cetyltrimethylammonium bromide (CTAB) may work considerably better. Methods for electrophoresis and staining of such gels are presented.

Key words Disk electrophoresis, Cationic electrophoresis, CTAB electrophoresis, Detergent, CTAB

1 Introduction

Sodium dodecyl sulfate polyacrylamide gel electrophoresis (SDS-PAGE) [1] and Western blotting [2, 3] have become indispensable in the protein lab to separate and detect proteins with high resolution. Separation is usually performed on an analytical scale, but preparative equipment is available commercially. With most proteins the R_f-value in SDS-PAGE is proportional to size because proteins bind about 1.4 g SDS per g of protein, equivalent to one molecule of SDS per three amino acids [4]. The negative charges introduced by the detergent far outweigh the charges on the protein itself; thus the charge/mass ratio—and hence the acceleration in an electrical field—is almost identical for all proteins. The restriction by the gel matrix however increases with the size of the protein.

There are, however, some situations where SDS-PAGE performs less well:

- Very hydrophobic proteins (*i.e.*, transmembrane proteins) bind more than the usual 1.4 g SDS per g of protein, increasing the charge/mass ratio of the protein. Thus these proteins run faster in SDS-PAGE than expected for their molecular mass. For

Biji T. Kurien and R. Hal Scofield (eds.), *Electrophoretic Separation of Proteins: Methods and Protocols*, Methods in Molecular Biology, vol. 1855, https://doi.org/10.1007/978-1-4939-8793-1_12,

example, the α-subunit of Na/K-ATPase has a molecular mass of 112 kDa but runs like an 87 kDa protein in SDS-PAGE [5].

- Proteins with a large number of charged amino acids (*e.g.*, histones [6]) also run faster than expected for their molecular mass.
- Glycoproteins contain a highly variable number of negative charges in their sugar side chains. In addition, they bind less SDS than proteins of similar mass [7]. Thus they do not run as crisp bands in SDS-PAGE, but as broad smears, reducing the achievable resolution of the method, and mass determination is unreliable.

In the latter two cases, it should be possible to circumvent the problems by using a positively charged detergent in lieu of the negatively charged SDS. Of course the proteins then run from the positive to the negative electrode. Such attempts have been reported several times in the literature [8–14], but resolution of the gels was usually low, caused by relatively broad protein bands.

The high resolution of SDS-PAGE is the result of band stacking in a discontinuous (multiphasic) buffer system, an effect first described by Ornstein [15] and later theoretically elaborated by Jovin and others [16–19]. While the protein moves through the stacking gel, it is electrophoretically concentrated from below 1 mg/mL (as present in usual samples) to several 100 mg/mL. Since the amount of protein cannot change due to mass conservation, the only way in which this can happen is by reducing the volume of the protein band, *i.e.*, its height. Changing from an anionic to a cationic detergent requires a change in buffer composition so that stacking is still possible [18]. With such selected buffer systems, high-resolution electrophoresis of proteins in cationic detergents is possible [20]. As with SDS-PAGE separation in CTAB (cetyltrimethylammonium bromide)-PAGE is based on protein size, as noted also by others [8, 9, 12, 14].

An additional advantage of CTAB compared to SDS is that it efficiently solubilizes membrane proteins, often without damaging their structure [10, 20–23]. Thus one can use the same detergent for electrophoresis that is also used for solubilization and purification, an advantage since extraneous detergent can interfere in PAGE.

Recently, Kramer [24] has published a similar system based on BAC (benzyldimethyl-N-hexadecylammonium chloride) as detergent with a methoxyacetate buffer; the relative merits of both systems have not been evaluated yet. Alkyl-dimethylamine-N-oxides, alkylpyridinium salts, or alkyltrimethylammonium salts of different chain lengths also still need to be evaluated for their utility in cationic electrophoresis. Two-dimensional gel electrophoresis using cationic detergents in the first and SDS in the second dimension [13, 25] has been used to separate membrane proteins, which are notoriously difficult to analyze by IEF/SDS-PAGE in proteomics studies. Spots are found on a diagonal ellipsoid as the two methods are not orthogonal.

This is an updated version of the paper published in [26].

2 Materials

All chemicals were of the highest purity available and were obtained mostly from Fluka (Buchs, Switzerland). Antibodies were from Accurate (Westbury, NY). Water came from a Milli-Q system (Millipore, Billerica, MA).

2.1 Casting of CTAB Gels

1. 40% acrylamide/bis (37:1): 1.08 g bisacrylamide and 38.9 g acrylamide made to 100 mL with water (*see* **Note 1**). Stable for months at 4 °C, especially when stored over an anion exchanger.
2. 40% acrylamide/bis (19:1): 2.11 g bisacrylamide and 37.9 g acrylamide made to 100 mL with water. Stable for months at 4 °C, especially when stored over an anion exchanger.
3. Potassium hydroxide (KOH; 1 M): 5.611 g KOH made to 100 mL with water. Stable at room temperature (RT) if protected from air.
4. 16.6 M acetic acid (commercial 99.5% glacial acetic acid): Stable at RT. Exact molarity is determined once by titration and noted onto the bottle.
5. 10% CTAB: 10 g CTAB made to 100 mL with water. Store at 37 °C to increase solubility.
6. Malachite green (1%): 10 mg/mL malachite green in water, stable at 4 °C.
7. Water-saturated butanol: *n*-butanol shaken with some water; after phase separation the upper, organic phase is used. Stable at RT.

2.2 For Photopolymerization

1. 100 mM methylene blue: 780 mg methylene blue made to 20 mL with water. Stable for months at 4 °C.
2. 100 mM sodium toluene 4-sulfinate (T4S): 356 mg T4S (anhydrous) made to 20 mL with water. Stable for months at 4 °C.
3. 1 mM diphenyl iodonium chloride (DPIC): 6.3 mg DPIC made to 20 mL with water. Stable for 1 week at 4 °C.

2.3 For Fenton System

1. 10 mM ferrous sulfate ($FeSO_4$): 27.8 mg $FeSO_4 \times 7\ H_2O$ made to 10 mL with water, make fresh daily.
2. 40 mM ascorbic acid: 70.5 mg ascorbic acid made to 10 mL with water, make fresh daily.

3. 30% hydrogen peroxide: commercially available, store at 4 °C (*see* **Note 2**).

The recipes for both photopolymerization [6] and a Fenton-system [27] are given; obviously only one needs to be prepared.

2.4 Running of CTAB Gels

1. Upper tank buffer: 40 mM (3.56 g/L) β-alanine, 70 mM (2.29 mL/L) acetic acid, 0.1% CTAB. Make fresh each time.
2. Lower tank buffer: 50 mM KOH, 187 mM (3.18 mL/L) acetic acid, 0.1% CTAB. Make fresh each time.
3. Sample buffer (2×): 1.27 mL 1 M KOH (127 mM final), 107 μL acetic acid (187 mM final), 2 mL 10% CTAB (2% final), 100 μL β-mercaptoethanol (βME) (1% final), 1 mL glycerol (10% final), 7.21 g urea (12 M final), and 50 μL 1% basic fuchsin (0.005% final), make to 10 mL with water. Stable at RT for a week.

2.5 Staining of CTAB Gels with Ponceau S

1. Fixative: 100 mL glacial acetic acid and 400 mL methanol made to 1 L with water. Stable at RT.
2. Ponceau S solution: 0.1 g Ponceau S and 10 mL glacial acetic acid made to 1 L with water. The solution is stable at RT and may be reused several times.

3 Methods

3.1 Casting of CTAB Gels

1. Mixing table for CTAB-PAGE gels (all volumes in mL). Note that either the reagents for photopolymerization or those for the Fenton-system should be used.

Solution	5%	7.5%	10%	12.5%	15%	17.5%	20%	Stack
Acrylamide 19:1	–	–	–	–	–	–	–	3.0
Acrylamide 37:1	7.5	11.3	15.0	18.8	22.5	26.3	30.0	–
1 M KOH	2.6	2.6	2.6	2.6	2.6	2.6	2.6.	1.91
Glacial acetic acid	0.962	0.962	0.962	0.962	0.962	0.962	0.962	0.161
Urea	10.8 g	10.8 g	10.8 g	10.8 g	10.8 g	10.8 g	10.8 g	5.4 g
10% CTAB	0.6	0.6	0.6	0.6	0.6	0.6	0.6	0.3
100 mM T4S	0.3	0.3	0.3	0.3	0.3	0.3	0.3	0.15
1 mM DPIC	1.5	1.5	1.5	1.5	1.5	1.5	1.5	0.75
10 mM $FeSO_4$	0.048	0.048	0.048	0.048	0.048	0.048	0.048	0.024
40 mM ascorbate	6.0	6.0	6.0	6.0	6.0	6.0	6.0	3.0
Water to	60	60	60	60	60	60	60	30

2. Pour the mixture, containing either the reagents for photo-crosslinking or the FENTON system, into an Erlenmeyer flask with magnetic stirrer. The stirrer is adjusted for vigorous movement without foam production. Then apply a vacuum for at least 10 min to remove dissolved oxygen. This is essential as oxygen inhibits the polymerization reaction. Do not apply the vacuum before starting the stirrer!
3. Add the catalyst: 18 μL 10 mM methylene blue or 2 μL hydrogen peroxide depending on whether the photopolymerization or the FENTON system is used. Mix by gentle inversion; do not shake oxygen into the solution again. At this stage do *not* add the catalyst to the stacking gel.
4. Pour the separating gel into the casting sandwich, and overlay with water-saturated *n*-butanol. The FENTON system will polymerize by itself; the photopolymerization system requires exposure to a strong source of white light. A sun-exposed window or a halogen lamp may be used. Polymerization is finished when the interface between gel and butanol becomes prominent. Methylene blue will become colorless during polymerization. Normally polymerization should be complete within 15 min.
5. Pour the butanol from the top of the gel, add catalyst to the stacking gel mixture, and cast the stacking gel. Insert combs immediately, and allow the gel to polymerize. Leave the gel in the fridge overnight (humid chamber), so that reactive intermediates of the polymerization chemistry can disintegrate.

3.2 Running of CTAB Gels

1. Mount the gel in the running chamber according to manufacturers' instruction; add upper and lower tank buffer. Sometimes air bubbles get trapped in the wells; these can be rinsed out using a tuberculin syringe with 27G needle. This process and the loading of samples are aided by malachite green in the stacking gel, which makes the wells easier to see.
2. Load the sample with a 25 μL Hamilton syringe; the needle should have a flat point. Between samples rinse the syringe with upper tank buffer, finally with water.
3. For a standard minigel, electrophoresis is performed at 20 mA per gel (10 mA during stacking), with a maximum voltage of 200 V. Do not forget to reverse the electrode polarity compared to SDS-PAGE (*see* **Notes 3–8**).

3.3 Staining of CTAB Gels with Ponceau S

1. Fix the gel twice for 15 min on an orbital shaker; then replace the fixative with dye solution for 5 min. The gel can be differentiated by incubation with several changes of fixative (in the same way as is commonly done with CBB-R250), but this procedure is time-consuming (*see* **Notes 9–17**).

4 Notes

1. Acrylamide is neurotoxic; handle with great care. Ready-made acrylamide/bisacrylamide solutions are commercially available; these avoid the development of dust during weighing. Store the solution over an anion exchanger to remove any acrylic acid which may form to prevent electroendosmosis during runs.
2. Store hydrogen peroxide solutions in the bottles supplied by the manufacturer which have release valves to prevent the buildup of pressure from decomposition.
3. FENTON polymerization results in gels which are somewhat more brittle than those produced by photopolymerization. However, if the Hoefer multi-casting stand is used, the non-transparent aluminia back plates prevent the use of a photopolymerization system. In that case the FENTON system must be used. The separation achievable does not depend on the polymerization reaction used, however.
4. Both systems are even more sensitive to the presence of oxygen than the TEMED/APS system used for LAEMMLI gels. Proper degassing of the gel mixture is essential. In addition it is important that the combs prevent access of air to the polymerizing stacking gel [20].
5. Gel concentration depends on the molecular mass range of the proteins of interest [28, 29]. I have found 5–15% gradient gels most convenient for proteins of 10–200 kDa. If the 10 gel multicaster is used, the gradient can be easily formed with a gradient maker, but for single gels the volume required is too small. In that case, cast step gradients by mixing:

% desired	5% (μL)	15% (μL)
15.0	0	392
14.0	36	356
13.1	71	320
12.3	107	285
11.3	142	249
10.5	178	214
9.5	214	178
8.6	249	142
7.7	285	107
6.8	320	71
5.9	356	36
5.0	392	0

These solutions are carefully layered on top of each other; this is made easier when the heavy solution contains 10% glycerol (which does not otherwise interfere with electrophoresis). The total volume of that gradient is 2.35 mL, enough for one minigel. With the photopolymerization chemistry, you can start polymerization once the gel is cast.

6. Proteins with very high molecular mass (>200 kDa) require low acrylamide concentration that results in very soft, difficult-to-handle gels. Add 0.5% agarose to stabilize them without any effect on separation.
7. CTAB has a relatively high KRAFFT point and precipitates if the temperature drops below 18 °C. If lower temperatures are desired during electrophoresis, consider replacing CTAB with a detergent with lower cmc and/or KRAFFT point like 16-BAC. Note that the stacking properties of buffer systems are temperature dependent; on http://www.buffers.nichd.nih.gov systems for use at room temperature and at 0 °C are available.
8. Pre-stained molecular weight markers, available for SDS-PAGE from several manufacturers, are unsuitable for CTAB-PAGE even after detergent exchange. The bound dye influences the R_f-value, and separation is no longer according to molecular mass. It can only be hoped that such standards will also become available for CTAB-PAGE.
9. The staining of gels with Ponceau S is sensitive and fast, but staining with CBB-R250 is also possible. Phenol red can be used in the same way as Ponceau S; the same is probably true for other acidic dyes. Phenol red is fluorescent under acidic conditions, making very sensitive detection of proteins possible.
10. Alternatively destain gels by putting them between filter papers (three sheets of Whatman No. 3 on both sides of the gel), and place it in a tank blotter with blotting buffer. Thirty minutes at 40 V (≈ 200 mA) removes the background stain, while the protein/stain complex is immobile. If need be, proteins become mobile again after incubating the gel in 1 mM KOH for 30 min.
11. Silver staining of CTAB gels is possible; the method of Heukeshoven and Dernick [30] achieves a higher sensitivity than that of Merril et al. [31]. You can execute the former method at a constant temperature of 37 °C for all steps; replace the glutardialdehyde with formaldehyde for maximum sensitivity.
12. Blotting of CTAB gels ("Eastern blotting") is possible with 10 mM KOH and 11 mM acetic acid as buffer. The procedure is similar to Western blotting, except for the reversed polarity [20]. If the blots are to be immunostained, incubate them

first with 0.05% SDS in methanol, to prevent unspecific binding of antibodies to CTAB/protein complexes.

13. After careful removal of CTAB with several changes of fixative, gels may also be stained with "stains all" (3,3'diethyl-9-methyl-4,5,4',5'-dibenzothiacarbocyanine [32]). This is interesting both because different types of protein (acidic, phosphor- and Ca^{2+} -binding) give different colors with "stains all" and because this dye increases the sensitivity of subsequent silver staining.
14. PAS staining of CTAB gels [33] works, but considerable savings in time and chemicals are possible if the staining is performed on Eastern blots rather than on gels [34].
15. Gels can also be stained with fluorescent transition metal complexes like "dye hard" (RuBPS). *See* [35] for details.
16. CTAB is a very mild detergent, which can retain the enzymatic activity of enzymes solubilized with it. Try zymograms [14], proteins in gels specifically stained by their enzymatic activity, at least with monomeric enzymes.
17. Gels can be dried after incubation with 1% glycerol between two sheets of cellophane.

Acknowledgment

This work was supported in part by Kuwait University grant MPB029 and by Ross University.

References

1. Laemmli U (1970) Cleavage of structural proteins during the assembly of the head of bacteriophage T4. Nature 227:680–685. https://doi.org/10.1038/227680a0
2. Towbin H, Staehelin T, Gordon J (1979) Electrophoretic transfer of proteins from polyacrylamide gels to nitrocellulose sheets: procedure and some applications. Proc Natl Acad Sci USA 76(9):4350–4354. https://doi.org/10.1073/pnas.76.9.4350
3. Dunn S (1986) Effects of the modification of transfer buffer composition and the renaturation of proteins in gels on the recognition of proteins on western blots by monoclonal antibodies. Anal Biochem 157(1):144–153. https://doi.org/10.1016/0003-2697(86)90207-1
4. Shirahama K, Tsujii K, Takagi T (1974) Free boundary electrophoresis of sodium dodecyl sulphate-protein polypeptide complexes with special reference to SDS-polyacrylamide gel electrophoresis. J Biochem 75(2):309–319 URL http://jb.oxfordjournals.org/content/75/2/309.full.pdf+html
5. Lane L, Copenhaver J, Lindenmayer G et al (1973) Purification and characterization of and [3H]ouabain binding to the transport adenosine triphosphatase from outer medulla of canine kidney. J Biol Chem 248 (20):7197–7200 URL http://www.jbc.org/content/248/20/7197.full.pdf+html
6. Rabilloud T, Girardot V, Lawrence JJ (1996) One- and two-dimensional histone separation in acidic gels: usefulness of methylene blue-driven photopolymerization. Electrophoresis 17(1):67–73. https://doi.org/10.1002/elps.1150170112
7. Segrest J, Jackson R, Andrews E et al (1971) Human erythrocyte membrane glycoprotein: a re-evaluation of the molecular weight as determined by sds polyacrylamide gel electrophoresis. Biochem Biophys Res Commun 44

(2):390–395. https://doi.org/10.1016/0006-291X(71)90612-7

8. Panyim S, Thitipongpanich R, Supatimusro D (1977) A simplified gel electrophoretic system and its validity for molecular weight determinations of protein-cetyltrimethylammonium complexes. Anal Biochem 81(2):320–327. https://doi.org/10.1016/0003-2697(77)90703-5
9. Eley M, Burns P, Kannapell C et al (1979) Cetyltrimethylammonium bromide polyacrylamide gel electrophoresis: estimation of protein subunit molecular weights using cationic detergents. Anal Biochem 92(2):411–419. https://doi.org/10.1016/0003-2697(79)90679-1
10. Macfarlane D (1983) Use of benzyldimethyl-n-hexadecylammonium chloride ("16-BAC"), a cationic detergent, in an acidic polyacrylamide gel electrophoresis system to detect base labile protein methylation in intact cells. Anal Biochem 132(2):231–235. https://doi.org/10.1016/0003-2697(83)90001-5
11. Mócz G, Bálint M (1984) Use of cationic detergents for polyacrylamide gel electrophoresis in multiphasic buffer systems. Anal Biochem 143(2):283–292. https://doi.org/10.1016/0003-2697(84)90664-X
12. Akin D, Shapira R, Kinkade J (1985) The determination of molecular weights of biologically active proteins by cetyltrimethylammonium bromide-polyacrylamide gel electrophoresis. Anal Biochem 145(1):170–176. https://doi.org/10.1016/0003-2697(85)90343-4
13. Hartinger J, Stenius K, Högemann D et al (1996) 16-BAC/SDS–PAGE: a two-dimensional gel electrophoresis system suitable for the separation of integral membrane proteins. Anal Biochem 240(1):126–133. https://doi.org/10.1006/abio.1996.0339
14. Akins R, Levin P, Tuan R (1992) Cetyltrimethylammonium bromide discontinuous gel electrophoresis: Mr-based separation of proteins with retention of enzymatic activity. Anal Biochem 202(1):172–178. https://doi.org/10.1016/0003-2697(92)90224-U
15. Ornstein L (1962) Disk electrophoresis: I. Background and theory. Ann N Y Acad Sci 121:321–351. https://doi.org/10.1111/j.1749-6632.1964.tb14207.x
16. Jovin T (1973) Multiphasic zone electrophoresis. I. Steady-state moving-boundary systems formed by different electrolyte combinations. Biochemistry 12(5):871–879. https://doi.org/10.1021/bi00729a014
17. Jovin T (1973) Multiphasic zone electrophoresis. II. design of integrated discontinuous buffer systems for analytical and preparative fractionation. Biochemistry 12(5):879–890. https://doi.org/10.1021/bi00729a015
18. Jovin T (1973) Multiphasic zone electrophoresis. III. Further analysis and new forms of discontinuous buffer systems. Biochemistry 12(5):890–898. https://doi.org/10.1021/bi00729a016
19. Jovin T (1973d) Multiphasic zone electrophoresis. IV design and analysis of discontinuous buffer systems with a digital computer. Ann NY Acad Sci 209:477–496. https://doi.org/10.1111/j.1749-6632.1973.tb47551.x
20. Buxbaum E (2003) Cationic electrophoresis and electrotransfer of membrane glycoproteins. Anal Biochem 314(1):70–76. https://doi.org/10.1016/S0003-2697(02)00639-5
21. Pritchard D, Crawford C, Duce I et al (1985) Antigen stripping from the nematode epicuticle using the cationic detergent cetyltrimethylammonium bromide (CTAB). Parasite Immunol 7(6):575–585. https://doi.org/10.1111/j.1365-3024.1985.tb00101.x
22. Freedman D, Nutman T, Ottesen E (1988) Enhanced solubilization of immunoreactive proteins from Brugia malayi adult parasites using cetyltrimethylammonium bromide. Exp Parasitol 65(2):244–250. https://doi.org/10.1016/0014-4894(88)90128-2
23. Maki K, Sagara J, Kawai A (1991) A cationic detergent, cetyltrimethylammonium bromide (CTAB), selectively dissociates the intermediate filament of the fibroblast. Biochem Biophys Res Commun 175(3):768–774. https://doi.org/10.1016/0006-291X(91)91632-M
24. Kramer M (2006) A new multiphasic buffer system for benzyldimethyl-n-hexadecylammonium chloride polyacrylamide gel electrophoresis of proteins providing efficient stacking. Electrophoresis 27(2):347–356. https://doi.org/10.1002/elps.200500563
25. Wenge B, Bönisch H, Grabitzki J et al (2008) Separation of membrane proteins by two-dimensional electrophoresis using cationic rehydrated strips. Electrophoresis 29(7):1511–1517. https://doi.org/10.1002/elps.200700546
26. Buxbaum E (2012) Cationic electrophoresis. In: Kurien and scofield, p. 55–63, doi: https://doi.org/10.1007/978-1-59745-542-8_14
27. Fenton H (1894) Oxidation of tartaric acid in presence of iron. J Chem Soc Trans 65:899–910. https://doi.org/10.1039/CT8946500899

28. Rodbard D, Chrambach A (1971) Estimation of molecular radius, free mobility, and valence using polyacrylamide gel electrophoresis. Anal Biochem 40(1):95–134. https://doi.org/10.1016/0003-2697(71)90086-8
29. Stellwagen N (1998) Apparent pore size of polyacrylamide gels. Electrophoresis 19 (10):1542–1547. https://doi.org/10.1002/elps.1150191004
30. Heukeshoven J, Dernick R (1988) Improved silver staining procedure for fast staining in PhastSystem development unit i. staining of sodium dodecyl sulfate gels. Electrophoresis 9 (1):28–32. https://doi.org/10.1002/elps.1150090106
31. Merril C, Goldman D, Sedman S et al (1981) Ultrasensitive stain for proteins in polyacrylamide gels shows regional variations in cerebrospinal fluid proteins. Science 211 (4489):1437–1438. https://doi.org/10.1126/science.6162199
32. Goldberg H, Warner K (1997) The staining of acidic proteins on polyacrylamide gels: Enhanced sensitivity and stability of "stains-all" staining in combination with silver nitrate. Anal Biochem 251(2):227–233. https://doi.org/10.1006/abio.1997.2252
33. Segrest J, Jackson R (1972) Molecular weight determination of glycoproteins by polyacrylamide gel electrophoresis in sodium dodecyl sulfate. Meth Enzymol 28:54–63
34. Thornton D, Holmes D, Sheehan J et al (1989) Quantitation of mucus glycoproteins blotted onto nitrocellulose membranes. Anal Biochem 182(1):160–164. https://doi.org/10.1016/0003-2697(89)90735-5
35. Buxbaum E (2012) Fluorescent staining of gels. In: [36], 543–550, doi:https://doi.org/10.1007/978-1-59745-542-8_14
36. Kurien B, Scofield R (eds) (2012) Protein electrophoresis—methods and protocols, Meth Mol Biol, vol vol. 869. Humana, Dordrecht ISBN 978-1-61779-820-7

Chapter 13

Two-Dimensional Gel Electrophoresis with Immobilized pH Gradients

Bre'Ana Byrd and Huyen Tran

Abstract

Isoelectric focusing (IEF) is a technique in protein research that has been used since 1975 to separate proteins based on their isoelectric point. When combined with sodium dodecyl sulfate polyacrylamide gel electrophoresis, this procedure allows for high-resolution separation of cellular proteins for analytical purposes. Laboratories perform IEF by (a) using carrier ampholytes that migrate through a gel to create the pH gradient or (b) using immobilized pH gradients (IPG) that contain ampholytes bound covalently to a gel. Here we describe an IEF system that uses immobilized pH gradient (IPG) strips that undergo the desired current and voltage setting to separate proteins based on its charge in the first dimension followed by SDS PAGE to generate a two-dimensional map of serum proteins.

Key words Isoelectric focusing, Two-dimensional electrophoresis, Immobilized pH gradients, High resolution

1 Introduction

Proteins are important for facilitating function, signal transduction, and structure [1]. Previous work has shown that it is important to have a high-resolution separation of proteins in electrophoresis because it allows for a more accurate look at the individual proteins rather than looking at the whole set of proteins. Single-dimension gel electrophoresis has been used in the past for identifying proteins and is a technique that generates multiple protein patterns and is discerned through molecular weight differences. Essentially, the technique facilitates protein subunits with similar molecular weights to migrate together. While this technique is accepted and used, it does not yield proteins in a manner that offers high resolution. High resolution of proteins in this regard refers to being able to look at individual proteins and their respective molecular weight. Two-dimensional electrophoresis (2-DE) has become a popular technique and alternative to quantify and identify proteins

Biji T. Kurien and R. Hal Scofield (eds.), *Electrophoretic Separation of Proteins: Methods and Protocols*,
Methods in Molecular Biology, vol. 1855, https://doi.org/10.1007/978-1-4939-8793-1_13,

[1–5]. This offers high resolution of proteins that the single-dimension electrophoresis did not. Isoelectric focusing (IEF) with the use of immobilized pH gradient (IPG) has become a newer advancement in protein separation because of the manipulability of the current and voltage control on the machine [5]. IPG offers better resolution and reproducibility while allowing higher protein loads [6]. IEF is an electrophoretic technique used for separating proteins based on their isoelectric point (pI) with an IPG strip. This is the first step in breaking down dimensions in a 2-D electrophoresis analysis. The 2-D analysis delivers a map of intact proteins, which reflects changes in protein expression level [3] and enables a powerful method for studying the profiles of proteins [5]. The coupling of IEF and SDS PAGE allows the separation of proteins based on their molecular weight as well as their respective pI thus giving a high resolution of an immunoblot. The 2-D gel electrophoresis technique involves two primary steps: (a) isoelectric focusing and (b) gel electrophoresis. It is important to note that IEF separates based on pI and 2-DE separates based on molecular weight. These differences are important concepts because the molecular weight and pH gradient becomes localized to a single protein. In this chapter, the protocol for the use of the Protean i12 IEF System and 2-DE will be discussed.

2 Materials

1. i12™ Focusing Tray, length 11 cm.
2. Gel-side up electrode wicks.
3. Electrode assemblies (one size fits all).
4. Rehydration/equilibration tray, 11 cm.
5. Mineral oil: Any standard mineral oil can be used.
6. Skinny, long blunt forceps.
7. Rehydration buffer: 8 M urea, 2% CHAPS, 50 mM DTT, 0.2% BioLyte® 3/10 ampholyte, and 0.001% bromophenol blue.
8. Filter paper.
9. IPG strips pH 3–10: We used 11 cm strips.
10. Cleaning brushes: Any standard toothbrush and cleaning brush will work.
11. Cleaning concentrate: 50× alkaline detergent (*see* **Note 1**).
12. Protean i12 IEF System.
13. One percent agarose containing bromophenol blue: Weigh 1 g agarose and suspend in 100 mL water.

    Microwave to dissolve. Add few drops of 0.1% bromophenol blue. Redissolve by heating after it solidifies, prior to use.

Fig. 1 Schematic for Criterion gel unit with IPG strip

14. Criterion gel electrophoresis unit (*see* Fig. 1).
15. Criterion precast gel (4–20% gradient gel).
16. Equilibration buffer: 31.25 mL 0.5 M Tris–HCl (pH 6.8), 5 g SDS, 90 g urea, 12.5 mL β-mercaptoethanol.

3 Methods

3.1 Strip Rehydration

1. Centrifuge the sample at 12,000 × *g* for 1 min to 5 min. Remove 4 μL of the supernatant.
2. Pipette 300 μL of rehydration buffer into a microcentrifuge tube, and then add the 4 μL of the supernatant and mix well.
3. Add the rehydration buffer and supernatant mixture into the rehydration tray.
4. Using forceps, place the IPG strip gel-side down into the rehydration tray (*see* **Notes 2** and **3**).
5. Cover the IPG strip with 4 mL of mineral oil (*see* **Note 4**).
6. Leave the rehydration tray with the IPG strips overnight; we recommend up to 16 h for rehydration.
7. Remove the mineral oil from the rehydration tray with the use of a 1-mL pipette; ensure the strip is not distressed during this process.
8. Using forceps, move the IPG strips from the rehydration tray, and place them gel-side up onto wet filter paper (*see* **Note 5**).
9. Wet the gel-side up wicks in deionized water. Wet two wicks for every IPG strip. Place the wicks on filter paper until ready for use.
10. Place the IPG strip into the focusing tray wells with the positive side of the strip facing the anode (*see* **Note 6**).

11. Place the wicks at the ends of the IPG strips (*see* **Note 7**).
12. Pipette 4 mL of mineral oil onto the strips.
13. Position the electrode assemblies on the focusing tray, and press down to securely attach the electrode assemblies.
14. Place the focusing tray into the instrument to run a protocol.

3.2 Running a Protocol

1. After placing the focusing tray into the instrument and pressing run, the machine will promote assigning a protocol.
2. Assign the protocol matching the strip length, pH 3–10 G (*see* **Note 8**).
3. Press run, and then press start on the following page to begin the protocol.

3.3 Gel Electrophoresis

1. After the protocol has been run, remove the mineral oil, again with a 1-mL pipette.
2. Rinse the strips in 15 mL of equilibration buffer for 15 min.
3. Remove the equilibration buffer, and add another 15 mL of equilibration buffer for 15 min (*see* **Note 9**).
4. Repeat the equilibration buffer step one more time.
5. Place the strip into the well of the Criterion gel. The gel side should be facing outward, toward the individual.
6. Pipette the molecular weight marker into the well beside the gel (*see* **Note 10**).
7. Pour 1% agarose containing bromophenol blue on top of the strip, and start the electrophoresis run.
8. Stain gel with Coomassie Brilliant Blue after the run is complete (Fig. 2).

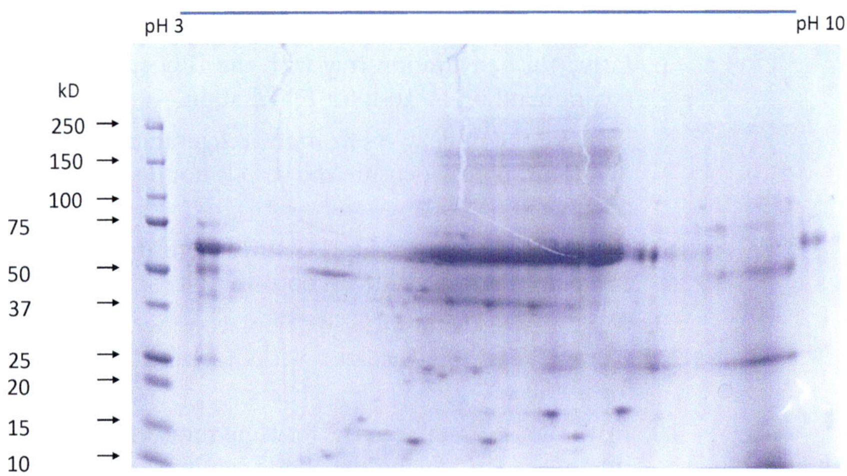

Fig. 2 Mouse serum proteins separated by 2-D gel electrophoresis and stained with Coomassie Brilliant Blue

4 Notes

1. Used for cleaning the glass plates and other equipment.
2. Remove the plastic backing from IPG strip.
3. Be careful not to introduce air bubbles and remove any existing air bubbles.
4. The mineral oil is to protect the strips from desiccation during the rehydration period.
5. This will help remove any mineral oil and keep the gel wet while the focusing tray is prepared.
6. Immediately load the IPG strip into the focusing tray after the excess mineral oil has been removed. The voltage in the instrument will be affected if the strip is left out longer than needed.
7. The wicks need to go between the IPG strip and the electrode assemblies to conduct the electricity from the machine through the strip.
8. The IPG strips used were 11 cm.
9. The preset protocol we used took 8 h to complete.
10. The equilibration buffer is used to remove urea from the strips and absorb SDS into the strips.
11. Criterion gels already have a marker well placed inside the gel.

References

1. Reinders J, Sickmann A (2009) Proteomics methods and protocols. Humana Press, London
2. Friedman DB, Hoving S, Westermeier R (2009) Isoelectric focusing and two dimensional gel electrophoresis. Methods Enzymol 463:515–540
3. Gorg A, Weiss W, Dunn MJ (2009) High resolution two dimensional electrophoresis. Proteomics 4:13–32
4. Roepstorff P (2012) Mass spectrometry based on proteomics, background, status, and future needs. Protein Cell 3:641–647
5. Posch A, Franz T, Hartwig S, Knebel B, Al-Hasani H, Passlack W, Kunz N, Hinze Y, Li X, Kotzka J, Lehr S (2013) 2D- Togo workflow: increasing feasibility and reproducibility of 2-dimensional gel electrophoresis. Arch Physiol Biochem 119:108–113
6. Bjellqvist B, Pasquali C, Ravier F, Sanchez J-C, Hochstrasser D (1993) A nonlinear wide range immobilized pH gradient for two dimensional electrophoresis and its definition in a relevant pH scale. Electrophoresis 14:1357–1367

Chapter 14

SARCOSYL-PAGE: Optimized Protocols for the Separation and Immunological Detection of PEGylated Proteins

Christian Reichel, Günter Gmeiner, Philipp Reihlen, Mario Thevis, and Wilhelm Schänzer

Abstract

PEGylation of recombinant proteins and synthetic peptides aims to generate biopharmaceuticals with altered physical properties. The modification may lead to a prolonged serum half-life caused by decreased receptor-mediated endocytosis and/or delay in renal clearance caused by the increased hydrodynamic volume of the pharmaceutical. MIRCERA, a PEGylated recombinant erythropoietin (rhEPO) used in the treatment of anemia due to chronic kidney disease, has also been abused by athletes as performance-enhancing drug. While it can be detected by sodium dodecylsulfate polyacrylamide gel electrophoresis (SDS-PAGE) and immunoblotting, the sensitivity of the test is significantly lower compared to other epoetins. By replacing SDS with sarcosyl in the sample and running buffers, the interaction between SDS and the PEG group of the protein no longer reduces the affinity of the monoclonal anti-EPO antibody (*clone AE7A5*) to the protein chain. Contrary to SDS, sarcosyl only binds to the amino acid chain of the PEGylated protein and thus leads to a sharper electrophoretic band and enhanced antibody binding. While the method was originally developed for anti-doping purposes, it may also be useful for the electrophoretic separation and immunological detection of other PEGylated proteins. Protocols for urine and serum are presented. They are also applicable for the general detection of EPO-based erythropoiesis-stimulating agents (ESA) in these matrices.

Key words MIRCERA, PEGylated proteins, SDS-PAGE, SAR-PAGE, Sarcosyl, Immunoblotting, Erythropoietin (EPO), Doping control, BlotCycler

1 Introduction

Due to the prolonged serum half-life of MIRCERA (ca. 130 h) and its decreased excretion in urine, the abuse of MIRCERA by athletes is preferably detected in serum or plasma samples. However, it was also shown to be excreted in urine [1]. MIRCERA, a PEGylated epoetin beta, contains one methoxy polyethylene glycol group (ca. 30 kDa) in covalent linkage to either the N-terminal amino group or the ε-amino group of Lys 52 or Lys 45 of the 165 amino acid chain of human recombinant erythropoietin (rhEPO) beta

Biji T. Kurien and R. Hal Scofield (eds.), *Electrophoretic Separation of Proteins: Methods and Protocols*, Methods in Molecular Biology, vol. 1855, https://doi.org/10.1007/978-1-4939-8793-1_14,

[2]. The resulting molecule has an average molecular mass of ca. 60 kDa and was mainly developed for treating anemia-related diseases (e.g., chronic kidney disease, CKD). The PEGylation not only leads to an increase in serum half-life but also to a repeated activation of the EPO-receptor by the molecule, due to a weakened interaction. Therefore, MIRCERA is also known as "continuous erythropoietin receptor activator (CERA)." Since erythropoietin stimulates the body's red blood cell (RBC) production, EPO pharmaceuticals have been mostly abused by athletes in endurance sports. Doping with rhEPO—including analogs, biosimilar epoetins, and MIRCERA—is prohibited according to the World Anti-Doping Agency (WADA) [3]. Its detection has been regulated by a technical document [4]. Since the molecular mass of MIRCERA is profoundly different to the mass of endogenous and recombinant EPOs (ca. 30 kDa mass difference), sodium dodecylsulfate polyacrylamide gel electrophoresis (SDS-PAGE) in combination with Western blotting appeared to be an ideal detection method. However, it turned out that SDS-PAGE is less sensitive for the PEGylated analog due to the interaction of SDS with both the PEG and amino acid chain of MIRCERA. By replacing SDS with N-lauroylsarcosinate (sarcosyl)—an anionic methyl glycine-based detergent—a significant increase in immunoblotting sensitivity was obtained [5]. It was shown that sarcosyl is not able to solubilize PEG the way SDS does. Contrary to SDS, sarcosyl does only bind to the protein part of PEGylated EPO, which leads to a sharper electrophoretic band and an enhanced antibody interaction compared to SDS-PAGE. The increased sensitivity of SARCOSYL-PAGE (SAR-PAGE) was explained by an inability of the monoclonal anti-EPO antibody to bind to the fully (i.e., PEG and amino acid chain) SDS-solubilized PEGylated protein. Instead, the antibody only interacted with molecules with SDS mainly bound to the protein chain. The result was a "pseudo-sharp" band on Western blot but with low sensitivity, as the majority of the MIRCERA molecules were fully solubilized by SDS. Since sarcosyl binds only to the protein chain, no interference with the antibody interaction occurred. This explained the appearance of a "truly sharp" band of the PEGylated protein in addition to a high sensitivity on Western blot (Fig. 1).

In 2015 the original SAR-PAGE protocol was optimized in particular for detecting CERA and other ESAs in urine [6]. It now uses automatized membrane incubation steps (BlotCycler), which lead to extremely low background and consequently higher sensitivity.

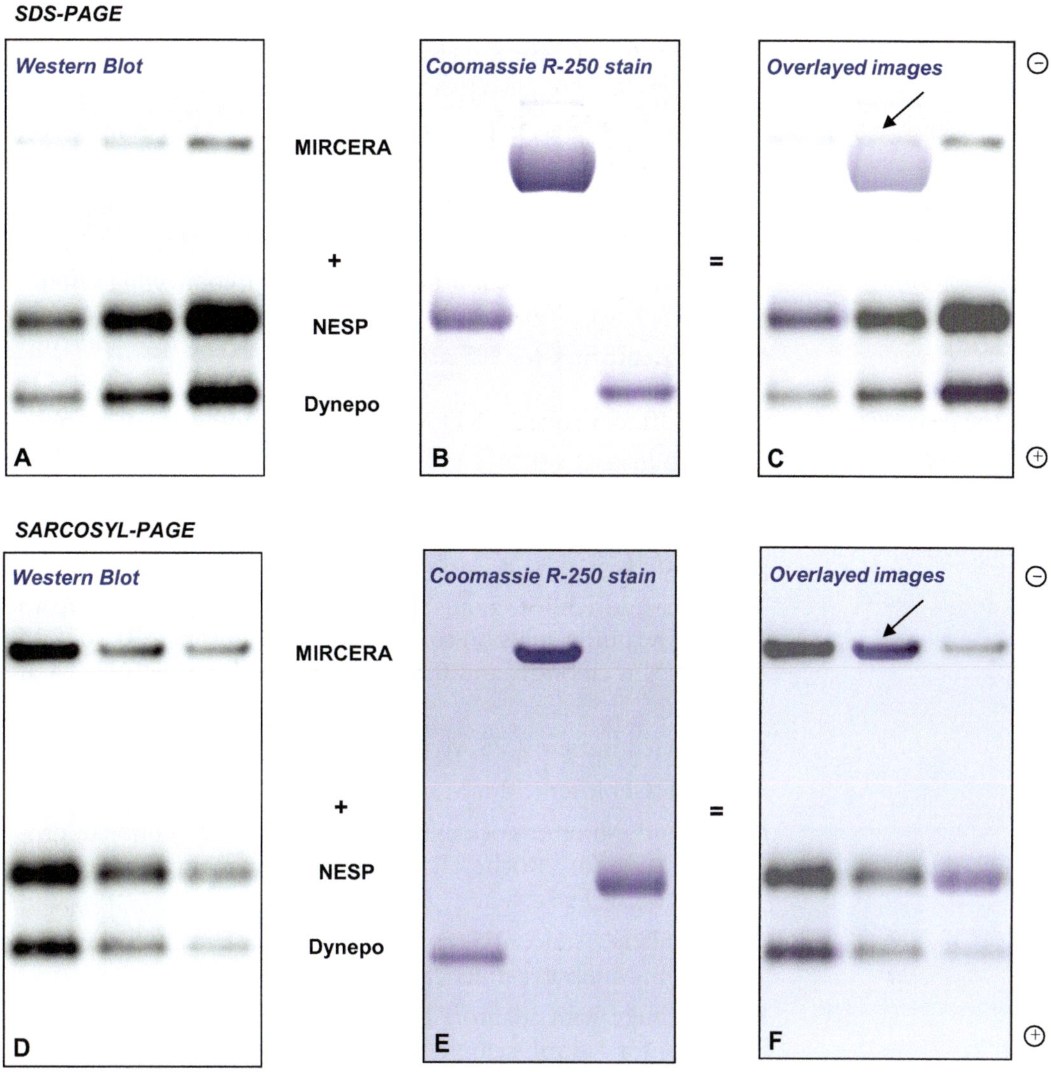

Fig. 1 Relationship between "true band sharpness" on gel and "pseudo band sharpness" on Western blots for MIRCERA. SDS-PAGE (**a–c**) and SAR-PAGE (**d–f**): Immunoblot with monoclonal anti-EPO antibody *(clone AE7A5)*, Coomassie R-250 stained identical second gel, and virtual overlay of both images. Note that the majority of the PEGylated protein molecules were not recognized by the antibody on SDS-PAGE *(arrows)*. Reproduced from ref. 5 with permission from Wiley

2 Materials

Use Milli-Q (MQ) water for preparing all buffers and solutions. Chemicals should be analytical or electrophoresis grade except otherwise noted. Wear non-powdered gloves during the entire procedure (*see* **Note 1**).

2.1 Immunoaffinity Purification of Serum Samples

1. EPO Purification Kit (MAIIA Diagnostics; Uppsala, Sweden).
2. Negative and positive (MIRCERA, rhEPO) control serum samples, test samples (typically 200 μL serum)
3. Microfilters (e.g., 0.22 μm; Nanosep MF; PALL; Ann Arbor, MI; or Steriflip, Millipore; Billerica, MA).
4. Ultrafilters (Microcon YM-30, MWCO 30,000; Millipore; Billerica, MA).
5. QIAvac 24 Plus system with vacuum pump, connecting system, and VacValves (QIAGEN; Hilden, Germany).
6. Microcentrifuge for 1.5 mL sample tubes.

2.2 Immunoaffinity Purification of Urine Samples

1. EPO-ELISA plate (STEMCELL Technologies; Grenoble, France).
2. Negative and positive (MIRCERA, rhEPO) control urine samples, test samples (typically 15 mL urine)
3. Phosphate-buffered saline (PBS): Dissolve 5 PBS tablets (Sigma-Aldrich, St. Louis, MO) in MQ water, and fill up to 1 L. The buffer solution contains 0.01 M phosphate, 0.0027 M potassium chloride, and 0.137 M sodium chloride (pH 7.4 at 25 °C).
4. Tris–HCl buffer (3.75 M, pH 7.4).
5. Tris–HCl buffer (50 mM, pH 7.4).
6. Protease inhibitor cocktail: dissolve one tablet cOmplete protease inhibitor (Roche Diagnostics; Mannheim, Germany) in 2 mL of MQ water.
7. Ultrafilters (Amicon Ultra-15 and Amicon Ultra-0.5, MWCO 30,000; Millipore; Billerica, MA).
8. Centrifuge with adaptors for 50 mL Falcon tubes, Microcentrifuge for 1.5 mL sample tubes.
9. PCR film.

2.3 Sarcosyl-PAGE (SAR-PAGE)

1. Polyacrylamide gels: Pre-cast BisTris gels (NuPAGE; Invitrogen/Thermo Fisher Scientific; Carlsbad, CA), e.g., 10% T, 1.0 mm, 20 wells, and Midi (*see* **Note 2**).
2. Sarcosyl (SAR) sample buffer (SB) (4×): 424 mM Tris hydrochloride, 564 mM Tris base, 8% SAR, 40% glycerol, and 2.04 mM EDTA (pH 8.5; do not adjust) [7]. Dissolve 0.666 g Tris(hydroxymethyl)aminomethane hydrochloride (Tris–HCl), 0.682 g Tris base, 0.8 g sodium N-lauroylsarcosinate (sarcosyl, SAR; #61743, Sigma-Aldrich; St. Louis, MO), 0.006 g ethylenediaminetetraacetic acid (EDTA, free acid), 4 g glycerol (≥99%), and ca. 3 mg of phenol red in 10 mL MQ water. Start with first dissolving Tris–HCl, Tris base, and SAR in ca. 5 mL of water (vortex), briefly

centrifuge to destroy excess foam, then add glycerol and phenol red, vortex again, and fill up with MQ water to 10 mL (*see* **Note 3**). The buffer can be stored for several months at room temperature.

3. Sarcosyl (SAR) running buffer (RB): 50 mM MOPS, 50 mM Tris base, 0.1% sarcosyl, 1 mM EDTA (pH 7.7; do not adjust). Dissolve 10.46 g MOPS, 6.06 g Tris base, 1 g sarcosyl, and 0.3 g EDTA (free acid) in MQ water, and fill up to 1 L.
4. Reducing agent (10×): 1 M DL-dithiothreitol (DTT) in MQ water. Prepare immediately before use (e.g., dissolve 154 mg DTT in 1 mL MQ water).
5. SAR elution buffer: mix 100 μL SAR-SB (4×) with 260 μL MQ water and 40 μL DTT (1 M). Prepare immediately before use.
6. Antioxidant: Dissolve 380 mg sodium metabisulfite ($Na_2S_2O_5$) in 1 mL of MQ water. Prepare immediately before use. The antioxidant is used only for the catholyte. Alternatively, the antioxidant for SDS-PAGE as supplied by Invitrogen/Thermo Fisher Scientific (Carlsbad, CA) may also be used.
7. Phosphate buffered saline (PBS): *see* above.
8. Insulin dilution buffer: Prepare 0.1% (w/v) insulin solution by diluting human insulin stock solution (10 mg/mL; Sigma-Aldrich #I9278; St. Louis, MO) 1:10 with PBS.
9. BSA dilution buffer: Prepare 0.05% (w/v) bovine serum albumin (BSA, ELISA grade) in PBS.
10. PEGylated protein standard: MIRCERA (50 μg/0.3 mL; Roche; Mannheim, Germany). For sensitivity testing prepare a serial dilution down to low pg/medium fg level (absolute amount on gel). Dilute with 0.1% insulin solution or 0.05% BSA/PBS (*see* **Note 4**).
11. Recombinant human EPO standards, e.g., Erypo (Janssen-Cilag; Vienna, Austria), NeoRecormon (Roche; Mannheim, Germany), Dynepo (Shire; Hampshire, UK), and Aranesp (NESP; Amgen; Thousand Oaks, CA). For sensitivity testing prepare a serial dilution down to medium/low fg level (absolute amount on gel). Dilute with 0.1% insulin solution or 0.05% BSA/PBS.
12. ThermoMixer C with SmartBlock for microtiter and DeepWell plates and SmartBlock for 1.5 mL sample tubes (Eppendorf; Hamburg, Germany).
13. Sample tubes (1.5 mL; Eppendorf; Hamburg, Germany).
14. MμltiFlex Round Tips (1–200 μL) (Sorenson BioScience; Salt Lake City, UT).
15. Electrophoresis cell (XCell4 SureLock Midi-Cell; Invitrogen/Thermo Fisher Scientific; Carlsbad, CA) (*see* **Note 5**).

16. Power supply.
17. Gel knife (Invitrogen/Thermo Fisher Scientific; Carlsbad, CA).

2.4 Immunoblotting

1. Transfer membranes: Immobilon-P (0.45 μm) and Durapore (0.65 μm) polyvinylidene fluoride (PVDF) membranes (Millipore; Billerica, MA).
2. Methanol.
3. Western blot transfer buffer I: 48 mM Tris base, 39 mM glycine, 0.00375% SAR, and 20% ethanol [8]. Dissolve 5.81 g Tris base, 2.93 g glycine, and 0.0375 g N-lauroylsarcosine sodium salt (SAR; Sigma-Aldrich #61743; St. Louis, MO) in MQ water, add 200 mL of ethanol, and fill up to 1 L with MQ water (*see* **Note 6**).
4. Western blot transfer buffer II: 0.7% acetic acid. Dilute 7 mL of glacial acetic acid with MQ water to 1 L.
5. Reducing buffer: 10 mM DTT. Dissolve 154 mg DTT in 100 mL PBS. Prepare immediately before use.
6. Blocking buffer: 5% (w/v) nonfat milk (NFM). Dissolve 5 g NFM in 100 mL PBS.
7. Washing buffer: 0.5% (w/v) NFM in PBS.
8. Incubation buffer: 1% (w/v) NFM in PBS.
9. Incubation vessels: Polypropylene, melamine, and/or glass containers of appropriate size for performing reducing, blocking, washing, and incubation steps.
10. Primary antibody: Monoclonal mouse anti-EPO antibody (*clone AE7A5*; R&D Systems; Minneapolis, MN).
11. Secondary antibody for single-blotting: Polyclonal goat anti-mouse IgG (H + L) antibody, cross-adsorbed, HRP-conjugated (#31432, Pierce/Thermo Scientific; Rockford, IL).
12. Secondary antibody for double-blotting: Biotinylated polyclonal goat anti-mouse IgG (H + L) antibody (#31800, Pierce/Thermo Scientific; Rockford, IL).
13. Streptavidin-horseradish peroxidase (HRP) complex (#G014-61, Biospa; Milan, Italy).
14. Blotting paper for single-blotting: Extra-thick blot paper (8.6 × 13.5 cm; Bio-Rad; Hercules, CA).
15. Blotting paper for double-blotting: Electrode paper NovaBlot (GE Healthcare; Uppsala, Sweden).
16. Blotter for semidry transfer: Trans-Blot SD Semi-Dry Electrophoretic Transfer Cell (Bio-Rad; Hercules, CA).
17. Power supply.
18. Incubator (37 °C).

19. Shaker (e.g., Stuart seesaw rocker SSL4; Bibby Scientific; Stone Staffordshire, UK).
20. Tweezers (flat tips).
21. Rubber roller.
22. Parafilm.
23. Sponge for cleaning the blotter.
24. BlotCycler (Precision Biosystems; Mansfield, MA).

2.5 Chemiluminescent Detection

1. Enhanced chemiluminescent (ECL) substrate: SuperSignal West Femto (Thermo Fisher Scientific; Rockford, IL).
2. CCD camera: LAS-4000 (GE Healthcare; Uppsala, Sweden) or equivalent.
3. Glass plate.
4. Ethanol (absolute).
5. Pipette tips (10 mL).
6. Image analysis software, e.g., GASepo *v2.3* [9] (Seibersdorf Laboratories; Seibersdorf, Austria).

3 Methods

Due to the high protein content of human serum (ca. 60–80 μg/μL) [10] and the low amount of EPO (ca. 26 pg/mL to 1 ng/mL), serum samples cannot be directly applied on electrophoresis gels for detecting EPO by Western blotting. Immunoaffinity purification is the most efficient way to remove the majority of these otherwise interfering proteins (*see* **Note** 7). The purified samples are then separated with SAR-PAGE, immunoblotted, and detected via chemiluminescence. Contrary to that, the protein content of urine is much lower and greatly varies with the specific gravity. Typical EPO-concentrations are in the low ng/L range [11]. In order to detect EPO in urine, the urine needs to be concentrated by ultrafiltration and subsequently also immunoaffinity purified to remove interfering proteins.

3.1 Immunoaffinity Purification of Serum Samples

1. Dilute 200 μL of serum sample with 1800 μL sample dilution buffer (mix 1 mL buffer for plasma/serum and 1 mL of exposure aid (both from the kit) with 20 mL of MQ water). The final serum concentration is 10% [12]. Incubate for 10 min at room temperature.
2. Filter the diluted sample through a 0.2 μm centrifugal filter in 500 μL aliquots (14,000 RCF/5 min). Alternatively, vacuum microfiltration can also be used (Steriflip filters).
3. Mount one dummy column of the kit on the QIAvac system with the VacValve, close valve, fill in 1 mL of PBS, and adjust flowrate to ca. 0.5 mL/min on the vacuum pump.

4. Remove the dummy column, and mount the EPO immunoaffinity purification column on the QIAvac system using the VacValve, close the valve, add 1 mL of PBS to the column, open the valve, let the buffer pass through with ca. 0.5 mL/min, and then close the valve.
5. Add the diluted serum sample (2 × 1 mL) to the column, open the valve and let the first mL pass through, then close the valve, refill, and open the valve again. The flow rate should be 0.5 mL/min or below.
6. Wash the column with 1 mL of washing buffer (ca. 0.5 mL/min).
7. Remove the column from the vacuum system, put it in a 1.5 mL tube, and spin in the microcentrifuge at 2000 RCF for 1 min.
8. Place the column in a new 1.5 mL tube containing 5 μL of adjustment buffer A (provided with the kit), add 50 μL of desorption buffer to the column, insert the assembly in the microcentrifuge, and spin again for 1 min at 2000 RCF.
9. Concentrate the eluate down to <10 μL by ultrafiltration (Microcon YM-30, 14,000 RCF/5 min). Recover the retentate by reversed spinning at 1000 RCF/2 min (*see* **Note 8**).
10. Add 3.75 μL SAR-SB (4×) and 1.5 μL DTT (1 M). Fill up with MQ water to 15 μL.
11. Proceed with SARCOSYL-PAGE or store sample at −20 °C or below.

3.2 Immunoaffinity Purification of Urine Samples

1. Fill 15 mL of urine into a 50 mL Falcon tube.
2. Add 1.5 mL Tris–HCl buffer (3.75 M, pH 7.4) and 300 μL cOmplete protease inhibitor solution.
3. Centrifuge at 4000 RCF/12 min.
4. Transfer supernatant to an Amicon Ultra-15 filter and spin at 4000 RCF/25 min.
5. Discard filtrate, and wash retentate with 15 mL Tris–HCL buffer (50 mM, pH 7.4) supplemented with 300 μL complete solution (4000 RCF/20 min).
6. Repeat washing step.
7. Transfer the retentate to an Amicon Ultra-0.5 filter and centrifuge at 14,000 RCF/ca. 15 min (*see* **Note 9**).
8. Recover retentate by reversed spinning at 1000 RCF/2 min. The final volume should be ca. 20–40 μL.
9. Transfer retentate to a well of the EPO-ELISA plate, cover with PCR film, and incubate in refrigerator without shaking

overnight (alternatively, incubation can also be performed in a ThermoMixer at 37 °C for 1–2 h).

10. Remove retentate and wash the well five times with 300 μL PBS.
11. Tap dry and add 15 μL SAR elution buffer.
12. Seal with PCR film and heat in ThermoMixer at 95 °C/5 min/ 650 rpm.
13. Spin down at 1300 RCF/1 min. The sample is now ready for SAR-PAGE.

3.3 Sarcosyl-PAGE (SAR-PAGE)

1. Heat an appropriate number of standards for 5 min at 95 °C under shaking (650 rpm; ThermoMixer). Cool down, and briefly centrifuge to remove condensate from lid (*see* **Note 10**).
2. Take one NuPAGE BisTris Midi gel (10% T, 1.0 mm, 20 wells) out of its plastic bag; remove the tape on the foot of the mold and the comb.
3. Put the gel into the XCell4 SureLock Midi-Cell, insert the buffer dam (or a second gel), and lock the sandwich consisting of gel, buffer core, and buffer dam (or second gel) with the SureLock assembly and displacement dam. This sandwich builds the inner buffer chamber and holds the catholyte.
4. Prepare the catholyte by mixing 400 mL of SAR-RB with 1000 μL of antioxidant.
5. Fill the catholyte in the inner chamber until the wells of the gel are completely covered with liquid. Check the inner buffer chamber for leaking.
6. Rinse the wells with catholyte.
7. Fill samples and standards (15 μL each) with MμltiFlex Tips into the wells of the gel.
8. Fill the anolyte (SAR-RB without antioxidant) in the outer buffer chamber until ca. 1 cm below the upper edge of the cell (this rather large liquid volume serves as a heat sink).
9. Put the electrophoresis cell in a plastic box (ca. 33 × 23 × 13 cm). Close the cell with the lid, and connect it to the power supply. Fill the box with ice-cold water and add frozen cooling packs.
10. Run for 180 min at constant voltage (125 V).
11. Stop electrophoresis as soon as the leading phenol red front has reached the bottom of the gel.

3.4 Immunoblotting

Depending on the sample type, either single- or double-blotting can be performed. Due to non-specific interactions of the secondary antibody with remaining serum proteins, double-blotting is mandatory for doping analyses of serum/plasma samples. For urine samples, single-blotting can be performed for most of the samples [4].

3.4.1 Single-Blotting

1. Cut one sheet of Immobilon-P membrane and two sheets of extra-thick blotting paper (7 × 13 cm). Activate membrane in methanol (ca. 15 s), then wash in MQ water (ca. 2 min), and finally equilibrate it in Western blot transfer buffer I for ca. 5 min.
2. Soak the two sheets of extra-thick blotting paper with transfer buffer I.
3. After electrophoresis, release the gel from the mold using the gel knife.
4. Place the gel in a vessel containing Western blot transfer buffer I and equilibrate it for ca. 2 min.
5. Clean anode and cathode of the Trans-Blot SD Semi-Dry Transfer Cell first with methanol or ethanol, and then with MQ water.
6. Prepare a sandwich consisting of one sheet of extra-thick blotting paper, PVDF membrane, gel, and the second sheet of extra-thick blotting paper. Assemble directly on the anode of the blotter and roll out air bubbles with the rubber roller. Close the blotter with the cathode and lid.
7. Connect the blotter with the power supply and blot for 45 min at 1.54 mA/cm^2 constant current. Blotting direction is from cathode to anode.
8. After blotting, disassemble the sandwich, and place the membrane with the side, which was next to the gel, facing upward in PBS and shake for ca. 5 min. Clean the blotter with a sponge and hot water, and rinse with MQ water.
9. Discard PBS and block the membrane in 100 mL 5% NFM/PBS solution for 60 min at room temperature under slow shaking.
10. Discard blocking solution and wash membrane briefly in PBS to remove excess milk.
11. Transfer the membrane into reducing buffer (10 mM DTT in PBS) and incubate at 37 °C for 20 min (make sure that the membrane does not swim up during the incubation period).
12. Discard reducing buffer and wash the membrane briefly in PBS, and then transfer it to the BlotCycler (fill the tray with ca. 40 mL PBS before; *see* **Note 11**).
13. Dilute primary antibody (*clone AE7A5*) 1:1000 in 1% NFM/PBS (25 mL).
14. Dilute secondary antibody (HRP-conjugated, Pierce #31432) 1:100,000 in 1% NFM/PBS (25 mL).

Table 1
BlotCycler program for single-blotting

Blocking (PBS only)	5 min
Wash 1	0 min
First antibody	180 min
Wash 2	9 × 5 min
Second antibody	810 min
Wash 3	9 × 5 min

15. Fill primary and secondary antibody solutions in tubes "P" and "S" of the BlotCycler, respectively. Fill ca. 3 L PBS into the tank (*see* **Note 12**).
16. Program BlotCycler as follows (Table 1):
17. Start BlotCycler.

3.4.2 Double-Blotting

1. Perform **steps 1** to **8** of as described under Subheading 3.4.1.
2. Transfer the membrane into reducing buffer (10 mM DTT in PBS) and incubate it at 37 °C for 60 min (make sure that the membrane does not swim up during the incubation period).
3. Discard reducing buffer and wash the membrane briefly in PBS.
4. Block the membrane in 100 mL 5% NFM/PBS solution.
5. Discard blocking buffer and wash the membrane briefly in PBS.
6. Incubate the membrane in primary antibody solution. Slowly shake for 60 min at room temperature or overnight in a cold room or refrigerator.
7. Wash the membrane 3 × 10 min in 0.5% NFM/PBS (shake faster than before).
8. Briefly rinse membrane in PBS in order to remove milk foam generated during washing.
9. Cut another sheet of Immobilon-P membrane (7 × 13 cm), and activate it in methanol for ca. 15 s while the first membrane is washing in 0.5% NFM/PBS. Rinse the membrane in MQ water, and equilibrate it in Western blot transfer buffer II (0.7% acetic acid) for ca. 5 min. Also cut one sheet of Durapore membrane (7 × 13 cm) and soak it directly in transfer buffer II (*see* **Note 13**).
10. Cut 2 × 6 sheets of NovaBlot blotting paper (7 × 13 cm), and soak in transfer buffer II for ca. 10–15 min.

11. Clean anode and cathode of the Trans-Blot SD Semi-Dry Transfer Cell first with methanol or ethanol and then with MQ water or transfer buffer II.
12. Assemble the double-blotting sandwich after rinsing the first membrane with PBS. The sandwich consists of six sheets of NovaBlot paper, the first Immobilon-P membrane, the Durapore membrane, the second Immobilon-P membrane, and another six sheets of blotting paper (assemble the sandwich directly on the anode and roll out air bubbles as described above). Close the blotter with the cathode and lid.
13. Connect the blotter with the power supply and blot for 10 min at 0.8 mA/cm^2 constant current (*see* **Note 14**).
14. Remove the second membrane immediately after the blot, and rinse ca. 5 min under shaking in PBS to neutralize the acid. Store the first membrane in PBS in the refrigerator or cold room until image analysis is completed (*see* **Note 15**). Clean the blotter with a sponge and hot water; rinse with MQ water.
15. Discard PBS and block membrane in 5% NFM/PBS for 60 min and under slow shaking at room temperature.
16. Discard blocking solution and rinse with PBS to remove excess milk.
17. Incubate the membrane in secondary antibody solution (diluted 1:2000 in 1% NFM/PBS) for 60 min at room temperature and under slow shaking (ca. 20–40 mL of solution depending on vessel size).
18. Discard antibody solution, and wash 3 × 10 min in 0.5% NFM/PBS (shake faster than before), and then rinse briefly in PBS to remove milk foam.
19. Incubate the membrane in streptavidin-HRP complex diluted 1:2000 in 1% NFM/PBS (60 min, room temperature, shake slowly).
20. Discard streptavidin-HRP solution, and wash the membrane in PBS (3 × 10 min), and then continue with chemiluminescent detection.

3.5 Chemiluminescent Detection

1. Prepare the substrate working solution by mixing both substrate components 1:1 (total 3 mL).
2. Clean the glass plate and tweezers with both ethanol and MQ water; wipe dry (*see* **Note 16**).
3. Take the membrane out of the PBS solution with tweezers; remove excess PBS by holding the membrane in vertical position and with the lower edge touching a sheet of blotting paper.

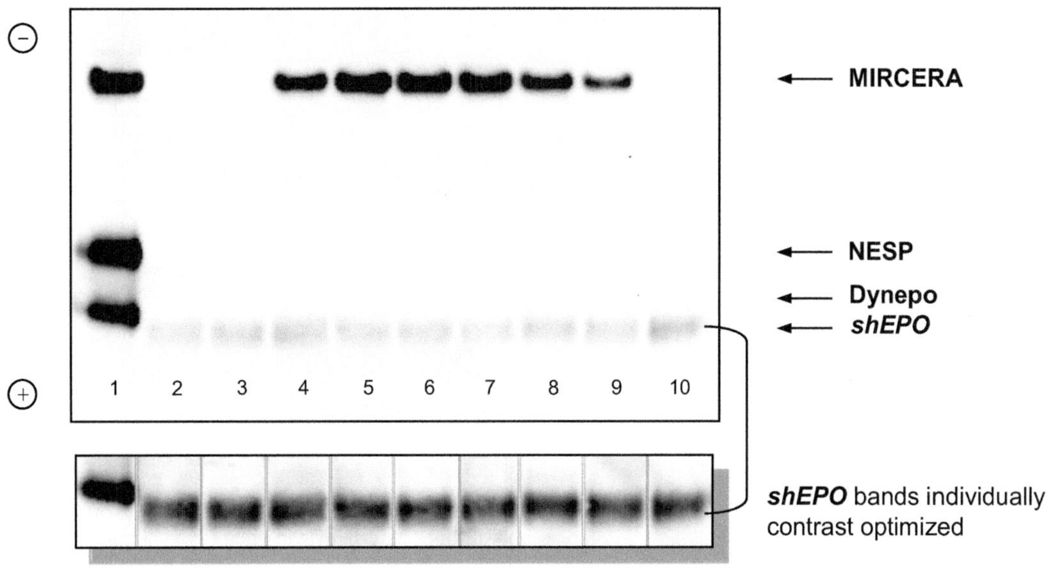

Fig. 2 Detection of MIRCERA in 200 μL of serum by SAR-PAGE. Lane sequence: (1) EPO standards (MIRCERA, NESP, Dynepo), (2 and 3) MIRCERA negative serum samples, (4–9) MIRCERA positive samples, (10) negative control serum. Note the lower molecular mass of endogenous serum EPO (shEPO) compared to the other epoetins. Reproduced from ref. 5 with permission from Wiley

4. Place the membrane on the glass plate, and transfer the plate into the CCD camera.
5. Distribute the substrate evenly on the membrane using the long side of a 10 mL pipette tip (do not directly touch the surface of the membrane).
6. Adjust membrane position and focusing of the camera. Immediately start with acquiring images (either by using the automatic mode of the camera or by manually selecting the exposure time (*see* **Note 17**).
7. Evaluate images using image analysis software (*see* **Note 18**). The presence of a band matching the position of the standards and the positive control sample is indicative for the presence of MIRCERA or another rhEPO in the test sample (Figs. 2 and 3) [5].

4 Notes

1. The powder of the gloves might lead to interference with the chemiluminescent substrate.
2. Instead of the BisTris buffer system (e.g., MOPS-chloride boundary), the standard Laemmli buffer system (glycine-chloride boundary) [13] can be used. SAR-PAGE was also demonstrated to be compatible with the Neville (borate-sulfate

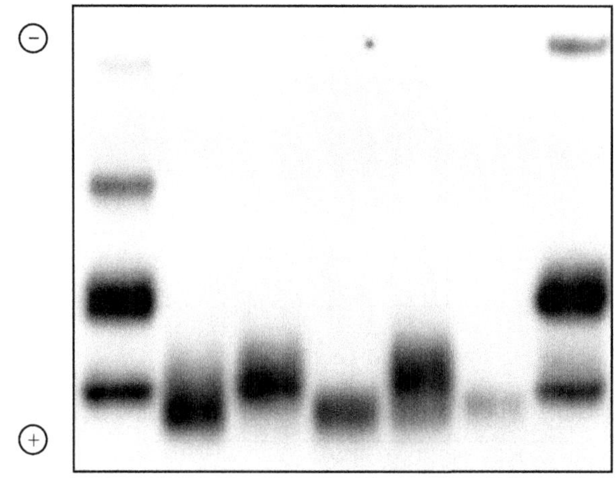

Fig. 3 Testing for doping with MIRCERA and rhEPOs in urine using single-blotting and BlotCycler. Lane sequence: (1) EPO standards (EPO-Fc, NESP, Dynepo), (2) positive control urine (low rhEPO, high uhEPO), (3) positive control urine (high rhEPO, low uhEPO, (4) negative control urine, (5) rhEPO-positive urine sample, (6) negative urine sample, and (7) EPO standards (CERA, NESP, Dynepo). Note the higher molecular mass of the positive controls and positive sample compared to the negative control and negative samples. Also note that EPO-Fc typically generates two bands (*see* **Note 10**)

boundary) [14], and Allen-Moore (e.g., borate-citrate boundary) [15] stacking systems. In case you decide to use these systems, make sure that you adapt all gel, sample, and running buffers accordingly by replacing the SDS with sarcosyl.

3. Use only highest purity sarcosyl as specified ($\geq$ 99%). The method will not work properly with a less pure detergent. Phenol red is used as front marker dye. The final color of the sample buffer should be reddish.

4. Since the effect of sarcosyl on the separation of PEGylated proteins is based on the non-interaction of sarcosyl with the covalently linked PEG molecule(s), the negative net charge is primarily dependent on the length of the amino acid chain and its interaction with the negatively charged sarcosyl molecules. Consequently, small PEGylated peptides will migrate much slower or hardly at all on SAR-PAGE than PEGylated proteins, given that the size of the attached PEG group(s) is identical. The effect of sarcosyl on the migration behavior can be visualized by comparing the results obtained on SDS-PAGE and SAR-PAGE after staining the gels with a PEG-specific reagent (Fig. 4). Coomassie R-250 on the other hand does only interact with the amino acid chain—hence it does not stain PEG [5]. For specifically staining the PEG group(s), first incubate

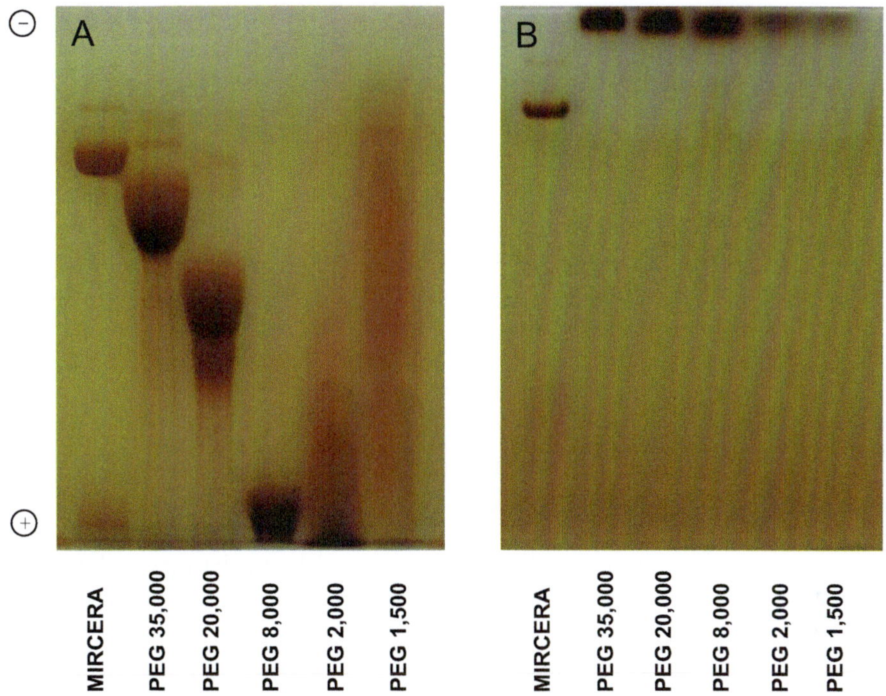

Fig. 4 Comparison of SDS-PAGE (**a**) and SAR-PAGE (**b**) of MIRCERA and polyethylene glycols of different average molecular masses (PEGs) after an iodine-based PEG-specific stain. Applied amounts on gels (BisTris, 10% T, 1.5 mm): MIRCERA (ca.2.2 μg), PEGs (ca. 5 μg each). Reproduced from ref. 5 with permission from Wiley

the gel for 15 min in a 5% (w/v) solution of barium chloride in Milli-Q water, then incubate for 15 min in an aqueous solution of 1% iodine/1.5% potassium iodide (w/v), and finally wash the gel until an optimal contrast between the brown PEG band and the still yellow gel is obtained (ca. 30 min) [16]. Avoid prolonged destaining as the PEG band will then also disappear (if this happens, stain again with the iodine solution). For documentation purposes scan or photograph the destained gel as soon as possible, since the iodine complex is instable. Proteins without PEG groups are not stained by this method (Fig. 4).

5. It is recommended to use a new electrophoresis cell for SAR-PAGE and to then use this cell only with sarcosyl. Otherwise make sure that you extensively clean your SDS-PAGE cell in order to completely remove SDS residues in the cell.

6. It is not necessary to replace the SDS in transfer buffer I for sarcosyl. Sarcosyl is only necessary during the electrophoretic separation for enhanced band sharpness and interaction of MIRCERA with the monoclonal anti-EPO antibody (*clone AE7A5*). Hence, the transfer can also be performed with the

regular Bjerrum buffer [8]. However, it was observed that using 0.00375% SAR improves the background of the BlotCycler method. For toxicity reasons methanol was replaced by ethanol.

7. Aside from the described immunoaffinity purification procedure [17, 18], which uses disposable monolithic columns containing an immobilized monoclonal anti-EPO antibody (*clone 3F6*), alternative procedures were also described (e.g., the removal of high abundant serum proteins by depletion with a multi-affinity column followed by immunoaffinity enrichment of EPO with a commercial an EPO-ELISA plate) (for additional details *see* ref. 5).
8. Contrary to filters of the Amicon Ultra type, Microcon filters run dry when centrifuged too long. Amicon Ultra-0.5 filters are not useful for this step, since the final retentate volume is around 20 μL regardless how long they are centrifuged (dead-stop filtration). For increased band sharpness, the sample volume applied on the gel should not be at the maximum of the well capacity.
9. Alternatively, the retentate of the first ultrafiltration step may be directly applied to the ELISA well. Make sure that in this case you individually centrifuge the samples as long as necessary to obtain a retentate volume $\leq$ 200 μL. Otherwise the sensitivity of the method may be compromised.
10. It is recommended to use two mixtures of recombinant erythropoietins as standards, namely, Dynepo/NESP/CERA and Dynepo/NESP/EPO-Fc. EPO-Fc typically generates two characteristic bands on SAR-PAGE. While the lower mass band is more intense and well separated from NESP, the band with the higher mass is weaker and just below the CERA band. Hence, it might be difficult to observe it in a Dynepo/NESP/EPO-Fc/CERA mixture [19].
11. The BlotCycler is available with trays for Mini- and Midi-gels. Select your instrument accordingly.
12. Do not fill buffers containing milk into the wash tank. Use only PBS to obtain images with low background. Put the BlotCycler in a refrigerator or cold room, and perform all steps under cooled conditions.
13. The original procedure of the Western double blot was described by Lasne [20, 21] and uses a Durapore membrane between the first and second Immobilon-P membranes. The Durapore membrane is a hydrophilic low protein-binding PVDF membrane in contrast to the hydrophobic high protein-binding PVDF Immobilon-P membrane and acts as a separator membrane between the two Immobilon-P membranes. However, the Durapore membrane might be omitted

in this step. The double-blotting technique was originally developed for detecting EPO in urinary retentates by isoelectric focusing in polyacrylamide slab gels and without preceding immunoaffinity purification. An additional explanation for the usefulness of the Durapore membrane during the double-blotting procedure is given in reference [22].

14. Blotting direction is from anode to cathode this time. Under acidic conditions the interaction between primary antibody and EPO is disrupted. During blotting the antibody migrates toward the cathode. Connect blotter as usual with the power supply (do not reverse polarity).
15. The first membrane can be re-incubated with primary antibody starting with the reduction (DTT) step. After completing image analysis, the first membrane can be dried on blotting paper overnight on air and stored at −20 °C for later re-incubation.
16. Make sure that no remnants of previous reagents and milk are on the tweezers. Otherwise, dark spots might occur on the membrane during incubation with the substrate and on the position where it was picked with the tweezers.
17. Typically, it is not necessary to use a pre-incubation period (e.g., 5 min) after the substrate was put on the membrane and before the acquisition of images. Also, it is normally not necessary to remove the substrate and then cover the membrane with a transparent plastic film (e.g., Saran wrap). A pre-incubation period is recommended when high sensitivity is required already for the first image. However, pre-incubation bears the risk that in case of a high antigen concentration, the signal is already gone before the first image is acquired due to excessive substrate reaction with HRP (identifiable by the appearance of brown bands on the membrane). Typical exposure times range from a few seconds to minutes.
18. To test the sensitivity of the entire system (i.e., immunoblotting, chemiluminescent detection, CCD camera), use a dilution series of MIRCERA and other EPO standards. The typical limit of detection (LOD) for MIRCERA on SAR-PAGE in combination with the described antibodies and a medium sensitivity ECL substrate (e.g., SuperSignal West Pico) is ca. 3 pg (around 50 amol). With a higher sensitivity substrate (e.g., SuperSignal West Femto), the achievable LOD is in medium fg range (Fig. 5).

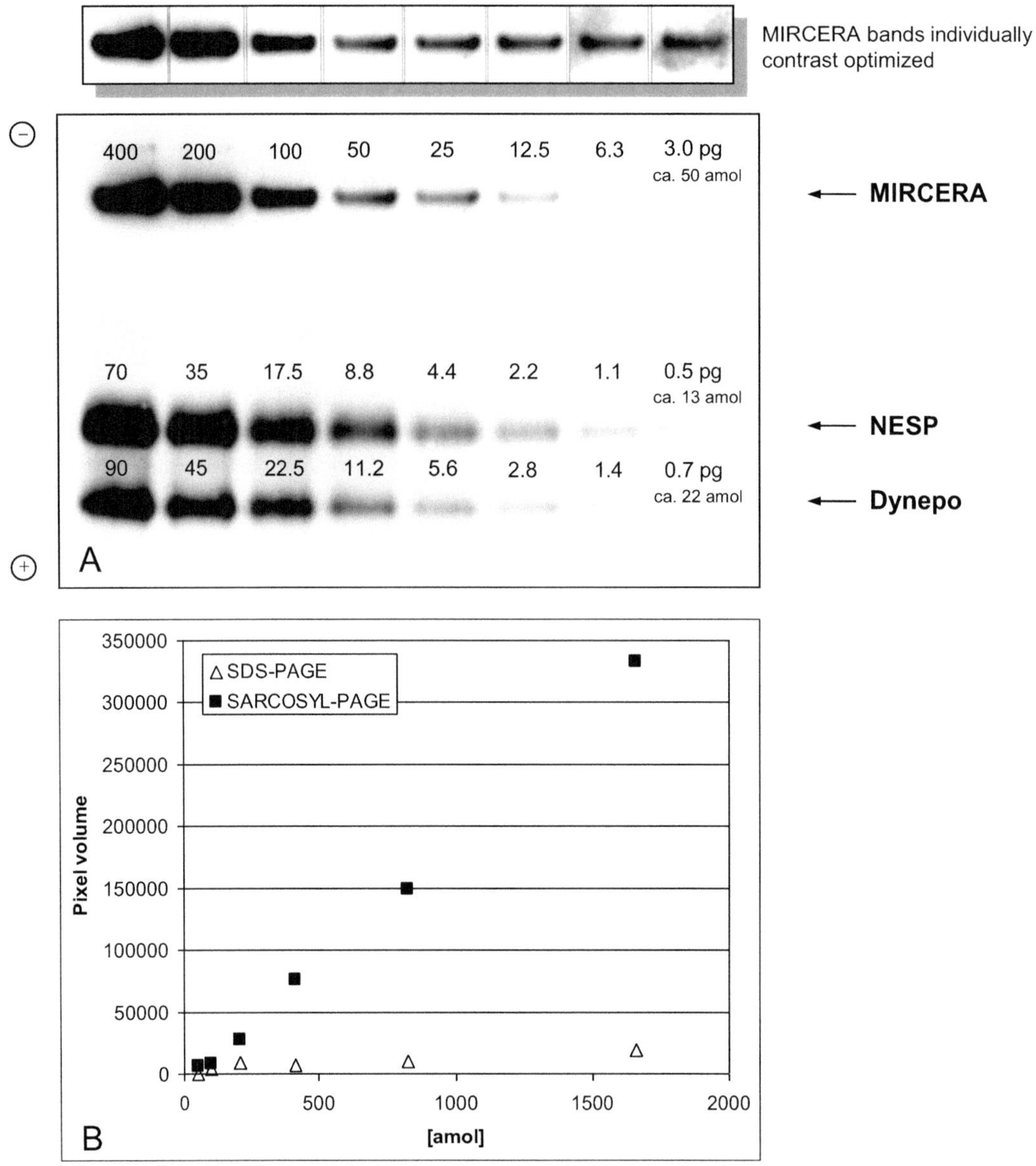

Fig. 5 (**a**) Sensitivity test of SAR-PAGE with a twofold serial dilution of three EPO standards (MIRCERA, NESP, Dynepo). Typically, low pg to fg sensitivity is obtained. (**b**) Densitometric comparison of SAR-PAGE and SDS-PAGE under identical test conditions. Note the significant increase in sensitivity for SAR-PAGE. Reproduced from ref. 5 with permission from Wiley

Acknowledgment

This project has been carried out with the financial support of WADA (grant 08B11CR).

References

1. Dehnes Y, Hemmersbach P (2011) Effect of single doses of methoxypolyethylene glycol-epoetin beta (CERA, Mircera) and epoetin delta (Dynepo) on isoelectric erythropoietin profiles and haematological parameters. Drug Test Anal 3(5):291–299
2. European Medicines Agency (EMEA) (2007) MIRCERA scientific discussion, http://www.ema.europa.eu/docs/en_GB/document_library/EPAR_-_Scientific_Discussion/human/000739/WC500033669.pdf. Accessed 29 Aug 2017, pp 1–4.
3. World Anti-Doping Agency (2016) The world anti-doping code. International Standard. Prohibited list January 2017. Montreal. https://www.wada-ama.org/sites/default/files/resources/files/2016-09-29_-_wada_prohibited_list_2017_eng_final.pdf. Accessed 29 Aug 2017.
4. World Anti-Doping Agency (2014) Technical document TD2014EPO. Harmonization of analysis and reporting of ESAs by electrophoretic techniques. Montreal. https://www.wada-ama.org/sites/default/files/resources/files/WADA-TD2014EPO-v1-Harmonization-of-Analysis-and-Reporting-of-ESAs-by-Electrophoretic-Techniques-EN.pdf. Accessed 29 Aug 2017.
5. Reichel C, Abzieher F, Geisendorfer T (2009) SARCOSYL-PAGE: a new method for the detection of MIRCERA- and EPO-doping in blood. Drug Test Anal 1(11–12):494–504
6. Reihlen P, Blobel M, Kempkes R, Reichel C, Völker Schänzer E, Majer B, Schänzer W (2015) Optimizing SAR-PAGE. Drug Test Anal 7(11–12):1014–1016
7. Invitrogen (2003). In: NuPAGE Technical Guide Version E (IM-1001), Carlsbad, CA, p 49.
8. Bjerrum OJ, Schafer-Nielsen C (1986) Buffer systems and transfer parameters for semidry electroblotting with horizontal apparatus. In: Dunn MJ (ed) Electrophoresis, 86th edn. Wiley-VCH, Weinheim, pp 315–327
9. Bajla I, Holländer I, Minichmayr M, Gmeiner G, Reichel C (2005) GASepo—a software solution for quantitative analysis of digital images in EPO doping control. Comput Methods Programs Biomed 80:246–270
10. Lasne F, Martin L, Martin JA, de Ceaurriz J (2007) Isoelectric profiles of human erythropoietin are different in serum and urine. Int J Biol Macromol 41:354–357
11. Berglund B, Wide L (2002) Erythropoietin concentrations and isoforms in urine of anonymous Olympic athletes during the Nagano Olympic games. Scand J Med Sci Sports 12(6):354–357
12. MAIIA Diagnostics (2016) EPO Purification Kit, Uppsala. http://maiiadiagnostics.com/gammal/files/Directions%20for%20Use%20-%20EPO%20Purification%20Kit%20%28new%21%29.pdf. Accessed 29 Aug 2017.
13. Laemmli UK (1970) Cleavage of structural proteins during the assembly of the head of bacteriophage T4. Nature 227:680–685
14. Neville DM Jr (1971) Molecular weight determination of protein-dodecyl sulfate complexes by gel electrophoresis in a discontinuous buffer system. J Biol Chem 246:6328–6334
15. Allen RC, Moore DJ (1971) Electrophoretic system. U.S. Patent 3620947.
16. Zheng C, Ma G, Su Z (2007) Native PAGE eliminates the problem of PEG-SDS interaction in SDS-PAGE and provides an alternative to HPLC in characterization of protein PEGylation. Electrophoresis 28:2801–2807
17. Lönnberg M, Dehnes Y, Drevin M, Garle M, Lamon S, Leuenberger N, Quach T, Carlsson J (2010) Rapid affinity purification of erythropoietin from biological samples using disposable monoliths. J Chromatogr A 1217:7031–7037
18. Dehnes Y, Lamon S, Lönnberg M (2010) Erythropoietin (EPO) immunoaffinity columns—a powerful tool for purifying EPO and its recombinant analogues. J Pharm Biomed Anal 53:1028–1032
19. Reichel C, Thevis M (2012) Detection of EPO-Fc fusion protein in human blood: screening and confirmation protocols for sports drug testing. Drug Test Anal 4 (11):818–829
20. Lasne F (2001) Double-blotting: a solution to the problem of non-specific binding of secondary antibodies in immunoblotting procedures. J Immunol Methods 253:125–131
21. Lasne F (2003) Double-blotting: a solution to the problem of nonspecific binding of secondary antibodies in immunoblotting procedures. J Immunol Methods 276:223–226
22. Lasne F (2009) Double-Blotting: a solution to the problem of nonspecific binding of secondary antibodies in immunoblotting procedures. In: Kurien BT, Scofield RH (eds) Methods in molecular biology, protein blotting and detection, vol 536. Humana Press, New York, p 217

Chapter 15

Tricine-SDS-PAGE

Syed R. Haider, Helen J. Reid, and Barry L. Sharp

Abstract

Tricine-sodium dodecyl sulfate-polyacrylamide gel electrophoresis (tricine-SDS-PAGE) is an efficient way of separating low molecular mass proteins. However, the standard system is quite complicated and specifically may not be useful when the separated proteins are to be recovered from the gel for quantitative analysis. Here, we describe a simplified system whereby these smaller proteins can be resolved in comparatively low percentage gels which have high compatibility with modern detectors such as UV and inductively coupled plasma-mass spectrometry (ICP-MS).

Key words Tricine-SDS-PAGE, Low molecular mass proteins, Running buffer, Gel buffer, Gel mixture, Low percentage gel, Protein standard, Polyacrylamide

1 Introduction

The separation of low molecular mass proteins (1–30 kDa), or fragmented peptides derived from enzymatic digestion, is a difficult task. To achieve such separations with high resolution, slab SDS-PAGE-based methods are frequently used and allow several lanes to be run on a single gel under identical conditions. Thus it is easy to compare and to determine the molecular masses of the separated species by calculating the relative mobility of each band relative to standards run on the same gel. The Laemmli-SDS-PAGE method [1] is widely used for the separation of the proteins; however, for the separation of low mass proteins, this system requires a very high percentage gel (>16%) or a gradient gel [2, 3]. Such gels have several disadvantages including difficulties in casting, irreproducibility, fragility, and short shelf life [4]. Further, it is difficult to quantitatively recover proteins in blotting and electroelution. The high background signals of sulfur and phosphorus in high percentage gels also makes them unsuitable for laser ablation (LA)-inductively coupled plasma-mass spectrometry (ICP-MS) detection of proteins [5–7].

Biji T. Kurien and R. Hal Scofield (eds.), *Electrophoretic Separation of Proteins: Methods and Protocols*, Methods in Molecular Biology, vol. 1855, https://doi.org/10.1007/978-1-4939-8793-1_15,

Tricine-SDS-PAGE is a remarkably simple and efficient alternative way of separating low molecular mass proteins/peptides in a single polyacrylamide gel with high resolution. Usually, a 10% gel (with pH 8.45) is employed to separate proteins in the range of 1–100 kDa using two running buffers: a cathode buffer, e.g., 100 mM Tris, 100 mM tricine, and 0.1% (w/v) SDS, pH 8.25, and an anode buffer, e.g., 100 mM Tris, pH 8.9. However, to obtain highly resolved bands, this method requires three gels (stacking, spacer, and resolving gels) [8–10] and addition of urea [4] in the resolving gel. The use of three gels is tedious, potentially troublesome, and always requires a freshly prepared gel mixture [10]. Addition of urea may also create problems in amino acid sequencing [11], and although it is useful for the analysis of low mass proteins, it crystallizes at low temperature and sometimes decomposes during sample preparation [4]. Recently, we have described an improved form of the tricine-SDS-PAGE method employing a simplified procedure which does not require two different electrode buffers, a spacer gel, or addition of urea [12]. This modified system is well suited for quantitative analysis and shows excellent compatibility with detectors such as ICP-MS or UV. For such analyses (and where the retention of metals is desired), the softest conditions are preferred, that is, low percentage gels and low concentration buffers.

For comparative purposes, the following provides descriptions of separations based on the modified reagent system and the original method reagent system. These show that using the modified system, a different buffer and pH are required and that the original reagent system does not work efficiently under these altered conditions. The user does not need to follow the steps for the original method; they are included only for information.

2 Materials

Prepare all the solutions in ultrapure water (18 M-Ω, e.g., from a Milli-Q water purification system, Millipore Corporation, Bedford, MA). Store at room temperature (if required), unless otherwise stated.

2.1 Tricine-SDS-PAGE Components

1. 2.5 M Tris buffer, pH 8.8 for the modified system, and pH 8.45 for original system (*see* **Notes 1** and **2**): Weigh 302.85 g of Tris-base and transfer it to a 1 L graduated cylinder or glass beaker. Dissolve the base by adding 600 mL of deionized water and adjust the pH using HCl. Make up to 1 L with water. This gel buffer can be stored at 4–8 °C.
2. Polyacrylamide gel solution: 30% (w/v) solution of acrylamide—bisacrylamide 29:1. Store this polyacrylamide solution at

4–8 °C (it is highly recommended to purchase the ready-made polymer solution from the suppliers instead of preparing it in the laboratory because acrylamide is recognized as a neurotoxin and suspected carcinogen).

3. Running buffer for modified system: 25 mM Tris, 25 mM tricine, and 0.05% (w/v) SDS. Dissolve 3.03 g Tris-base, 4.5 g tricine, and 0.5 g SDS in 1 L of deionized water (*see* **Note 3**). There is no need to adjust the pH. Store at 4–8 °C (*see* **Note 4**).
4. Running buffer for original system: 100 mM Tris and 100 mM tricine 0.1% (w/v) SDS. Weigh 12.1 g Tris-base, 17.9 g tricine, and 1 g SDS, and dissolve in 1 L of deionized water (*see* **Notes 3–5**). There is no need to adjust the pH. Store at 4–8 °C (*see* **Note 4**).
5. Sample or loading buffer: 2× sample buffer containing 100 mM Tris–HCl (pH 6.8) (*see* **Note 6**), 1% (w/v) SDS, 4% (v/v) 2-mercaptoethanol, 0.02% (w/v) Coomassie Brilliant Blue (CBB), and 24% (w/v) glycerol. Store at −20 °C (*see* **Note 4**).
6. Ammonium persulfate (APS) solution: Weigh 0.03 g of APS and dissolve in 1 mL of deionized water (*see* **Note 7**).
7. *N,N,N′,N′*-tetramethylethylenediamine (TEMED) (*see* **Note 8**).
8. Coomassie Brilliant Blue (CBB) staining solution: Dissolve 0.025–0.030 g of CBB in 100 mL of 10% (v/v) acetic acid solution (*see* **Note 9**).
9. Acetic acid (10% v/v) destaining solution: Prepare 100 mL of 10% (v/v) acetic solution (*see* **Note 9**).
10. Glutaraldehyde (5% v/v) fixing solution: Prepare 5% glutaraldehyde solution by diluting 5 times the 25% (v/v) glutaraldehyde stock solution (*see* **Note 9**). Both stock and diluted solutions should be stored at −20 °C (*see* **Note 4**).

2.2 Protein Mass Ladder

Protein mass ladder: (a) with a mass range of 2.5–200 kDa and (b) 1.0–26.6 kDa Ultra Low Range Molecular Weight Marker™. Store the protein mass ladder (a) at 4–8 °C and protein mass ladder (b) at −20 °C (*see* **Note 4**). For details of the protein mass ladders (a) and (b), *see* Table 1.

2.3 Protein Digestion

1. 1, 4-dithio-DL-threitol (DTT) (100 mM): Weigh 0.015 g DTT and dissolve in 1 mL of deionized water. Store at 4–8 °C.
2. Iodoacetamide (IAA) (100 mM): Weigh 0.02 g IAA and dissolve in 1 mL of deionized water. Store at 4–8 °C.
3. Ammonium bicarbonate (NH_4HCO_3) (50 mM): Dissolve 0.04 g of NH_4HCO_3 in 10 mL of deionized water (*see* **Note 10**).

Table 1
Detail of protein mass ladders (a) and (b). Reproduced by permission of Springer-Verlag © 2010

Protein mass ladder (a)	Protein mass ladder (b)	Molecular mass kDa
Myosin	–	200.0
β-galactosidase	–	116.3
Phosphorylase B	–	97.4
Bovine serum albumin (BSA)	–	66.3
Glutamic dehydrogenase	–	55.4
Lactate dehydrogenase	–	36.5
Carbonic anhydrase	–	31.0
–	Triosephophate isomerase	26.6
Trpsin inhibitor	–	21.5
–	Myoglobin (from horse heart)	17.0
Lysozyme	–	14.4
–	α-lactalbumin (from bovine milk)	14.2
–	Aprotinin (from bovine lung)	6.5
Aprotinin	–	6.0
Insulin B chain	Insulin B chain (oxidised bovine)	3.5
Insulin A chain	–	2.5
–	Bradykinin	1.06

4. Sequencing grade modified trypsin (0.1 μg/μL): Dissolve 100 μg of the lyophilized powder of modified trypsin in 1 mL of deionized water (*see* **Note 11**). Store trypsin solution in aliquots at −20 °C. For long-term storage use −80 °C.
5. Prepare sample for protein digestion by dissolving 1 mg of β-casein in 1 mL of deionized water.

3 Methods

All procedures should be carried out at room temperature.

3.1 Protein Digestion

Perform protein digestion in solution for a comparative study according to the following method. Here β-casein is used as a model protein.

1. Add 10 μL of 1 mg/mL β-casein solution in the mixture of 15 μL of 50 mM NH_4HCO_3 and 1.5 μL of 100 mM DTT solution and incubate at 90 °C for 5–10 min.
2. After cooling to room temperature, add 3 μL of 100 mM IAA to this solution and incubate for 25 min in the dark.
3. Add 1 μL of 0.1 μg/μL sequencing grade modified trypsin and incubate at 37 °C for 4 h.
4. Finally, add an additional 1 μL of trypsin and incubate overnight at 30 °C. Next day, reduce digest mixture to 10 μL using vacuum.

3.2 Gel Casting

1. Prepare all stacking (4% w/v) and resolving gels mixture for 7, 9, and 10% (w/v) gels for both the original and modified methods as shown in Table 2 (*see* **Notes 12–14**).
2. Cast all the gels in gel cassettes at 7.3 cm height × 8 cm width and 0.75 mm thickness using a mini-gel casting apparatus (Bio-Rad, Hemel Hempstead, UK) (*see* **Notes 15** and **16**).
3. Firstly, inject the resolving gel mixture into the gel cassette (*see* **Notes 17** and **18**).
4. Do not fill the resolving gel mixture to the top of the gel plates. Leave a space of approximately 1–2 cm from the top. Quickly fill this space with butanol or 70% ethanol (*see* **Notes 19** and **20**).
5. The resolving gel usually takes 10–15 min for polymerization. Check for polymerization of the gel in the centrifuge tube, and if the gel has been polymerized, then pour off the butanol

Table 2
Tricine-SDS-PAGE protocol for the preparation of the polyacrylamide gels. Reproduced by permission of Springer-Verlag © 2010

Percentage of acrylamide in the gel		Acrylamide/bis-acrylamide 29:1 30% (w/v) solution mL	2.5 Tris-buffer (pH 8.8 for modified and pH 8.45 for original method) mL	Deionized water mL	TEMED μL	APS 30 mg/mL μL	Total volume mL
Stacking Gel	4	0.66	0.76	3.42	5.0	150	~5
Resolving Gel	7	2.33	5.6	1.91	7.0	150	~10
	9	3.00	5.6	1.24	6.0	150	~10
	10	3.33	5.6	0.90	6.0	150	~10
	12	4.00	5.0	0.89	6.0	100	~10
	15	5.00	4.6	0.29	6.0	100	~10
	16	5.33	4.3	0.22	6.0	100	~10

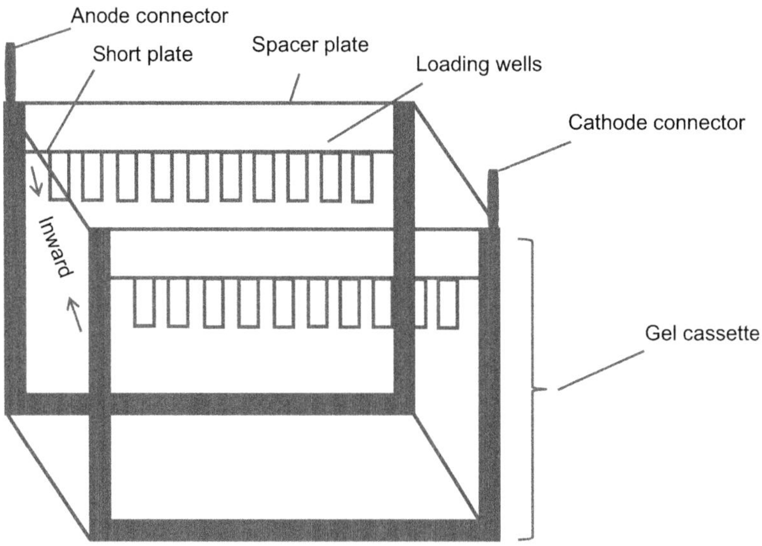

Fig. 1 Dual plate electrode assembly showing the gel cassette sandwich. The short plates must be pointing inward

(or 70% ethanol) from the top of the resolving gel. Start injecting the stacking gel mixture onto the polymerized resolving gel in the gel cassette, and fill up to the top of the short plates, and then quickly insert the gel combs for the loading wells. Check the polymerization of the stacking gel in the centrifuge tube, or by just slightly lifting the combs, and if the stacking gel is polymerized, then remove the gel cassette sandwich from gel casting frame and set this into the electrode assembly. The short plates must be pointing inward in the dual plate electrode assembly (*see* Fig. 1).

3.3 Sample Preparation

Mix 10 μL of the sample buffer with 10 μL of the digest mixture. Dilute 20-fold the protein mass ladder (b) with the 1× sample buffer and heat at 65 °C (*see* **Note 21**). There is no need to dilute or heat the protein mass ladder (a).

3.4 Sample Loading and Gel Running

1. Place the electrode assembly containing the two gel cassettes in the running buffer tank.
2. Fill the tank with the running buffer and gently remove the combs. Wash the loading wells with the running buffer using 1 mL pipette tips to remove any unpolymerized acrylamide.
3. Load 7 μL of protein mass ladder (a), 5 μL of protein mass ladder (b), and 10 μL of β-casein digest (as prepared above) in each well of the 7, 9, and 10% (w/v) slab gels for the modified and original tricine-SDS-PAGE methods.
4. Apply a constant voltage of 125–150 V. Do not stop until the dye front touches the bottom (*see* **Notes 22** and **23**).

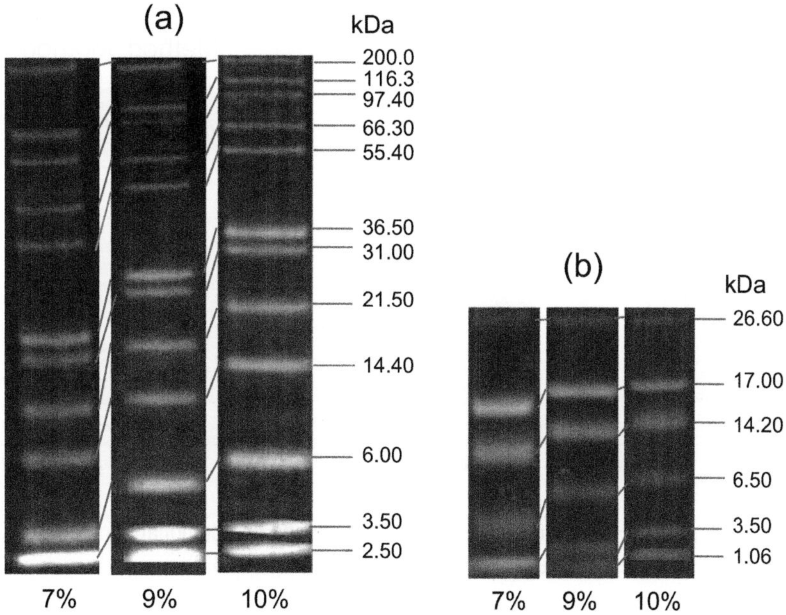

Fig. 2 Modified tricine-SDS-PAGE separation of low mass proteins in the range of (**a**) 2.5 to 200 kDa (**b**) 1.0 to 26.6 kDa in 7, 9, and 10% gels. For proteins corresponding to the listed masses, *see* Table 1. Reproduced by permission of Springer-Verlag © 2010

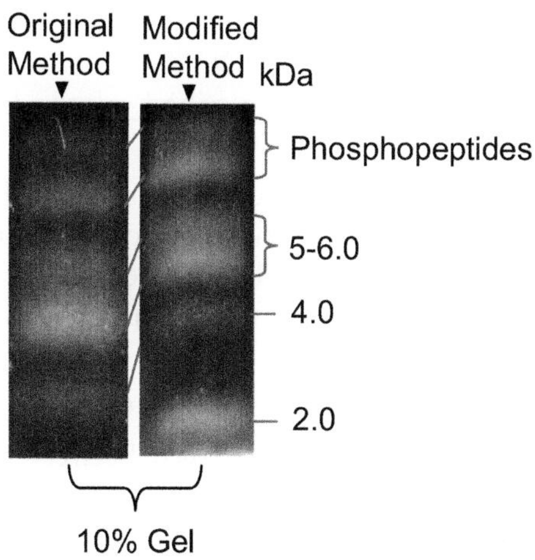

Fig. 3 Head-to-head comparative results for the β-casein digest using a 10% polyacrylamide slab gel compared with the original and modified tricine-SDS-PAGE systems (*see* **Note 26**). Reproduced by permission of Springer-Verlag © 2010

3.5 Gel Fixing, Staining, and Destaining

1. Fix the gels for 25 min (using a shaker) with the fixing solution of 5% (v/v) glutaraldehyde.
2. Stain each gel for 20 min (using a shaker) using CBB staining solution.
3. Perform destaining for 20 min (using a shaker) using 10% (v/v) acetic acid solution (*see* **Notes 24** and **25**) (*see* Figs. 2, 3, and 4 for typical results).

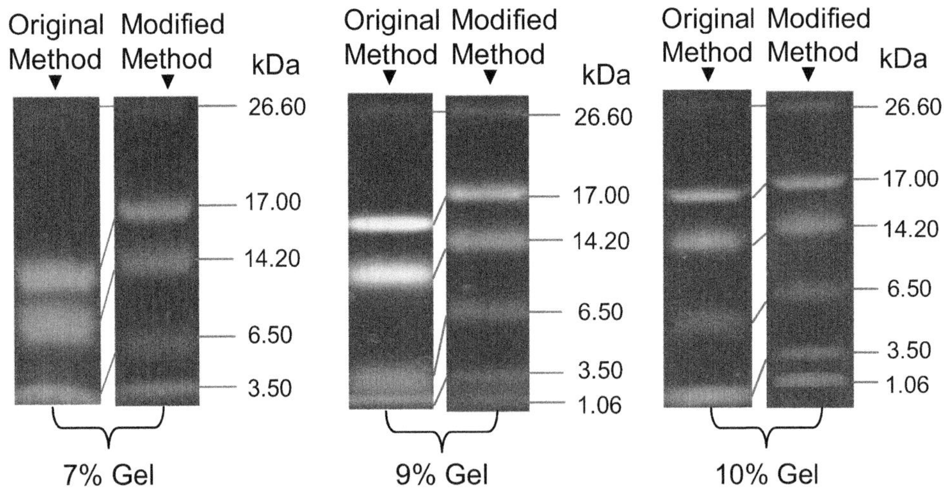

Fig. 4 Effect of increasing the pH of the gel buffer to pH 8.8 showing an additional band in the 7, 9, and 10% modified gel systems. Reproduced by permission of Springer-Verlag © 2010

4 Notes

1. The "original method" refers to the previously described [8, 9] tricine-SDS-PAGE, and "modified method" [12] refers to the modification here described.
2. The previously described methods for tricine-SDS-PAGE employed 3.0 M

   Tris-base, while the modified method uses 2.5 M—So here, for the comparative study, the concentration of the gel buffer was kept to 2.5 M for both systems.
3. The running buffer can be prepared at 10× concentration and can be diluted accordingly.
4. All the solutions should be used at room temperature.
5. For the comparative study between the modified and original tricine-SDS-PAGE system, only one running buffer was used.
6. Tris-base with pH of 7.0 can also be used instead of pH 6.8.
7. APS must always be freshly prepared.
8. TEMED is a very odorous neurotoxin, so extra care is required when handling it.
9. Avoid using methanol in staining, destaining, and fixing solutions as this may cause loss of the low molecular mass proteins.
10. There is no need to adjust the pH of the ammonium bicarbonate solution.
11. Sequencing grade modified trypsin does not suffer from self-autolysis and minimizes mis-cleavages during the digestion.

12. We prefer a 10–50 mL centrifuge tube for preparing the gel mixture.
13. Do not include SDS, glycerol, or urea in any of the gels.
14. TEMED and APS must be added last in the gel mixture which requires quick injection into the gel cassettes; otherwise, gel will be polymerized in the tube.
15. Acrylamide is recognized as a neurotoxin, so gloves must be worn all the time and the work must be done in a properly ventilated area.
16. For the purpose of comparing the original and modified methods, the same gel casting protocol was used for both the systems.
17. Inject the gel mixture slowly into the gel cassettes using 5 mL pipette tips, and avoid bubble formation which may increase the electrical resistance between the electrodes during electrophoresis.
18. In case of direct contact of the gel mixture with the skin, wash the skin thoroughly with water, and seek medical advice if the contact is severe because acrylamide can readily be absorbed by the skin.
19. Butanol or 70% ethanol provides a smooth resolving gel surface.
20. The butanol or 70% ethanol must be removed before casting the stacking gel on the top of the resolving gel.
21. For example, sigma Aldrich (Poole, UK) provides 10 mL of 2× sample buffer free with the protein mass ladder (b).
22. Do not overfill the wells; otherwise, they will contaminate the neighboring wells.
23. For rapid separation, a higher voltage of 200–250 can be applied: However, it is not preferred because it increases the temperature of the buffer which may cause a significant change in the pH of the running buffer and/or resolving gel.
24. Wash the gel with deionized water after each of the fixing, staining, and destaining steps.
25. For best results, do not reuse fixing, staining, and destaining solutions.
26. Confirmation of the phosphopeptide bands was obtained by whole gel elution and ICP-MS detection of ^{31}P (for details, *see* ref. 12).

Acknowledgments

This work was supported by a grant from Loughborough University, Loughborough, Leicestershire, LE11 3TU, UK.

References

1. Laemmli UK (1970) Cleavage of the structural proteins during the assembly of the head of bacteriophage. Nature 227:680–685
2. Hashimoto F, Horigome T, Kanbayashi M, Yoshida K, Sugano H (1983) An improved method for separation of low-molecular-weight polypeptides by electrophoresis in sodium dodecyl sulfate-polyacrylamide gel. Anal Biochem 129:192–199
3. Bothe D, Simonis M, Dohren H (1985) A sodium dodecyl sulphate-gradient gel electrophoresis system that separates polypeptides in the molecular weight range of 1500-100,000. Anal Biochem 151:49–54
4. Okajima T, Tanabe T, Yasuda T (1993) Non-urea sodium dodecyl sulfate-polyacrylamide gel electrophoresis with high-molarity buffers for the separation of proteins and peptides. Anal Biochem 211:293–300
5. Marshall P, Heudi O, Bains S, Freeman HN, Abou-Shakra F, Reardon K (2002) The determination of protein phosphorylation on electrophoresis gel blots by laser ablation inductively coupled plasma-mass spectrometry. Analyst 127:459–461
6. Elliot VL, McLeod C, Marshall P (2005) Combination of gel electrophoresis and ICP-mass spectrometry—novel strategies for phosphoprotein measurement. Anal Bioanal Chem 383(3):416–423
7. Wind M, Feldmann I, Jakubowski N, Lehmann WD (2003) Spotting and quantification of phosphoproteins purified by gel electrophoresis and laser ablation-element mass spectrometry with phosphorus-31 detection. Electrophoresis 7:1276–1280
8. Schägger H (2006) Tricine-SDS-PAGE. Nature 1:16–22
9. Schägger H, von Jagow G (1987) Tricine sodium dodecyl sulfate-polyacrylamide gel electrophoresis for the separation of proteins in the range from 1 to 100 kDa. Anal Biochem 166:368–379
10. Yim SK, Ahn T, Kim JK, Yun CH (2002) Polyacrylamide gel electrophoresis without a stacking gel: application for separation of peptides. Anal Biochem 305:279–281
11. Hames BD (1998) Gel electrophoresis of proteins. Oxford University Press, New York
12. Haider SR, Reid HJ, Sharp BL (2010) Modification of tricine-SDS-PAGE for online and offline analysis of phosphoproteins by ICP-MS. Anal Bioanal Chem 397:655–664

Chapter 16

Analysis of Protein Glycation Using Phenylboronate Acrylamide Gel Electrophoresis

Marta P. Pereira Morais, Omar Kassaar, Stephen E. Flower, Robert J. Williams, Tony D. James, and Jean M. H. van den Elsen

Abstract

Carbohydrate modification of proteins adds complexity and diversity to the proteome. However, undesired carbohydrate modifications also occur in the form of glycation, which have been implicated in diseases such as diabetes, Alzheimer's disease, autoimmune diseases, and cancer. The analysis of glycated proteins is challenging due to their complexity and variability. Numerous analytical techniques have been developed that require expensive specialized equipment and complex data analysis. In this chapter, we describe two easy-to-use electrophoresis-based methods that will enable researchers to detect, identify, and analyze these posttranslational modifications. This new cost-effective methodology will aid the detection of unwanted glycation products in processed foods and may lead to new diagnostics and therapeutics for age-related chronic diseases.

Key words Protein glycation, Gel electrophoresis, Boronic acid, mP-AGE, Flu-PAGE

1 Introduction

1.1 Protein Glycation

Over the past decades, there has been considerable interest in the development of simple and efficient techniques to detect glycated proteins as indicators for chronic age-related diseases such as diabetes, Alzheimer's disease, and cancer [1]. Protein glycation is a nonenzymatic posttranslational modification whereby saccharides react in a nonrandom manner with free amino groups at the N-terminus of proteins and with lysine and arginine side chains to form Schiff base or aldimine conjugates. The Schiff base intermediate rearranges to form a more stable ketoamine called the Amadori product, which can undergo a series of reactions including oxidation, dehydration, fragmentation, and degradation to form heterogeneous nonfluorescent and fluorescent yellow-brown products collectively known as advanced glycation end products or AGEs [2] (Fig. 1). Despite extensive research, the glycation process

Biji T. Kurien and R. Hal Scofield (eds.), *Electrophoretic Separation of Proteins: Methods and Protocols*, Methods in Molecular Biology, vol. 1855, https://doi.org/10.1007/978-1-4939-8793-1_16,

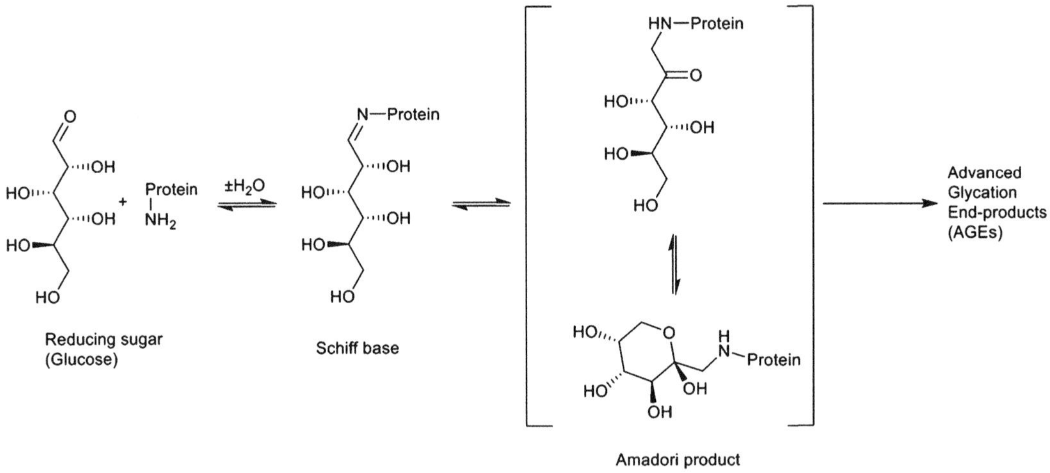

Fig. 1 Schematic diagram of the protein glycation reaction. The condensation reaction between reducing sugar (glucose) and free amino groups in proteins leads to the formation of a Schiff base, which rearranges to a more stable Amadori product (fructosamine), ultimately forming advanced glycation end products (AGEs)

is still poorly understood due to its complexity and variability [3]. Since AGE formation is irreversible, these products accumulate over the lifetime of proteins, thereby affecting their structure and function [1, 4, 5]. Protein glycation is a slow process and although it takes hours for Schiff bases to form, the formation of the more advanced products is in the order of weeks, depending on physiological conditions such as blood glucose levels [1]. The rapid turnover of most proteins, therefore, protects them against potential damage [3]. As a result, AGE formation mainly affects long-lived proteins such as albumin, collagen, and lens crystallins [6].

Glycation also occurs during cooking and storage of food, affecting its flavor, color, texture, digestibility, as well as nutritional value [7]. Foerster and Henle have shown that the high temperatures involved in food preparation result in the formation of complexes that are different from those observed in biological systems. Only little is known about the resorption and elimination of these dietary glycated products.

1.2 Analysis of Glycated Proteins

The implication of protein glycation in disease has in recent years led to the development of numerous analytical methods aimed at identifying and quantifying heterogeneous glycated protein products. Glycation analysis techniques include mass spectroscopy (MS) [8], boronate affinity chromatography (BAC) [9], high-performance liquid chromatography (HPLC) [10], and fluorescence spectroscopy [11]. Each of these methods has its advantages and limitations. MS, for example, only detects the most abundant proteins, and any information regarding the glycation state is lost when proteins are modified or digested prior to analysis, as part of the sample preparation. While fluorescence spectroscopy, on the other hand, allows monitoring of the course of the glycation reaction, it is biased toward fluorescent AGE products and does not provide any structural information of these adducts. Often the cost and availability of specialized instruments, such as MS, limit their general use. We chose the widely available protein analysis tool, polyacrylamide gel electrophoresis [12], and modified it to facilitate the visualization, identification, and quantification of glycated proteins [13].

1.3 Uses of Polyacrylamide Gel Electrophoresis

Polyacrylamide gel electrophoresis was first described by Orstein and Davis and by Raymond and Weintraub in 1959 [14, 15]. This technique utilizes the characteristic of proteins that, when they are charged at any pH other than their isoelectric point (pI), they will migrate in the presence of an electric current. The polyacrylamide gel matrix thereby acts as a molecular sieve to separate proteins, providing information about the molecular weight, charge, subunit composition, and purity of proteins. Polyacrylamide gel properties can be altered by varying the amount of acrylamide and cross-linkers and can be cast into a range of shapes for sensing and detection purposes. Because of its inertness and biocompatibility, polyacrylamide gels can be used in numerous medical applications [16]. Gel electrophoresis is simple, highly reproducible [17], cost effective, and sensitive, detecting proteins at picomole levels when using silver staining [18]. Multiple samples can be run on a single gel, ensuring identical running conditions and the potential for high-throughput analysis. Until recently, this technique could not differentiate between various posttranslationally modified proteins due to their similarity in molecular weight. In order to utilize this analytical system for glycated proteins, we used a special class of compounds in combination with polyacrylamide gel electrophoresis that reversibly interact with cis-1,2- and 1,3-diol moieties of the carbohydrate protein adducts [19].

1.4 Utilizing Boronic Acid-Diol Interaction

The interaction between boronic acids and diols, first described by Lorand and Edwards [20], is unusual because it involves a rapid and reversible covalent interaction (Fig. 2). Boronic ester formation is more favorable at high pH, since boronic acid molecules in the tetrahedral anionic state are more reactive than the trigonal analogues. Boronic acids have different affinities for different diols, and this property has been utilized in a number of applications in affinity support from sample separation to sensors [21, 22]. Boronic acids were first used in affinity chromatography in 1970 to separate sugars and nucleic acids [23] and were subsequently employed to separate a variety of cis-diol compounds including proteins [9, 24]. Protein purification [25] and clinical analysis of fructosamine content are some of the reported uses of boronic acid affinity chromatography (BAC) due to its low cost, speed, and accuracy [26]. In this technique, enrichment and isolation of glycated

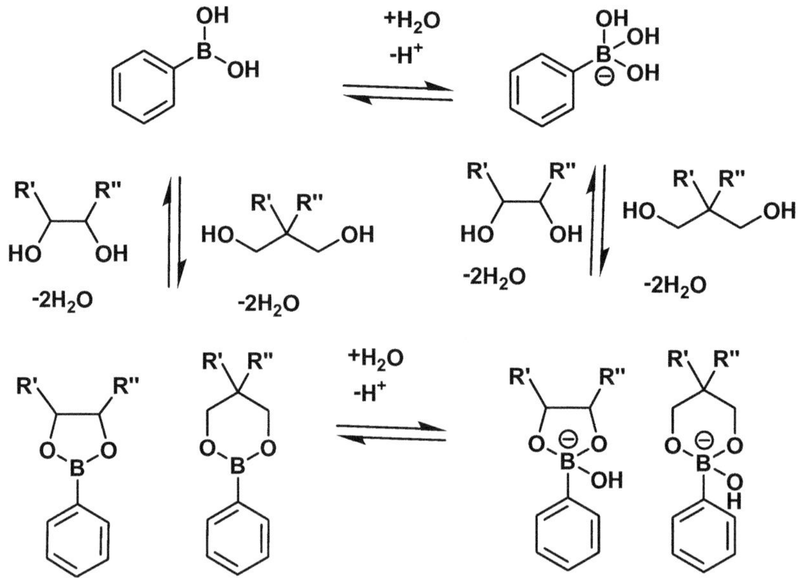

Fig. 2 The interaction between boronic acids and diols. Boronic acids exist in a trigonal state (top left) under neutral conditions or a tetrahedral state (top right) under basic conditions. Both states form esters with *cis*-1,2- and 1,3-diols, giving five- and six-membered rings, respectively. Five-membered cyclic boronic esters are more stable than six-membered ones. Furthermore, the condensation reactions are in equilibrium, with the interaction between anionic boronic acids and diols kinetically faster than the neutral boronic acid analogues

proteins [10] is achieved by passing samples through an agarose matrix containing covalently attached m-aminophenylboronic acid. Glycated proteins are bound to the matrix, while unglycated counterparts are removed. The bound fraction of glycated proteins are subsequently eluted using sorbitol containing buffer, which has a higher affinity for boronic acid. This enrichment of glycated proteins are often used prior to non-chromatographic analyses, such as MS [27].

1.5 The Use of Boronic Acids in Gel Electrophoresis

We first explored the application of boronic acid functionalized polyacrylamide gels for the separation of saccharides [28]. Direct analysis of carbohydrates via gel electrophoresis is not possible because these molecules are electroneutral in nature and lack detectable chromophores and fluorophores. To overcome this problem, fluorophore-assisted carbohydrate electrophoresis (FACE) [29] was developed, allowing simple, rapid, and sensitive analysis of carbohydrates by labeling them with fluorophores. However, some of the fluorophores used in this method, such as 8-aminonaphthalene-1,3,6-trisulfonic acid (ANTS), affect the true nature of carbohydrates by the introduction of a number of negative charges that can result in unexpected migration profiles [29–31].

The incorporation of small quantities (0.5% w/v) of methacrylamido phenylboronic acid (MPBA), in boronate-assisted saccharide electrophoresis (BASE), resulted not only in the correction of aberrant migration profiles of carbohydrates labeled with the neutral fluorophore 2-aminoacridone (AMAC) but also enabled the separation of monosaccharides from disaccharides [28].

Encouraged by our results from the abovementioned BASE analyses, we hypothesized that sugar-modified proteins, such as glycated proteins, should also interact with gel-incorporated boronic acids, enabling their analysis by gel electrophoresis [13]. This concept will be further explored below, using a number of model proteins for glycation analysis.

1.6 Analysis of Glycated Proteins Using Methacrylamido Phenylboronate Acrylamide Gel Electrophoresis (mP-AGE)

Human serum albumin (HSA) is the most abundant protein in human plasma and binds a wide variety of compounds, acting as transport proteins for substrates like fatty acids, amino acids, hormones, and drugs [32]. HSA is an ideal model system for studying protein glycation because of its abundance (40 mg/mL) and long half-life (approximately 20 days). In healthy adults, 10% of HSA is glycated affecting mainly lysine and arginine residues, with a two- to threefold increase under hyperglycemic conditions [33]. Hemoglobin (Hb) is another protein commonly affected by glycation and is used clinically as a diabetic control index. However, various investigations have indicated that glycated HSA is a better glycemic indicator than glycated Hb [34].

While normal SDS-PAGE analysis cannot distinguish between glycated and unglycated proteins, incorporation of methacrylamido phenylboronic acid (MPBA) in polyacrylamide gel electrophoresis, mP-AGE, results in the retention of glycated proteins, while the mobility of unmodified proteins remains unchanged [13] (Figure 3).

1.7 The Molecular Interactions Between Glycation Adducts and Boronic Acid

With our results, we can now provide a molecular model explaining the behavior of glycated protein in mP-AGE. Boronic acids have different affinities for different carbohydrates and form the strongest interaction with sugars containing an anomeric *cis*-1,2-diol (Fig. 2). Glucose, the dominant contributor to glycation in vivo, contains such an anomeric *cis*-1,2-diol in the protein adduct and an additional 4,5-diol that also interacts with MPBA. The anomeric diol interaction can also be stabilized by electrostatic interaction between the negatively charged boronate and protonated amino group, forming a trident interaction (Fig. 4) [13]. In contrast, the pyranose structure from the Heyns rearrangement of fructose does not contain any *cis*-diols that can interact strongly with the MPBA incorporated in the gel. Although the furanose fructose adduct contains anomeric 3,4- and 4,5-diols, its inability to interact with boronic acid in the gel leads us to conclude that the majority of this adduct exist in the pyranose form in HSA, or the absence of amino group stabilization results in weaker boronic acid binding.

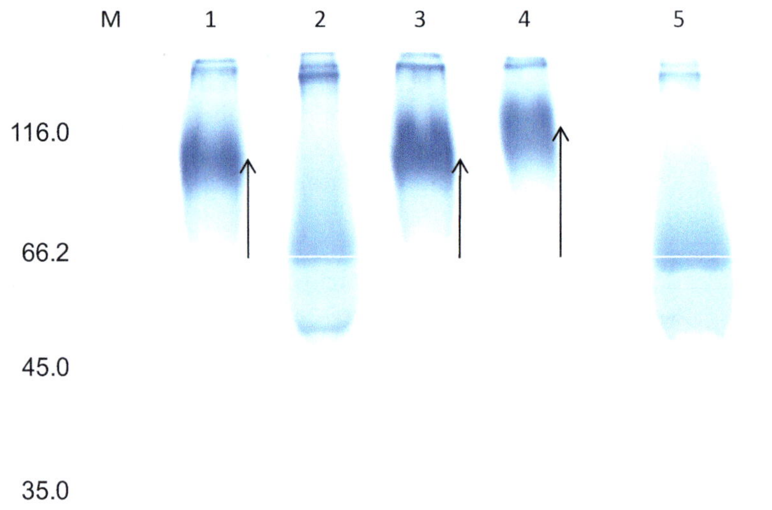

Fig. 3 Methacrylamido phenylboronic acid gel electrophoresis (mP-AGE) analysis of glycated human serum albumin (HSA). SDS-PAGE analysis of glycated HSA on an 8% polyacrylamide (PA) gel with 0.5% w/v MPBA. For this experiment, HSA samples were incubated at 37 °C for 21 days in the presence of glucose (lane 1), fructose (lane 2), mannose (lane 3), maltose (lane 4), and PBS as control (lane 5), prior to electrophoresis. HSA appears as a single band at ~ 67 kDa in the PBS in the control lane (lane 5) while the glucose-, mannose-, and maltose-incubated protein bands are retained at higher position in the gel around 80–90 kDa (indicated by arrows). The gel migration pattern of fructose incubated HSA is similar to that of the PBS control. This observation can be explained by differences in affinity of MPBA for the Amadori adducts formed with the different sugars [13], as explained in Fig. 4

This molecular model can also be used to explain the results obtained with HSA that is glycated with other sugars, such as maltose, which as a disaccharide has additional cis-diols and is retained higher in the mP-AGE gel compared to glucose and mannose glycated HSA (Fig. 3). In addition, it elucidates the differences in retention of sugar adducts caused by glycation or from the natural enzymatic process of glycosylation [13]. Glycosylated proteins contain *N*- and *O*-linked glycans that do not have anomeric *cis*-diols that can interact strongly with boronic acids, and as a result, their electrophoretic mobility is not altered in mP-AGE.

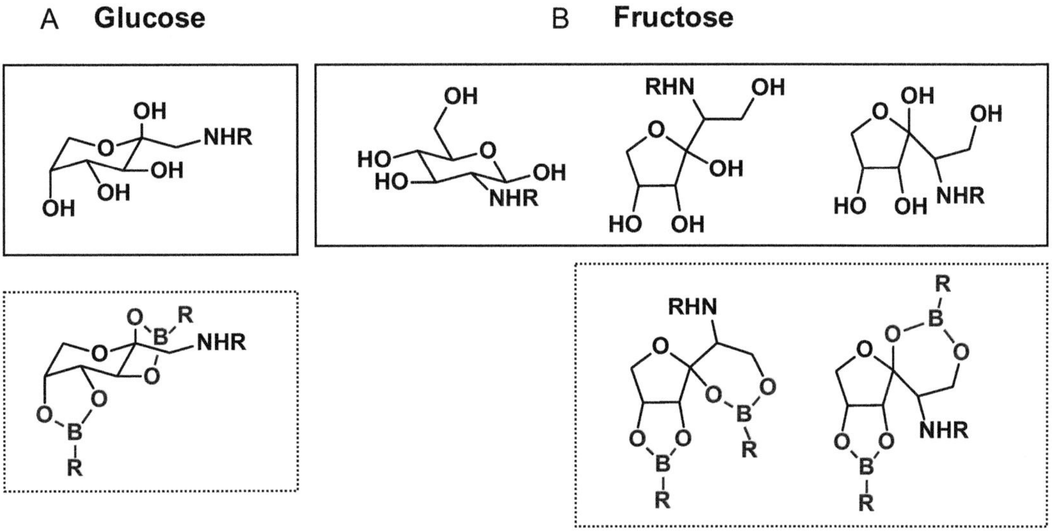

Fig. 4 Molecular model explaining the separation of glycated proteins in mP-AGE. Shown are the molecular interactions between phenylboronate and the Amadori product formed when HSA is incubated with (**A**) glucose (fructosamine) and (**B**) fructose. Structures in black represent the sugar-modified Amadori adducts, while their boronic acid interactions are shown in blue below. Both glucose and fructose predominantly exist in a pyranose form, but it is the linear form that reacts with amino groups of proteins (Fig. 1). Interaction between boronic acid and anomeric *cis*-1,2-diols is preferential, as five-membered cyclic boronic esters are more stable than six-membered ones. The Amadori product resulting from *glucose* incubation contains anomeric *cis*-1,2- (positions 2,3) and an additional *cis*-1,2-diol at positions 4,5 available for interaction with MPBA. The location of the *cis*-1,2-diol at position 2,3, in close proximity to the Amadori amino group, allows for an additional interaction between the negatively charged tetrahedral phenylboronate and the Amadori amino group [13]. In contrast, the pyranose and furanose configurations of the *fructose* Amadori product do not have *cis*-diols in close proximity to the Amadori amino group, required for such a trident interaction. Molecular carbohydrate structures were produced using ChemDraw (CambridgeSoft)

1.8 Detection of Glycation in Recombinant Protein Production Using mP-AGE

In addition to the analysis of glycated HSA, we would like to mention the potential use of mP-AGE for the detection and separation of glycated protein products that are formed during the production of recombinant proteins in *E. coli* expression systems. Recombinant proteins containing an N-terminal poly-histidine tag are prone to modification by 6-phosphogluconolactone (6PGL), an intermediate of the pentose phosphate pathway. The 6PGL intermediate reacts directly with the N-terminal amino group of proteins resulting in a 259 dalton molecular addition. The 6PGL adduct can subsequently be dephosphorylated by a phosphatase to give a single gluconolactone modification of 178 daltons [4, 5]. Such modifications affect protein purity and activity, which are unwanted characteristics of recombinant proteins that are used as therapeutics. We have shown that a single gluconolactone modification can be detected in recombinantly expressed *Staphylococcus aureus* protein Sbi [35] using mP-AGE, resulting in a dramatic retention of the carbohydrate-modified fraction of this protein in the gel

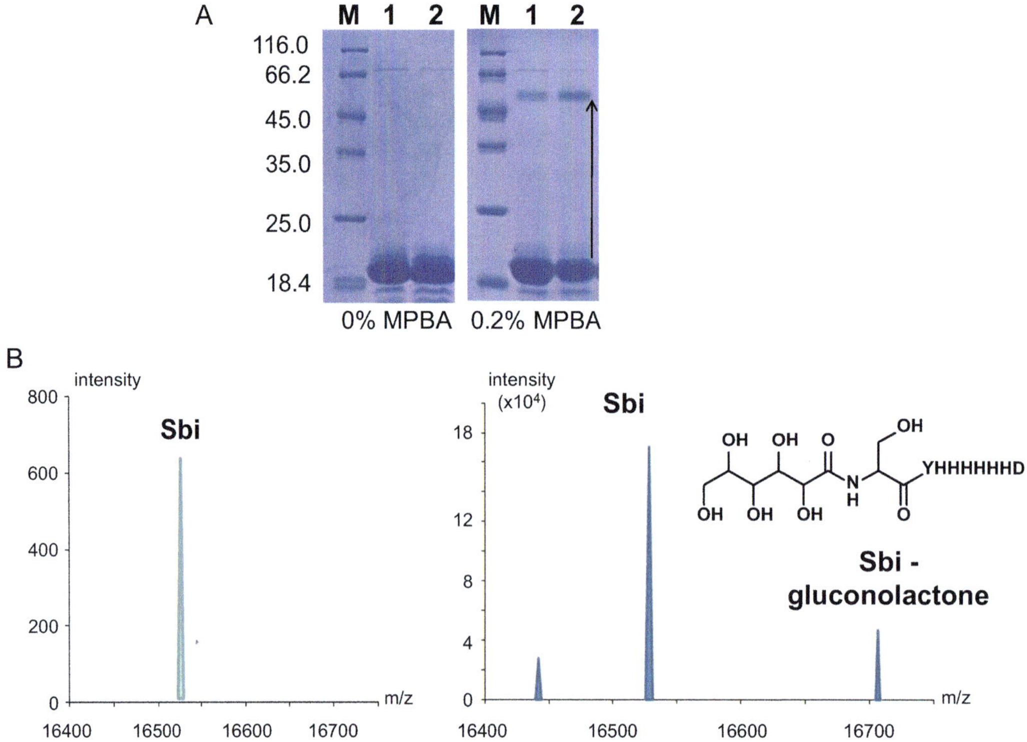

Fig. 5 mP-AGE analysis of a gluconoylated recombinant protein. (**A**) SDS-PAGE analysis of freshly purified recombinant Sbi (lane 1) and the same protein after 1 h exogenous incubation with 50 mM δ-gluconolactone (lane 2) on gels without (left) and with MPBA (right). Normal SDS-PAGE shows identical bands for both freshly purified and exogenously gluconoylated protein. The faint band of gluconoylated protein at around 20 kDa in the normal SDS-PAGE (left) is shifted to a position almost three times its molecular weight in the gel containing 0.2% w/v MPBA (right), while the electrophoretic mobility of the unmodified fraction of the protein remains unaffected. The dramatic enhancement in protein retention indicates a strong affinity between the gluconoylated protein and the gel-incorporated boronic acid. (**B**) Mass spectra of freshly purified Sbi (left) and exogenously gluconoylated protein (right) with the additional peak corresponding to the addition of a gluconolactone adduct (+178 Da). The molecular structure of this adduct is shown in the inset

(Fig. 5) [13]. Our results show that even low levels of gluconoylation can be detected using mP-AGE, while remaining undetected with MS (Fig 5B, left panel). Sbi is gluconoylated at the N-terminal histidine-tag during protein expression in E. coli and the linearity of the gluconoyl adduct, and the number of hydroxyl groups present in a resonance structure, makes the adduct very suitable for interactions with MPBA.

1.9 Analysis of Glycated Proteins Using Fluorescent Phenylboronate Acrylamide Gel Electrophoresis (Flu-PAGE)

While mP-AGE is a useful method to aid the detection and separation of glycated proteins, it is not suitable for the analysis of carbohydrate-modified proteins in complex biological samples. For this reason, we developed fluorescent phenylboronate gel electrophoresis (Flu-PAGE) [36]. Flu-PAGE can be used to directly visualize glycated proteins in normal SDS-PAGE and Western blots, because the samples are labeled with the fluorophore-appended boronate prior to electrophoresis. The labeling does not affect the electrophoretic mobility of the proteins, nor does it hinder the subsequent identification of the glycated proteins using mass spectrometry. Similar to mP-AGE, Flu-PAGE exploits the reversible covalent interaction between boronic acid and *cis*-diols that are present in fructosamine-protein adducts in glycated proteins, strengthened by the additional charge interaction between the boronate and the fructoselysine amino group [36] as shown in Fig. 4. This technique allows for specific labeling of glycated proteins over glycosylated and unmodified proteins in complex biological samples such as plasma and brain homogenates [36]. An example of a Flu-PAGE analysis of normal human serum is shown in Fig. 6, resulting in the specific fluorescent staining of glycated HSA. More recently, we have used Flu-PAGE to describe

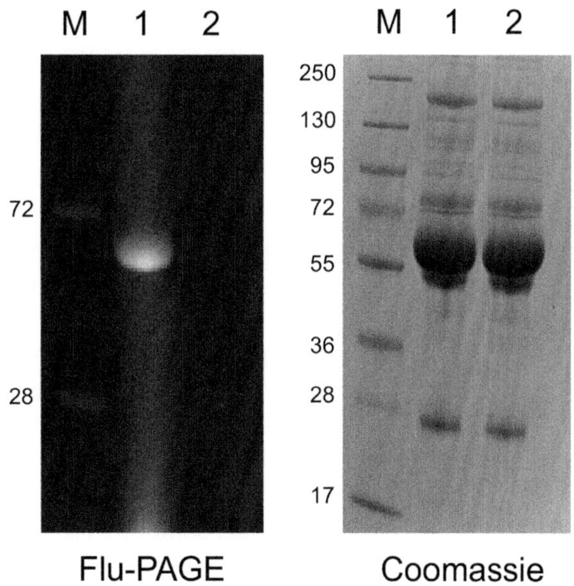

Fig. 6 Flu-PAGE analysis of normal human serum. When visualized with a blue light transilluminator, using an orange filter (595 nm) (left panel), a band of glycated human serum albumin (HSA) can be easily observed in the fluorescein-boronic acid-labeled sample (lane 1) but not in the fluorescein-labeled control (lane 2). The panel on the right shows the same gel subsequently stained with Coomassie Brilliant Blue. The Thermo Scientific PageRuler Plus Prestained Protein Ladder was used as molecular weight marker (M)

the glycation profile in human Alzheimer's disease brain homogenates and identified macrophage migration inhibitory factor (MIF) as being glycated and oxidized [37].

In summary, we have demonstrated that gel electrophoresis can be used to differentiate glycated proteins from unmodified ones by the incorporation of boronic acid in polyacrylamide gels (mP-AGE) and by labeling of glucose-modified proteins in complex biological samples using fluorescent boronic acids (Flu-PAGE). Since the degree of protein glycation has been shown to be an important causal factor and indicator in aging and age-related chronic disease states such as diabetes, cardiovascular disease, autoimmune disease, cancer, and Alzheimer's disease, both mP-AGE and Flu-PAGE therefore have the potential to be used in disease diagnosis and the screening of glycation inhibitors.

2 Materials

2.1 For Casting Gels

1. Gel cassette.
2. Comb.
3. 40% acrylamide solution (29:1 acrylamide: N,N′-methylene-bisacrylamide).
4. Tris(hydroxymethyl)aminomethane (Tris).
5. N,N,N′,N′-tetramethylethane-1,2-diamine (TEMED).
6. 10% ammonium persulfate (APS).
7. Methacrylamido phenylboronic acid (MPBA, Sigma-Aldrich 771465).

2.1.1 Gel Solutions

1. Resolving gel solution: 8 or 15% polyacrylamide (acrylamide: bisacrylamide 29:1) in 375 mM Tris pH 8.8 (*see* **Note 1**).

   Example: For an 8% polyacrylamide mini gel (dimension height (100 mm) × width (100 mm) × thickness (1.0 mm)), prepare 1.25 mL of 40% acrylamide solution, 1.56 mL 1.5 M Tris pH 8.8, and 3.40 mL water. For 15% polyacrylamide mini gel, use 2.34 mL of 40% acrylamide solution, 1.56 mL of 1.5 M Tris pH 8.8, and 2.31 mL water.
2. Stacking gel solution: 4% polyacrylamide (acrylamide/bisacrylamide 29:1) in 125 mM Tris pH 6.8. *Example*: For each mini gel, mix 0.78 mL of 0.5 M Tris pH 6.8, 0.31 mL of 40% acrylamide solution, and 2 mL water.

2.2 For Electrophoresis

1. Electrophoresis gel tank.
2. Power pack.
3. Protein marker.

2.2.1 Buffers

1. Tris-Glycine SDS running buffer (1×): 25 mM Tris, 190 mM glycine, and 0.1% SDS.
2. SDS reducing sample buffer (2×): 3% w/v SDS, 5% β-mercaptoethanol, 10% glycerol, 62.5 mM Tris pH 6.8, and 0.01% bromophenol blue.

2.3 For Visualization

1. Fluorescein-boronic acid (abcam ab219802), Coomassie Brilliant Blue or silver stain.
2. Light box.
3. Blue light transilluminator (wavelength range 420–520 nm, with amber (530 nm) or orange (595 nm) filters).
4. UV transilluminator (wavelength 365 nm, with orange (595 nm) or green (537 nm) filters).

3 Method

3.1 Sample Preparation

1. For mP-AGE: Add an equal volume of 2× sample buffer to the protein sample.
2. For Flu-PAGE: Add fluorescein-boronic acid sample incubation solution (1 mg/mL) to your sample in a 1:10 ratio (*see* **Note 2**) and incubate for 1 h at room temperature (*see* **Note 3**).

   Example: Add 1 μL fluorescein-boronic acid solution to 9 μL of your sample solution (i.e., blood plasma, serum, or tissue homogenate).
3. Heat the mixture at 100 °C for 5–10 min for protein denaturation and incorporation of SDS.
4. For Coomassie staining, load approximately 1 μg of protein on the gel depending on sample complexity. In silver stained gels, the sample loading should be decreased by 20-fold to obtain an interpretable gel profile (*see* **Note 4**).

3.2 Casting Gels

3.2.1 Normal Polyacrylamide Gel

1. Polymerize resolving gel solution (*see* Subheading 2.1.1) with 3 μL TEMED and 31 μL 10% APS. Add a few drops of saturated butanol or water to the top of the gel solution to level it and prevent atmospheric oxygen from inhibiting the polymerization process. Gel polymerizes in approximately 20 min (*see* **Notes 5–7**).
2. Once the gel is set, pour away the butanol, rinse the gel with water, and dry the inside of the gel cassette with a piece of filter paper without touching the resolving gel.
3. Pour the stacking gel on top of the resolving gel and polymerize by adding 3 μL TEMED and 16 μL of 10% APS. Insert a gel comb before this gel polymerizes to generate sample wells (*see* **Note 8**). This gel can be stored at 4 °C for about a month.

4. For electrophoresis, remove gel comb and insert the gel cassette into the gel tank of the electrophoresis apparatus according to the manufacturer's instructions. After adding running buffer (*see* Subheading 2.2.1), protein samples can be loaded onto the gel and electrophoresis commence. Stop the run when the dye front reaches the bottom of the gel (*see* **Note 9**).

3.2.2 Methacrylamido Phenylboronic Acid Gel

1. Dissolve the required amount of methacrylamido phenylboronic acid (MPBA) in the resolving gel solution prior to polymerization.
2. Once dissolved, MPBA can be polymerized and will be incorporated into the backbone of the gel matrix ready for use in electrophoresis (*see* **Note 10**). Boronic acids are not needed in the stacking gel.

3.3 Visualization of Glycated Proteins

1. Protein bands in normal SDS-PAGE and mP-AGE can be visualized by Coomassie Brilliant Blue or silver staining and analyzed using a light box. Please follow the manufacturer's instructions for these subsequent analyses.
2. The fluorescein-boronic acid-labeled glycated protein bands in Flu-PAGE can be visualized immediately after electrophoresis by placing them on a blue light transilluminator with an amber or orange filter screen. Alternatively a UV transilluminator can be used, with an orange or green filter (*see* **Note 11**).
3. Flu-PAGE gels can subsequently be used for further analyses such as anti-AGE or anti-CML Western blotting or total protein staining with Coomassie Brilliant Blue or silver stain. We recommend following the manufacturer's instructions for these subsequent analyses.

4 Notes

1. Gels are often identified with their percentages of acrylamide (%T) and cross-linker (%C) that can be calculated as follows:

   $$\%T = [\text{mass of acrylamide (g)} + \text{bisacrylamide (g)}] \times 100/\text{volume (mL)}.$$

   $$\%C = \text{mass of bisacrylamide (g)} \times 100/[\text{mass of acrylamide (g)} + \text{bisacrylamide (g)}]$$
2. The fluorescein-boronic acid sample incubation solution (1 mg/mL) can be prepared from the stock solution by adding in water (as a saturated solution) or methanol. Methanol improves fluorescein-boronic acid solubility and can help intensify glycated protein labeling.
3. It may be preferable for some samples to be labeled overnight at 4 °C for improved labeling.

4. Due to the stacking effect, where the rapid migration of proteins in the stacking gel allows them to be stacked into a thin zone before resolution, large volumes of 15–20 μL of dilute sample can be loaded onto each lane of the gel.
5. APS and radical initiator TEMED should only be added just before polymerization is required.
6. Polymerization is temperature dependent. Therefore, gels should be cast above 20 °C. Lower temperatures will result in longer polymerization time and may cause incomplete polymerization. Always allow the gels to polymerize at least 2 h prior to electrophoresis to ensure complete polymerization.
7. To determine if the gel is set, gently tilt the gel cassette to see whether the gel is in solid or liquid state.
8. Mark the bottom of each sample well in the gel cassette with a marker to ease sample loading.
9. For Flu-PAGE the best results are obtained with Tris-Glycine/SDS gels and running buffer at pH 8.3. We do not recommend using Bis-Tris, MES, MOPS, or other neutral- and low-pH gels and running buffer conditions. Please ensure the electrophoresis process is not interrupted until after most of the free fluorescein-boronic acid has left the bottom of gel together with the dye front. Alternatively, excess fluorescein-boronic acid can be removed from the samples after incubation by treatment with desalting column, prior to electrophoresis.
10. Maximum dissolution of MPBA in Tris pH 8.8 gel solution is about 1% w/v. The required amount of MPBA is added to polyacrylamide gel solution in a falcon tube and shaken until MPBA completely dissolves. Do not heat the solution as this might result in unwanted polymerization. The solubility of MPBA varies with the pH and composition of gel buffer.
11. When imaging the glycated bands in Flu-PAGE, it is recommended to cut off any parts at the bottom of the gel containing excess fluorescein-boronic acid and rinse the gel in water before imaging. By adjusting the contrast and brightness of the gel images, the visibility of the fluorescein-boronic acid-labeled glycated protein bands can be optimized.

Acknowledgments

We thank Alzheimer's Research UK (ARUK-PPG2011B-17) and the Dunhill Medical Trust (DMT research grant R320/1113) for supporting this research.

References

1. Ulrich P, Cerami A (2001) Protein glycation, diabetes, and aging. Recent Prog Horm Res 56:1–22
2. Lapolla A, Traldi P, Fedele D (2005) Importance of measuring products of non-enzymatic glycation of proteins. Clin Biochem 38:103–115
3. Pokharna H, Pottenger L (1997) Nonenzymatic glycation of cartilage proteoglycans: an in vivo and in vitro study. Glycoconj J 14:917–923
4. Geoghegan KF, Dixon HBF, Rosner PJ et al (1999) Spontaneous [alpha]-N-6-Phosphogluconoylation of a "His Tag" in Escherichia coli: the cause of extra mass of 258 or 178 Da in fusion proteins. Anal Biochem 267:169–184
5. Yan Z, Caldwell GW, McDonell PA (1999) Identification of a gluconic acid derivative attached to the N-terminus of histidine-tagged proteins expressed in bacteria. Biochem Biophys Res Commun 262:793–800
6. Monnier VM, Cerami A (1981) Nonenzymatic browning in vivo: possible process for aging of long-lived proteins. Science 211:491–493
7. Foerster A, Henle T (2003) Glycation in food and metabolic transit of dietary AGEs (advanced glycation end-products): studies on the urinary excretion of pyrraline. Biochem Soc Trans 31:1383–1385
8. Montgomery H, Tanaka K, Belgacem O (2010) Glycation pattern of peptides condensed with maltose, lactose and glucose determined by ultraviolet matrix-assisted laser desorption/ionization tandem mass spectrometry. Rapid Commun Mass Spectrom 24:841–848
9. Klenk DC, Hermanson GT, Krohn RI et al (1982) Determination of glycosylated hemoglobin by affinity chromatography: comparison with colorimetric and ion-exchange methods, and effects of common interferences. Clin Chem 28:2088–2094
10. Zhang Q, Ames JM, Smith RD et al (2008) A perspective on the Maillard reaction and the analysis of protein glycation by mass spectrometry: probing the pathogenesis of chronic disease. J Proteome Res 8:754–769
11. Wu JT, Tu M-C, Zhung P (1996) Advanced glycation end product (AGE): characterization of the products from the reaction between D-glucose and serum albumin. J Clin Lab Anal 10:21–34
12. Weber K, Osborn M (1969) The reliability of molecular weight determinations by dodecyl sulfate-polyacrylamide gel electrophoresis. J Biol Chem 244:4406–4412
13. Pereira Morais MP, Mackay JD, Bhamra SK et al (2009) Analysis of protein glycation using phenylboronate acrylamide gel electrophoresis. Proteomics 10:48–58
14. Davis BJ, Ornstein L (1959) A new high resolution electrophoresis method. In: Society for the study of blood. New York Academy of Medicine, New York, NY
15. Raymond S, Weintraub L (1959) Acrylamide gel as a supporting medium for zone electrophoresis. Science 130:711
16. Miksík I, Deyl Z (1997) Post-translational non-enzymatic modification of proteins II. Separation of selected protein species after glycation and other carbonyl-mediated modifications. J Chromatogr B Biomed Sci Appl 699:311–345
17. Garfin DE (2003) Gel electrophoresis of proteins. In: Davey J, Lord M (eds) Essential cell biology: a practical approach. Oxford University Press, Oxford
18. Switzer RC, Merril CR, Shifrin S (1979) A highly sensitive silver stain for detecting proteins and peptides in polyacrylamide gels. Anal Biochem 98:231–237
19. Springsteen G, Wang B (2002) A detailed examination of boronic acid-diol complexation. Tetrahedron 58:5291–5300
20. Lorand JP, Edwards JO (1959) Polyol complexes and structure of the benzeneboronate ion. J Org Chem 24:769–774
21. Nishiyabu R, Kubo Y, James TD et al (2010) Boronic acid building blocks: tools for sensing and separation. Chem Commun 47:1106–1123
22. Ma WMJ, Pereira Morais MP, D'Hooge F et al (2009) Dye displacement assay for saccharide detection with boronate hydrogels. Chem Commun 5:532–534
23. Weith HL, Wiebers JL, Gilham PT (1970) Synthesis of cellulose derivatives containing the dihydroxyboryl group and a study of their capacity to form specific complexes with sugars and nucleic acid components. Biochemistry 9:4396–4401
24. Liu X-C (2006) Boronic acids as ligands for affinity chromatography. Chin J Chromatogr 24:73–80
25. Cartwright SJ, Waley SG (1984) Purification of beta-lactamases by affinity chromatography on phenylboronic acid-agarose. Biochem J 221:505–512

26. Armbruster DA (1987) Fructosamine: structure, analysis, and clinical usefulness. Clin Chem 33:2153–2163
27. Priego Capote F, Sanchez J-C (2009) Strategies for proteomic analysis of non-enzymatically glycated proteins. Mass Spectrom Rev 28:135–146
28. Jackson TR, Springall JS, Rogalle D et al (2008) Boronate affinity saccharide electrophoresis: a novel carbohydrate analysis tool. Electrophoresis 29:4185–4191
29. Jackson P (1990) The use of polyacrylamide-gel electrophoresis for the high-resolution separation of reducing saccharides labelled with the fluorophore 8-aminonaphthalene-1,3,6-trisulphonic acid. Detection of picomolar quantities by an imaging system based on a cooled charge-coupled device. Biochem J 270:705–713
30. Mahoney DJ, Aplin RT, Calabro A et al (2001) Novel methods for the preparation and characterization of hyaluronan oligosaccharides of defined length. Glycobiology 11: 1025–1033
31. Calabro A, Benavides M, Tammi M et al (2000) Microanalysis of enzyme digests of hyaluronan and chondroitin/dermatan sulfate by fluorophore-assisted carbohydrate electrophoresis (FACE). Glycobiology 10:273–281
32. Curry S, Mandelkow H, Brick P et al (1998) Crystal structure of human serum albumin complexed with fatty acid reveals an asymmetric distribution of binding sites. Nat Struct Mol Biol 5:827–835
33. Sattarahmady N, Moosavi-Movahedi AA, Ahmad F et al (2007) Formation of the molten globule-like state during prolonged glycation of human serum albumin. Biochim Biophys Acta Gen Subj 1770:933–942
34. Inaba M, Okuno S, Kumeda Y, et al., and the Osaka CKD Expert Research Group (2007) Glycated albumin is a better glycemic indicator than glycated hemoglobin values in hemodialysis patients with diabetes: effect of anemia and erythropoietin injection, J Am Soc Nephrol 18, 896–903
35. Burman JD, Leung E, Atkins KL et al (2008) Interaction of human complement with Sbi, a staphylococcal immunoglobulin-binding protein: indications of a novel mechanism of complement evasion by Staphylococcus aureus. J Biol Chem 283:17579–17593
36. Pereira Morais MP, Marshall D, Flower SE et al (2013) Analysis of protein glycation using fluorescent phenylboronate gel electrophoresis. Sci Rep 3:1437
37. Kassaar O, Pereira Morais MP, Xu S et al (2017) Macrophage migration inhibitory factor is subjected to glucose modification and oxidation in Alzheimer's disease. Sci Rep 7:42874

Chapter 17

Immunofixation Electrophoresis for Identification of Proteins and Specific Antibodies

Gyorgy Csako

Abstract

Immunofixation electrophoresis (IFE) is a technique for the identification of proteins within complex mixtures after separation by either conventional zone electrophoresis or isoelectric focusing. Most commonly antigens (which are often immunoglobulins) are separated by electrophoresis followed by precipitation with specific antibodies in situ. However, immunoglobulins with specific reactivity can be also precipitated with the proper antigens after electrophoresis in reverse or reversed IFE. Because of its great versatility, potentially high sensitivity, ease to perform and customize, and relatively low cost with no requirement for expensive instrumentation, manual IFE remains a valuable tool for both clinical diagnostic testing and research. Any low-viscosity body fluid specimen or, possibly, culture fluid could be tested with IFE if proper antibodies (or antigens in reverse[d] IFE) are available. After pretreatment with chaotropic and/or reducing agents, even high-viscosity specimens might be amenable to testing with IFE.

Key words Amido black, Agarose gel, Immunoglobulins, Immunoprint, Isoelectric focusing, Immunoprecipitation, Lipoproteins, Plasma proteins, Reverse(d) immunofixation electrophoresis, Zone electrophoresis

1 Introduction

Immunofixation electrophoresis (IFE) is an alternative technique to immunoelectrophoresis (IEP) for visualizing specific proteins in situ through an immunologic reaction, namely, antigen-antibody precipitin formation, following separation by electrophoresis in semisolid (gels) or solid media (membranes). Although most commonly antigens (which are often immunoglobulins [Ig's]) are separated by electrophoresis followed by precipitation with their corresponding antibodies in situ, immunoglobulins with specific reactivity can be also precipitated with the proper antigens after electrophoresis in situ, and this is called reverse or reversed IFE. The principle of IFE is simple. The antibodies (or antigens in reverse[d] IFE) are applied to the surface of a gel or onto a cellulose acetate membrane after electrophoresis, and, at antigen-antibody

Biji T. Kurien and R. Hal Scofield (eds.), *Electrophoretic Separation of Proteins: Methods and Protocols*, Methods in Molecular Biology, vol. 1855, https://doi.org/10.1007/978-1-4939-8793-1_17,

equivalence or in moderate antibody excess, they form large precipitates with their counterparts in the gel or membrane. Upon washing, uncomplexed antibodies (or specific antigens in reverse [d] IFE) and other proteins are removed from the gel or membrane, leaving only the immunoprecipitates which are too large and insoluble to be washed out from the pores. The immunoprecipitates remaining in the gel or membrane can be then directly stained or identified by other techniques such as by means of fluorescein-, enzyme-, or radiolabeled primary or secondary antibodies or using a biotin-avidin detection system. In short, IFE permits the identification of proteins within complex mixtures after separation by either conventional zone electrophoresis or isoelectric focusing (IEF) and is suitable for localization of both discrete protein bands and diffuse zones. Further, reverse(d) IFE with (labeled) antigen addition allows for detecting the function of the immunoglobulins separated by electrophoresis.

The concept of IFE was reported independently by Afonso [1] and Wilson [2] as a modified technique of IEP for the quantitation of low protein concentrations in 1964. However, the method became practical only when Alper and Johnson substituted monospecific antisera for polyvalent antisera in 1969 [3]. They called this method IFE and used it for improved detection of protein polymorphism. Interestingly, in a commentary published along with reprint of the original article in 1983 [3], the authors recall that their manuscript was rejected by one (unnamed) journal before being accepted for publication in the *Vox Sanguinis*. In addition to protein polymorphisms [4–10], IFE applications also have been reported for the study of protein conversion products of enzymes [3, 11, 12], for the detection and identification of abnormal B-cell-clone-produced immunoglobulins (paraproteins) [13, 14], and even for protein quantitation [11]. In these studies, IFE was shown to be applicable to a variety of gel types and cellulose acetate membrane as well. In a pivotal series of articles in 1976, Ritchie and Smith [15–17] concluded that "Immunofixation offers the worker an economical means of physically locating a protein in an electrophoretic strip and is ideally suited to forensic medicine, genetic studies, or research. The method is as simple and economical as the commonly used one- or two-dimensional immunoelectrophoresis, yet yields considerably more information." Increasing availability of commercial kits since the mid-1980s greatly contributed to the increasing popularity of IFE for paraprotein detection and identification. IFE in agarose gel gradually replaced IEP and became the new standard for the detection and identification of paraproteins in human body fluids (serum, urine, and cerebrospinal fluid [CSF]) [18–20]. In combination with isoelectric focusing, IFE became a "gold standard" for detecting oligoclonal bands in CSF [21]. Finally, zone electrophoresis coupled with IFE is used for distinguishing serum transferrin from CSF asialo-(β_2) transferrin (previously known as β-trace protein) after head trauma or surgery, allowing the diagnosis of CSF leak [22–24]. Reverse(d) IFE proved to be a useful tool for identifying the specificity of electrophoretically separated antibodies [13, 25–28]. Immunosubtraction methods (also called capillary IFE) in which immunoabsorption of the samples with solid-phase-bound monospecific antibodies (Paragon 2000, Beckman) or formation of slow-migrating immune complexes by mixing the samples with antibodies in liquid phase (CAPILLARYS2, Sebia) is followed by capillary zone electrophoresis (CZE) became recently available for testing for paraproteins and other proteins [29–31]. Although capillary immunosubtraction is relatively fast (no stains and drying) and can be fully automated as a walk-away system, it is less sensitive and generally considered inferior to IFE for paraprotein detection [29–31]. The utility of a method proposed for detection of specific antibodies using immunosubtraction and capillary electrophoresis, called immunocapture-immunosubtraction or ICIS [32], requires additional studies. Thus, because of its great versatility, ease to perform and customize, potentially high sensitivity, and relatively low cost with no requirement for expensive instrumentation, manual IFE remains a valuable tool for both basic and clinical research. The following is the description of a simple and practical manual IFE method. Note that, unless stated otherwise, antiserum and antibody variably refer to the same primary reagent of direct immunofixation. Further, depending on the context, specific antibodies may refer to the primary antibodies in direct IFE or to the antibodies identified by their respective antigens in reverse(d) IFE.

2 Materials

Prepare all solutions with distilled or deionized water and high-grade analytical reagents at room temperature (15–30 °C). Except stated otherwise, reagents and other materials are stored at room temperature. Unless obtained commercially, prepare and store all reagents and accessories used in these methods without adding sodium azide. Follow all waste disposal regulations for discarding reagents and disposables. As a safe laboratory practice, always wear clean gloves to minimize or eliminate exposure to infectious agents and hazardous chemicals.

2.1 Agarose Gel Plates for Zone Electrophoresis (See Note 1)

1. Hydragel 7 and Hydragel 15/30 Protein(e) alkaline-buffered precast 0.8% agarose gels (Sebia, Norcross, GA, USA). These gels contain Tris-barbital buffer (pH 9.2 ± 0.1) and proprietary additives which are claimed to be "nonhazardous" at the concentrations used. Store horizontally at room temperature or refrigerated (2–8 °C) until stated expiration date (*see* **Note 2**).

2. Hydragel 7 and Hydragel 15 Lipoprotein(e) alkaline-buffered precast 0.8% agarose gels (Sebia). These gels contain "buffer" (pH 8.5 ± 0.1) and proprietary additives which are claimed to be "nonhazardous" at the concentrations used. Store horizontally at room temperature or refrigerated until stated expiration date (*see* **Note 2**).

2.2 Apparatus and Accessories for Zone Electrophoresis in Agarose Gel

1. Hydrasys® semiautomated electrophoresis system (Sebia) (*see* **Note 3**).
2. Buffered, ready-to-use sponge strips for protein electrophoresis (Sebia). Each contains Tris-barbital buffer pH 9.2 ± 0.3, sodium azide, and proprietary additives which are claimed to be "nonhazardous" at the concentrations used. Store horizontally at room temperature or refrigerated until stated expiration date (*see* **Note 4**).
3. Buffered, ready-to-use sponge strips for lipoprotein electrophoresis (Sebia). Each contains "buffer" pH 8.5 ± 0.1, sodium azide, and proprietary additives which are claimed to be "nonhazardous" at the concentrations used. Store horizontally at room temperature or refrigerated until stated expiration date (*see* **Note 4**).
4. Precut, single-use applicators (ready to use). (Sebia). Store in a dry place at room temperature or refrigerated.
5. Precut, single-use filter papers. Store in a dry place at room temperature or refrigerated (*see* **Note 5**).
6. Wet storage chamber supplied with Hydrasys. (Sebia).
7. Pipettes or syringe for delivering volumes between 10 and 200 μL.

2.3 Accessories for Immunofixation and Staining in Agarose Gel

1. Thick filter papers. Single-use, thick absorbent paper for removing unprecipitated proteins off the gel surface after immunofixation step. Store in a dry plate at room temperature or refrigerated (*see* **Note 6**).
2. Lens paper strips for antiserum overlay, 5 × 30 mm each. Cut by means of a scissor from lens paper (VWR Scientific, Radnor, PA, USA) proportionate to the location of the expected plasma protein bands in the electrophoretic lanes (Fig. 1a) (*see* **Note 7**).
3. Template for antiserum application. Custom cut from X-ray film with tracks matching the position of the expected plasma protein electrophoretic patterns (Fig. 1b) (*see* **Note 7**).
4. Pipettes or syringes for delivering volumes between 0.5 and 100 μL.
5. Wet storage chamber supplied with Hydrasys (*see* **Note 8**).
6. Staining pans.
7. Paragon dryer (Beckman Coulter, Fullerton, CA, USA).

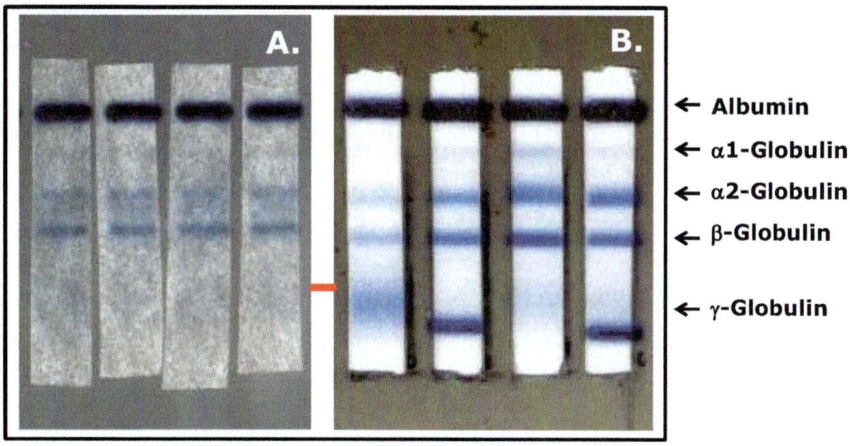

Fig. 1 Two methods of antibody (or in "reverse[d]" IFE: antigen) application for immunofixation following electrophoresis. Applying the antibody (or antigen) solution via cellulose acetate or paper strips (**a**) and placing the antibody (or antigen) solution into the tracks of an antiserum application template (**b**) positioned over the target area of the gel (in this case, electrophoretically separated human serum proteins). For illustration, the methods are shown as overlays on amido black-stained normal and abnormal (monoclonal band present) human serum protein patterns after electrophoresis in agarose gel. In the case of (**a**), strips of a simple lens paper were used, while in the case of (**b**), the template was prepared by cutting a previously developed X-ray film with a scalpel. Short horizontal line in the center indicates approximate point of sample application for electrophoresis

2.4 Protein Fixative for Reference Lane(s) in Agarose Gel IFE (See Note 9)

For convenience, a commercial protein fixative was used in these IFE runs. However, a simple "home-made" alternative also is described (*see* **Note 10**).

1. IFE fixative solution, ready to use (Sebia). This is an acidic fixative of undisclosed composition and contains "nonhazardous" additives at concentrations used. Store at room temperature or refrigerated until stated expiration date.

2.5 Antibodies for Immunofixation (See Notes 11–15)

1. Monospecific animal antisera to human α_1-acid glycoprotein, albumin, α_1-antitrypsin, α_2-macroglobulin, complement factor C3c, and haptoglobin, fibrinogen, and transferrin.
2. Monospecific "mammalian immunoglobulins" (antisera) to human IgG, IgA, and IgM, κ and λ [free and bound] light chains, and free κ and free λ light chains.
3. Anti-whole human serum, monospecific antisera to human apo AI and apo B (Chemicon International Inc. [Millipore], Billerica, MA, USA) and apo E (Biodesign International, Inc. [Meridian Life Science, Inc.], Saco, ME, USA).
4. Reconstitute lyophilized preparations according to the manufacturer's instructions. Store all antisera refrigerated until stated expiration date.

2.6 Protein Staining and Destaining Solutions for Agarose Gel IFE (See Note 16)

For convenience, a commercial amido black stain was used in these IFE runs. However, a simple "home-made" alternative also is described (*see* **Notes 17** and **18**).

1. Amido black stain. Reconstitute and store according to the manufacturer's instructions. The working solution is acidic (pH 2) and contains 0.4 g/dL amido black, 6.7% (v/v) ethylene glycol, and additives which are claimed to be "nonhazardous" at concentrations used. Store working solution in a closed container to prevent evaporation for 1 month.
2. Destaining solution for amidoblack staining: 5% acetic acid [v/v]): To 950 mL distilled or deionized water add 50 mL of glacial acetic acid and mix thoroughly using a magnetic stirrer. Store in a closed container for 1 week.

2.7 Sudan Black Staining and Destaining Solutions for Agarose Gel IFE (See Note 19)

For convenience, a commercial Sudan Black stain was used in these IFE runs. However, a simple "home-made" alternative also is described (*see* **Note 20**).

1. Sudan Black stain. Prepare working solution 30 min before use according to the manufacturer's instructions. Working solution contains ~0.04 g/dL Sudan Black, ~52% (v/v) absolute ethanol, ~48% water, and < 0.01% dimethylformamide (v/v). Discard solution after use.
2. Destaining solution and wash solution 1 for Sudan Black staining.
3. Wash solution 2 for Sudan Black staining.

2.8 Enzymatic Staining for Lipoprotein Cholesterol in Agarose Gel IFE

1. Reagents from an enzymatic cholesterol staining kit: SPIFE^R Vis Cholesterol (Helena).
2. Store, reconstitute, and prepare reagents according to the manufacturer's instructions.

2.9 Specimens and Positive Controls for IFE in Agarose Gel (See Note 21)

1. For illustration of the method, use human serum, plasma, and concentrated (30- to 50-fold) urine specimens (*see* **Notes 22** and **23**).
2. Collect serum, plasma, and random urine (with no preservatives!) specimens with standard medical procedures. Store refrigerated for up to 1 week (preferably use within 48 h), freeze at −20 °C for up to 3 months or at −70 °C for up to 12 months.
3. For electrophoresis, use serum and plasma samples straight or dilute with 0.85% saline to the concentration range established to be appropriate for target detection in trial runs with different antigen-antibody ratios (i.e., different combinations of sample vs. antiserum dilutions) (*see* **Notes 13, 24**, and **25**).

4. For electrophoresis, use concentrated urine samples according to the protein concentration of the original (unconcentrated) urine either straight or dilute with 0.85% saline to the concentration range established to be appropriate for target detection in trial runs with different antigen-antibody ratios (i.e., different combinations of sample vs. antiserum dilutions).
5. Positive control specimens for human paraprotein testing: IFE Control Kit (IgG/K, IgA/1, IgM) (Helena). Store controls according to the manufacturer's instructions (*see* **Note 26**).

3 Methods

Carry out all procedures at room temperature unless otherwise specified. For general guidance and safety issues, consult the operating manual of the electrophoretic system used. Estimated total time needed to complete one gel with any of the electrophoretic and detection methods described below is ~3 h.

3.1 Zone Electrophoresis in Agarose Gel for Proteins and Lipoproteins

Switch on the Hydrasys instrument and proceed according to the Hydragel 7 or 15/30 Protein(e) method or, as appropriate, to the Hydragel 7 or 15 Lipoprotein(e) method for migration. Briefly:

1. Place one applicator for 7 Protein(e) and Lipoprotein(e) plate or two applicators for 15/30 Protein(e) or 15 Lipoprotein (e) plate.
2. Apply 10 μL of sample in each well. Load each applicator within 2 min.
3. Place the applicator(s) into the wet storage chamber with the teeth up. Let the samples diffuse into the teeth for 5 min after the last sample application.
4. Attach buffered strips as appropriate for protein or lipoprotein run to the pins on the uplifted electrode carrier.
5. Unpack the appropriate Hydragel plate and roll quickly and uniformly one thin filter paper onto the gel surface to absorb excess of liquid. Remove and discard the paper immediately.
6. Complete migration setup per manufacturer's instructions, including placement of the gel in the migration module and applications of samples onto the gel.
7. After selecting the proper migration program, carry out electrophoresis.
   (a) under 10 W constant for serum/plasma and urine proteins using Hydragel 7 Protein(e) or 20 W constant using Hydragel 15/30 Protein(e) at 20 °C, controlled by Peltier effect, until 33 Vh have accumulated (~7 min), and,

(b) under 160 V constant voltage for serum lipoproteins using Hydragel 7 and 15/30 Lipoprotein(e) at 20 °C, controlled by Peltier effect, until 65 Vh have accumulated (~25 min).

3.2 Immunofixation

1. After stopping electrophoresis, remove gel from the migration module and wipe excess solution from back of gel.
2. Gently blot the gel surface with thin filter paper. Discard paper.
3. When available, apply 0.5 μL of positive control materials as "dots" to designated locations at either end of the electrophoretic tracks (except for the acid fixative lane) and allow absorption into gel for 2–3 min.
4. (a) For overlay of acid fixative and antisera by means of lens paper strips, carefully place each strip over the protein zones, including the control "spots," using the position numbers of the plate as a guide (Fig. 2a). A 1:1 size print of the stained protein/lipoprotein electrophoretic gel placed under the plate helps in positioning of the strips (Fig. 2a) (*see* **Notes 27–29**).

   (b) Using a micropipetter, apply 35 μL of protein fixative onto the strip corresponding to the reference lane and the same amount of appropriate antisera onto each of the other strips. If needed for a protocol, the same amount of antisera can be reapplied after 3–5 min onto the strips.

   (c) Carefully spread acid fixative and antisera over the entire surface of the strips by means of the tip of the pipet without damaging the gel surface.
5. (a) For using template for antiserum delivery, align the IFE antiserum template with the position of the electrophoretic tracks identified by the printed position numbers of the plate (Fig. 2b). Apply the template to the gel by slightly bending it in a "U-shape." This minimizes or avoids getting air bubbles under the template, especially when very thin plastic templates are used. If bubbles form, gently lift the edge of the template with the edge of a thin spatula, and then lay the template back down on the gel to remove the bubble(s).

   (b) Gently rub around the tracks to ensure complete seal between template and gel surface.

   (c) Apply 35 μL of IFE protein fixative to the reference track and the same amount of each antiserum to the designated tracks, including the position of the control "spots." Avoid spillage over tracks. Do not touch gel surface with pipet or syringe tip (*see* **Notes 27–29**).
6. Place gel in wet incubation chamber.
7. Cover chamber and incubate gel at room temperature for 30 min (*see* **Note 30**).

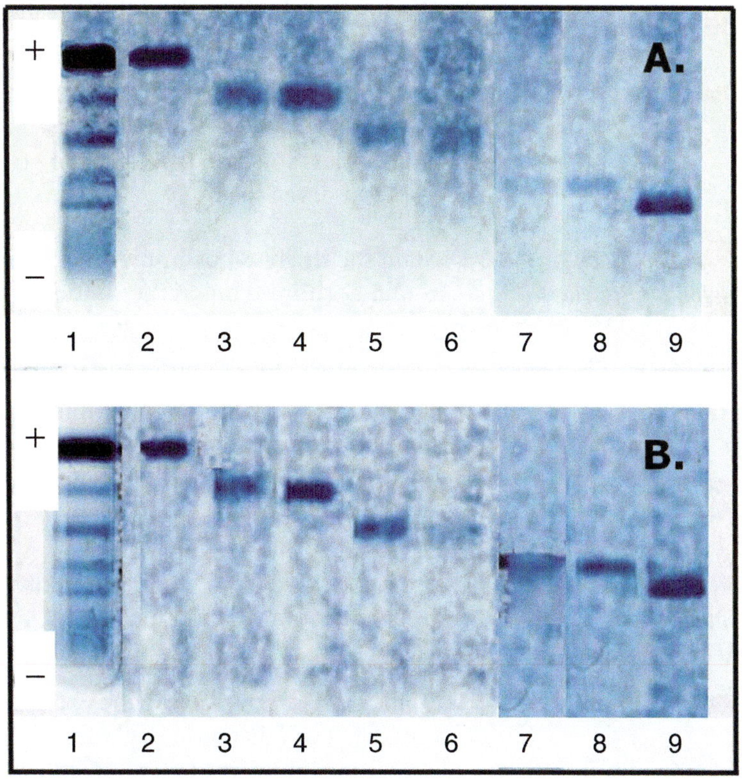

Fig. 2 Detection of various human plasma proteins with IFE in a protein electrophoresis agarose gel using two different methods of immunofixation with monospecific antibodies, followed by protein staining with amido black. Application of antibodies via lens paper (**a**) and antiserum template (**b**) as shown in Fig. 1. Lane 1, protein fixation with acid fixative as a reference, while all other lanes represent immunofixation of various plasma proteins with monospecific antibodies as follows: lane 2, albumin; lane 3, α_1-acid glycoprotein; lane 4, α_1-antitrypsin; lane 5, haptoglobin; lane 6, α_2-macroglobulin; lane 7, complement factor C3c; lane 8, transferrin; and lane 9, fibrinogen

3.3 Processing of Agarose Gels for Staining

1. Following incubation period, remove gel from wet incubation chamber.
2. (a) In case of overlay with lens paper strips, wash gel with ~75 mL of 0.85% saline with gentle shaking on a platform (or just by hands) for 5 min. Remove and discard strips.

   (b) In case of cut IFE antiserum template, soak gel in ~75 mL of 0.85% saline with gentle shaking on a platform (or just by hands) for 5 min and gently peel off the template (*see* **Note 31**).
3. Remove gel from saline and wipe excess solution from back of gel.
4. Press dry by placing two thick absorbent papers ("thick filter paper") on the top of the gel, followed by adding a weight of ~75 g for 5 min (*see* **Note 32**).

5. Wash gel with ~75 mL of 0.85% saline for 5 min.
6. Press dry by placing two thick absorbent papers ("thick filter paper") on the top of the gel, followed by adding a weight of 7 ~ 5 g for 5 min (*see* **Note 32**).
7. Wipe excess solution from back of gel and dry gel for 5 min in a Paragon dryer at 56 °C.

3.4 Amido Black Staining for Proteins

1. Stain the dried gel by immersing into the amido black stain for 5 min at room temperature under gentle agitation.
2. Destain gel with 5% (v/v) acetic acid solution for 3 min at room temperature under gentle agitation.
3. Destain gel with water for 3 min at room temperature under gentle agitation.
4. Remove gel from water, wipe excess solution from back of gel, and dry it in a Paragon dryer for 5 min at 56 °C.

3.5 Sudan Black Staining for Lipoproteins (See Note 33)

1. Stain the dried gel by immersing into the Sudan Black stain for 5 min at room temperature under gentle agitation.
2. Briefly (~1 min) destain gel with water under gentle agitation.
3. Remove gel from water, wipe excess solution from back of gel, and dry it in a Paragon dryer for 5 min at 56 °C.

3.6 Enzymatic Staining for Lipoprotein Cholesterol (See Note 33)

1. Pour cholesterol substrate into incubation chamber and place agarose gel down.
2. Incubate gel with substrate for 15 min at 30 °C.
3. Wash gel briefly (~1 min) with citric acid destaining solution at room temperature.
4. Briefly (~1 min) destain gel in water at room temperature.
5. Remove gel from water, wipe excess solution from back of gel, and dry it in a Paragon dryer for 5 min at 56 °C.

3.7 Interpretation of IFE Patterns (See Note 34)

3.7.1 General Comments:

1. Evaluate gel visually by observing location of specific precipitin bands in relationship to (acid-fixed) reference protein electrophoretic pattern (*see* **Note 35**).
2. A stained immunofixed band represents a reaction between a specific antiserum and a protein (antigen) in the sample in both conventional IFE and reverse(d) IFE (*see* **Notes 36** and **37**).
3. By comparing the location of a band in the acid-fixed reference pattern with the stained immunofixed band in the same location, determine the identity of a particular protein in the sample (*see* **Notes 38–40**).

Specific comments on IFE patterns produced with the manual method on human serum, plasma, and urine specimens:

1. Figure 2: Even in a minimally optimized IFE method with home-made antiserum applicators, various plasma proteins were satisfactorily identified with either method of antibody application for immunofixation after electrophoresis of a human plasma specimen. The "dirty" background apparently is due to inadequate removal of the heavy load of nonreactive proteins from the undiluted or minimally diluted, relatively low-avidity antisera.
2. Figure 3: The same three paraproteins (IgG[κ], IgA[λ], and IgM[κ]), signifying a triclonal gammopathy, were correctly identified by two major semiautomated commercial methods and the manual IFE method in a human serum specimen. Note the clarity of background with the manual IFE method when highly optimized and apparently "prepurified" antibodies (IgG fraction?) were used versus the "dirty" background in a minimally optimized system in Fig. 2.
3. Figure 4: Based on the IFE patterns, the first urine specimen (**a**) contains a strong band of free κ light chain, confirmed by testing with anti-free κ antibody, whereas the second urine (**b**) contains both IgG(λ) and free λ light chain, again confirmed by testing with anti-free λ antibody. As in Fig. 3, note the clearness of background when highly optimized and apparently "prepurified" antibodies (IgG fraction?) are used in the manual method. These observations underscore the importance of antiserum quality: a critical need for high-avidity and high-affinity antibodies in the presence of minimal amounts of extraneous proteins to reduce or virtually eliminate background staining, especially when protein stains are used for detection. Obviously, this becomes less of an issue when probes are involved in detection of the immunoprecipitates.
4. Figure 5: The results obtained with plasma lipoproteins demonstrate the versatility of the IFE method. It is interesting to note that more apolipoprotein is carried in the α- lipoprotein fraction than the combined pre-β and β-lipoprotein fraction, while the reverse is true for the transport of both major lipids, triglycerides and cholesterol. The biochemical results are consistent with these observations.

4 Notes

1. Besides alkaline agarose, applications have been also reported for agar, acid-starch, acidic agarose, agarose-acrylamide, polyacrylamide and cellulose acetate (Cellogel) gels (gel thickness

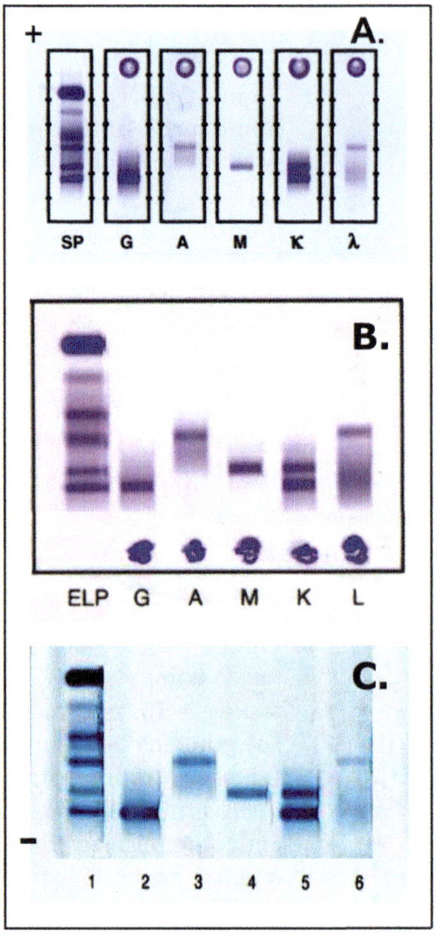

Fig. 3 Comparison of the manual IFE method with two major semiautomated commercial IFE methods for human serum paraprotein detection and identification in agarose gel. (**a**) SPIFE[R] ImmunoFix kit on the SPIFE 3000 system, acid violet staining (Courtesy of Dr. Rita Ellerbrook of Helena Laboratories, Beaumont, TX, USA); (**b**) Hydragel 2 IF on Hydrasys[R], acid violet staining (Sebia Inc., Norcross, GA, USA); and (**c**) manual IFE method on a protein electrophoresis plate and using a hand-cut IFE antiserum template for manually applying acid fixative and antisera, followed by amido black staining. The acid fixative and monospecific antibodies with the manual method are the same as those in the semiautomated Sebia method. Note presence of "dots" of positive controls (Helena) for antibody identification in both commercial IFEs. Codes: lane 1, SP, and ELP, acid fixative; lane 2, G, and G, antihuman IgG heavy chain-specific antibody; lane 3, A, and A, antihuman IgA heavy chain-specific antibody; lane 4, M, and M, antihuman IgM heavy chain-specific antibody; lane 5, κ, and K, antihuman κ (free and bound) light chain-specific antibody; and lane 6, λ, L, antihuman λ (free and bound) light chain-specific antibody

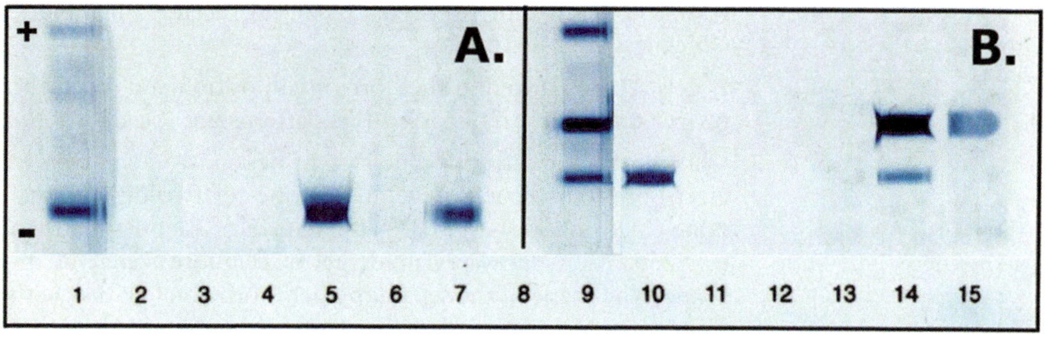

Fig. 4 Detection of paraproteins in two different human urine specimens (**a** and **b**) with the manual IFE method in protein electrophoresis agarose gel using a home-cut IFE antiserum template for acid fixative and antibody application, followed by amido black staining. Both urine specimens were preconcentrated ~50-fold using Amicon Ultra-4 10 K microconcentrators (Millipore, Billerica, MA, USA) for electrophoresis. The acid fixative and monospecific antibodies are the same as those in the semiautomated Sebia method. Codes: lanes 1 and 9, acid fixative; all other lanes received monospecific antibodies to human Ig heavy and light chains as follows: lanes 2 and 10, IgG heavy chain; lanes 3 and 11, IgA heavy chain; lanes 4 and 12, IgM heavy chain; lanes 5 and 13, κ (free and bound) light chain; lanes 6 and 14, λ (free and bound) light chain; lane 7, free κ light chain; and lane 15, free λ light chain

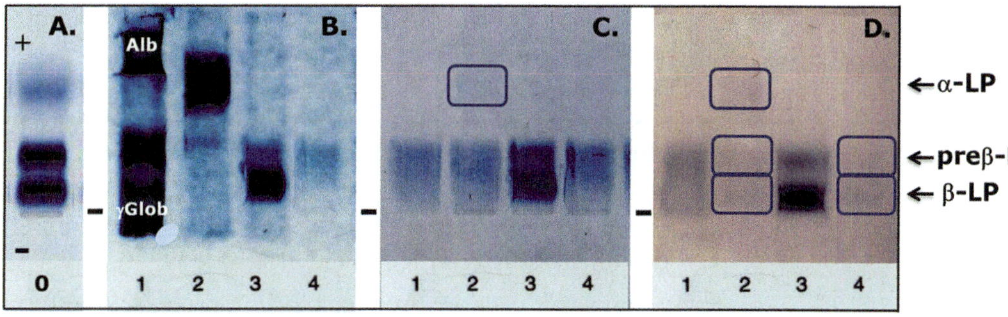

Fig. 5 Manual IFE method used for detecting various lipoprotein constituents via immunofixation of human whole serum and apolipoproteins A-I, B, and E in an agarose gel for lipoprotein electrophoresis. (**a**) Standard Sudan Black staining for lipids (Sebia), (**b**) through (**d**), staining after immunofixation with antibodies against human whole serum (lane 1), apo A-I (lane 2), apo B (lane 3), and apo E (lane 4), using a hand-cut antiserum template for application: (**b**) amido black staining for proteins (Sebia method), (**c**) Sudan Black staining for lipids (Sebia method), (**d**) enzymatic staining for cholesterol (Helena method). Short horizontal lines indicate point of sample application. The serum specimen tested contained total cholesterol 289 mg/dL (desirable, <200; borderline high cardiovascular risk, 200–239), triglycerides 219 mg/dL (*N*, <150; high, 200–499), high-density lipoprotein cholesterol (HDL-C) 35 mg/dL (high risk, <40), low-density lipoprotein cholesterol ($LDL\text{-}C_{calculated}$) 210 mg/dL (optimal, <100; very high cardiovascular risk, ≥190), apolipoprotein A-I 118 mg/dL (*N*, 110–205), apolipoprotein B 146 mg/dL (*N*, 55–140), lipoprotein electrophoresis α-lipoprotein 14% (*N*, 17–46), pre-β-lipoprotein 42% (*N*, 2–32), β-lipoprotein 43% (*N*, 39–66), and chylomicrons 1% (*N*, 0–2)

commonly $\leq$1 mm), and dry cellulose acetate membrane (thickness 0.12–0.16 mm).

2. If gels are not stored in the horizontal position and/or at the proper temperature, distorted IFE patterns may occur.
3. Choice of the electrophoretic system depends on the type of electrophoretic separation required (one- or two-dimensional zone electrophoresis vs. IEF), the nature of support medium (e.g., agarose vs. polyacrylamide gel vs. cellulose acetate membrane), the shape of the separating gel (horizontal vs. disk with some polyacrylamide gels), the degree of required resolution (e.g., low- vs. high-resolution agarose gel electrophoresis [HRAGE]), availability of plate cooling (e.g., a must for IEF and HRAGE), and, obviously, the size of the gel or membrane (fit into the electrophoretic chamber). Any electrophoretic system that satisfies these requirements is suitable for generation of a gel or cellulose acetate membrane with the target proteins separated for subsequent immunofixation. Currently, there are at least four major commercial kits with dedicated instrumentation for clinical diagnostic testing of human body fluid specimens (serum, neat and/or concentrated urine, and/or CSF) for paraprotein detection and identification. Semiautomated systems include the SPIFE® 2000/3000, Helena Laboratories, Beaumont, TX, USA; Interlab G26, Interlab Sri, Rome, Italy; and Hydrasys-Hyrys® system, Sebia, Norcross, GA, USA. Manual methods include the Standard and Minifix™ IFE kits (both from The Binding Site Ltd., London, England).
4. These strips constitute the electrophoresis buffer reservoir and provide contact between the gel and electrodes.
5. These thin paper pads are used for blotting excessive moisture off the gel surface before sample application.
6. These single-use, thick absorbent papers are used for removing unprecipitated proteins off the gel surface after immunofixation.
7. Immunofixation of electrophoretic gels or membranes can be carried out by either immersing into an antibody solution (all polyacrylamide disks and often cellulose acetate membranes) or overlaying with an antibody solution (horizontal gels and membranes). In this latter approach, either the entire surface is spread by an antibody solution or different electrophoretic zones are exposed to different antibody solutions by means of semipermeable membranes or (filter) papers (Fig. 1). Application of antiserum directly to the gel surface may be advantageous over application via cellulose acetate and paper strips because (a) much smaller volumes (i.e., less antibody or less antigen in reverse[d]) IEF) are needed to develop the same

intensity IFE patterns, (b) more rapid diffusion of antiserum into the gel is less likely to be associated with antigen excess in most cases, and (c) the range of antigen concentrations detectable with a given concentration of antibodies is wider [19]. On the other hand, direct application of the antiserum to the surface of a gel requires availability of proper application templates which may be less accessible than cellulose acetate or paper strips. Nonetheless, as shown in Fig. 1, both types of IFE antiserum application devices can be made from materials that either can be purchased or are already available in the laboratory. Proportionate strips of cellulose acetate membrane or filter papers such as Whatman no. 1 or, simply, a lens paper (e.g., VWR Scientific, Radnor, PA, USA) for the area of gel or membrane containing the target protein band(s) can be cut using scissors (Fig. 1a). Likewise, antiserum application templates can be custom cut from thin, pliable plastic sheets or X-ray film using a scalpel (Fig. 1b).

8. Any level humid incubation chamber can substitute this chamber for room temperature incubations. Incubation at higher temperatures (up to 45 °C) will shorten the incubation time for immunoprecipitation.
9. A variety of protein fixatives, often with undisclosed composition in commercial IFE kits, have been proposed and used over the years. Apart from the type of electrophoretic medium (e.g., agarose and polyacrylamide gels or cellulose acetate membrane), choice of the fixative often is arbitrary, based on personal experience and practicability.
10. Alternative "home-made" protein fixative (10% acetic acid [v/v] and 5% sulfosalicylic acid [w/v] in water): Weigh out 5.8 g of 5-sulfosalicylic acid dehydrate (MW 254.2), add it to a graded beaker, dissolve in 80 mL of distilled water under continuous mixing using a magnetic stirrer, and then add 10 mL of glacial acetic acid, mix again, and add distilled water up to 100 mL. Store in a closed container at room temperature for 1 month.
11. High-affinity and high-avidity antisera with optimized dilutions to avoid prozoning will produce maximal immunoprecipitation. These attributes, along with their relatively low price, make polyclonal antibodies the reagent of choice for most immunoprecipitation-based applications. High titer has the advantage of allowing for higher antiserum dilutions which, in turn, reduce the undesirable background protein "load" in the gel or membrane. Because of their generally low affinity, monoclonal antibodies are rarely suitable for IFE. This is in contrast with their high applicability to and widespread use in the alternative immunodetection method, immunoblotting (Western blotting).

12. In order to minimize the protein "load" on the gel or membrane during the immunofixation step, instead of using full antisera, the immunoglobulin fraction (IgG) of antisera is preferred. This is a common approach in commercial kits for human paraprotein testing to reduce the background due to incomplete removal of nonreactive serum proteins. Alternatively, high dilution of high-titered full antisera helps to minimize the protein "load" in the gel or membrane and, hence, to reduce the background. In reverse(d) IFE, the same considerations apply: use of partially or completely purified antigens will minimize the background staining due to incompletely removed nonreactive proteins.
13. For each antigen-antibody combination, the optimal ratios should be determined in trial runs using several combinations of dilutions.
14. For enhancement of both speed and intensity of immune complex formation, inclusion of 4% polyethylene glycol (PEG, M_r 6000) in the antiserum has been proposed [33], but it did not gain wide acceptance.
15. While replacing polyvalent antisera for monospecific ones was revolutionary at the time when IFE has been introduced in 1969 [3], polyvalent antisera with well-defined specificities became recently advantageous at least in one application. Several semiautomated commercial methods for paraprotein detection and identification now offer the option of screening serum and/or concentrated urine samples with pentavalent antisera consisting of a pool of monospecific antibodies to IgG, IgA, and IgM heavy chains and κ and λ free and bound light chains (e.g., SPIFE[R] IFE Pentavalent Kit for screening serum, Helena Laboratories; serum and concentrated urine pentavalent immunofixation, Interlab G26, Interlab Sri; and Hydragel 6/12 IF Penta for screening serum, Hydrasys, Sebia). The concept may be applicable to other protein antigen or specific antibody systems as well.
16. Besides fluorescein and radioactive or enzyme probe-labeled primary or secondary antibodies, a variety of stains have been used for detecting protein in the immunoprecipitates of IFE. Major differences among these stains involve detection sensitivity (ranging from a few ng to 500 ng [often 50–100 ng] of target protein per band), electrophoretic media compatibility (e.g., agarose vs. polyacrylamide gel vs. cellulose acetate membrane), and "reversibility" (e.g., both Ponceau S and Nigrosin [e] stainings are completely reversible). Comprehensive direct comparisons of all of these stains are not available, and band detection sensitivity also is greatly affected by unequal affinity of stains toward various proteins and by factors unrelated to the

stain (e.g., optimization of the precipitin reaction). Nevertheless, an approximate order of detection sensitivity for commonly used stains is estimated as follows: silver > Coomassie Brilliant Blue R-350 > Coomassie Brilliant Blue R-250 or G-250 > Ponceau S > Nigrosin(e) > acid violet > amido black. Note that the use of Ponceau S is limited to cellulose acetate membrane (and nitrocellulose membrane in immunoblotting). Finally, practicability (e.g., stability on storage, speed of staining and destaining, concern for toxicity, etc.) and simplicity of use also play a major role in selection of stains for IFE. These considerations explain why amido black, a relatively insensitive dye, is so widely used.

17. Amido black (MW 616.5), a dark red to black powder, stains proteins with a blue-black color. Of the many synonyms used for amido black, a short list is as follows: Acid Black 1, Acidal Black 10B, Acidal Navy Blue 3BR, Amidoschwarz, Amido Black 10B, Buffalo Black NBR, Eriosin Blue Black B, Naphthalene Blue Black, Naphthalene (Black) 12B, Naphthol Blue Black, and C.I. 20,470. For protein detection in electrophoretic gels and membranes, amido black stains have been prepared in a number of ways, and specific formulations of the stain often are not disclosed in publications and IFE kits for paraproteins. While the nature of the electrophoretic matrix needs to be considered, the choice of a particular solution is often based on personal preference. Note that after staining of lipoproteins with Sudan Black, the proteins no longer take amido black.
18. Alternative "home-made" amido black staining solution: Prepare a saturated solution of the dye (~1 g/dL) in 10% acetic acid in methanol (v/v) by dissolving 1 g of amido black in 450 mL of methanol in a glass beaker, and then add 50 mL of glacial acetic acid. Mix well using a magnetic stirrer for at least 30 min and then filter. Store in a closed container for up to 2 months.
19. Sudan Black (MW 457), also known as Fat Black HB, Solvent Black, Sudan Black B, and C.I. 26,150, is a dark brown to black powder. It is slightly basic and combines with acidic groups in compound lipids, resulting in a blue-black color. Sudan Black is thought to stain mostly triglycerides in human plasma lipoproteins.
20. Alternative "home-made" Sudan Black staining solution: Prepare a saturated solution of Sudan Black in 60% ethanol by adding an excess of the dye (100 mg) to 100 mL of warm (37 °C) 60% ethanol solution under constant mixing with a magnetic stirrer. Keep it at 37 °C for 16 h, allow to cool, let it stand for few days at room temperature, filter, and then store in

a dark bottle. Stable for up to 2 months. Alternative "home-made" destaining solutions: 50% (v/v) ethanol and then 40% (v/v) ethanol. Store in a closed container for up to 2 months.

21. Any low-viscosity body fluid specimen or culture fluid could be tested with IFE if proper antibodies (or antigens in reverse [d] IFE) are available. High-viscosity specimens (e.g., sera from certain gammopathy cases, most often of the IgM, and, less commonly, of the IgA and IgG type) can be pretreated with special chaotropic agents (such as the Fluidil with undisclosed composition in the Sebia system) and/or with the reducing agent 2-mercaptoethanol (2-ME). Cryoproteins may cause gelation of the serum specimens at room temperature and, hence, interfere with sample application for electrophoresis. Pretreatment with a chaotropic agent such as the Fluidil (Sebia) and/or preheating the specimen at 37 °C for 30–60 min would reverse the "gelation" and allow for proper sample application for electrophoresis.

22. In order to get insight into the impact of IFE on clinical practice and research, PubMed searches were performed on July 14, 2011. While these retrievals obviously underestimated the true usage of this method, they still show IFE is a widely used laboratory technique. Using the search term "immunofixation," a total of 1128 articles were retrieved, of which 852 or 64% (852/1128) were diagnosis related based on adding "diagnosis" as a second search term. Using the search terms "immunofixation electrophoresis," a total of 715 articles were retrieved from PubMed, of which 543 or 64% (543/715) were diagnosis related based on adding "diagnosis" as a third search term. These data indicate that most commonly (~two-thirds of the cases) IFE is used as a diagnostic test. A quick review of the title (and, occasionally, the abstract) of the articles retrieved by adding the search term "diagnosis" to either retrieval approach revealed that practically all are related to paraprotein detection and identification in hematologic malignancies. Further, the identical proportions of diagnosis-related immunofixation (electrophoresis) articles with the two retrieval approaches suggest that the word "immunofixation" is used virtually identically to IFE in the literature.

23. Most clinical urine specimens (dipstick protein $\leq$3+ or protein concentration $<$ 200 mg/dL), virtually all CSF specimens, and, likely, many culture fluid specimens require concentration for IFE testing after zone electrophoresis. Up to 50-fold enrichments with yields of 60–120 μL of concentrate could be obtained by centrifuging 2 mL samples in Amicon Ultra-4 microconcentrators (Millipore, Billerica, MA) (10 K for urines and 30 K for CSF at 3100 × *g* for 14 min and 2600 × *g* for 8 min, respectively, at 20 °C in a swing-out rotor with a radius

of ~11.5 cm). Retentate recovery is claimed to be high (>90%) with these devices. However, it is noteworthy that concentrating specimens may lead to artifacts, including protein losses that differ from sample to sample and may be substantial (up to 50%) with some other devices.

24. To avoid antigen excess and minimize unwanted diffusion, samples generally need to be adjusted to contain approximately 5–50 mg/dL or 50–500 ng/μL of the protein under study.

25. Because of the concentrating (focusing) effects of IEF, the protein concentrations required for band detection are only approximately one-tenth of those with conventional zone electrophoresis [19].

26. The use of positive controls ensures that the proper antibodies (or, in case of reverse[d] IFE, antigens) are used in the proper locations in the gel or membrane for immunofixation. Along with a commercial kit for IFE testing of paraproteins, one manufacturer markets such controls (IFE Control Kit [IgG/K, IgA/1, IgM], Helena). If commercially not accessible, use solutions of "purified" target antigens (or, if available, specific antibodies in reverse[d] IFE) to confirm proper usage of each antibody (or antigen in reverse[d] IFE). While high purity is desirable to reduce the total protein "load" (and, hence, risk of false positivity), greatly "enriched" preparations might be satisfactory if their reactivity pattern with the corresponding monospecific antibody (or antigen in reverse [d] IFE) has been already established. Commercially available purified or recombinant preparations could be a major source of positive controls. Alternatively, specimens available in the laboratory either as a home-purified product or without further purification from natural sources can be considered for use as positive controls. For instance, for gammopathy testing of any body fluid specimen, urine specimens with a single strong paraprotein band (IgG, IgA, IgM, free κ, or free λ) might be appropriate for this purpose. The controls are applied after electrophoresis either to a preformed well (Helena) or simply to the surface of the gel and allowed to be absorbed into the gel as a "dot" (this manual method and Sebia IFE in Fig. 3). After protein staining, a positive antigen-antibody reaction affirming that the right antibody was used in the right place will be indicated by the development of a ring or dot (Fig. 3). Target concentrations of the positive controls usually are in the range of 5–50 mg/dL (corresponding to 25–250 ng of target protein assuming 0.5 μL is being applied to each control spot). For obvious reasons, performance of controls should be verified before routine use.

27. Since improper antibodies (or antigens in reverse[d] IFE) obviously would cause inaccurate results, great care should be taken to use the "right" antibodies in the "right" position of the gel. In order to minimize or avoid this type of errors, manufacturers of commercial kits for paraprotein detection and identification use differently colored antibody reagents. Color is removed from the gel during the wash steps.
28. Application of antibodies (or antigens in reverse[d] IFE) should be no later than 10 min after completion of electrophoresis.
29. The volume of antiserum (or antigen in reverse[d] IFE) needed for immunofixation varies with the type and character of electrophoretic matrix (e.g., gel vs. membrane, thin vs. thick gels), mode of antiserum application (immersion into the antiserum vs. direct surface overlay, overlay using cellulose acetate membrane or thick filter paper vs. thin lens paper), and, obviously, size of the target area (defined by width and length of the tracks [lanes] and expected distribution of target proteins). Because of the loss due to uptake by the delivery device itself, direct overlay of an electrophoretic track generally requires less antiserum (or antigen in reverse[d] IFE) than overlay via a membrane or filter paper (*see* **Note** 7). Probably, searching for paraproteins represents the far extreme in antiserum volume needs because it requires covering virtually the entire area of serum electrophoretic separation from the anodal end (albumin) to the cathodal end of the normal gamma globulin zone. In fact, in order to assure that no paraproteins are missed, immunofixation (~1 cm) beyond the cathodal end of the normal gamma globulin zone is necessary to detect rare cases of far-cathodal migrating paraprotein variants. Most target proteins separated in the electrophoretic matrix are localized in a relatively short segment of a track and, therefore, require smaller volumes for immunofixation. Antiserum (or, in reverse[d] IFE, antigen) volume needed for immunofixation is variably reported as total per gel (e.g., 0.25 mL for a 10-position gel), total per track (lane) (e.g., 25–80 μL/track [lane]), or volume per unit of surface area (e.g., 10.2 μL/cm^2).
30. In order to speed up and enhance the antigen-antibody reaction, higher than room temperature incubations up to 45 °C for 10–30 min have been suggested and used for paraprotein IFE. Since high temperatures will increasingly denature proteins, heat stability of both antigens and antibodies needs to be considered for establishing optimal incubation with a given system.

31. Unlike membranes or paper strips, IFE antiserum application templates may be reused after washing with mild detergents and, when indicated, disinfecting.
32. Many protocols call for using multiple layers of filter papers and adding more weight (as much as 1 to 1.5 kg) for increasing the efficiency of gel preparation for subsequent staining. Increasing the weight is particularly advantageous for "cleaning up" the gel after immunofixation. On the other hand, the present protocol with a "minimalist" approach shows that even ~one-tenth of the "optimal" weight load is able to generate usable IFE patterns if IgG fractions of the antisera are being used (Figs. 3 and 4). Additional cycle(s) of wash and press dry can be also considered for improving background "cleanup."
33. In case of complex particles such as lipoproteins, each containing several different apolipoproteins, a series of monospecific antibodies targeting the same immunoprecipitate location can be used to" dissect" the various protein constituents (Fig. 5b). Further, in addition to using stains or probes to detect the protein content of immunoprecipitates formed in the electrophoretic matrix, special stains or enzyme reactions can be applied to detect the nonprotein components. Examples include but are not limited to the detection of various lipid components of lipoproteins (Fig. 5c, d). A common lipid stain, Sudan Black, detects mostly triglycerides (Fig. 5c). For more specific identification and quantification of lipids in electrophoretically separated lipoproteins, enzymatic cholesterol, triglyceride, and phospholipid assays have been developed [34–37]. These assays can be readily adapted for the detection of specific lipid components such as cholesterol in immunoprecipitates and allow for the study of co-localization of various apolipoproteins and lipids (Fig. 5d). Using a similar concept, calcium, iron, esterase and oxidase activity, etc. can be also detected in the immunoprecipitates.
34. Although the original intention of IFE was to identify proteins with antigenic property in the electrophoretic media in situ, it was later discovered that diffusion of the immune reactants can generate interpretable patterns in another way as well. When antibodies are applied with a cellulose acetate overlay, protein antigens from the separatory gel also diffuse into the cellulose acetate membrane and form precipitates there. This membrane can be then washed and stained for these immunoprecipitates, resulting in an "immunofixation print" or "immunoprint" [10]. However, it was also observed that, while any cellulose acetate membrane is suitable for applying the antiserum for precipitation in the separating gel, those for generating good immunoprints require certain brands of membranes such as the ones used initially (Separphore III, Gelman Sciences, Ann

Arbor, MI, USA, or Millipore Corp., Bedford, MA, USA). Details of the immunoprint technique have been summarized by Johnson [19].

35. Since IFE is suitable for localization of both discrete protein bands and diffuse zones, in the common patterns of paraproteinemic cases, the monoclonal or oligoclonal bands are present as discrete band(s) against the background of polyclonal Ig's as a diffuse zone in the gamma region. Detection sensitivity of IFE for paraproteins is generally higher in specimens with low background of polyclonal Ig's (e.g., urines with no or low level proteinuria and most CSF specimens). Because the co-occurrence of antigenically similar diffusely migrating and restricted mobility proteins in the same specimen is rather unique to the combination of polyclonal and monoclonal Ig's, similar phenomena are less likely with other proteins.

36. Antigen excess in direct IFE results in increased diffusion and wider band. In extreme cases, this excess is associated with the presence of an unstained area in the center of an immunoprecipitin band, variably described as "central clear spot," "hole," or "doughnut." Rerunning the sample at a higher dilution (i.e., lowering the antigen concentration) will eliminate this prozoning artifact.

37. Because of the increased detectability of immunofixed protein bands with increased protein mass due to the antigen-antibody reaction, proteins occurring at low concentration in the electrophoresed sample may not always be visible in the acid-fixed reference lane. Consequently, if artifactual bands can be ruled out (*see* **Notes 38** and **39**), weak IFE bands in the absence of matching bands in the reference lane should be interpreted as being consistent with low target protein concentrations.

38. Multimeric paraproteins, especially IgM, and immune complexes may occasionally adhere to the gel matrix. During IFE testing for paraproteins, these bands will appear in all five antisera reaction areas of the gel at or around the point of application. In most cases, true IgM or, possibly, other monoclonal bands still can be identified with a marked increase in size and staining intensity where they react with the specific antisera for their heavy chain and light chain. Rerunning the sample after pretreatment with the reducing agent 2-ME or a chaotropic agent (e.g., Fluidil, Sebia) usually eliminates or markedly weakens the artifactual bands and confirms the initial assessment for the type of the monoclonal band.

39. Care should be taken to avoid overconcentration of urine samples because this can cause protein aggregation and consequent decrease of solubility. Consequently, artifactual IFE bands may appear with all anti-Ig antibodies when tested for

paraproteins. Similar phenomena may be also observed in urine with other antibodies and with other type of specimens that require preconcentration. Re-electrophoresis of such samples on dilutions usually eliminates these artifacts.

40. Routine IFE methods, including commercial kits for paraproteins, do not test initially for IgD and IgE heavy chains and free (unbound) κ and λ light chains. However, manufacturers of commercial IFE kits provide antisera for testing for IgD and IgE heavy chains and free κ and free λ light chains as well. There is even a dedicated kit available for testing with a semiautomated method for free light chains in neat urine (Bence-Jones Immunofixation, Interlab Sri). When light chains without corresponding heavy chains are identified in a serum IFE pattern and inappropriate antigen-antibody ratios, deteriorated or low-avidity antisera, or possible errors in antibody application (antibody omitted or wrong) can be ruled out, additional testing with antisera against free κ and free λ light chains should be carried out. If these are negative, monospecific antibodies to IgD and IgE heavy chains should be used for identification of rare cases of the respective monoclonal Ig's. If even these tests are negative, then the possibility of "unreactive" light chains needs to be entertained [38–40]. Some IgA and IgD paraproteins have been found to have either a configurational change in their light chain or an enzymic cleavage in their molecule which prevents the precipitation reaction with anti-light chain antisera, hence the name "unreactive" light chains [38–40]. In such very rare cases of immunologically inaccessible light chains, it is often necessary to isolate the abnormal Ig and use special techniques for its proper characterization, thereby ruling out true heavy chain disease [38–40]. Since light chains without corresponding heavy chains are relatively common in urine IFE, retesting the samples with anti-free κ and anti-free λ usually is sufficient for diagnosis.

References

1. Afonso E (1964) Quantitative immunoelectrophoresis of serum proteins. Clin Chim Acta 10:114–122
2. Wilson AT (1964) Direct immunoelectrophoresis. J Immunol 92:31–34
3. Alper CA, Johnson AM (1969 and 1993) Immunofixation electrophoresis: a technique for the study of protein polymorphism. Vox Sang 17:445–452, Vox Sang 65:76
4. Johnson AM, Cleve H, Alper C (1975) Variants of the group-specific component system as demonstrated by immunofixation electrophoresis. Report of a new variant, Gc Boston (Gc B). Am J Hum Genet 27:728–736
5. Johnson AM (1976) Genetic typing of alpha1-antitrypsin by immunofixation electrophoresis, identification of subtypes of Pi M. J Lab Clin Med 87:152–163
6. Lieberman J, Gaidulis L (1976) Simplified alpha1-antitrypsin phenotyping by immunofixation of acid-starch gels. J Lab Clin Med 87:710–716
7. Grunbaum BW, Zajac PL (1977) Rapid phenotyping of the group specific component by

immunofixation on cellulose acetate. J Forensic Sci 22:586–589
8. Cleve H, Constans J, Berg S et al (1981) Gc revisited: six further Gc-phenotypes delineated by isoelectric focusing and by polyacrylamide gel electrophoresis. Hum Genet 57:312–316
9. Mauff G, Hummel K, Pulverer G (1975) Properdin factor B (glycine-rich beta-glycoprotein or C3 proactivator)-polymorphism: genetic and biochemical aspects. First application to paternity cases. Z Immunitatsforsch Exp Klin Immunol 150:327–338
10. Arnaud P, Wilson GB, Koistinen J et al (1977) Immunofixation after electrofocusing: improved method for specific detection of serum proteins with determination of isoelectric points. I. Immunofixation print technique for detection of alpha-1-protease inhibitor. J Immunol Methods 16:221–231
11. Baumstark JS (1978) Quantitative immunofixation of proteins following zone electrophoresis in agarose gel: application to the determination of the stoichiometry of the alpha1-antitrypsin-elastase interaction. J Immunol Methods 23:79–89
12. Morrison R, Noppinger K, Brown MG Jr (1985) Immunofixation of complement component C3 phenotypes in bloodstains after cellulose acetate electrophoresis. J Forensic Sci 30:1221–1225
13. Cawley LP, Minard BJ, Tourtellotte WW et al (1976) Immunofixation electrophoretic techniques applied to identification of proteins in serum and cerebrospinal fluid. Clin Chem 22:1262–1268
14. Chang CH, Inglis NR (1979) Convenient immunofixation electrophoresis on cellulose acetate membrane. Clin Chim Acta 65:91–97
15. Ritchie RF, Smith R (1976) Immunofixation. I. General principles and application to agarose gel electrophoresis. Clin Chem 22:497–499
16. Ritchie RF, Smith R (1976) Immunofixation. II. Application to typing of alpha1-antitrypsin at acid pH. Clin Chem 22:1735–1737
17. Ritchie RF, Smith R (1976) Immunofixation. III. Application to the study of monoclonal proteins. Clin Chem 22:1982–1985
18. Kyle RA, Greipp PR (1978) The laboratory investigation of monoclonal gammopathies. Mayo Clin Proc 53:719–739
19. Johnson AM (1982) Immunofixation electrophoresis and electrofocusing. Clin Chem 28:1797–1800
20. Keren DF (1999) Procedures for the evaluation of monoclonal immunoglobulins. Arch Pathol Lab Med 123:126–132
21. Freedman MS, Thompson EJ, Deisenhammer F et al (2005) Recommended standard of cerebrospinal fluid analysis in the diagnosis of multiple sclerosis: a consensus statement. Arch Neurol 62:865–870
22. Laurenzi MA, Link H (1978) Localization of the immunoglobulins G, A and M, beta-trace protein and gamma-trace protein on isoelectric focusing of serum and cerebrospinal fluid by immunofixation. Acta Neurol Scand 58:141–147
23. Bateman N, Jones NS (2000) Rhinorrhoea feigning cerebrospinal fluid leak: nine illustrative cases. J Laryngol Otol 114:462–464
24. Bleier BS, Debnath I, O'Connell BP et al (2011) Preliminary study on the stability of beta-2 transferrin in extracorporeal cerebrospinal fluid. Otolaryngol Head Neck Surg 144:101–103
25. Ottó S (1982) Reversed immunofixation agar gel electrophoresis. Immunol Lett 4:85–86
26. Mehta PD, Patrick BA, Thormar H et al (1982) Oligoclonal IgG bands with and without measles antibody activity in sera of patients with subacute sclerosing panencephalitis (SSPE). J Immunol 129:1983–1985
27. Storstein A, Monstad SE, Honnorat J et al (2004) Paraneoplastic antibodies detected by isoelectric focusing of cerebrospinal fluid and serum. J Neuroimmunol 155:150–154
28. Stich O, Rauer S (2007) Antigen-specific oligoclonal bands in cerebrospinal fluid and serum from patients with anti-amphiphysin- and anti-CV2/CRMP5 associated paraneoplastic neurological syndromes. Eur J Neurol 14:650–653
29. Bossuyt X, Bogaerts A, Schiettekatte G et al (1998) Detection and classification of paraproteins by capillary immunofixation/subtraction. Clin Chem 44:760–764
30. Litwin CM, Anderson SK, Philipps G et al (1999) Comparison of capillary zone and immunosubtraction with agarose gel and immunofixation electrophoresis for detecting and identifying monoclonal gammopathies. Am J Clin Pathol 112:411–417
31. Yang Z, Harrison K, Park YA et al (2007) Performance of the Sebia CAPILLARYS 2 for detection and immunotyping of serum monoclonal paraproteins. Am J Clin Pathol 128:293–299
32. Paquette DM, Banks PR (2001) Detection of specific antibodies using immunosubtraction and capillary electrophoresis instrumentation. Electrophoresis 22:2391–2397
33. Pascali E, Pezzoli A, Chiarandini A (1982) Immunofixation: application to the

identification of "difficult" monoclonal components. Clin Chem 28:1404–1405

34. Leglise D, Menez JF, Person B et al (1982) A detailed lipidograph: enzymatic determination of cholesterol, phospholipids and glycerides in plasma lipoprotein after a cellulose acetate electrophoretic procedure (author's transl). [Article in French]. Clin Chim Acta 118:265–277
35. Nauck M, Winkler K, März W et al (1995) Quantitative determination of high-, low-, and very-low-density lipoproteins and lipoprotein(a) by agarose gel electrophoresis and enzymatic cholesterol staining. Clin Chem 41:1761–1767
36. Winkler K, Nauck M, Siekmeier R et al (1995) Determination of triglycerides in lipoproteins separated by agarose gel electrophoresis. J Lipid Res 36:839–847
37. García-Sánchez C, Torres-Tamayo M, Juárez-Meavepeña M et al (2011) Lipid plasma concentrations of HDL subclasses determined by enzymatic staining on polyacrylamide electrophoresis gels in children with metabolic syndrome. Clin Chim Acta 412:292–298
38. Cejka J, Kithier K (1979) IgD myeloma protein with "unreactive" light chain determinants. Clin Chem 25:1495–1498
39. Netto D, Vladutiu AO (1981) A simple technique for identification of "unreactive" light chains of immunoglobulins. Clin Chim Acta 116:253–260
40. Rabhi H, Ghaffor M, Abbadi MC (1989) Spontaneous enzymatic cleavage of IgD myeloma protein giving a pattern of delta heavy chain disease. Arch Inst Pasteur Alger 57:135–140

Chapter 18

Electrophoretic Separation of Very Large Molecular Weight Proteins in SDS Agarose

Marion L. Greaser and Chad M. Warren

Abstract

Very large proteins (subunit sizes, >200 kDa) are difficult to electrophoretically separate on polyacrylamide gels. A SDS vertical agarose gel system has been developed that has vastly improved resolving power for very large proteins. Proteins with molecular masses between 200 and 4000 kDa can be clearly separated. Inclusion of a reducing agent in the upper reservoir buffer and use of a large pore-sized agarose have been found to be key technical procedures for obtaining optimum protein migration and resolution.

Key words SeaKem Gold agarose, Titin, DATD, Large proteins

1 Introduction

Laemmli SDS (sodium dodecyl sulfate) polyacrylamide systems perform poorly in resolving proteins with sizes greater than 200–500 kDa [1]. Protein migration in SDS gels is linear with the log of the molecular weight [2], so the larger the protein, the more poorly it is separated from other big proteins. Some workers have attempted to improve protein separation by using very low concentration acrylamide gels [3], acrylamide mixed with agarose [4], or acrylamide gradients [5]. However, low concentration acrylamide gels are difficult to use because of their mechanical fragility and tendency to distort during handling. A new electrophoresis system using SDS agarose for protein electrophoresis and blotting has been described [6]. Although the system was developed and has been most extensively used with muscle samples, it has also been applied in work with other large proteins such as nesprin/SNAP (subunit size 1100 kDa) [7], von Willebrand factor multimers [8–12], and huntingtin aggregates [13]. Cross-linked polymers of peanut proteins [14, 15] and casein [16] have been examined. In addition agarose has served as a component of a two-dimensional proteomics protocol [17]. An example showing the resolution for

Biji T. Kurien and R. Hal Scofield (eds.), *Electrophoretic Separation of Proteins: Methods and Protocols*, Methods in Molecular Biology, vol. 1855, https://doi.org/10.1007/978-1-4939-8793-1_18,

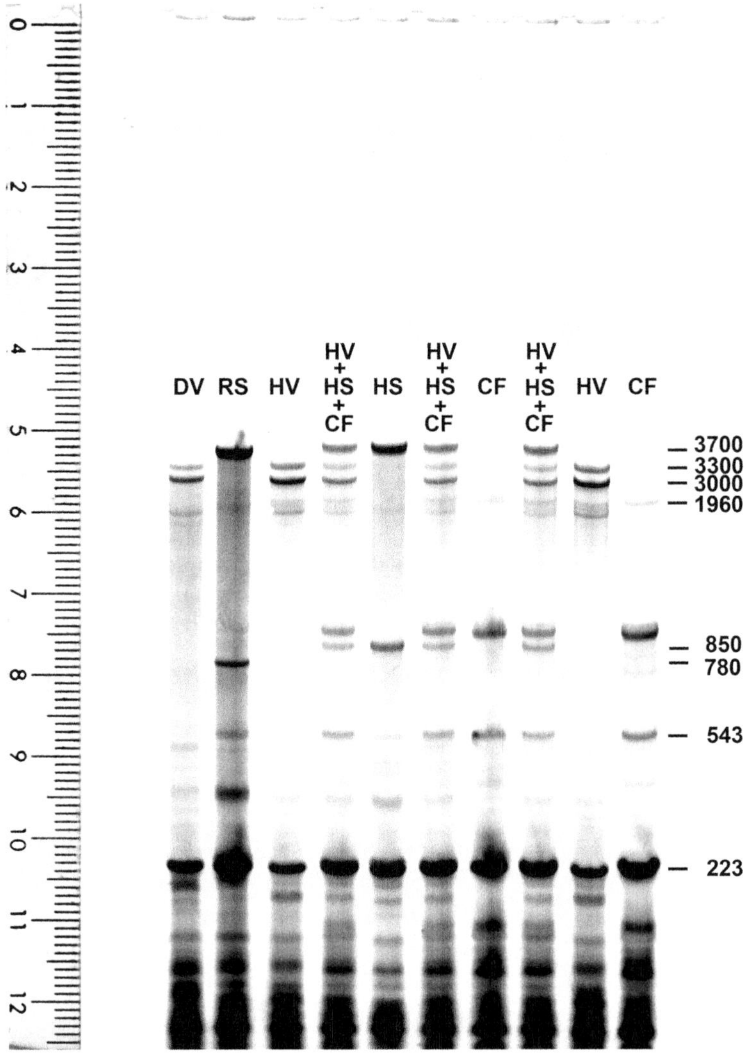

Fig. 1 SDS 1% agarose gel stained with silver. A centimeter ruler is shown on the left and the sizes of the various protein bands in kDa are listed on the right. Abbreviations: *DV* dog ventricle, *RS* rat soleus, *HV* human ventricle, *HS* human soleus, *CF* crayfish claw muscle. Human soleus titin is 3700 kDa, and the human ventricle has two titin bands of 3300 and 3000 kDa. The bands at 780 and 850 kDa in the skeletal muscle samples are rat and human nebulin, respectively. The myosin heavy chain is 223 kDa

several muscle samples containing large proteins is shown in Fig. 1. Migration distance shows a linear relationship with the log of the molecular weight [6]. This system also allows more quantitative transfer of proteins from the gel for Western blots and achieves much higher reproducibility than can be obtained with methods using low percentage acrylamide.

2 Materials

2.1 Apparatus

1. SE 600 Slab Gel Unit with 16 × 18 cm glass plates (Hoefer, Holliston, MA) or a similar commercial gel unit (*see* **Note 1**).
2. 65 °C oven.
3. A constant current power supply.
4. Circulating cooler.

2.2 Stock Solutions

1. Acrylamide gel for plug: 38.5% acrylamide. Weigh 37.5 g of acrylamide and 1 g DATD (N,N′-diallyl-tartardiamide) (Biorad, Hercules, CA) into a beaker, add about 50 mL of water, stir till dissolved, and dilute to 100 mL. Filter through a 0.45 micron filter (such as a Millex-HA, Millipore Corporation, Billerica, MA). Store in a brown bottle in the cold room (4 °C). Danger! Avoid skin contact.
2. Reservoir and agarose gel buffer concentrate (5×): 0.25 M Trizma base, 1.92 M glycine, 0.5% SDS. Store at room temperature.
3. Ammonium persulfate: Prepare a 100 mg/mL solution in water; store frozen in 0.5 mL aliquots (stable indefinitely at −20 °C).
4. Sample buffer: 8 M urea, 2 M thiourea, 0.05 M Tris–HCl (pH 6.8), 75 mM DTT, 3% SDS, 0.05% bromophenol blue (adapted from Ref. 18). (Dissolve urea and thiourea and treat with mixed bed resin to remove ionic constituents; then add remaining ingredients. Store at −20 °C).
5. 50% v/v glycerol.
6. Acrylamide plug solution: In a 15 mL plastic beaker, add 1.924 mL deionized water, 1.7 mL 50% glycerol, 2.12 mL 3 M Tris–HCl (pH 9.3), 2.72 mL acrylamide plug solution (38.5%), 24 μL 10% ammonium persulfate, and 13 μL TEMED (tetramethylethylenediamine). Mix by gently pipetting a few times. Prepare complete solution only after gel plates are assembled.
7. Agarose gel solution: Weigh 0.8 g of SeaKem Gold agarose (Lonza Group Ltd., Basel, Switzerland) into a 600 mL beaker (*see* **Note 2**). To a 100 mL graduated cylinder, add 48 mL of 50% v/v glycerol (*see* **Note 3**) and 16 mL 5× electrophoresis buffer, and bring volume up to 80 mL with deionized water. Place parafilm over top of the graduated cylinder, mix by inverting a few times, and pour solution into the 600 mL beaker containing the agarose. Place saran wrap over top of beaker and poke a few holes in the saran wrap. Weigh beaker with contents. Place beaker in a microwave oven along with a separate beaker of deionized water. Heat for a total of 2 min

(stop every 30 s to swirl—protect hand with an insulated glove) (*see* **Note 4**). Prepare agarose solution after the polyacrylamide plug has been inserted.

8. Coomassie blue protein stain:
   (a) Stain: Dissolve 0.05% (w/v) Coomassie blue R-250 in 50% (v/v) methanol-10% (v/v) acetic acid. Store at room temperature in a closed container (*see* **Note 5**).
   (b) Destain: 10% (v/v) methanol-7.5%(v/v) acetic acid.
9. Silver protein stain:
   (a) Fixing solution: 50% v/v methanol-5% v/v glycerol-12% v/v acetic acid. Prepare fixing solution by adding, in the following order, 500 mL methanol, 50 mL glycerol, and water to 880 mL mark in a graduated cylinder, mix, and then add 120 mL acetic acid and mix. This makes enough for two agarose gels.
   (b) Potassium ferrocyanide solution: potassium ferrocyanide (20 g/L) (500 mL per gel).
   (c) Staining solutions: Solution A – dissolve 25 g sodium carbonate in deionized water and bring up to 500 mL (prepare solution A in 1000 mL beaker or larger).

Solution B—dissolve 5 g silicon tungstic acid, 1 g ammonium nitrate, 1 g silver nitrate, and 3.35 mL 37% formaldehyde in deionized water; bring final volume to 500 mL.

3 Methods

3.1 Gel Preparation

1. Volumes listed will provide enough solution for two 16 × 18 cm gels with 1.5 mm spacers.
2. Clean plates and spacers with soap, and rinse with distilled water and finally with ethanol.
3. Assemble gel plates. Place plate on clean bench top. Place spacers hanging half the way off each side of plate. Place second plate on top. Stand the plates and place one side into the clamp. Align spacer with side of plates and clamp and push spacer down so that bottom is flush with the glass plates (top buffer will leak if spacers are not flush with plates).
4. Pour acrylamide plugs in bottom of gel plate assembly (*see* **Note 6**). Mix by gently pipetting a few times. Immediately add 2.5 mL to each gel assembly. Add a small amount of water on top of each plug to level the upper surface and provide an oxygen barrier. Allow gel to polymerize for 20–30 min. Drain off water layer by inverting gel plate assembly on a paper towel.

5. Place assembly, 20 lane sample combs, and 60 mL plastic syringe in a 65 °C oven for 10 min (*see* **Note 7**).
6. Allow freshly prepared agarose to cool for a few minutes at room temperature. Re-weigh, and add sufficient heated deionized water to replace that lost by evaporation.
7. Draw up about 40 mL of agarose in the pre-warmed 60 mL Luer-Lock syringe, and pour each gel slowly until it just overflows the top of the plates. Try to avoid formation of bubbles (if bubbles are present, allow them to migrate to the top of the gel and remove them with the back of the sample comb by pinching off the top bead of agarose. Insert sample combs and allow unit to cool at room temperature for about 45 min (*see* **Note 8**).

3.2 Electrophoresis Setup and Sample Loading

1. Add 4 L of buffer to lower chamber (3200 mL deionized water plus 800 mL 5× electrophoresis buffer). Start cooling unit and stir bar (gels run at 6 °C).
2. Prepare 600 mL upper chamber electrophoresis buffer (same concentration as lower chamber buffer). Add 2-mercaptoethanol (final concentration of 10 mM) [19] (*see* **Note 9**). Buffer will be poured into top chamber after samples are loaded and assembly placed in unit.
3. Take combs out of gels by bending them back and forth to detach from gel, and slowly pull them up. Pour a small amount of upper chamber buffer into a 15 mL beaker and pipette buffer into first and last wells (the rest will fill over). Add buffer to remove any trapped bubbles. Insert pipette tip to deposit sample in bottom of the sample well. Skip the first and last lanes (*see* **Note 10**).
4. Running gels. Once samples are loaded, put upper chamber on the assembly. Pour upper chamber buffer into upper chamber from corners (don't pour buffer directly over wells). Place lid on unit, and connect to power supply. Turn electrophoresis unit on and run at 30 mAmps (2 gels) for 3 h.

3.3 Staining and Destaining

1. After the tracking dye reaches the bottom of the acrylamide plug, turn off the power and disassemble the plates. Cut off sample wells and acrylamide plug and discard.
2. Gels can be stained with Coomassie blue R-250 or silver (*see* **steps 3–11** below) depending on the sensitivity needed. Place agarose gel in a plastic or glass dish with the Coomassie stain and gently agitate for 1 h. Discard stain, and soak gel in methanol-acetic acid destain with gentle shaking until the background clears (usually overnight). Include a Chemwipe in the destain to trap the Coomassie.

3. For more sensitive protein detection, use a silver staining procedure adapted from Ref. [20].
4. Add 500 mL of 50% v/v methanol-10% v/v acetic acid fixing solution to a pre-cleaned polypropylene bin (*see* **Note 11**), and slide agarose gel into the solution. Cover and gently agitate on a shaker for 1 h. Discard fixing solution, and then transfer bin with gel to a 37 °C oven overnight.
5. Add 500 mL deionized water to gel, agitate gently for 20 min, and discard solution.
6. Repeat **step 5** two more times (three washes, 20 min each).
7. Add 500 mL potassium ferrocyanide solution to each gel and shake for 5 min, and discard solution.
8. Add 500 mL of deionized water to each gel and shake for 5 min, and discard water.
9. Repeat gel wash two more times with 500 mL deionized water for 5 min (total of three washes, 5 min each).
10. Slowly add solution B to a stirring solution A just prior to staining. Shake gels in staining solution (500 mL/gel) until bands appear (5–10 min; do not overstain). Decant staining solution, and add 500 mL of 1% v/v acetic acid to each gel, and shake for 5 min to stop staining. Decant acetic acid and wash gels with 500 mL deionized water for 5 min.
11. Dry gels between two sheets of wet Mylar, and add a couple ml of glycerol to make the sandwich less brittle.

4 Notes

1. The agarose gel procedure works equally well with small format gels (i.e., 8 × 10 cm).
2. It is essential to use SeaKem Gold agarose for optimal migration of high molecular weight proteins. This type has large pore size and excellent mechanical stability. Other types of agarose may be used, but the protein mobility will be significantly reduced.
3. Glycerol is included in the mixture to increase the solution viscosity inside the gel and thus sharpen the protein bands.
4. Periodic swirling during the heating step eliminates non-hydrated agarose granules in the final gel.
5. Dissolving the Coomassie in methanol first and then adding the water and acetic acid reduce problems with dye aggregates.
6. The acrylamide plug is used to prevent the agarose from slipping out of the vertical gel plate assembly. Use of DATD as the cross-linker provides an acrylamide that sticks better to the

glass plates than if a conventional bisacrylamide cross-linker is used. Plugs can be poured a day before making the agarose gel (place tape or parafilm over the top of the plates to prevent drying and store in cold room).

7. Preheating the glass plate assembly, well comb, and syringe prevents premature agarose gelling when the solution touches the colder surfaces. In addition the plates are less likely to crack during pouring if they are closer to the temperature of the hot agarose.
8. Sample combs should extend no longer than 1 cm into agarose; otherwise they may be difficult to remove. Gels can be used right away or stored overnight in a cold room.
9. The disulfide bond formation of large proteins during electrophoresis also retards their migration and may lead to smearing (*see* Ref. 21). Thus inclusion of 2-mercaptoethanol in the upper buffer improves resolution of high molecular weight proteins. Alternatively, protein can be alkylated to prevent disulfide bond formation [21].
10. Conventional sample buffers may not be dense enough for the sample to stay at the bottom of the well. If necessary add additional glycerol (up to 30% v/v final concentration) to increase sample density.
11. Bins may be cleaned with a 50% nitric acid solution to remove silver deposits and then given a final rinse with water.

Acknowledgments

This work was supported by the College of Agricultural and Life Sciences, University of Wisconsin-Madison, and from grants MLG-NIH HL77196 and Hatch NC1131.

References

1. Laemmli UK (1970) Cleavage of structural proteins during the assembly of the head of bacteriophage T4. Nature 227:680–685
2. Weber K, Osborn M (1969) The reliability of molecular weight determinations by dodecyl sulfate-polyacrylamide gel electrophoresis. J Biol Chem 244:4406–4412
3. Granzier H, Wang K (1993) Gel electrophoresis of giant proteins: solubilization and silver-staining of titin and nebulin from single muscle fiber segments. Electrophoresis 14:56–64
4. Tatsumi R, Hattori A (1995) Detection of giant myofibrillar proteins connectin and nebulin by electrophoresis in 2% polyacrylamide slab gels strengthened with agarose. Anal Biochem 224:28–31
5. Cazorla O, Freiburg A, Helmes M et al (2000) Differential expression of cardiac titin isoforms and modulation of cellular stiffness. Circ Res 86:59–67
6. Warren CM, Krzesinski PR, Greaser ML (2003) Vertical agarose gel electrophoresis and electroblotting of high-molecular-weight proteins. Electrophoresis 24:1695–1702
7. Razafsky D, Hodzic D (2015) A variant of Nesprin1 giant devoid of KASH domain underlies the molecular etiology of autosomal recessive cerebellar ataxia type I. Neurobiol Dis 15:57–67

8. Wu JJ, Fujikawa K, McMullen BA et al (2006) Characterization of a core binding site for ADAMTS-13 in the A2 domain of von Willebrand factor. Proc Natl Acad Sci U S A 103:18470–18474
9. Ott HW, Griesmacher A, Schnapka-Koepf M et al (2010) Analysis of von Willebrand Factor multimers by simultaneous high- and low-resolution vertical SDS-agarose gel electrophoresis and Cy5-labeled antibody high-sensitivity fluorescence detection. Am J Clin Pathol 133:322–330
10. Yamashita K, Yagi H, Hayakawa M et al (2016) Rapid restoration of thrombus formation and high-molecular-weight von Willebrand factor multimers in patients with severe aortic stenosis after valve replacement. J Atheroscler Thromb 23:1150–1158
11. Tsujii N, Nogami K, Yoshizawa H et al (2016) Influenza-associated thrombotic microangiopathy with unbalanced von Willebrand factor and a disintegrin and metalloproteinase with a thrombospondin type 1 motif, member 13 levels in a heterozygous protein S-deficient boy. Pediatr Int 58:926–929
12. Nishigori N, Matsumoto M, Koyama F et al (2015) von Willebrand factor-rich platelet thrombi in the liver cause sinusoidal obstruction syndrome following oxaliplatin-based chemotherapy. PLoS One 10:e0143136
13. Hoffner G, Island ML, Djian P (2005) Purification of neuronal inclusions of patients with Huntington's disease reveals a broad range of N-terminal fragments of expanded huntingtin and insoluble polymers. J Neurochem 95:125–136
14. Radosavljevic J, Nordlund E, Mihajlovic L et al (2014) Sensitizing potential of enzymatically cross-linked peanut proteins in a mouse model of peanut allergy. Mol Nutr Food Res 58:635–646
15. Mihajlovic L, Radosavljevic J, Nordlund E et al (2016) Peanut protein structure, polyphenol content and immune response to peanut proteins in vivo are modulated by laccase. Food Funct 7:2357–2366
16. Stanic D, Monogioudi E, Dilek E et al (2010) Digestibility and allergenicity assessment of enzymatically crosslinked beta-casein. Mol Nutr Food Res 54:1273–1284
17. Oh-Ishi M, Maeda T (2007) Disease proteomics of high-molecular-mass proteins by two-dimensional gel electrophoresis with agarose gels in the first dimension (agarose 2-DE). J Chromat B-Anal Tech Biomed Life Sci 849:211–222
18. Yates LD, Greaser ML (1983) Quantitative determination of myosin and actin in rabbit skeletal muscle. J Mol Biol 168:123–141
19. Fritz JD, Swartz DR, Greaser ML (1989) Factors affecting polyacrylamide gel electrophoresis and electroblotting of high-molecular-weight myofibrillar proteins. Anal Biochem 180:205–210
20. Peats S (1984) Quantitation of protein and DNA in silver-stained agarose gels. Anal Biochem 140:178–182
21. Sechi S, Chait BT (1998) Modification of cysteine residues by alkylation. A tool in peptide mapping and protein identification. Anal Chem 70:5150–5158

Chapter 19

Increase in Local Protein Concentration by Field-Inversion Gel Electrophoresis

Henghang Tsai and Hon-Chiu Eastwood Leung

Abstract

Proteins that migrate through cross-linked polyacrylamide gels (PAGs) under the influence of a constant electric field experience negative factors, such as diffusion and nonspecific trapping in the gel matrix. These negative factors reduce protein concentrations within a defined gel volume with increasing migration distance and, therefore, decrease protein recovery efficiency. Here, we describe the enhancement of protein separation efficiency for up to twofold in conventional one-dimensional PAG electrophoresis (1D PAGE), two-dimensional (2D) PAGE, and native PAGE by implementing pulses of inverted electric field during gel electrophoresis.

Key words Field-inversion gel electrophoresis, Pulsed-field gel electrophoresis, Two-dimensional polyacrylamide gel electrophoresis, Forward pulse time, Reverse pulse time, Separation efficiency

1 Introduction

Sodium dodecyl sulfate-polyacrylamide gel electrophoresis (SDS-PAGE) is an indispensable technique in protein separation. This technique has only changed marginally over the past three decades [1]. Despite its popularity, SDS-PAGE, as well as native PAGE for protein separation, suffers the basic limitation of band broadening by diffusion and trapping of biomolecules in gel matrices. Nevertheless, protein separation by SDS-PAGE that interfaces with mass spectrometry (MS) has currently emerged as the method of choice in the forefront of proteomics. Thus, new tools for upstream gel electrophoresis that can improve protein separation efficiency and recovery will possibly lead to new discoveries in downstream processes.

Pulsed-field gel electrophoresis (PFGE) is an elegant, simple, and universally accepted technique for the separation of large DNA molecules [2]. Several modifications of PFGE with different electrophoretic configurations exist [3–5]. One modification is the

Biji T. Kurien and R. Hal Scofield (eds.), *Electrophoretic Separation of Proteins: Methods and Protocols*, Methods in Molecular Biology, vol. 1855, https://doi.org/10.1007/978-1-4939-8793-1_19,

field-inversion gel electrophoresis (FIGE) in which net forward molecular migration is achieved by either employing a longer net forward field time or a higher forward field strength compared with the reverse direction [4]. Among the various pulsed-field gel electrophoresis techniques, FIGE is likely the easiest to perform with minimal special equipment that generates a highly uniform electric field across the gel [4]. Systematic investigations of FIGE on the separation of DNA molecules in agarose gels were previously reported [6–10]. However, published studies of the use of FIGE for protein separation are still lacking. Another modification is the orthogonal-field-alternation gel electrophoresis (OFAGE) techniques [3]. Both of these methods were optimized to maximize separation efficiency (band width in the dimension of separation) and selectivity (distance between the center of two bands) for large DNA molecules.

Attempts were made to enhance the separation of proteins by means of PFGE [11]. Subsequent applications of PFGE were used to resolve either specific model protein species [12, 13] or high molecular weight muscle myosin heavy chain isoforms [14]. However, these approaches were limited to the application of alternating cycles of on-and-off electric fields across a slab gel. This approach inevitably allowed diffusion to occur during the off times. Unwanted band broadening as a result of diffusion compromises general separation efficiency.

There are currently three models to explain molecular migration during PAGE: (I) The extended Ogston (EO) model assumes an overall sphere-like conformation for native or small proteins in which mobility is a function of available gel pores in a regular lattice fashion ([15]; II). The reptation model assumes that molecules go through a rather disordered matrix, such as polyacrylamide [16], accounting for the snake-like movement of pearl necklace-shaped polymers, such as protein-SDS complexes. These two models only apply to polypeptides within a certain molecular weight (MW) range at a given cross-linked polyacrylamide concentration. Any deviation from this linearity implies a change or transition in molecular shape (i.e., the radius, and net charge III). The door-corridor (DC) model explains the behavior of polypeptides above a critical MW where the mobility of a protein becomes independent of the cross-linked polyacrylamide concentration [17]. Effective trapping of migrating molecules by the matrix predominates in this model, and electrokinetic energy is required to overcome the trapping effect [17].

We report here the engineering of a simple field-inversion device (Fig. 1) and an extensive study of protein recovery enhancement using FIGE. Thus, in general FIGE enhances protein separation by improving local protein concentrations during SDS-PAGE or under native gel conditions. The increased local protein concentration thus improves the observable intensity of protein species in

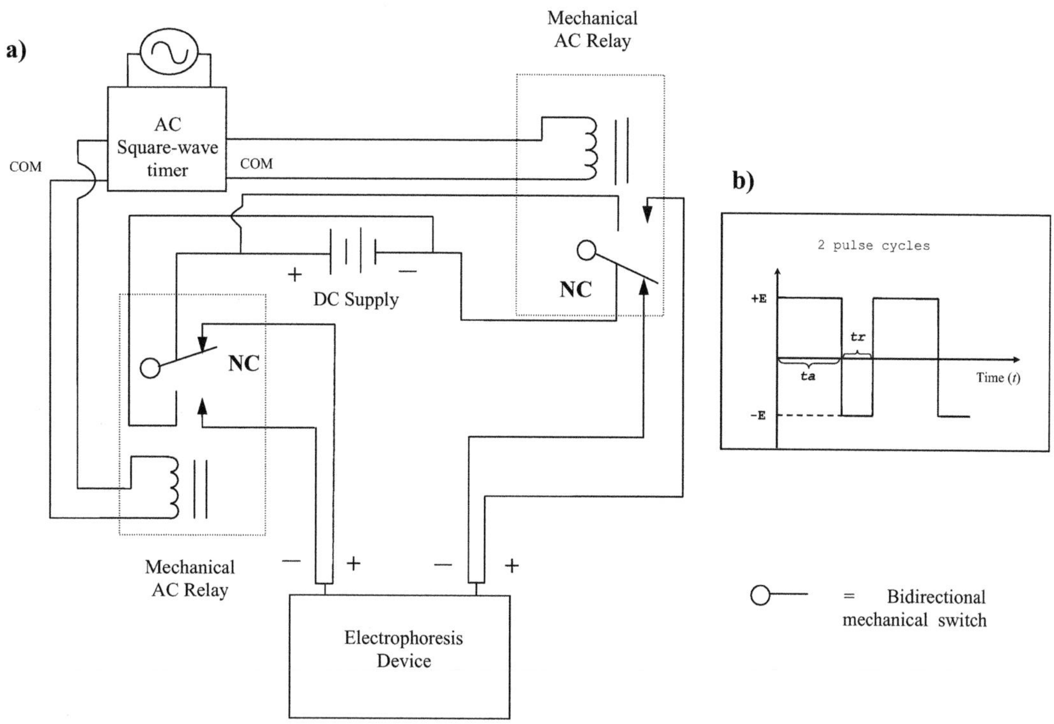

Fig. 1 Electric circuit of pulse generator and field diagram. (**a**) Electric circuit diagram for generating positive (+E) and negative (−E) square-wave electric fields during field inversion experiments. Direct current (DC) supply is from an external source. NC, normally closed switch; AC relay, alternate current relay; Com, common outlet. Inset (**b**) shows a profile of the electric field during a typical FIGE experiment where +E = −E and *ta* (forward field time) is longer than *tr* (reverse field time) (reproduced from [18] with permission from BioMed Central)

PAGs and also improves the success of downstream peptide sequencing using MS. Taken together, FIGE can be used to complement constant field gel electrophoresis for better protein separation and recovery.

2 Materials

All solutions used should be made by using milliQ water; if possible, filter all solution, and store in appropriate DURAN® bottles (Schott, Elmsford, NY, USA) unless otherwise mentioned. It is also important to ensure proper pH prior to gel casting and electrophoresis. If protein sequencing using mass spectrometry is to be conducted after electrophoresis, all solutions used should be in molecular biology grade (DNase, RNase, and protease-free). All experiments, if possible, are to be conducted in the fume hood. We preferred the use of precast gels if protein sequencing is involved in subsequent steps. Staining of gels can be done using Coomassie blue staining such as the SimplyBlue™ SafeStain (Invitrogen Inc., Carlsbad, CA, USA) or mass spectrometry-compatible silver stain, such as the SilverQuest™ silver staining kit (Invitrogen Inc., Carlsbad, CA, USA).

2.1 1D SDS Polyacrylamide Gel Components

1. Resolving gel buffer: 1.5 M Tris–HCl, pH 8.8. Weigh 181.7 g Tris-base and transfer to a 1 L glass measuring cylinder. Add milliQ water to a volume of 900 mL. Mix and adjust pH to 8.8 with HCl. Make up to 1 L with autoclaved milliQ water and store at 4 °C.
2. Stacking gel buffer: 0.5 M Tris–HCl, pH 6.8. Weigh 60.6 g Tris-base and prepare to 1 L solution as in **step 1** and store at 4 °C.
3. Thirty percent acrylamide/Bis solution (37.5:1 acrylamide/Bis): Weigh 146.1 g acrylamide monomer and 3.9 g Bis-acrylamide, and transfer to cylinder. Add milliQ water to a volume of 500 mL. Mix to completely dissolve. Store at 4 °C.
4. Ten percent SDS solution: Dissolve 10 g SDS in 100 mL milliQ water. Store at room temperature.
5. Ten percent ammonium persulfate: Add 1 g ammonium persulfate to 10 mL milliQ water. Store at −20 °C (*see* **Note 1**).
6. N,N,N,N'-Tetramethyl-ethylenediamine (TEMED). Store at 4 °C.
7. Running buffer (5×): Weigh 15 g Tris-base; 72 g glycine; 5 g SDS. Add milliQ water to 1 L; it is not necessary to adjust the pH of the running buffer. It should have a pH of 8.3. Store at room temperature.
8. Laemmli sample buffer (4×): 0.3 M Tris–HCl (pH 6.8), 8% SDS, 40% glycerol, and trace amount of bromophenol blue (BPB). Store at room temperature.
9. 2 M dithiothreitol (DTT) solution: Dissolve 30.8 mg DTT in 100 μL milliQ water (*see* **Note 2**).
10. Protein standards.

2.2 2D PAGE Components

1. Rehydration buffer: 7 M urea, 2 M thiourea, 4% CHAPS, and trace amount of bromophenol blue. Weigh 10.5 g of urea, 3.8 g of thiourea, 10 g of CHAPS, 500 μL of IPG buffer (ampholytes), and 154 mg of DTT to a 100 mL glass cylinder. Make up to 25 mL with milliQ water and dissolve all components thoroughly. Store at −20 °C.
2. SDS equilibration buffer solution: Weigh 72.1 g of urea, 69 mL of glycerol, 10 mL of 1.5 M Tris buffer pH 8.8, 4 g of SDS, and trace amount of BPB to cylinder. Make up to 200 mL. Store at −20 °C.

3. Electrophoresis buffer: 1.5 M Tris buffer pH 8.8—dissolve 181.7 g of Tris-base in 1 L of milliQ water. Adjust pH to 8.8 with HCl. Store at 4 °C.
4. Agarose sealing solution: Weigh 0.5 g of agarose in 100 mL electrophoresis buffer (*see* **Note 3**).
5. SDS running buffer (1×): Weigh 3.03 g of Tris-base, 14.4 g of glycine, and 1 g of SDS in 1 L cylinder, and top up to 1 L with milliQ water.
6. Ettan™ IPGphor™ II Isoelectric Focusing System (GE Healthcare, Piscataway, NJ, USA).

2.3 Native PAGE Components

1. 40% acrylamide/Bis solution (37.5:1): Weigh 100 g of acrylamide and 2.65 g of Bis-acrylamide in 250 mL of milliQ water. Store at 4 °C (*see* **Note 4**).
2. Resolving gel buffer (4×): Weigh 36.3 g of Tris and add milliQ water to dissolve. Adjust pH to 8.8 with HCl. Add milliQ water to a final volume of 200 mL. Store at 4 °C.
3. Stacking gel buffer (4×): Weigh 15.1 g of Tris and add milliQ water to dissolve. Adjust pH to 6.8 with HCl and make up final volume to 50 mL. Store at 4 °C.
4. Ammonium persulfate (10%): Weigh 1.0 g of ammonium persulfate in a 15 mL Falcon tube, and add milliQ water to a final volume of 10 mL. Store at 4 °C.
5. Electrophoresis buffer (1×): Weigh 28.8 g of glycine and 6.0 g of Tris in a 2 L cylinder, and add milliQ water to obtain a final volume of 2 L.
6. Water-saturated *n*-butanol (55 mL): Mix 50 mL of *n*-butanol with 5 mL of milliQ water. Store at room temperature (*see* **Note 5**).
7. 2× sample buffer (10 mL): Mix 2.5 mL of 4× stacking gel buffer (pH 6.8) with 0.4 mL of glycerol, and trace amount of BPB with 5.5 mL of milliQ water. Store 0.5 mL aliquots at −20 °C for a maximum period of 6 months.

2.4 FIGE Components

A schematic representation of the pulsing circuitry in conjunction with an external electrophoresis unit is provided in Fig. 1. A picture of the prototype can be found elsewhere (*see* Supplementary data in Ref. [18]). Forward and reverse switching of the electric field supply to the gel was achieved by interfacing the voltage supply (Power Pac 1000 power supply, Bio-Rad, Hercules, CA, USA) with an 240 V alternate current (AC) relay (MY2, OMRON, Japan) that could handle a current of 5 Amp. The rate of the forward and reverse switching was controlled by a MD4E-AP programmable 110–240 V AC switching device (Fuji Electric, Japan) that was able to deliver pulses as short as 1 ms. This simple instrumentation generates square wave form. Amplitude, length, and stability of the pulses were checked and ascertained by a Tektronix oscilloscope model TDS 1000 (*see* Supplementary data in Ref. [18]). Expensive systems for generating inverse pulsed wave can also be purchased from Bio-Rad or GE Healthcare.

3 Methods

All procedures are done at room temperature unless otherwise stated.

3.1 1D PAGE

1. Use 70% alcohol and Kimwipes to wipe the glass plates used for casting gel, and then set up the rest of the apparatus as per manufacturer's instructions.
2. Mix 1.9 mL of separating gel buffer, 2.5 mL of acrylamide/Bis solution, 75 μL of 10% SDS solution, and 3.1 mL of milliQ water in a 15 mL conical tube. Add 20 μL of 10% ammonium persulfate and 10 μL of TEMED (10% polyacrylamide gel in mini cassette: 7.5 cm × 10 cm × 1.0 mm).
3. Invert the tube gently after adding all components.
4. Pour the gel solution into the glass cassette. Allow space for stacking gel and gently overlay with *n*-butanol or water.
5. Remove *n*-butanol, wash with water, and wipe with filter paper.
6. Add the stacking gel solution and put in comb without introducing air bubbles. Prepare the stacking gel solution by mixing 1.25 mL of stacking gel buffer, 0.5 mL acrylamide/Bis solution, 50 μL of 10% SDS, and 3.25 mL milliQ water in a 15 mL conical tube. Add 20 μL of 10% ammonium persulfate, 10 μL of TEMED.
7. Set up the gel in the tank and make sure that the running buffer is not leaking.
8. For each sample, use ¼ the volume of 4× Laemmli sample buffer. Heat the sample at 95 °C for 5 min. Cool the samples to room temperature. Add DTT solution to 50 mM final concentration.
9. Load the samples into the wells using a pipette, ensuring that each well has been loaded with appropriate volume of sample.
10. Reassemble the apparatus and connect the leads to the power supply. Proceed to pulsing (*see* Subheading 3.3).
11. Stop gel running when dye front reaches the bottom of the gel. Take out the gel and process to gel staining.

3.2 2D PAGE

1. Select the strip holder(s) corresponding to the immobilized pH gradient (IPG) strip length (7, 11, 13, 18, or 24 cm) chosen for the experiment. Wash each holder with strip holder cleaning solution supplied to remove residual protein. Rinse thoroughly with double distilled water. Use a cotton swab or a lint-free paper to dry the holder or allow it to air-dry.
2. Prepare the required volumes per strip as summarized below:

IPG strip length (cm)	Total volume per strip holder (μL, including any sample volume)
7	125
11	200
13	250
18	350
24	450

3. Deliver the solution slowly with sample at a central point in the strip holder channel away from the sample application wells.
4. Remove the protective cover from the IPG strip. Position it with the gel side down and the pointed (anodic) end of the strip directed toward the pointed end of the strip holder. Align the pointed end first, and then lower the strip onto the rehydration solution. To help soaking the entire strip, one can gently lift and lower the strip, and slide it back and forth along the surface of the rehydration solution. Tilt the strip holder slightly as needed to assure complete and even wetting. Finally, lower the cathodic (square) end of the strip into the strip holder, making sure that the IPG gel contacts the strip holder electrodes at each end. (The gel can be visually identified once the rehydration solution begins to enter the gel.) Be careful not to trap bubbles under the strip.
5. Apply Immobiline DryStrip cover fluid to minimize evaporation and urea crystallization. Pipette the cover fluid dropwise into one end of the strip holder until one half of the strip is covered. Pipet the fluid dropwise into the other end of the strip holder, adding fluid until the entire IPG strip is covered.
6. Place the cover on the holder. Pressure foams on the underside of the cover are to assure that the strip maintains good contact with the electrodes as the gel rehydrates.
7. A minimum of 10 h is required for rehydration; overnight is recommended.
8. Start focusing by increasing the voltage stepwise, to a maximum of 8000 V. (Different manufacturers have different focusing protocols for immobilized pH gradients strips. We used

Ettan™ IPGphor™ II Isoelectric Focusing System. Detailed protocol can be found in GE's 2D electrophoresis handbook). Stop focusing when the voltage hour reaches the desired value.

9. Proceed to two-dimensional electrophoresis.
10. The two-dimensional vertical gel must be ready for use prior to the equilibration of the Immobiline DryStrip after the one-dimensional separation.
11. Equilibration is carried out in a two-step process using capped tubes.
12. Place the IPG strips in individual capped tubes, with the support film toward the tube wall.
13. Add the appropriate volume of SDS equilibration buffer (add DTT to 1% final concentration) to each strip. Cap or seal the tubes with flexible paraffin film, and place them on their sides on a rocker for the equilibration process. Equilibrate for 15 min.
14. Pour off buffer from above step, and add the appropriate volume of SDS equilibration buffer (add iodoacetamide to 4% final concentration) to each strip. Again cap or seal the tubes with flexible paraffin film, and place them on their sides on a rocker for the equilibration process. Equilibrate for an additional 15 min.
15. Apply the Immobiline DryStrip gels on top of SDS-polyacrylamide gel. Push the strips gently down to touch the gel surface.
16. Seal the Immobiline DryStrip gel in place with melted agarose sealing solution.
    Proceed to Subheading 3.4.

3.3 Native PAGE

1. Thoroughly clean and dry the glass plates and three spacers, and then assemble them with bulldog clips. Clamp the chamber in an upright and leveled position.
2. Prepare 10 mL of separating gel mixture as follows.

	4% (mL)	5% (mL)	6% (mL)	8% (mL)	10% (mL)
40% acrylamide/Bis solution (37.5:1)	1	1.25	1.5	2	2.5
4× separating gel buffer	2.5	2.5	2.5	2.5	2.5
50% glycerol	2.5	–	–	–	–
ddH_2O	4	6.25	6	5.5	5

3. Degas the solution, then add 5 μL of 10% ammonium persulfate, and 10 μL of TEMED (N, N, N′, N′-tetramethylethylenediamine).
4. Mix gently and use immediately (because polymerization starts when TEMED is added). Carefully pour the freshly mixed solution into the gel chamber without generating air bubbles. Pour to a level about 1 cm below where the bottom of the well-forming comb will come when it is in position.
5. Carefully overlay the acrylamide solution with H_2O-saturated *n*-butanol without mixing to eliminate oxygen and generate a flat top to the gel.
6. Polymerize the acrylamide for 1 h.
7. Prepare the 4 mL stacking gel solution as follows. Mix the following:

40% acrylamide/Bis solution (37.5:1)	0.4 mL
4× stacking gel buffer	1.0 mL
ddH_2O	2.6 mL

8. Degas the stacking gel solution, and then add 20 μL of 10% ammonium persulfate and 5 μL of TEMED. Mix gently and use immediately.
9. Pour off the *n*-butanol from the polymerized resolving gel, wash the gel top with water, and fill the gap remaining in the gel chamber with the stacking gel mixture. Insert the comb at the top of the gel.
10. Polymerize the acrylamide for 1 h.
11. When the stacking gel has polymerized, remove the comb without distorting the shapes of the well. Remove the clips holding the plates together, and install the gel in the electrophoresis apparatus.
12. Fill the electrophoresis apparatus with electrophoresis running buffer. Push out the bottom spacer from the gel, and remove bubbles from both the top and underneath of the gel. Use the gel immediately.
13. While the gel is being polymerized, prepare samples for loading.
14. Dissolve the protein sample solution in a similar volume of 2× sample buffer, or dissolve a dry sample in 1× sample buffer. The concentration of sample in the solution should be such as to give a sufficient amount of protein in a volume not greater than the size of the sample well. (The BPB dye in sample buffer indicates when the sample solution is acidic, by turning yellow. If this happens, add a little NaOH for the color to turn blue again.)

15. Load the gel with 10–30 μL (20–50 μg) protein sample solution by pipette.
16. Start electrophoresis immediately by turning on power. Proceed to Subheading 3.4.
17. Remove the gel from between the glass plates once the BPB reaches the bottom of the gel.
18. Stain the gel in the staining solution according to the vendor's instruction.

3.4 FIGE

1. The two main reasons for conducting FIGE are [1] to increase the amount of retrievable proteins within a specific locality from the gel by means of detrapping of the migrating proteins; therefore higher local concentration of protein species can be achieved (Fig. 2); and [2] depending on the migrating property of proteins within a particular concentration of polyacrylamide gel, FIGE can be useful for better separating proteins of similar molecular weights (Fig. 3 and spot numbers 3 and 4 in Fig. 4).
2. Before conducting FIGE, it is essential to ensure that the anode and cathode polarities of the power source are connected to the correct electrodes of the timer and the electrophoresis system. A simple voltmeter can be employed to test that forward pulse time (*ta*) always give a positive reading and reverse pulse time (*tr*) always give a negative reading when the electrodes are connected correctly (*see* **Note 6**).
3. Percentages of slab gels used for protein separation should be based on the sizes of the different proteins under investigation with reference to the migration chart for the protein standards obtained from the vendor (*see* Subheading 2.1, **step 10**).
4. The tested *ta*/*tr* ratios effective for 1D PAGE are in the range of 1.07–3.75. Smaller ratio can be used for separating proteins with closed molecular weight; however, too small a ratio may result in overheating to the PAG system. Therefore smaller ratios are recommended for 1D PAGE when one intends to separate proteins with similar molecular weights (*see* Fig. 3, Table 1).
5. As for small gels (1 mm × 7 cm gel), we recommend a voltage of 200 V and an average run time of 12 h.
6. For FIGE in the second dimension in 2D PAGE, we recommend a *ta*/*tr* of 400/106 ms for a gel with 7–9 cm in width (*see* Fig. 4).
7. For native gel, we recommend a pulse rate of *ta*/*tr* at 400/100 ms. If gradient native gel is required for separating large complex native protein mixture, it is better to seek a commercial source. FIGE results are extremely sensitive to the consistency of the gel used (*see* Fig. 5).

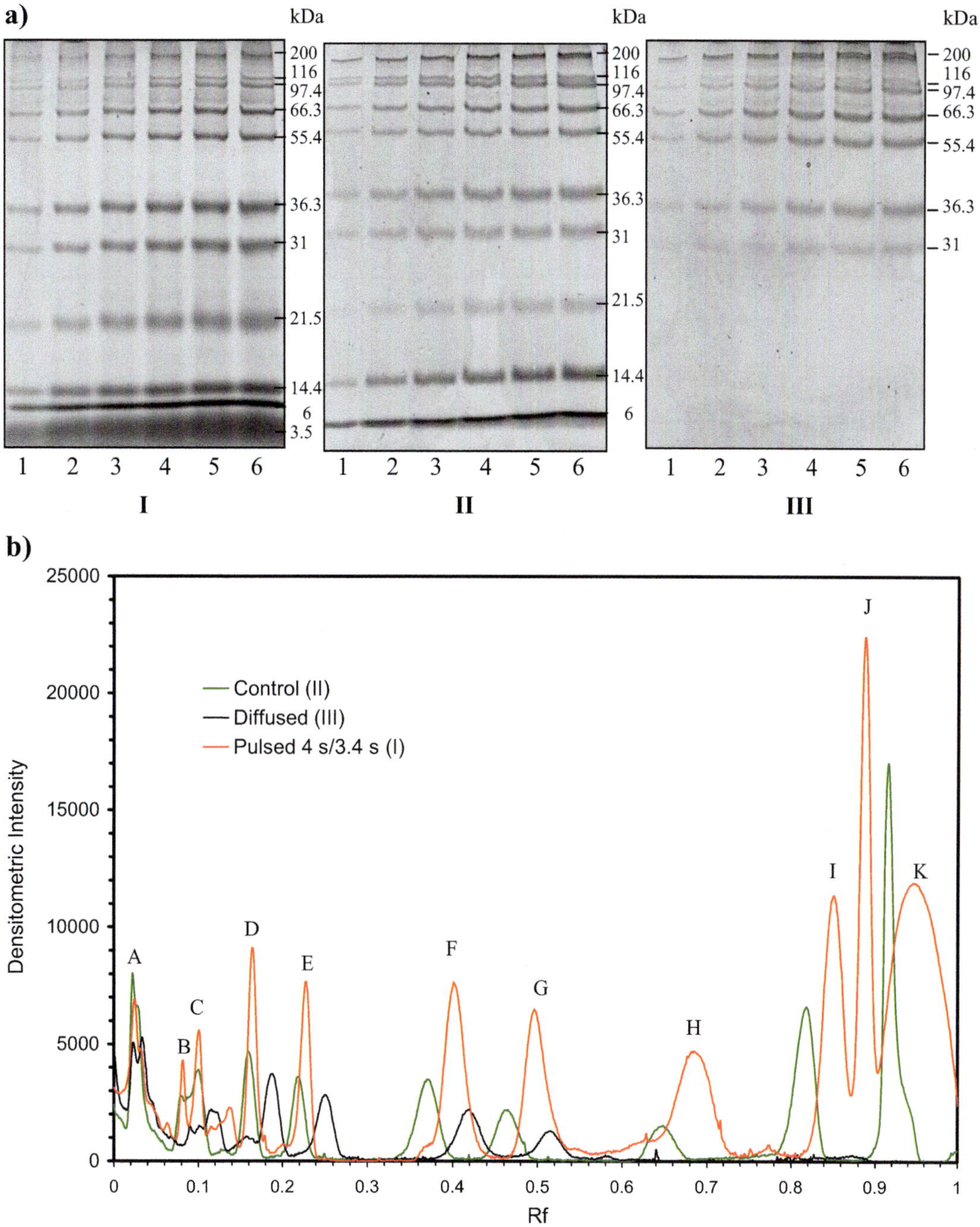

Fig. 2 Increased local concentrations of protein bands upon pulsing. Protein band intensity analyses in FIGE (**a** I), CFE (**a** II), and CFE followed by resting within glass plates in room temperature for 12 h (**a** III). Lanes 1–6 are 2 μL, 4 μL, 6 μL, 8 μL, 10 μL, and 12 μL of Mark12 protein standards, respectively, in a self-cast Bio-Rad 14% SDS-PAGE 1 mm × 7 cm gel followed by Coomassie blue staining. (**a** I) Gel was run with a pulsed field at (4 s/3.4 s) at 200 V for 13 h, with an average buffer temperature of 25 °C. (**a** II) Gel was run at a constant field of 200 V for 1 h and an average buffer temperature of 25 °C. (**a** III) Gel was run at a constant field of 200 V for 1 h and left at rest for another 12 h within the glass plates to permit diffusion prior to staining. (**b**) Densitometry

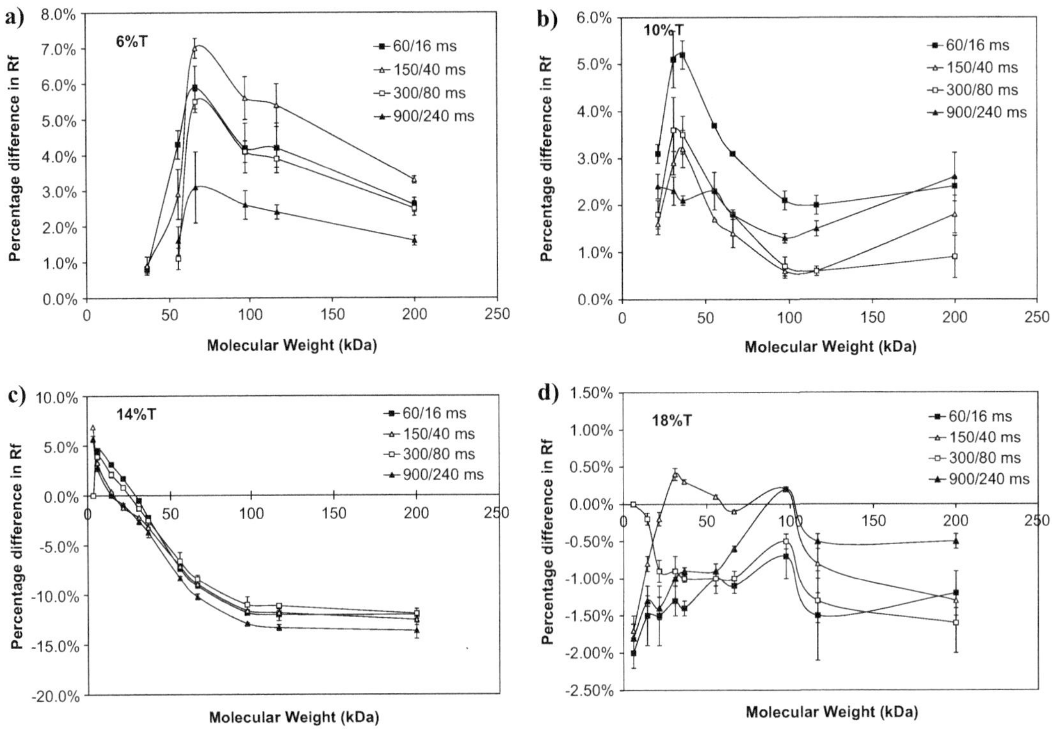

Fig. 3 Changes in mobility upon different pulsing conditions. Comparison of changes in protein mobility between FIGE and CFE conditions in 6% (**a**), 10% (**b**), 14% (**c**), and 18% (**d**) cross-linked polyacrylamide concentration self-cast Bio-Rad SDS-PAG (1 mm $\times$ 7 cm). Different concentrations of polyacrylamide were cast in a mini-Protean 3 apparatus. Five microliters of Mark12 protein standards were used. Graphs of mobility were generated based on relative mobilities to the resolving front (R_f) using Quantity One software (Bio-Rad) and expressed as percent differences in R_f with respect to CFE controls. Each data point was the average of two separate experiments. All gels were run at 200 V with the average buffer temperature of 10 °C. Positive values denote slower mobility, and negative values denote faster mobility with respect to CFE control (reproduced from [18] with permission from BioMed Central)

Fig. 2 (continued) analysis of protein bands in the gels of the three conditions tested. Molecular weight was represented by alphabet A to K, where $A = 200$ kDa, $B = 116.3$ kDa, $C = 97.4$ kDa, $D = 66.3$ kDa, $E = 55.4$ kDa, $F = 36.3$ kDa, $G = 31.0$ kDa, $H = 21.5$ kDa, $I = 14.4$ kDa, $J = 6.0$ kDa, and $K =$ unresolved 3.5/2/0 kDa bands, respectively. Migration distance relative to the dye front (R_f) and intensity of bands from lane 6 of all three gels were densitometrically analyzed by Bio-Rad quantity one software. The graph results were the average of two independent experiments. The graph results were subsequently employed in the calculation of peak variance, σ^2, in Table 1. Red line represents the sample run in pulsed condition. Green line represents the sample run in constant field. Black line represents the sample run in constant field and then left for 12 h before staining (reproduced from [18] with permission from BioMed Central)

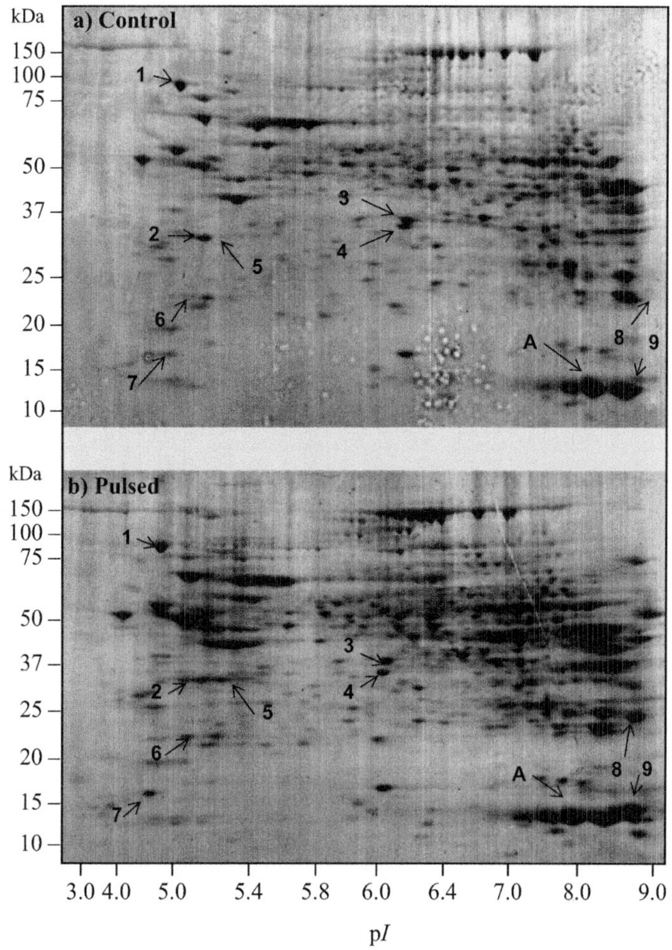

Fig. 4 Effects of FIGE on 2D PAGE analysis of rat liver lysates. Comparison of 2D PAGE images of rat liver lysate under CFE (**a**) and FIGE (**b**) conditions. Each gel represents 100 μg of rat liver lysate separated by isoelectric focusing (IEF) using a nonlinear pH 3–10 IPG strip in the first dimension and a Criterion precast SDS-10–20% PAG in the second dimension at room temperature. Control denotes CFE, and pulsed denotes FIGE with a *ta*/*tr* of 400/106 ms in the two-dimensional separation. Gels were stained with Coomassie blue. Spots selected for LC-MS/MS analysis are denoted by numbers (*see* Table 1), and spots IC and IP denote internal controls for equivalent sample loading for control and pulsing, respectively (reproduced from [18] with permission from BioMed Central)

4 Notes

1. Aliquots of 500 μL in Eppendorf tubes are prepared. Tubes are stored in −20 °C up to 6 months.
2. It is best to prepare this solution fresh each time.

Table 1
Peak variance (σ^2) of proteins separated as in Fig. 2aI by FIGE (pulsing) and Fig. 2aII by CFE (control)

MW (KDa)	Well	200 (A)	116.3 (B)	97.4 (C)	66.3(D)	55.4 (E)	36.5(F)	31.0(G)	21.5 (H)	14.4 (I)	6.0 (J)
σ^2 (Peak variance, mm^2)											
Control	–	1.62 ± 0.05[a]	8.83 ± 0.21	6.50 ± 0.18	1.62 ± 0.15	2.63 ± 0.13	5.86 ± 0.28	5.86 ± 0.45	6.94 ± 0.42	4.51 ± 0.20	1.52 ± 0.11
Pulsed	–	1.85 ± 0.07	0.41 ± 0.05	0.85 ± 0.12	0.87 ± 0.07	1.32 ± 0.15	4.51 ± 0.24	4.51 ± 0.38	8.51 ± 1.00	4.51 ± 0.15	1.52 ± 0.05

Peak variance (σ^2) = (Full-width-half-maximum)^2/5.54

[a]Results are the average of two experiments ± SEM (reproduced from [18] with permission from BioMed Central)

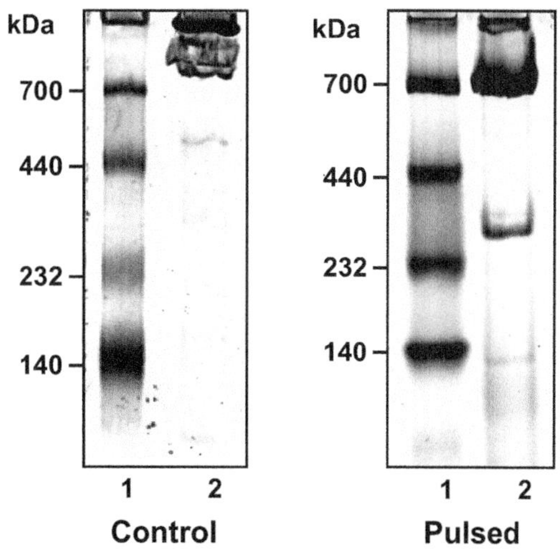

Fig. 5 Effects of FIGE on protein separation under native gel conditions. Comparison of GroEL 14-mer complex (840 kDa) in native PAG between CFE and FIGE conditions. CFE control was the left panel. FIGE was at the right panel. The gels used were 1 mm × 7 cm native 6% PAG cast in a mini-Protean 3 apparatus. Run time was 2 h 15 min in control condition. Pulsed condition run time was 5 h 30 min with a *ta/tr* of 400/100 msec. Lane 1, native MW markers (10 μg total); Lane 2, 10 μg purified *E. coli* GroEL native complex (14-mer, 840 kDa). The band at 300 kDa could be a minor cofactor associated with GroEL during purification. The gel was stained with Coomassie blue. Proteins were purposefully overloaded to best represent the effect of "detrapping" of native proteins under pulsing conditions (reproduced from [18] with permission from BioMed Central)

3. Dissolve 0.5 g agarose in 50 mL milliQ water in a 250 mL DURAN® bottle. Put the solution in microwave at high power until it boils. Swirl the solution to ensure complete dissolving. It may take several rounds of microwave boiling and swirling. Add room temperature milliQ water to a total volume of 75 mL. Add 25 mL Laemmli sample buffer (4×) to total 100 mL and mix well. Aliquots of 1 mL agarose solution in Eppendorf tubes are prepared. Store solidified agarose solution in −20 °C for up to 1 year. Heat the agarose tube to 90 °C for 5 min, and then keep at 45 °C heat block till use.
4. Wear a mask when weighing acrylamide. Put the scale in fume hood when weighing. Mix the solution with a magnetic stir bar on top of a stirrer. Unpolymerized acrylamide is a neurotoxin, and care should be exercised to avoid skin contact. The acrylamide solution can be stored at 4 °C for 1 month.
5. Shake the water and *n*-butanol for 1 min. Two layers will be seen. The top is the butanol layer.

6. To ensure a proper *ta/tr* ratio, that is, >1, especially when pulse rate is relatively fast at > 5 Hz (number of total *ta* and *tr* per second), a voltmeter is recommended. When the voltmeter reading becomes unstable, it means that the pulse rate (hertz) is too fast for any accurate detection. In this situation, change the timing to lower hertz (e.g., 0.2 Hz or *ta* = 4 s and *tr* = 1 s) first. Following the proper voltmeter reading, one can increase the pulse rate accordingly. For long period of pulsing, large volume of running buffer should be used to prevent buffer capacity deterioration. At pulse rate of >5 Hz, an electrophoresis system with built-in cooling system is recommended. Alternatively, this can be achieved with large volume of running buffer with the submerged gel during the run. If it is possible, the heat sinking effect of the running buffer can be enhanced through the use of a magnetic stirrer.

Acknowledgment

This work was supported by the Agency for Science Technology and Research (ASTAR) of Singapore. BioMed Central is the original publisher of the original research article [18]. Figures are used with permission of BioMed Central.

References

1. Laemmli UK (1970) Cleavage of structural proteins during the assembly of the head of bacteriophage T4. Nature 227:680–685
2. Schwartz DC, Cantor CR (1984) Separation of yeast chromosome sized DNA fragments by pulsed field gradient gel electrophoresis. Cell 37:67–75
3. Bancroft I, Wolk CP (1988) Pulsed homogeneous orthogonal field gel electrophoresis (PHOGE). Nucleic Acids Res 16:7405–7418
4. Carle GF, Frank M, Olson MV (1986) Electrophoretic separations of large DNA molecules by periodic inversion of the electric field. Science 232:65–68
5. Chu G, Vollrath D, Davis RW (1986) Separation of large DNA molecules by contour-clamped homogeneous electric fields. Science 234:1582–1585
6. Bostock CJ (1988) Parameters of field inversion gel electrophoresis for the analysis of pox virus genomes. Nucleic Acids Res 16:4239–4252
7. Heller C, Pohl FM (1989) A systematic study of field inversion gel electrophoresis. Nucleic Acids Res 17:5989–6003
8. Mathew MK, Hui CF, Smith CL et al (1988) High-resolution separation and accurate size determination in pulsed-field gel electrophoresis of DNA. 4. Influence of DNA topology. Biochemistry 27:9222–9226
9. Mathew MK, Smith CL, Cantor CR (1988) High-resolution separation and accurate size determination in pulsed-field gel electrophoresis of DNA. 2. Effect of pulse time and electric field strength and implications for models of the separation process. Biochemistry 27:9210–9216
10. Mathew MK, Smith CL, Cantor CR (1988) High-resolution separation and accurate size determination in pulsed-field gel electrophoresis of DNA. 1. DNA size standards and the effect of agarose and temperature. Biochemistry 27:9204–9210
11. Tischfield JA, Bernhard HP, Ruddle FH (1973) A new electrophoretic-autoradiographic method for the visual detection of phosphotransferases. Anal Biochem 53:545–554
12. Brassard E, Turmel C, Noolandi J (1991) Observation of orientation and relaxation of protein-sodium dodecyl sulfate complexes

during pulsed intermittent field polyacrylamide gel electrophoresis. Electrophoresis 12:373–375
13. Houri A, Starita-Geribaldi M (1994) Pulsed field electrophoresis for the separation of protein-sodium dodecyl sulfate-complexes in polyacrylamide gels. Electrophoresis 15:1032–1039
14. Sant'Ana Pereira JA, Greaser M, Moss RL (2001) Pulse electrophoresis of muscle myosin heavy chains in sodium dodecyl sulfate-polyacrylamide gels. Anal Biochem 291:229–236
15. Rodbard D, Chrambach A (1970) Unified theory for gel electrophoresis and gel filtration. Proc Natl Acad Sci U S A 65:970–977
16. Guo XH, Chen SH (1990) Reptation mechanism in protein-sodium-dodecylsulfate (SDS) polyacrylamide-gel electrophoresis. Phys Rev Lett 64:2579–2582
17. Kozulic B (1994) On the "door-corridor" model of gel electrophoresis. I. Equations describing the relationship between mobility and size of DNA fragments and protein-SDS complexes. Appl Theor Electrophor 4:125–136
18. Tsai H, Low TY, Freeby S et al (2007) Increase in local protein concentration by field-inversion gel electrophoresis. Proteome Sci 5:18

Chapter 20

Two-Dimensional Difference Gel Electrophoresis

Malachi Blundon, Vinitha Ganesan, Brendan Redler, Phu T. Van, and Jonathan S. Minden

Abstract

Two-dimensional difference gel electrophoresis (2D DIGE) is a modified form of 2D electrophoresis (2D E) that allows one to compare two or three protein samples simultaneously on the same gel. The proteins in each sample are covalently tagged with different color fluorescent dyes that are designed to have no effect on the relative migration of proteins during electrophoresis. Proteins that are common to the samples appear as "spots" with a fixed ratio of fluorescent signals, whereas proteins that differ between the samples have different fluorescence ratios. With conventional imaging systems, DIGE is capable of reliably detecting as little as 0.2 fmol of protein, and protein differences down to ± 15%, over a ~10,000-fold protein concentration range. DIGE combined with digital image analysis therefore greatly improves the statistical assessment of proteome variation. Here we describe a protocol for conducting DIGE experiments, which takes 2–3 days to complete. We have further improved upon 2D DIGE by introducing in-gel equilibration to improve protein retention during transfer between the first and second dimensions of electrophoresis and by developing a fluorescent gel imaging system with a millionfold dynamic range.

Key words Proteomics, Difference gel electrophoresis (DIGE), Digital fluorescent gel imaging, IPG strips

1 Introductions

The central goals of proteomics include identifying protein changes that differentiate normal and diseased states in cells, tissues, or organisms and examining how protein changes correlate with developmental age and environment. The first stage in comparative proteomics is to separate complex mixtures of protein into individual components; this is typically done using gel electrophoresis (at the whole protein level) or column chromatography (at the peptide level). Both of these separation schemes have advantages and disadvantages. We have focused on two-dimensional electrophoresis (2D E) because of its accessibility to most laboratories. This approach was described simultaneously by several groups in 1975 [1–3]. Despite the substantial advances in the technology

Biji T. Kurien and R. Hal Scofield (eds.), *Electrophoretic Separation of Proteins: Methods and Protocols*, Methods in Molecular Biology, vol. 1855, https://doi.org/10.1007/978-1-4939-8793-1_20,

since its launch—the most notable of which was the introduction of immobilized pH gradients in the first dimension [4, 5]—some of the more significant systemic shortcomings have remained unsolved. The most troublesome of these is the inherent lack of reproducibility between gels. Efforts to surmount this limitation have mostly focused on developing computational methods for gel matching. These approaches have had limited success because the sources of gel-to-gel variation are numerous, complex, and difficult to model [6–8].

Difference gel electrophoresis (DIGE) was developed to overcome the irreproducibility problem in the 2D E methodology by labeling two samples each with a different fluorescent dye prior to running them on the same gel (Fig. 1) [9, 10]. The fluorescent dyes used in DIGE, Cy3-NHS and Cy5-NHS (Fig. 2), are cyanine based, molecular weight matched, amine reactive, and positively charged. These characteristics, coupled with sub-stoichiometric labeling, result in no electrophoretic mobility shifts arising between the two differentially labeled samples when they are co-electrophoresed. Therefore, in DIGE, every identical protein in one sample superimposes with its differentially labeled counterpart in the other sample, allowing for more reproducible and facile detection of differences. Furthermore, DIGE is a sensitive technique, capable of detecting as little as 0.2 fmol of protein, and this detection system is linear over a ~10,000-fold concentration range [9, 11, 12].

The most important considerations in performing DIGE experiments are experimental design and sample preparation [13]. DIGE has been used to analyze proteome changes from a wide variety of cell types, tissue types, model organisms, and bodily fluids including serum [14–20]. The sample preparation protocol depends on the sample type. Most samples require mild homogenization in lysis buffer to extract protein. DIGE is an extremely sensitive method, in which a 15% change in protein abundance is more than two standard deviations (SDs) above the normal variation [12]. One must take great care in deciding which samples to compare while bearing in mind the sources of variation. If specific tissues are to be compared, one must carefully dissect the tissue to avoid variation. Hypothetically, if one has 10% contamination of neighboring tissue in one sample and 5% variation in another, the contamination could be sufficient to cause a false-positive conclusion about protein differences in the experiment. One way to avoid tissue contamination is to use laser microdissection, which provides a precise method for capturing specific populations of cells [21–23]. Another source of variation can arise during sample cleanup or fractionation. If sample cleanup or fractionation is planned, it is best to label the samples independently and then combine them prior to cleanup or fractionation. This will alleviate variation due to handling. Simply measuring total protein after

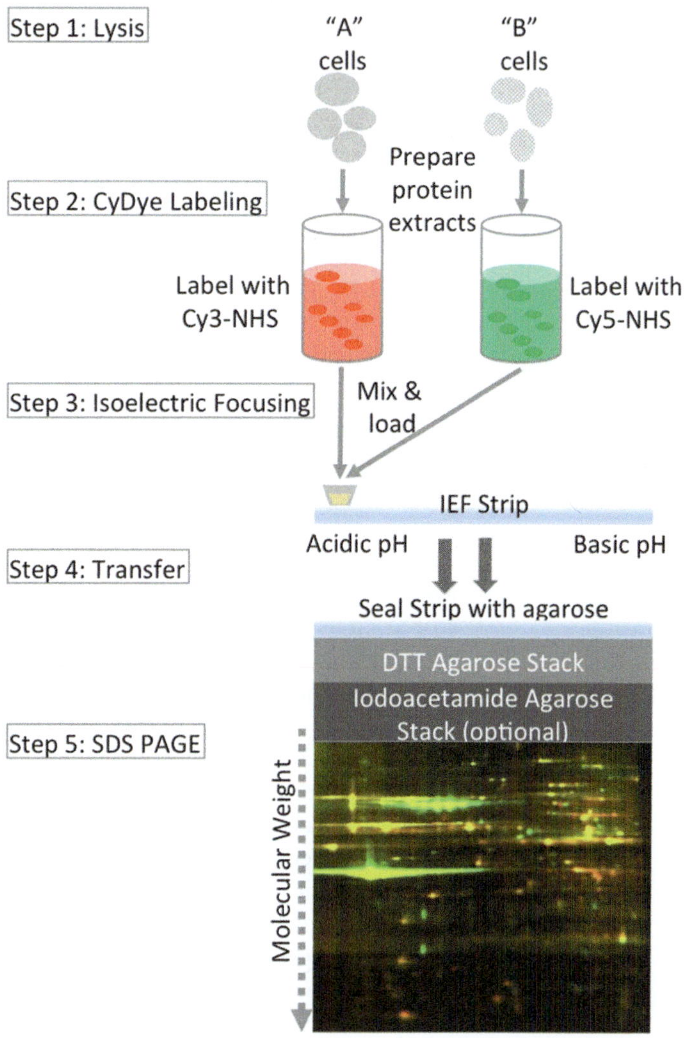

Fig. 1 Schematic of DIGE analysis. Extracts are made of two cell samples, denoted as "A" and "B." These extracts are separately labeled with Cy3-NHS and Cy5-NHS, which covalently link to lysine residues. A low stoichiometry of labeling is used, where ~5% of all proteins carry a single dye molecule. The labeled protein extracts are then combined and co-electrophoresed on an isoelectric focusing strip gel. Agarose stacking gels containing iodoacetamide and DTT are layered on top of a polyacrylamide gel. The IEF strip is then placed on top of the stacks and held in place with sealing gel. Following transfer of the IEF strip, proteins are separated based on their size in the second dimension–SDS PAGE. The 2D gel is then imaged on a fluorescent gel imager at the Cy3 and Cy5 emission wavelengths. Shown here is a color overlay of Cy3 (green) and Cy5 (red) images of *mitochondria* extracts from mouse cell line. Regions of equal Cy3 and Cy5 signals appear yellow. MWt, molecular weight

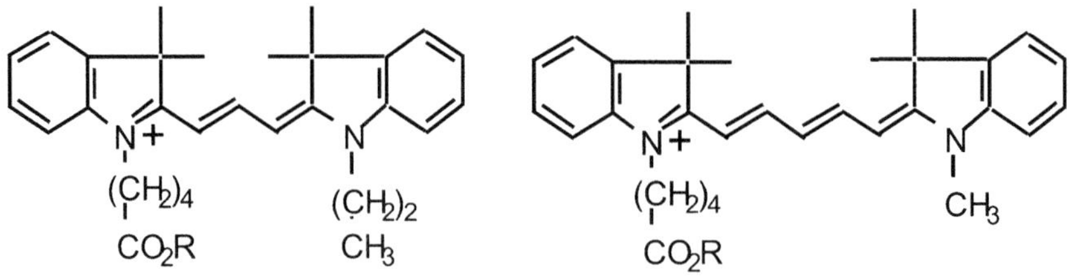

Cy-3

Cy-5

R = NHS

Fig. 2 Chemical structure of DIGE dyes. Propyl-Cy3-NHS and methyl-Cy5-NHS are shown here. These compounds are charge and mass matched. The fluorescent characteristics of Cy3 and Cy5 are dictated by the three-carbon and five-carbon polyene chains, respectively, linking the two indoline rings

sample cleanup or fractionation will not guard against variation in loss of specific proteins during processing. If samples are to be stored, it is extremely important to avoid multiple freeze-thaw cycles. As a general rule of thumb, protein samples should not experience more than one or two freeze-thaw cycles.

The protocol described here outlines the steps we use in performing 2D DIGE experiments. We typically use a two-dye approach with Cy3 and Cy5. For comparing model systems, such as *Drosophila* or yeast, the two-dye approach is more reproducible than the three-dye method, which utilizes a Cy2-labeled pooled sample for normalization [24]. The three-dye method is better suited to multiple sample comparisons where there is considerable genetic background variation, as in the analysis of human samples [25]. The precursor to the three-dye method was a two-dye method comparing a Cy3-labeled pooled control + test sample and Cy5-labeled control or test samples [26]. We always run two gels for each comparison, in which the order of labeling is reversed (referred to as reciprocal labeling). This allows one to differentiate between sample-dependent differences and rare dye-dependent differences. The latter are presumably due to incomplete solubilization of proteins, allowing the two dyes to associate differently with the protein. Loading equal amounts of each protein sample is advisable, but exact loading is not required as slight load differences can be normalized during image acquisition with specific imaging technologies or image analysis. All steps should be performed on ice or in a cold room if possible to minimize unwanted side reactions. As soon as lysis is complete, samples can be stored at –80 °C in aliquots of 100–200 μg of total protein where the concentration is >1 mg/mL. When comparing whole-cell extracts, we generally load 100 μg of each sample (200 μg total protein) for pH 3–10 strips. One can load up to 400 μg total protein on narrower pH range strips (e.g., 4–7).

Recently, we have streamlined the workflow at the transfer step between IEF and SDS-PAGE [27]. This improved step using a stacking gel instead of liquid equilibration has reduced protein loss by ~45% in the overall process. An agarose stacking gel containing dithiothreitol (DTT) is used to reduce proteins electrophoresed out of the first-dimension IEF strips. If alkylation is desired, a second agarose stacking gel containing iodoacetamide is employed.

High-quality digital imaging of DIGE gels is essential for detecting proteome changes. To probe the proteome as deep as possible, it is important to use an imager that is capable of true 16-bit data collection. Fluorescent gel imagers come in two formats: scanner-based or CCD camera-based. There are several commercial gel imagers, such as the Typhoon imager (Amersham Biosciences/GE Healthcare), which is commonly used for DIGE gel imaging. The Typhoon imager is based on a scanning laser illumination system and photomultiplier detector. The gel-scanning systems that are suitable for DIGE analysis require a separate device for spot picking. We use a homemade CCD-based fluorescence imager that has an integrated spot-picking robot (Fig. 3) [28, 29].

The main consideration in gel imaging is utilizing the full dynamic range of the detector. The dynamic range of the detector is measured by the number of discrete gray levels each pixel can accommodate, represented digitally as bits. The CCD camera in our imager has a 16-bit dynamic range or 65,536 gray levels. Ideally, one would like to load enough fluorescently labeled protein and set the exposure parameters so that the brightest protein spot nearly saturates the full dynamic range of the detector. This would maximize the detection of dim spots without obscuring the fluorescence ratios between protein spots on the gel. We have recently developed an imaging system that has a millionfold dynamic range. This was accomplished by using an iterative imaging method using structured illumination. Rather than using wide-field illumination or laser scanning, a video projector was used to modulate gel illumination. After each imaging cycle, the brightest regions of the gel are masked by turning off the appropriate pixels of the video projector, thus allowing for longer exposure times without ever saturating the detector. Individual subsaturation 16-bit gel images are automatically combined by our imaging software to create high-dynamic-range 32-bit images. To further improve the detectability of proteins, we also increased the extent of fluorescent labeling up to tenfold over the minimal labeling standard conditions [29].

Another important issue is the fluorescent background. Many of the materials and substances that are placed in the imager can have intrinsic fluorescence. For example, borosilicate glass is highly fluorescent, so any glass used to hold the gels in the imager must be made of quartz or fused silica. Although background fluorescent

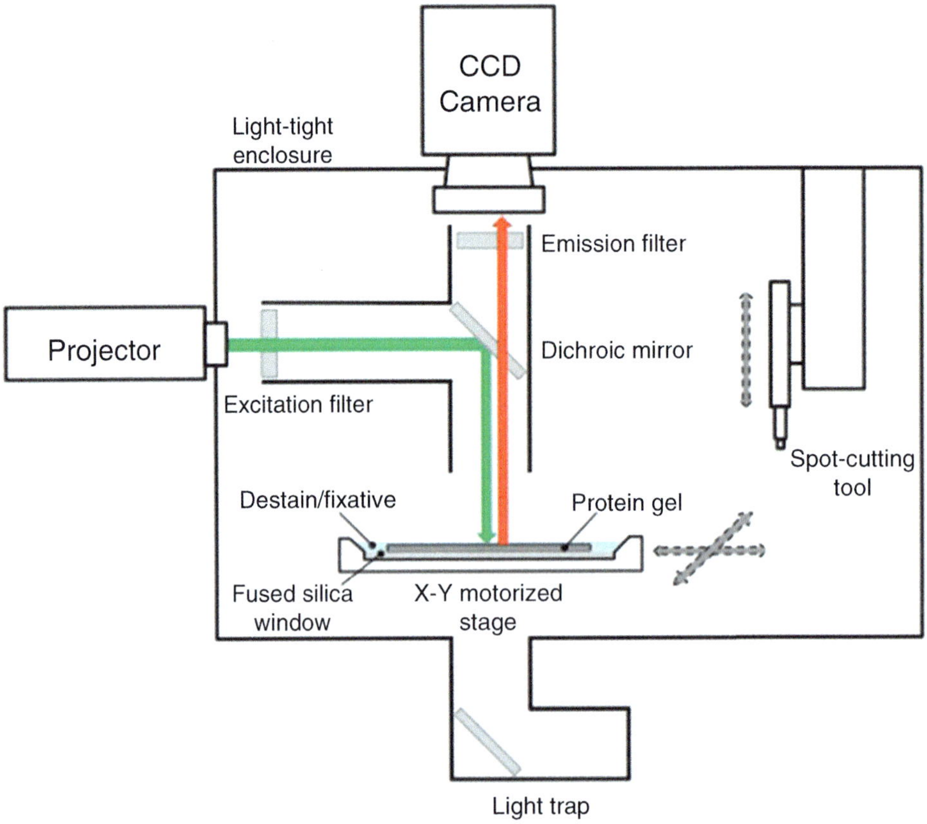

Fig. 3 Fluorescence gel imager/spot picker. This diagram illustrates the components of the DIGE gel imager. This device comprises the following components: a scientific-grade Peltier-cooled 16-bit CCD camera (Photometrics/Roper Scientific); an 105 mm macro lens (Nikon); a computer-driven stage (New England Affiliated Technology); and a spot-picker drive (Applied Precision). All electronic components are controlled by a computer workstation running Windows and custom software. The illumination system is video projector (Hitachi) that directs light through motorized filter wheels (CVI Laser Corporation). The fluorescence wavelengths are selected by band-pass filters (Chroma Technology)

signals can be removed computationally, they limit the dynamic range of the gel images and mask signals from low-abundance proteins. Our new gel-imaging system is capable of detecting as little as 0.02 fmol of protein and can detect proteins over a ~1,000,000-fold concentration range. As there are a limited number of such gel-imaging systems currently in use, we will not describe our imaging protocol here in detail, but it has been made available to the public [29].

Image analysis in our laboratory is done using several software applications, including ImageJ, QuickTime, and SExtractor. There are also integrated image analysis packages available for DIGE gels, most notably DeCyder (Amersham Biosciences/GE Healthcare). We use ImageJ for image visualization and annotation. Additionally, we use SExtractor, an astrophysics freeware application that has been adapted to quantify protein-spot intensities and ratios (https://sourceforge.net/projects/sextractor) [30]. Although these separate applications are more time-consuming than the commercial packages, we find the analysis to be as reliable as that achieved with DeCyder. None of the current image analysis applications are completely automatic in that the computer determines which protein spots are changing and the degree of change. A certain amount of editing of the detected spots by visual inspection is required. We find that visual inspection of a two-frame looping movie of the Cy3 and Cy5 gel images is the most reliable method for detecting protein differences and for spot-list editing. We find that two-frame looping movies are much more robust than superimposing two pseudo-color images of Cy3-labeled and Cy5-labeled proteins. While current automated spot detection applications are becoming more robust, visual inspection is still used as the final arbiter. The outcome of image analysis is a list of difference-protein spots that indicate putative differentially expressed or modified candidate proteins. This spot list can be used for large-scale protein-profiling studies using standard bioinformatics tools [31].

Despite DIGE being sensitive and reproducible, two caveats need to be mentioned. First, 2D E does not efficiently resolve integral membrane proteins. This is due to their hydrophobic domains causing precipitation during isoelectric focusing (IEF). Other laboratories continue to work on this problem. Second, labeling with the amine-reactive DIGE dyes limits one to sub-stoichiometric labeling (also known as minimal labeling), where less than 5% of all proteins carry a single bound dye molecule and the rest have no bound dye. For proteins that are >25 kDa, there is no appreciable molecular weight shift between labeled and unlabeled protein, while there is a slight but predictable shift for smaller proteins in which unlabeled proteins run about half a spot diameter faster than their labeled counterparts. This shift problem has been addressed with the development of cysteine-reactive dyes. These dyes allow one to saturate-label all available cysteines, which eliminates the shift between labeled and unlabeled proteins, as all proteins are maximally labeled [21, 22, 32]. Regardless of these limitations, DIGE combined with MS is an economical, sensitive, robust, and useful approach for comparative proteomics.

2 Materials

2.1 Reagents

Use double-distilled H_2O unless otherwise stated. The source and purity of ingredients are critical, especially for solutions used for labeling and IEF.

1. SDS: 20% w/v stock solution.
2. HEPES: 1M Na-HEPES, pH 8.0, stock.

3. Dithiothreitol (DTT): 1M stock, kept at −80°C.
4. Acrylamide: 30% T, 2.6% C stock solution.
5. Ammonium persulfate (APS): 10% w/v stock kept at –20 °C.
6. IEF strips (e.g., pH 3–10 NL, 18 cm Immobiline DryStrips, Amersham Biosciences/GE Healthcare, Piscataway, NJ).
7. IEF strip-holder cleaning detergent (Amersham Biosciences/GE Healthcare).
8. RBS 35 detergent (concentrate; Pierce, Thermo Fisher Scientific, Rockford, IL).
9. SeaPlaque UltraPure low-melting agarose (Lonza.com).
10. Kimwipes (Kimberly-Clark Professional, GA).
11. Lens paper.

2.2 Equipment

1. IPGphor IEF apparatus (Amersham Biosciences/GE Healthcare).
2. Bio-Rad Protean II XL.
3. Fluorescent gel imager/spot picker (custom made, Fig. 3).
4. Optional: Model 485 Gradient Former gradient maker (Bio-Rad, Hercules, CA).
5. DryStrip rehydration tray (Amersham Biosciences/GE Healthcare).
6. Image analysis software applications (e.g., ImageJ, QuickTime, SExtractor).

2.3 Reagent Setup (See Note 1)

1. Lysis buffer: 7 M urea, 2 M thiourea, 4% (wt/vol) CHAPS, 10 mM DTT, and 10 mM Na-HEPES (pH 8.0) (*see* **Note 2**).
2. Dye solutions: Add 5 μL DMF to each 5 nmol tube of Cy3-NHS (Amersham Biosciences/GE Healthcare). As the extinction coefficient of Cy5 is greater than that of Cy3, add 6 μL DMF to each 5 nmol tube of Cy5-NHS (Amersham Biosciences/GE Healthcare) (*see* **Note 3**).
3. Quencher: 5 M methylamine-HCl and 100 mM HEPES (pH 8.0). Dissolve 2.38 g HEPES in 38.8 mL of 40% vol/vol methylamine aqueous solution (*see* **Note 4**).
4. Rehydration buffer: 7 M urea, 2 M thiourea, 4% (wt/vol) CHAPS, 10 mM DTT, 2 mM acetic acid, 0.002% (wt/vol) bromophenol blue and 1% (wt/vol) IPG buffer. Use the appropriate IPG buffer (Amersham Biosciences/GE Healthcare) that corresponds to the pH range of the IEF strips in the experiment (*see* **Note 5**).
5. Gradient gel solutions: *See* Table 1 for the composition of the two acrylamide solutions needed to form a 10–15% gradient gel that is 14 × 24 cm (*see* **Note 6**).

Table 1
Composition of acrylamide solutions needed to form a 10–15% gradient gel

Stock solution	Light	Heavy
Acrylamide (30% T, 2.6% C)	8.25 mL	12.25 mL
1.5 M Tris, pH 8.8	6.25 mL	6.25 mL
Sucrose	–	3.75 g
SDS (20% w/v)	125 μL	125 μL
H_2O	10.375 mL	4.175 mL
APS (10% w/v)	82.5 μL	82.5 μL
TEMED	8.25 μL	8.25 μL

Table 2
Composition of agarose solutions needed to make the IAA stack, DTT stack, and sealing gels

Stocks	IAA stack (1 gel)	DTT stack (1 gel)	Sealing (1 gel)
Agarose	50 mg	50 mg	20 mg
water	4 mL	4 mL	1.732 mL
0.5M Tris	650 μL	650 μL	260 μL
20% SDS	25 μL	25 μL	10 μL
500× bromophenol blue	–	–	8 μL
Melt agarose into solution and refrigerate as 1 aliquot per gel			
1M IAA	650 μL	–	–
1M DTT	–	325 μL	–

6. Tank buffer: 25 mM Tris, 190 mM glycine, and 0.1% (wt/vol) SDS. Can be made or purchased as a 10× stock.
7. Agarose stacking gel solutions: *See* Table 2.

   DTT Stack: 65 mM Tris (pH 6.8), 0.1% SDS, 1% agarose, 65 mM dithiothreitol (DTT). Heat all of the components, except DTT, in a microwave until the agarose dissolves. DTT is added immediately before use when casting the stacking gel. Iodoacetamide stack (optional): 65 mM Tris (pH 6.8), 0.1% SDS, 1% agarose, 130 mM iodoacetamide (IAA). Heat all of the components, except IAA, gently in a microwave until the agarose dissolves. IAA is added after dissolving the agarose, just before casting the second stacking gel. Stacking gel solutions can be aliquoted and stored at 4 °C (without the DTT or IAA, add just before casting).

8. Agarose sealing solution: 65 mM Tris (pH 6.8), 0.1% (wt/vol) SDS, 1% (wt/vol) agarose, and 0.002% (wt/vol) bromophenol blue. Heat in a microwave until the agarose dissolves. Aliquot and store at 4 °C.
9. Fixative: 40% (vol/vol) methanol (HPLC grade) and 1% (vol/-vol) acetic acid (glacial).
10. IEF strips: IEF strips are available in different pH ranges and sizes. pH 3–10 NL strips are a good starting point for whole-cell extracts; "NL" indicates that the strip has a nonlinear pH gradient, with increased resolution between pH 5 and 7. The 13 cm strips fit well into the Hoeffer SE 660 system that we use for SDS-PAGE. If using 18 cm strips, the ends will have to be cut to fit. This does not lead to appreciable loss of proteins when using pH 3–10 NL strips, as most samples do not have many proteins focused with pI values near the extremes. In practice, using 18 cm pH 3–10 NL strips improves the resolution of proteins in the region where most proteins lie. pH 4–7 strips might be used instead of pH 3–10 NL if most proteins are within this pI range, which is the case for bacterial samples. Narrow pH-range strips (1 pH unit) are available for closer study of proteins within a region of interest.

3 Methods

3.1 Casting the 2D E Gel

1. Typically, we cast 2D E gels on the same day we perform the IEF. Assemble the gel cassette using cleaned and dried plates. We use homemade 10% or 12% acrylamide gels. Gradient gels using increasing amounts of acrylamide from 10 to 15% may be desirable for smoother resolution of protein spots. The spacers used are 1.5 mm in thickness (*see* **Note 7**).
2. Add SDS to acrylamide solutions, and mix well. Then add 10% (wt/vol) APS and TEMED when ready to pour the gel.
3. Overlay with ~750 μL water-saturated *n*-butanol (top layer) and allow the gel to polymerize for ≥1 h (*see* **Note 8**).

3.2 Rehydrating the IEF Strip

1. The day before the 2D DIGE experiment, rehydrate Immobiline DryStrips in a DryStrip rehydration tray according to the manufacturer's instructions (*see* **Note 9**).

3.3 Sample Preparation

1. Place the cells or tissue sample in a 1.5-mL centrifuge tube that has a fitted pestle. Glass homogenizers may be used for difficult tissues or cells. Sufficient material to yield 100–250 μg protein is needed. Rinse the cells with an ice-cold low-salt Na-HEPES (pH 8.0) buffer that does not contain primary amines. Remove

excess liquid and add lysis buffer desired. Homogenize the cells with a few passes or turns of the pestle. Centrifuge the sample in a microcentrifuge for 5–15 min at 15,000 × *g* at 4 °C to remove unbroken cells and debris (*see* **Note 10**).

2. Measure protein concentration using a Bradford or BCA assay. As urea, CHAPS, and DTT affect the Bradford assay, the standards and samples should all be made up in lysis buffer. Add 2 μL sample/standard to 800 μL water in a plastic test tube and mix. Add 200 μL Bradford reagent, mix, and measure the OD at 595 nm within 1 h. If the sample is too concentrated (>2.5 mg/mL), dilute with lysis buffer. A standard curve should be made with BSA dissolved in lysis buffer at a concentration range of 0.5–2.5 mg/mL. The blank should be made with 2 μL lysis buffer (*see* **Note 11**).

3.4 Protein Labeling

1. *See* Table 3. The following four steps describe a typical 2D DIGE experiment in which protein samples denoted as "A" and "B" are to be compared. Set up four 1.5 mL microcentrifuge tubes, numbered 1–4. Pipette 50 μL protein sample A (at 2.0 mg/mL), which contains 100 μg of protein in lysis buffer, into tubes 1 and 3. Pipette 50 μL protein sample B into tubes 2 and 4 (*see* **Note 12**).
2. Add 2 μL Cy3-NHS stock solution to tubes 1 and 4. Add 2 μL Cy5-NHS stock solution to tubes 2 and 3. Mix by vortexing and briefly spin in a microcentrifuge at top speed for a few seconds at 4 °C to consolidate all of the liquid. Incubate in the dark on ice for 15 min (*see* **Note 13**). On each tube, add 1 μL quenching solution and incubate on ice for 30 min (*see* **Note 14**).
3. In a new tube labeled as "gel 1," combine the entire contents of tubes 1 and 2. In a new tube labeled as "gel 2," combine the contents of tubes 3 and 4. Add 2.4 μL of the appropriate IPG buffer solution (2% of the total volume of sample) to the gel 1 and 2 tubes. Mix by vortexing and briefly spin in a microcentrifuge at top speed for a few seconds at 4 °C to consolidate all of the liquid.

Table 3
Typical protein labeling scheme

	Tube 1	Tube 2	Tube 3	Tube 4
Sample A	50 μL	–	50 μL	–
Sample B	–	50 μL	–	50 μL
Cy3-NHS	2 μL	–	–	2 μL
Cy4-NHS	–	2 μL	2 μL	–

Table 4
Recommended program for running 18 cm, pH 3–10 IEF strips

Step	Voltage	Voltage gradient type	Duration (h)	Duration (kVh)
1	500	Step-n-hold	1	
2	4000	Step-n-hold	1	
3	8000	Step-n-hold	2–4	
4	8000	Step-n-hold		30–40

3.5 First-Dimension Electrophoresis: IEF

IEF should be set up according to the manufacturer's instructions with a few modifications. Refer to the user manual for more details and precautions to be taken while setting up IEF. Below are instructions for GE's IPGphor machine. Slight modifications of the protocol are necessary for Bio-Rad's Protean II XL.

1. Turn on the IPGphor. Place ceramic strip holders on the IPGphor platform, with the pointed end toward the anode.
2. Open the rehydration cassette assembly. Remove a rehydrated IEF strip and rinse it briefly with HPLC water. Blot excess water with lens paper if necessary. Place the IEF strip in the holder with the acidic end toward the anode. Place water-saturated wicks at both ends of the strip. Place an electrode on each of the wicks. Place a sample cup near the electrode on the acidic end of the gel. Cover the surface of the gel with ~1 mL of DryStrip Cover Fluid (*see* **Note 15**).
3. Pipette the sample into the sample cup, taking care to avoid bubbles. Place the lid over the strip holder. Start the IEF program shown in Table 4, using 50 μA per strip at 18 °C (*see* **Note 16**).

3.6 Transfer

1. Drain the *n*-butanol from the top of the second-dimension gel and rinse well with water. Dry the glass completely with a Kimwipe. Allow to drip-dry upside down at an angle.
2. Melt agarose stacking and sealing solution in a microwave and place in a beaker of hot water, or hot block, to keep melted.
3. Add 325 μL DTT to agarose stack. Pour 4 mL of this stack on top of the acrylamide gel. Let this solidify for 20 min; refrigeration or performing this step in a cold room will speed the process and help prevent DTT degradation. Optional: If performing mass spec after electrophoresis, add IAA stack.
4. Remove and rinse strips with water, lightly blot excess oil (IPG fluid) with a lens paper sandwich. To create the sandwich, place a paper towel on the bench and place two pieces of lens paper over the paper towel. Create enough surface area for the size

IEF strip in use. Place the IEF strip on the lens paper, plastic side down. With a third piece of lens paper, blot off the gel side by gently placing the paper on the strip and peeling off the paper slowly, starting from one end. Slow steady motions are necessary to ensure the strip does not rip.

5. Place the second-dimension gel horizontal on the bench. Place the IEF strip on top of the second-dimension gel with the plastic touching the back glass and the acidic end of the strip toward the left. Gently push the IEF strip down until it contacts the stacking gel; a gel casting spacer works well for this purpose; take care not to scratch the IEF strip. Cover the IEF strip with melted agarose sealing solution until it just covers the IEF strip (*see* **Note 17**).

3.7 Second-Dimension Electrophoresis: SDS-PAGE

1. Place the gel in the electrophoresis unit and fill the upper and lower chambers with tank buffer. The lower tank should be filled with 3.5-L tank buffer and stirred constantly at 4 °C in a cold room.
2. Electrophorese at a constant voltage set at 60 V until the bromophenol blue passes the stack. Increase the voltage stepwise by ~100 V every 15 min to 300 V and run at constant voltage; takes ~6 h to complete (*see* **Note 18**).
3. At the completion of electrophoresis, remove the gels from the glass plates. The stacking gel, dye front, and IEF strip should be removed and the gel soaked in fixative for ≥2 h with gentle swirling when necessary (*see* **Note 19**).

3.8 Image Acquisition

1. Acquire images using your chosen imaging system. The 2D DIGE experiment described will yield four images: two each from gel 1 and reciprocal gel 2. Each image is 1,024 × 1,280 pixels, with a resolution of 135 μm per pixel. These images should be stored as raw unsigned 16-bit data.

3.9 Image Analysis

Perform image analysis. Image analysis in our laboratory is done using several software applications: ImageJ and SExtractor. Visually inspect a two-frame looping movie of the Cy3 and Cy5 gel images to detect protein differences. The outcome of image analysis is a list of difference-protein spots that indicate significant differences between the two protein samples being compared.

In general, we rely on the two-frame looping movies to identify the significant protein changes we want to identify by MS. Visual difference-protein detection requires <1 h for an experienced user to complete. Precise quantification for comparative proteomics is performed by our SExtractor pipeline [10, 12, 29, 33]. To accurately assess the level of protein change, the Cy3 and Cy5 images are summed and submitted to SExtractor for spot detection, which generates a list of elliptical objects that specify individual protein spots. This list is used as a mask to determine the protein spot intensities in background-subtracted Cy3 and Cy5 images. This approach typically yields an SD of 5–7% for unchanging spots. Manual quantitative difference-protein analysis usually requires 10–15 h for an experienced user to complete. Additional computational spot detection methods are available [34, 35] but require manual editing. The computational methods are particularly useful when comparing multiple gels.

3.10 Spot Picking

Spot-cutting should be performed in a clean room to avoid contamination of the spots. Use the integrated spot-cutting tool to "click" on the spots of interest, cut a 1.8-mm-diameter plug from the gel, and deposit it in a 1.5 mL sample tube containing 1% acetic acid. The imager software allows one to create a list of protein spots to be retrieved. After all the desired spots are cut, the acetic acid solution should be removed from the sample tube and the plate moved to the digestion robot (*see* **Note 20**).

3.11 Protein Identification

Identify the difference proteins using MS. This can be done by routine methods using a variety of instruments. DIGE labeling with Cy3-NHS or Cy5-NHS labels only a small fraction of proteins in the lysate: ~5% of all proteins have a single dye molecule bound, whereas the rest have none. Hence, DIGE labeling does not affect the MS identification of proteins. It is important to note, however, that the DIGE imaging system is capable of detecting <0.2 fmol of protein. To identify low-abundance difference proteins, one must either enrich for the proteins of interest or pool protein from multiple DIGE gels (*see* **Note 21**).

3.12 Timing

1. Gel setup: 1 h.
2. Rehydrating the IEF strip: 15 min setup with already prepped reagents or 1 h. for setup with prepping fresh reagents. 12–24 h for rehydration.
3. Sample preparation: 1 h.
4. Protein labeling: 1 h.
5. First-dimension IEF: 17–19 h.
6. Transfer of IEF strip for second dimension: 1 h.
7. Second-dimension SDS-PAGE: 6–12 h.
8. Image acquisition: 30 min per gel.
9. Image analysis: 1–5 h per 24 × 18 cm gel with ~1500 protein spots.
10. Spot picking: 1–2 h.

4 Notes

1. Wear gloves at all times to avoid keratin contamination, which is a key obstacle to further analysis by MS.
2. Never heat urea solutions above room temperature (18–21 °C) as this will cause the formation of cyanate, which carbamylates protein and produces charge trains in 2D E gels [23]. The pH of this solution is critical for the labeling reaction. We typically store the lysis buffer in 1 mL aliquots at –80 °C for no longer than 2 weeks for the most consistent 2D E gel results.
3. Use fresh, dry DMF of the best quality. Store dye solutions at –80 °C and allow them to equilibrate to ice temperature before opening the tube. If one dye is consistently brighter in your specific imager despite identical protein loads, the initial concentration of dye solution may be altered to account for this to achieve more comparable images.
4. Slowly add concentrated HCl with stirring, until the pH reaches 8.0. This step should be done on ice in a fume hood. You will need ~43 mL acid to reach the desired pH. Add water to a final volume of 100 mL. Aliquot and store at –80 °C.
5. Store rehydration buffer at 4 °C. Use within 7 days. Can be stored at −80 °C for extended storage of about 2 months.
6. Gradient gels are optional. Standard 10–12% gels may be used. Alternatively, one can bypass this step by purchasing premade gradient gels for 2D E.
7. If pouring several gels at a time, rinse the gradient maker with DI water between gels.

   Use a magnetic stirrer and stir bar in the front chamber to ensure sufficient mixing of the heavy and light solutions.
8. Acrylamide is a neurotoxin; wear gloves and use appropriate handling precautions.

   Gels should be allowed to polymerize for ≥8 h if one intends to cut protein spots for MS. Unpolymerized acrylamide causes side-chain and amino-terminal modification of proteins, and these might pose problems for MS. Gels can be allowed to polymerize overnight. They should be layered with sufficient water-saturated *n*-butanol that the gels do not dry out. It is alright to pour the stacking gel just 30 min–6 h before the electrophoresis is set up. Commercial gels might be suitable, provided they do not contain any fluorescent contaminants. Gels might be stored at 4 °C for 2 days, provided care is taken to prevent drying.
9. We prefer to use sample cups to load protein samples. We do not recommend including the protein sample in the

rehydration solution as this leads to vertical and horizontal streaks and protein loss [36].

Make sure the IPG buffer matches the pH profile of the IEF strip. The Immobiline DryStrips need to be rehydrated in rehydration buffer for 8–24 h.

10. The majority of our samples are *Drosophila* embryos, which lyse easily by this method. Keep the centrifuge tube on ice during homogenization or work in a cold room. Some cell types might require more vigorous homogenization, such as *Saccharomyces cerevisiae*, which requires vortexing at high speed in the presence of glass beads after addition of lysis buffer. A glass homogenizer may be used for larger samples, but the lysate needs to be centrifuged at top speed in a microcentrifuge for 5–15 min at 4 °C to remove debris. Magnetic beads need to be removed completely if used for sample preparation—purifying organelles such as mitochondria.
11. Ideally the protein concentration of the lysate should be 1–2 mg/mL. If the lysate concentration is too high, dilute with lysis buffer. If the concentration is too low, the sample may be concentrated by precipitation, but this could create problems with protein loss during precipitation, re-solubilization, and pH adjustment. It is preferable to start with sufficient material to yield a suitably concentrated lysate. Another important consideration is the removal of contaminating factors that affect IEF, such as salt, lipids, detergents, and nucleic acids. There are a number of commercial protein cleanup kits that are worth testing; however, one must be cautious about sample bias via protein loss and pH effects. We recommend labeling the protein samples first and then combining the samples before using a protein cleanup protocol.
12. Try to use equal amounts of protein in each sample. Optionally you can also add 1 μL of 1 mg/mL BSA in lysis buffer to each tube as a loading control. Mix by vortexing.
13. To synchronize the labeling time, add the dye to the walls of all reaction tubes. The Cy3 and Cy5 reaction tubes containing samples to be loaded on the same gel are briefly spun, vortexed, and spun again in a microcentrifuge at top speed for a few seconds at 4 °C. Allow the dye stocks to warm up to ice temperature before opening.
14. Again, the solution is dispensed onto the walls of the tubes, and the reaction is started as described in the previous step. Incubate in the dark on ice for 30 min. The duration of quenching can be extended up to 2 h if necessary. After labeling, the samples can be stored at –80 °C. Thaw on ice prior to applying to the IEF strips.

15. The samples should be loaded onto IEF strips immediately after combining. Avoid unnecessary delays during this step to prevent the IEF strip from drying. The IEF gel is fragile and requires careful handling.
16. We cover the IEF machine with an opaque cover to limit photobleaching. The time taken to reach 25–32 kVh using the above protocol is typically between 17 and 19 h for various sample types. The duration depends on the sample and can vary from run to run depending on the amount of salt and other contaminants in the sample.
17. Wash all IEF accessories on finishing IEF. Use double-distilled water for washing. Wash strip holders with the IEF strip-cleaning detergent. Use Q-Tips. Avoid scratching. Rinse thoroughly with water and then absolute ethanol. Wash electrodes with water and ethanol. Clean strip-holder lids and the IPG-phor platform.
18. Directly increasing gel voltage from 80 V to 300 V may result in cracking the glass plates from thermal stress.
19. Gels can be stored in fixative for several months without significant loss of fluorescent signal. For long-term storage, refrigeration is recommended. However, protein spots to be identified by MS should be cut out and frozen at –80 °C within 1 week.
20. A key difference between the gel-imaging system we use and the Typhoon imaging system is that we use an open-gel format that is amenable to imaging and spot picking within the same device, whereas the Typhoon imager can image gels in open-gel format, but spot picking is done on a different device. The open-gel format does not require the gel to adhere to one of the two electrophoresis plates.

    The cut gel fragments can be stored at –80 °C for at least several weeks.
21. There are two main reasons for poor results: insufficient protein labeling and incomplete IEF. Insufficient labeling is commonly due to the pH of the lysate being <8.0 or the presence of excess primary amines, such as Tris. If there is sufficient sample, one can test the pH using a small piece of pH paper. Tris <50 mM does not appear to interfere with protein labeling, and 30 mM is recommended. Incomplete focusing is usually the result of interfering contaminants, such as salts and nucleic acids. Salts, which can be removed using a protein cleanup kit, prevent the IEF from reaching the desired voltage. Nucleic acids make the sample viscous and difficult to handle and also interfere with focusing. Nucleic acids can be removed by commercially available cleanup kits or can be broken up by sonication or the addition of nuclease [1].

There will always be gel-to-gel variation for 2D E gels. This is due to a myriad of uncontrollable features in manufacturing, sample preparation, and electrophoretic conditions. DIGE was developed to circumvent some of this variability. The digital-imaging system and fluorescence labeling also provide a three or four orders of magnitude linear response, which provides a much more accurate estimation of protein abundance than silver or Coomassie blue staining of proteins. DIGE protein differences can be visualized in a variety of ways, including color overlay (Fig. 1) and two-frame looping movies.

Acknowledgments

The development and optimization of this DIGE protocol would not have been possible without the efforts and dedication of my students and staff: Mustafa Ünlü, Liz Morgan, Chris Lacenere, Surya Viswanathan, Lei Gong, Mamta Puri, Susan Dowd, and Amber Lucas, as well as current and past members of the undergraduate proteomics platoon.

References

1. O'Farrell PH (1975) High resolution two-dimensional electrophoresis of proteins. J Biol Chem 250:4007–4021
2. Scheele GA (1975) Two-dimensional gel analysis of soluble proteins. Characterization of guinea pig exocrine pancreatic proteins. J Biol Chem 250:5375–5385
3. Klose J (1975) Protein mapping by combined isoelectric focusing and electrophoresis of mouse tissues. A novel approach to testing for induced point mutations in mammals. Humangenetik 26:231–243
4. Gorg A, Postel W, Gunther S (1988) The current state of two-dimensional electrophoresis with immobilized pH gradients. Electrophoresis 9:531–546
5. Hamdan M, Righetti PG (2005) Proteomics today: protein assessment and biomarkers using mass spectrometry, 2D electrophoresis, and microarray technology. John Wiley & Sons, Hoboken, N.J.
6. Miller MJ, Olson AD, Thorgeirsson SS (1984) Computer analysis of two-dimensional gels: automatic matching. Electrophoresis 5:297–303
7. Han CM, Park JH, Chang CS, Ryoo MC (2014) A spot-matching method using cumulative frequency matrix in 2D gel images. Biotechnol Biotechnol Equip 28:S37–S43
8. Dowsey AW, Morris JS, Gutstein HB, Yang GZ (2010) Informatics and Statistics for Analyzing 2-D Gel Electrophoresis Images. Methods Mol Biol 604:239–255
9. Ünlü M, Morgan ME, Minden JS (1997) Difference gel electrophoresis: a single gel method for detecting changes in protein extracts. Electrophoresis 18:2071–2077
10. Viswanathan S, Ünlü M, Minden JS (2006) Two-dimensional difference gel electrophoresis. Nat Protoc 1:1351–1358
11. Lilley KS, Friedman DB (2004) All about DIGE: quantification technology for differential-display 2D-gel proteomics. Expert Rev Proteomics 1:401–409
12. Gong L, Puri M, Unlu M, Young M, Robertson K, Viswanathan S, Krishnaswamy A, Dowd SR, Minden JS (2004) Drosophila ventral furrow morphogenesis: a proteomic analysis. Development 131:643–656
13. Shaw MM, Riederer BM (2003) Sample preparation for two-dimensional gel electrophoresis. Proteomics 3:1408–1417
14. Lilley K (2002) Protein profiling using two-dimensional difference gel electrophoresis (2-D DIGE). In: Coligan JE, Dunn BM, Speicher DW, Wingfield PT (eds) Current protocols in protein science. John Wiley & Sons, Inc., New York, pp 22.22.21–22.22.14

15. Marouga R, David S, Hawkins E (2005) The development of the DIGE system: 2D fluorescence difference gel analysis technology. Anal Bioanal Chem 382:669–678
16. Wang H, Clouthier SG, Galchev V, Misek DE, Duffner U, Min CK, Zhao R, Tra J, Omenn GS, Ferrara JL, Hanash SM (2005) Intact-protein-based high-resolution three-dimensional quantitative analysis system for proteome profiling of biological fluids. Mol Cell Proteomics 4:618–625
17. Okano T, Kondo T, Kakisaka T, Fujii K, Yamada M, Kato H, Nishimura T, Gemma A, Kudoh S, Hirohashi S (2006) Plasma proteomics of lung cancer by a linkage of multidimensional liquid chromatography and two-dimensional difference gel electrophoresis. Proteomics 6:3938–3948
18. Diez R, Herbstreith M, Osorio C, Alzate O (2010) 2-D Fluorescence Difference Gel Electrophoresis (DIGE) in Neuroproteomics. In: Alzate O (ed) Neuroproteomics. CRC Press/Taylor & Francis, Boca Raton (FL)
19. Martyniuk CJ, Alvarez S, Denslow ND (2012) DIGE and iTRAQ as biomarker discovery tools in aquatic toxicology. Ecotoxicol Environ Saf 76:3–10
20. Arentz G, Weiland F, Oehler MK, Hoffmann P (2015) State of the art of 2D DIGE. Proteomics Clin Appl 9:277–288
21. Kondo T, Seike M, Mori Y, Fujii K, Yamada T, Hirohashi S (2003) Application of sensitive fluorescent dyes in linkage of laser microdissection and two-dimensional gel electrophoresis as a cancer proteomic study tool. Proteomics 3:1758–1766
22. Sitek B, Potthoff S, Schulenborg T, Stegbauer J, Vinke T, Rump LC, Meyer HE, Vonend O, Stuhler K (2006) Novel approaches to analyse glomerular proteins from smallest scale murine and human samples using DIGE saturation labelling. Proteomics 6:4506–4513
23. Zhou G, Li H, DeCamp D, Chen S, Shu H, Gong Y, Flaig M, Gillespie JW, Hu N, Taylor PR, Emmert-Buck MR, Liotta LA, Petricoin EF 3rd, Zhao Y (2002) 2D differential in-gel electrophoresis for the identification of esophageal scans cell cancer-specific protein markers. Mol Cell Proteomics 1:117–124
24. Karp NA, Lilley KS (2005) Maximising sensitivity for detecting changes in protein expression: experimental design using minimal CyDyes. Proteomics 5:3105–3115
25. Tonge R, Shaw J, Middleton B, Rowlinson R, Rayner S, Young J, Pognan F, Hawkins E, Currie I, Davison M (2001) Validation and development of fluorescence two-dimensional differential gel electrophoresis proteomics technology. Proteomics 1:377–396
26. Yan JX, Devenish AT, Wait R, Stone T, Lewis S, Fowler S (2002) Fluorescence two-dimensional difference gel electrophoresis and mass spectrometry based proteomic analysis of Escherichia coli. Proteomics 2:1682–1698
27. Van PT, Ganesan V, Bass V, Parthasarathy A, Schlesinger D, Minden JS (2014) In-gel equilibration for improved protein retention in 2DE-based proteomic workflows. Electrophoresis 35:3012–3017
28. Sellers KF, Miecznikowski J, Viswanathan S, Minden JS, Eddy WF (2007) Lights, Camera, Action! systematic variation in 2-D difference gel electrophoresis images. Electrophoresis 28:3324–3332
29. Van PT, Bass V, Shiwarski D, Lanni F, Minden J (2014) High dynamic range proteome imaging with the structured illumination gel imager. Electrophoresis 35:2642–2655
30. Bertin E, Arnouts S (1996) SExtractor: software for source extraction. Astron Astrophys 117:393–404
31. Suehara Y, Kondo T, Fujii K, Hasegawa T, Kawai A, Seki K, Beppu Y, Nishimura T, Kurosawa H, Hirohashi S (2006) Proteomic signatures corresponding to histological classification and grading of soft-tissue sarcomas. Proteomics 6:4402–4409
32. Shaw J, Rowlinson R, Nickson J, Stone T, Sweet A, Williams K, Tonge R (2003) Evaluation of saturation labelling two-dimensional difference gel electrophoresis fluorescent dyes. Proteomics 3:1181–1195
33. Blundon MA, Schlesinger DR, Parthasarathy A, Smith SL, Kolev HM, Vinson DA, Kunttas-Tatli E, McCartney BM, Minden JS (2016) Proteomic analysis reveals APC-dependent post-translational modifications and identifies a novel regulator of beta-catenin. Development 143:2629–2640
34. Millioni R, Miuzzo M, Sbrignadello S, Murphy E, Puricelli L, Tura A, Bertacco E, Rattazzi M, Iori E, Tessari P (2010) Delta2D and Proteomweaver: Performance evaluation of two different approaches for 2-DE analysis. Electrophoresis 31:1311–1317
35. Goldfarb M (2007) Computer analysis of two-dimensional gels. J Biomol Tech 18:143–146
36. Zhou S, Bailey MJ, Dunn MJ, Preedy VR, Emery PW (2005) A quantitative investigation into the losses of proteins at different stages of a two-dimensional gel electrophoresis procedure. Proteomics 5:2739–2747

Chapter 21

Immunoelectrophoresis: A Method with Many Faces

Gyorgy Csako

Abstract

Immunoelectrophoresis (IEP) was the first practical method that combined electrophoresis and immunoprecipitation for identifying and characterizing proteins within complex mixtures. Over the years, IEP has been extended to include a variety of techniques and, as a general name, has been applied to virtually any technique that involves electrophoresis and antigen-antibody precipitin reaction for proteins. Because of the diversity in technical details of different IEP versions, the method described here deals only with classic IEP. Although it requires some manual expertise, IEP is versatile, relatively easy to customize, and economical with no need for expensive instrumentation. Further, it can discern identity, partial identity, and nonidentity of the proteins. Any low-viscosity body fluid specimen or, possibly, culture fluid and tissue extract could be tested with IEP if proper antibodies are available. With these attributes, classic IEP remains a valuable tool for clinical diagnostic testing, purity checking of biochemical and pharmaceutical products, and research.

Key words Countercurrent immunoelectrophoresis, Counterimmunoelectrophoresis, Crossed immunoelectrophoresis, Double immunodiffusion, Immunoglobulins, Immunoprecipitation, Laurell crossed immunoelectrophoresis (two-dimensional electroimmunodiffusion), Reversed rocket immunoelectrophoresis, Rocket immunoelectrophoresis, Starch gel

1 Introduction

In its classic form, first devised by Grabar and William [1–3] and then modified as a micromethod by Scheidegger [4, 5], immunoelectrophoresis (IEP) combines separation of protein antigens by electrophoresis and immunodiffusion against an antiserum. The proteins are first separated by electrophoresis in semisolid media (gels), and then the specific proteins are visualized through an immunologic reaction, namely, antigen-antibody precipitin formation. Double immunodiffusion ensues after an antiserum prepared against the target proteins is applied into a trough cut into the gel parallel to the direction of electrophoretic separation and distinct elliptical precipitin arcs develop with the antisera for each protein antigen. Over the years, however, IEP has been extended to include

Biji T. Kurien and R. Hal Scofield (eds.), *Electrophoretic Separation of Proteins: Methods and Protocols*, Methods in Molecular Biology, vol. 1855, https://doi.org/10.1007/978-1-4939-8793-1_21,

a variety of techniques and, as a general name, has been applied to virtually any technique that involves electrophoresis and antigen-antibody precipitin reaction for separation, characterization, and, possibly, quantitation of proteins. In fact, the initial reports about the immunofixation electrophoresis (IFE) technique were misleadingly also called "quantitative IEP" [6] and "direct IEP" [7]. "True" IEP techniques now include two versions of the one-dimensional (classic) IEP [1–5], counter(-current) or cross-over (immuno) electrophoresis or electrosyneresis [8–10], crossed IEP ([two-dimensional] quantitative IEP, two-dimensional electro-immunodiffusion) [11–13], rocket IEP or electroimmunoassay [14–17], fused rocket IEP [18], affinity IEP or crossed immuno-affinoelectrophoresis [19], line immunoelectrophoresis [20], crossed radioimmunoelectrophoresis [21], crossed-line immuno-electrophoresis [22], rocket-line immunoelectrophoresis [23], and numerous other modifications of these techniques (e.g., ref. 24). Even reverse version of some of these assays has been developed (e.g., reverse rocket IEP [25]). Because of their shared concept, these methods are all capable of identifying (and, depending on assay design, quantifying) antigen(s) (or antibodies in reversed versions) in complex specimens.

With much improved resolution and quantitation of proteins in biological fluids and biological extracts, the introduction of the IEP analysis had a great impact on protein chemistry and laboratory diagnostics. Interestingly, the history of classic IEP for human paraprotein immunotyping follows an opposite pattern to IFE. IFE was introduced for improved detection of protein polymorphisms and only years later was applied to the study of paraproteins as a clinical diagnostic technique, but it became the preferred test in this area. Classic IEP was developed primarily as a clinical diagnostic test for the detection and identification of paraproteins, but then it was gradually replaced by IFE, so that now it is only selectively used for that purpose. On the other hand, classic IEP is still widely employed for purity checks and identification of biochemical and pharmaceutical products (e.g., in major official reference publications for drug specifications [26, 27]), and research. Further, with its many new versions, IEP re-invented itself for both clinical diagnostic and research applications. The following is the description of a classic IEP method for paraprotein immunotyping which can be readily modified for other diagnostic or research applications.

2 Materials

Prepare all solutions with distilled or deionized water and high-grade analytical reagents at room temperature (15–30 °C). Except stated otherwise, reagents and other materials are stored at room temperature. Unless obtained commercially, prepare and store all reagent and accessories used in these methods without adding sodium azide. Follow all waste disposal regulations for discarding reagents and disposables. As a safe laboratory practice, always wear clean gloves to minimize or eliminate exposure to infectious agents and hazardous chemicals.

2.1 Agarose Gel Plates for Protein Electrophoresis (See Notes 1 and 2)

1. Hydragel 7 and Hydragel 15/30 Protein(e) alkaline buffered precast 0.8% (w/v) agarose gels (Sebia, Norcross, GA, USA). These gels contain Tris-barbital buffer (pH 9.2 $\pm$ 0.1) and proprietary additives which are claimed to be "nonhazardous" at concentrations used. Store horizontally at room temperature or refrigerate (2–8°C) until the stated expiration date.

2.2 Accessories for Preparing Agarose Gel Plates for Electrophoresis and Immunodiffusion (See Note 3)

1. Templates for positioning wells and troughs (Fig. 1a) (*see* **Note 4**).
2. "Cutting templates" for antiserum application (Fig. 1a) (*see* **Notes 3** and **5**).
3. Suction needle as a well cutter (well diameter of ~1 mm) (*see* **Notes 3** and **6**).
4. Scalpel.
5. Glass Pasteur pipet connected by tubing to a filter flask (*see* **Note 7**).
6. Filter flask connected by tubing to a vacuum source (*see* **Notes 3** and **5**).

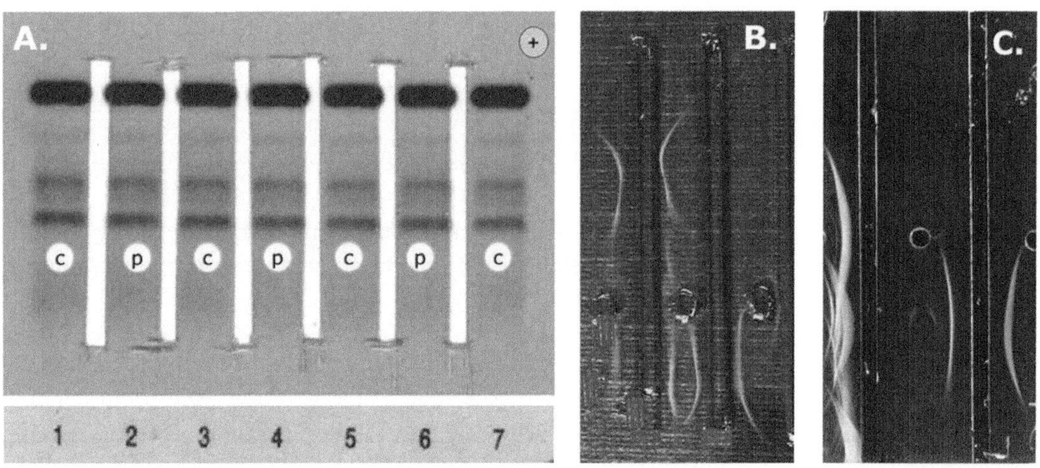

Fig. 1 Positioning template for wells and troughs, gel-cutting template for antiserum application, and photographs of wet agarose gel with precipitin arcs for IEP. (**a**) The "homemade" gel-cutting template for making troughs is shown on top of the "positioning" template (1:1 paper copy of amido black-stained human serum protein patterns after electrophoresis in agarose gel). Codes: c, control sample positions; p, patient sample positions. (**b**) Photograph of wet agarose gel that shows a developing human serum IEP pattern using "homemade" devices. (**c**) Photograph of wet agarose gel that shows part of a completed human serum IP pattern produced by a discontinued commercial kit

7. Leveling table.
8. Pipets or syringes for delivering volumes between 0.5 and 50 μL.
9. Wet incubation chamber (*see* **Note 8**).

2.3 Apparatus and Accessories for Protein Electrophoresis in Agarose Gel

1. Hydrasys® semiautomated electrophoresis system (Sebia) (*see* **Note 9**).
2. Buffered, ready to use sponge strips for protein electrophoresis (Sebia). Each contains Tris-barbital buffer pH 9.2 $\pm$ 0.3, sodium azide, and proprietary additives which are claimed to be "nonhazardous" at the concentrations used. Store horizontally at room temperature or refrigerated until stated expiration date (*see* **Note 10**).
3. Precut, single use filter papers. Store in a dry place at room temperature or refrigerated.
4. Pipet for delivering 200 μL.
5. Bromophenol blue sodium salt: 0.05% (w/v) solution in a Tris-barbital buffer pH 8.6 (first, prepare a 5× buffer in a measuring cylinder); dissolve 0.24 *g* 5,5'-diethyl-barbituric acid and 4.43 *g* Tris in distilled water to a final volume of 100 mL. Dilute by a factor of five before use. Store in a closed bottle for 6 months. Second, weigh 5 mg of the dye powder (3% difference due to the sodium content is negligible) and pour the dye into a 15 mL tube with a screw cap. Add 10 mL of the fivefold diluted stock buffer to the tube and cap it. Shake the tube well to dissolve the dye in the buffer. Store for up to 1 year (*see* **Note 11**).

2.4 Antibodies for Immunodiffusion (See Notes 12, 13, and 14)

1. Polyclonal antihuman whole serum and monospecific rabbit antibodies to human IgG, IgA, IgM, and κ and λ (free and bound) light chains.
2. Reconstitute lyophilized preparations according to the manufacturer's instructions. Store all antisera refrigerated until stated expiration date.

2.5 Solutions and Accessories for Protein Staining in Agarose Gel (See Note 15)

1. Amido black stain. Reconstitute and store according to the manufacturer's instructions. The working solution is acidic (pH 2) and contains 0.4 g/dL amido black, 6.7% (v/v) ethylene glycol, and additives which are claimed to be "nonhazardous" at concentrations used. Working solution in a closed container to prevent evaporation is stable for 1 month (*see* **Note 16**).
2. Destaining solution: distilled water.
3. Staining pans.
4. Paragon dryer (Beckman).

2.6 Specimens for IEP in Agarose Gel (See Note 17)

1. Human serum for illustration of the method (*see* **Note 18**).
2. Collect blood specimens with standard medical procedures. Store serum refrigerated for up to 1 week (preferably use within 48 h) and freeze at −20 °C for up to 3 months or at −70 °C for up to 12 months.
3. Use serum samples straight or dilute with 0.85% saline when antigen excess is suspected (*see* **Note 19**).

2.7 Control (Reference) Specimens for IEP in Agarose Gel (See Note 20)

1. Normal human serum or serum pool for paraprotein testing. Store serum refrigerated for up to 1 week (preferably use within 48 h) and freeze at −20 °C for up to 3 months or at −70 °C for up to 12 months.

3 Methods

Carry out all procedures at room temperature unless otherwise specified. For general guidance and safety issues, consult the operating manual of the electrophoretic system used. Estimated total time needed to complete one gel with the electrophoretic and detection method described below is ~48 h.

3.1 Preparation of Wells and Troughs in Agarose Gel

1. Unpack a Hydragel plate selected for IEP (Hydragel 7 or 15/30 Protein[e]) and roll quickly and uniformly one thin filter paper onto the gel surface to absorb excess liquid. Remove and discard the paper immediately.
2. Place the gel plate over the appropriate template showing the position of wells for sample application and troughs for antiserum application (Fig. 1a) (*see* **Note 4**).
3. Carefully overlay the gel with the "cutting template" (Fig. 1a) (*see* **Note 5**).
4. Cut the desired number of troughs with a scalpel and aspirate the gel from the troughs by suction using a glass Pasteur pipet connected to the side arm of a filter flask which, in turn, is connected to a vacuum line. Avoid damaging the walls of the trough. Clean blade often by immersion into distilled water and wiping it dry on paper towels laid down on the benchtop. For safety, do *not* hold towels in your hand while wiping!
5. Remove the cutting template by making sure that the gel is not damaged (e.g., not lifted up from its plastic support) (*see* **Note 21**).
6. Guided by the template placed under the gel for assay layout, punch sample application wells and simultaneously aspirate the agarose plugs with the suction needle connected to the side arm of a filter flask which, in turn, is connected to a vacuum line. Wash suction needle between punching wells by aspirating distilled water (*see* **Note 6**).

3.2 Sample Preparation and Loading onto Precut Agarose Gel

1. Add 2 μL of 0.05% bromophenol blue solution to 18 μL of control specimen and briefly (1–2 s) vortex it. This results in a final dye concentration of 0.005% (w/v).
2. Apply 2 μL of control serum sample to each application well as follows: positions 1, 3, 5, and 7 for testing one unknown specimen on a 7 Protein(e) plate and positions 1, 3, 5, 7, 9, 11, 13, and 15 for testing two unknown specimens on a 15 Protein(e) plate (*see* **Note 22**).
3. Apply 2 μL of unknown serum sample to each application well as follows: positions 2, 4, and 6 for testing one unknown specimen on a 7 Protein(e) plate and positions 2, 4, 6, 12, and 14 for testing two unknown specimens on a 15 Protein (e) plate. Note that position 8 receives no sample; this position separates the two unknown specimen runs (*see* **Note 22**).

3.3 Protein Electrophoresis in Agarose Gel

Switch on the Hydrasys instrument and proceed according to the Hydragel 7 or 15/30 Protein(e) method for migration. Briefly,

1. Attach buffered strips as appropriate for protein run to the pins on the uplifted electrode carrier.
2. Place the 7 Protein(e) or 15/30 Protein(e) plate preloaded with samples into the migration module by making sure that the sample application wells are toward the cathode.
3. Select migration program appropriate for the type of agarose plate.
4. Run electrophoresis under 10 W constant for serum proteins using Hydragel 7 Protein(e) plate (for 1 unknown IEP specimen) or 20 W constant using Hydragel 15/30 Protein(e) plate (for 2 unknown IEP specimens) at 20 °C, controlled by Peltier effect, until 33 Vh have accumulated (~7 min).
5. Interrupt electrophoretic program after migration has been completed.

3.4 Immunodiffusion in Agarose Gel

1. After stopping electrophoresis, remove gel from the migration module and wipe excess solution from back of gel.
2. Verify proper migration by observing the position of the tracking dye, bromophenol blue: the second fastest migrating blue spot showing the position of albumin should be slightly above the anodic end of the troughs.
3. Gently blot the gel surface with thin filter paper. Discard paper.
4. Place the plate on a leveling platform.
5. Slowly apply 15 μL of various antisera into the anodic end of the appropriate troughs. Although antiserum usually spreads along the trough by capillary action, make sure that it has evenly spread along the entire length of the trough (*see* **Note 23**).

6. Place gel in level humidity incubation chamber.
7. Cover chamber and incubate gel for 18–24 h in a refrigerator (4°°C) or until visible precipitin arcs develop at expected locations with special attention to the controls in the gel (Fig. 1b, c) (*see* **Note 24**).
8. After ~16 h of incubation or at the earliest convenient time, begin checking the progress of precipitin formation at every few hour. Visible precipitin arcs begin to appear after 6–8 h, and the immunoprecipitation pattern is fully developed after 24 h in this system.

3.5 Processing of Agarose Gel for Staining (see Note 25)

1. At the end of incubation, remove gel from wet incubation chamber.
2. Stop immunodiffusion by washing the gel in ~75 mL of 0.85% saline on a rocking platform at a low speed for 1 h.
3. Continue washing with saline for 6–24 h. During this time, change saline five to ten times to increase efficiency of washing (obviously, more frequently for shorter wash time).
4. Soak gel in ~75 mL of distilled water with gentle shaking on a platform (or just by hands) to desalt for ~1 h. Change water two to three times during this time.
5. Wipe excess solution from back of gel and dry gel in a Paragon dryer at 60 °C (5–10 min).

3.6 Amido Black Staining for Proteins (See Notes 26 and 27)

1. Stain the dried gel by immersing into the amido black stain for 3–5 min at room temperature under gentle agitation.
2. Destain gel with distilled water at room temperature under gentle agitation until the desired low background level is achieved.
3. Remove gel from water, wipe water from back of gel, and dry gel in a Paragon dryer for 5–10 min at 60 °C.

3.7 Interpretation of IEP Patterns (See Notes 28 and 29)

General comments:

1. Evaluate visually the gel by noting
   (a) The presence of precipitin arc(s) at the expected location with each antiserum in the control (reference) sample.
   (b) The location and shape of specific precipitin arc(s) with each antiserum in the unknown sample in relation to those in the control (reference) sample.
2. In case of paraprotein testing, distortion of an arc (broadening, bulging, or displacement) indicates paraprotein. Matching the location of distortions with the heavy chain-specific and light chain-specific antisera identifies the immunotype of paraprotein (s).

3. In case of paraprotein testing, if technical errors such as inappropriate or low-avidity antisera can be ruled out, then
   (a) The absence of heavy chain abnormality in the presence of light chain abnormality suggests the possibility of IgD or IgE paraprotein which requires further testing with specific antisera against these immunoglobulins (*see* **Note 30**).
   (b) The absence of light chain abnormality in the presence of heavy chain abnormality suggests the possibility of heavy chain disease.
   (c) The absence of abnormality in the routine IEP pattern but abnormal band in the gamma region on protein electrophoresis suggests the possibility of free light chain, IEP artifact (the so-called umbrella effect), or pseudoparaprotein (e.g., fibrinogen) (*see* **Note 29**).

Specific comments on IEP patterns produced with the manual method on human serum:

1. Fig. 2: Even this minimally optimized IEP method shows abnormalities (distortion exemplified by broadening) at the cathodal end of the IgG and κ precipitin arcs with the patient serum (**c**), indicating the presence of an IgG(κ) paraprotein. This paraprotein is even more apparent in the corresponding IFE pattern (**b**).
2. Fig. 3: IFE and IEP performed with optimized assays both show unequivocal presence of an IgG(κ) paraprotein. Note characteristic distortion (primarily broadening) of the precipitin arcs that correspond to the paraprotein.
3. Fig. 4: Since IEP assays do not routinely test for the presence of free light chains, testing with anti-free light chain antibodies may be indicated when either clinical suspicion arises or discordant routine IEP and protein electrophoretic results are obtained (no abnormality in the routine IEP pattern but abnormal band in the gamma region on protein electrophoresis). This figure shows a case where routine IEP panel suggested a normal pattern, but testing with anti-free light chain antibodies revealed a weak but distinct band of free κ light chain in the patient's serum.

4 Notes

1. Although agar and agarose gel have been used in the majority of classic IEP assays, applications have been also reported for starch, acrylamide, and cellulose acetate (Cellogel) gels and even for dry cellulose acetate membrane [28]. Agar, the

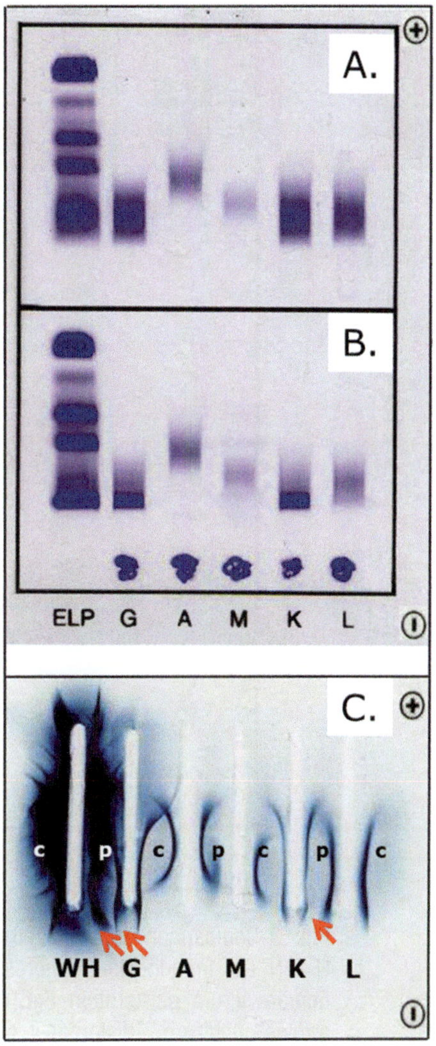

Fig. 2 Comparison of the manual IEP method with IFE obtained by a semiautomated commercial method for human serum paraprotein immunotyping in agarose gel. (**a**) Normal control human serum and (**b**) human serum with IgG(κ) paraprotein, both analyzed with Hydragel 2 IF kit on Hydrasys[R] followed by staining with acid violet (Sebia Inc., Norcross, GA, USA), and (**c**) minimally optimized manual EIP method on a protein electrophoresis plate using a hand-cut IEP antiserum template for cutting the troughs prior to electrophoresis and manually applying various antisera after electrophoresis, followed by amido black staining. Arrows indicate abnormal distortion and broadening of precipitin arcs. Codes: ELP, acid-fixed protein reference lanes; G, A, and M, monospecific antibodies against human IgG, IgA, and IgM heavy chains; K and L, monospecific antibodies against human κ and λ (free and bound) light chains; WH, antibodies against whole human serum; c, control serum (same as in A); p, patient serum (same as in B)

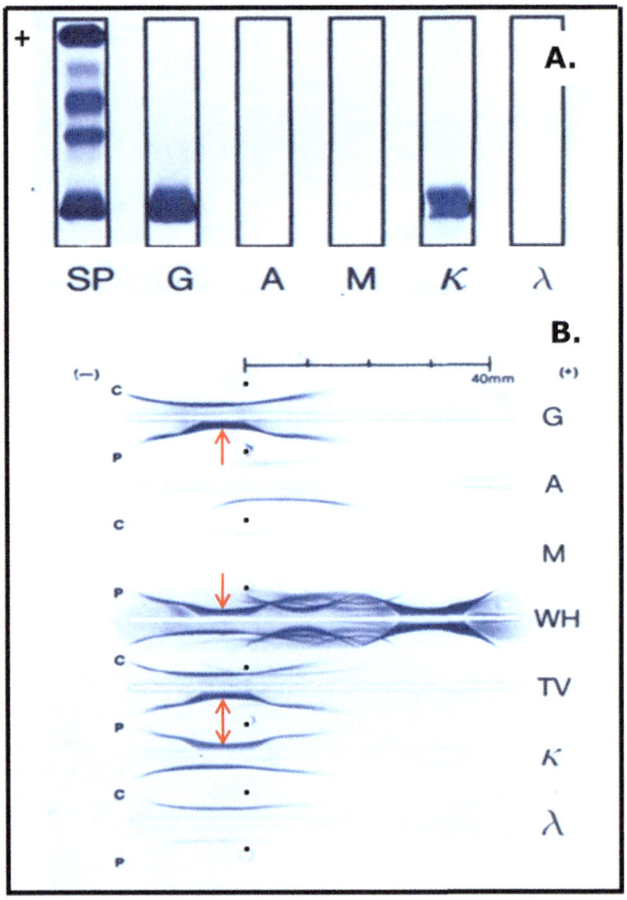

Fig. 3 Comparison of optimized IFE (TITAN Gel ImmunoFix, Helena) and IEP (TITAN Gel Immunoelectrophoresis, Helena) methods for immunotyping of a human serum paraprotein (IgG[κ]). Dots show sites of sample application; arrows indicate distortion (abnormal widening) of precipitin arcs. Immunoglobulin quantitation by radial immunodiffusion: IgG 3,524 mg/dL, IgA <40 mg/dL, and IgM 20 mg/dL. Codes: SP, acid-fixed protein reference lane; G, A, and M, monospecific antibodies against human IgG, IgA, and IgM heavy chains; K and L, monospecific antibodies against human κ and λ (free and bound) light chains; WH, polyclonal antibodies against whole human serum; TV, trivalent antiserum against IgG, IgA, and IgM heavy chains; c, control (reference) serum; p, patient serum. (Courtesy of Helena Laboratories, Beaumont, TX, USA)

original medium for IEP, consists of a mixture of agarose and agaropectin. Agarose, the main component of agar, is a linear polymer of the repeating monomeric unit of a disaccharide (agarobiose). Agaropectin is a similar polymer but contains numerous acidic side groups and gels poorly. The neutral charge and relatively simple chemical composition of agarose coupled with the high gel strength, low electroendosmosis, large pore size, and low background staining of agarose gel

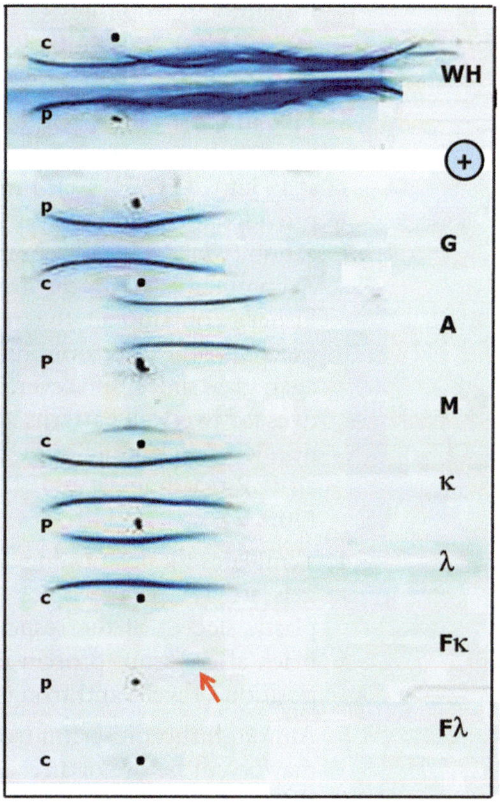

Fig. 4 Detection of serum free κ light chain with IEP (TITAN Gel Immunoelectrophoresis, Helena). Dots show sites of sample application; arrow indicates free κ light chain. Codes: WH, polyclonal antibodies against whole human serum; G, A, and M, monospecific antibodies against human IgG, IgA, and IgM heavy chains; κ and λ, monospecific antibodies against human κ and λ (free and bound) light chains; Fκ and Fλ, monospecific antibodies against human free κ and free λ light chains. (Courtesy of Helena Laboratories, Beaumont, TX, USA)

made this gel the most popular medium for IEP. Most agarose gels are prepared as 0.7–1.5% slabs of ~1 mm thickness. The large pores of agarose gels allow free passage and separation of proteins, while providing an anchor for the immunoprecipitates of protein antigens and their specific antibodies.

2. In a currently available diagnostic kit for human paraprotein immunotyping (TITAN Gel Immunoelectrophoresis, Helena Laboratories, Beaumont, TX, USA), the IEP plates contain 1% (w/v) agarose. A general purpose commercial kit for IEP includes agar (Immunoelectrophoresis kit, Thermo Fisher Scientific Inc., Waltham, MA, USA), while two different commercially available teaching kits for IEP both supply agarose for preparing gel-coated glass slides (EDVO-Kit #272 Immunoelectrophoresis, EDVOTEK^{R}, Inc., Rockville, MD, USA [kit also marketed by Thermo Fisher] and Manual

Immunoelectrophoresis Teaching Kit, Catalog #IMI-KIT-1022, IMGENEX India Pvt. Ltd., Bhubaneswar, India) (websites for all three kits accessed on July 3, 2011).

3. Like in most earlier IEP kits for human paraprotein immunotyping, the agarose plates in the currently available diagnostic IEP kit (TITAN Gel Immunoelectrophoresis, Helena) are provided as "ready-to-use", i.e., precut for both sample application (wells) and subsequent antiserum application (troughs). A commercial IEP kit (Immunoelectrophoresis kit, Thermo Fisher) contains all the required materials to perform IEP, except for the electrophoresis system and power supply. Besides agar, glass slides, and several accessories, these materials include dyes for two well patterns, cutters for 1–2 mm troughs, suction needle, and gel knife. Along with agarose and other accessories, an IEP teaching kit by EDVOTEK^{R} supplies well cutters (*see* **Note 2**).
4. Templates for positioning wells and troughs in 7 and 15 sample gels are 1:1 paper copies (preferably placed in a transparent plastic sleeve) of the respective gel plates stained with amido black after serum protein electrophoresis and marked for the position of wells and troughs (Fig. 1a).
5. Although the antiserum troughs (also called trenches and slits) may be cut before or after the electrophoresis, for convenience they are often cut before as was the case here. The "cutting templates" for 7 and 15 position gels used here were custom cut (~1 mm slits) with a scalpel from a previously exposed X-ray film using the appropriate templates for well and trough positioning as a guide (Fig. 1a). A Pasteur pipet connected to the side arm of a filter flask, serving as a trap to receive pieces of gel and rinse water for protection of the vacuum line, was used to "clear" the troughs. Historically, troughs have been cut, in the simplest way, by means of single-edge razors, scalpels, or sharp (gel) knives using only "positioning templates." Most commonly, however, troughs were prepared by means of "gel cutters" which included blades [29]. A simple trough cutter can be made with a pair of single-edge razor blades bolted together across metal spacers with machine screws [30]. A U-shaped trough cutter device constructed from an aluminum profile and scalpel blade for making ~2 mm wide troughs was described in detail in an early book about IEP [29]. In the "scoop method," a metal scoop was made of 24-gauge stainless steel bent in the form of a U along its long axis [31]. A commercial IEP kit includes cutters for 1–2 mm troughs along with suction needle (Thermo Fisher), whereas an IEP teaching kit reports a gel "displacement" method for making troughs (EDVOTEK^{R}). Most commonly, trough widths range from 1 to 2 mm (*see* **Note 2**).

6. A suction needle for making sample application wells in the gel can be "homemade" by cutting off the point of a hypodermic needle and sharpening the cut end on the inside surface.

   These "well cutters" are then connected to a filter flask which, in turn, is connected to a vacuum source. This suction needle may also be used to remove pieces of gel when cutting troughs (*see* **Note 5**). As a guide for choosing a hypodermic needle for gel punching, the nominal outer diameter is 1.1 mm for a 19-gauge, 1.5 mm for a 17-gauge, 2.1 mm for a 14-gauge, 3.0 mm for an 11-gauge, and 4.2 mm for an 8-gauge regular-wall needle. Gel punchers also can be made by cutting micro-pipette tips or using the back of these tips for making (large) sample application wells. Instead of using vacuum, the gel plugs can be removed with a toothpick or, in case of larger wells, with forceps. In an early book about IEP, a copper tube was used for cutting wells [29]. A commercial IEP kit provides "dies for two well patterns" and "suction needle" (Thermo Fisher), whereas an IEP teaching kit supplies "well cutters" (EDVOTEK^R) for making sample application wells (*see* **Note 3**). Note that small sample volumes (up to ~5 μL) may be applied directly to the gel surface with a micropipette and this approach may be even preferred in case of very small volumes (0.5–2 μL). In some experiments, we also used this technique. However, this technique requires steady hands for precise sample application. In general, samples >5 μL applied as a droplet spread too widely on the gel surface, and direct application is not advisable [30]. As another alternative to using wells, application templates made of thin plastic sheets either commercially or "handmade" in the laboratory may be used, especially for somewhat larger sample volumes (>2 μL). When directly applied to the surface, wait for 3–5 min for the serum to absorb into gel. The diameter of sample application wells in different applications varies from 1–2 mm to 8 mm but most often probably is ~2 mm. Since the volume of these wells also depends on the thickness of the gel, sample volumes ranging from 1–2 μL to 30–40 μL can be loaded into these wells. However, most often only a few μL (commonly 1–2 μL) of sample is electrophoresed. Besides the size of the application wells, their position relative to the troughs also is an important factor to consider. In case of precast gels, this distance generally is defined by the manufacturer's plate design, though deviation from that is possible. In case of custom cast gels, the user can freely optimize the system for IEP. Although special needs allow for deviations, the distance between the rim of the application well and the edge of the nearest trough should be at least ~2 mm and <5 mm for optimal immunoprecipitation to occur.

7. This is used to aspirate pieces of gel from the troughs after cutting with scalpel.
8. Any level humidifying incubation chamber of proper size to accommodate the gels (e.g., a plastic container with wet paper towel at the bottom and tightly covered with a lid or plastic foil) is suitable for incubation. Use distilled water to saturate towel.
9. Electrophoresis equipments with a horizontal cooling plate are usually recommended for the electrophoresis. Currently, there is only one major IEP kit with dedicated instrumentation for human paraprotein immunotyping (TITAN Gel Immunoelectrophoresis on semiautomated SPIFE® systems, Helena).
10. These strips constitute the electrophoresis buffer reservoir and provide contact between the gel and electrodes.
11. Bromophenol blue (3′,3″,5′,5″-tetrabromophenol sulfonephthalein, bromophenol blue sultone form) is a light pink to purple crystalline powder (MW 670). It is used as a tracking dye to monitor the progress and completion of electrophoretic separation for proteins (and also serves as a size marker for nucleic acids) in agarose gel. It also is a pH indicator with a yellow color below pH 3 and blue-purple at more basic pH. Usually a final concentration of 0.002–0.01% (w/v), bromophenol blue is recommended for tracking electrophoresis. While the dye is soluble in organic solvents such as alcohol, usually aqueous solutions are used. The sodium salt of bromophenol blue is readily soluble in alkaline electrophoresis buffers and is a convenient way to dissolve it. If a stock solution is made in water, it should be slowly titrated with 1 N NaOH solution until the powder dissolves and the solution turns dark blue.
12. Because polyclonal antisera (and their IgG fraction) contain antibodies to multiple recognition sites (epitopes) on an antigen, the apparent affinity and avidity of the resultant antigen-antibody interactions is very high. This explains that high affinity and high avidity polyclonal antisera with optimized dilutions to avoid prozoning will produce maximal immunoprecipitation. These attributes, along with a relatively low price, make polyclonal antibodies, commonly produced in rabbit, sheep, goat, or donkey, the reagent of choice for most immunoprecipitation-based applications, including IEP. High antibody titer has the advantage of allowing for higher antiserum dilutions which, in turn, reduce the undesirable background protein "load" in the gel. Because of their generally low affinity, monoclonal antibodies are rarely suitable for IEP.
13. While anti-whole human serum has been the initial reference and continues to be a widely used reference for identifying specific protein components in IEPs, polyvalent antisera with

well-defined specificities are increasingly used alternatives for human paraprotein testing. Trivalent (a mix of IgG, IgA, and IgM heavy chain-specific antibodies) and pentavalent (a mix of IgG, IgA IgM, IgD, and IgE heavy chain-specific antibodies) are being used in a commercial kit (TITAN Gel Immunoelectrophoresis, Helena). The concept may be applied to other protein systems as well.

14. Inclusion of 3–5% of dextran (average MW 70,000) or polyethylene glycol (PEG, MW 6000) in low-avidity antibody-containing agarose gel increased the detection sensitivity (i.e., immune complex formation) of two-dimensional IEP and electroimmunodiffusion for at least up to fivefold [32, 33]. Further, it was suggested that other high molecular weight polymers with good water solubility would have the same effect [32]. Nevertheless, inclusion of these polymers in IEP has not gained wide acceptance.

15. After successful immunodiffusion, most precipitin arcs are visible by a naked eye in the wet agarose gel by holding the plate over a dark background close to a diffuse light source or using indirect illumination, and these patterns can be documented by photography (Fig. 1b, c). However, the precipitin arcs almost always are stained with protein dyes in the dried gel for enhanced detection and for permanent records. Most commonly used stains in decreasing order of detection sensitivity include Coomassie brilliant blue R-250 > acid violet > amido black. Less commonly used or historic protein stains include light green SF, Ponceau red (only for cellulose acetate membrane), nigrosin(e), thiazine red, and azocarmine. Although silver staining of the immunoprecipitates is possible in agarose and acrylamide gels and is more sensitive than staining with dyes, it is rarely used for IEP.

16. For convenience, a commercial amido black stain was used in these IEP runs. Alternative "homemade" amido black 10B staining solution: prepare a saturated solution of the dye (~1 g/dL) in 10% acetic acid (v/v) in methanol by dissolving 1 g of amido black in 450 mL of methanol in a glass beaker, then adding 50 mL of glacial acetic acid. Mix well using a magnetic stirrer for at least 30 min, leave the solution at room temperature overnight, and then filter. Store in a closed container for up to 2 months. For further information regarding amido black, *see* **Note 17** in Chapter 15 (on IFE) of this book.

17. In order to assess the standing of IEP methods in the literature, PubMed searches were performed on July 21, 2011. Using the search term "immunoelectrophoresis," a total of 27,727 articles were retrieved, of which 95% (26,403/27,727) were

diagnosis-related based on adding "diagnosis" as a second search term. These data indicate that (a) the classic IEP method and its various versions have been widely used in laboratories, and (b) IEP with its various versions was likely used as a diagnostic test in an overwhelming majority of the reports.

18. Any low-viscosity body fluid specimen, culture fluid, or tissue extract could be tested with classic IEP if proper antibodies are available. Human serum specimens with high viscosity due to the presence of large amounts of polymeric paraproteins like IgM can be pretreated with the reducing agent 2-mercaptoethanol for IEP. Prewarming of serum specimens containing cryoglobulins allowed immunotyping of these cryoglobulins with IEP only in less than one-third of the cases [34]. Whether pretreatment of high-viscosity and cryo-gelating specimens with chaotropic agents would make them amenable to testing with IEP remains to be fully elucidated.
19. In general, IEP is much less sensitive to antigen excess than IFE. Consequently, human serum specimens initially are tested straight in IEP for paraproteins and diluted only when evidence for antigen excess (prozoning) emerges with incomplete precipitin reaction. Even in this case, an alternative approach includes retesting of the undiluted serum with doubling the amount of antiserum added to the trough immediately after electrophoresis or after the usual incubation period but before washing (TITAN Gel Immunoelectrophoresis method, Helena). Urine specimens are often preconcentrated for IEP testing, but using both native and concentrated urine samples also is proposed in the abovementioned diagnostic kit. It is noteworthy, however, that concentrating specimens may lead to artifacts, including protein losses that differ from sample to sample and may be substantial (up to 50%) with some concentrators.
20. Based on the presence, shape, intensity, and location of the corresponding precipitin arcs, control specimens are used to confirm the performance and correctness of antisera to human whole serum and to human immunoglobulin heavy and light chains when immunotyping for paraproteins. The controls also serve as a side-by-side reference for the evaluation of immunoprecipitin arcs developing with the unknown specimens. In case of purity checks, the presence or absence of contaminating proteins and degradation products of the protein of interest are evaluated against reference preparations and/or the "starting material."
21. The IEP antiserum application templates are reusable after washing with mild detergents and, when indicated, disinfection.

22. Sample volumes used in different IEPs vary widely from 0.5 to 40 μL.
23. It is desirable to fill up the troughs with the antisera. Depending on the volume of the troughs, antiserum volumes may vary from 10 μL to as much as 200 μL per trough.
24. Incubation time primarily depends on the type and concentration of the gel, the distance between the sample application well and the lateral wall of trough, and the incubation temperature. Room temperature incubations may be needed for up to 3 days in some gels. In order to shorten the time for immunoprecipitation, incubation at temperatures higher than room temperature (e.g., at 37 °C) can be used up to ~45 °C. Conversely, incubation in a refrigerator (2–8 °C) may be needed to slow down immunodiffusion when overnight room temperature incubation would result in partial or total dissolution of precipitin arcs.
25. When whole antisera and large volume of antibody (>20 μL/mL of agarose gel) are used in IEP, the background staining will likely increase due to inadequate removal of unbound proteins. This can be overcome by including repeat press-dry cycles in preparation of the gel for subsequent staining. Press-dry can also markedly reduce the time for washing gels in the absence of protein "overload." For press-dry, multiple layers of filter papers and weight are added on top of the gels for periods of 10–15 min each between wash steps. The first filter paper should be soaked with distilled water and, after completing the press step, should be removed last after soaking it again with distilled water. Regarding the amount of weight, at least 10 g/cm^2 pressure is suggested for optimal effect (Bio-Pheresis^R Horizontal Electrophoresis Cell [Cat. No. 170-2900] Instruction Manual, rev. 031193, Bio-Rad Laboratories, Hercules, CA, USA). This corresponds to ~0.5 kg for the smaller 7-position and ~1 kg for the larger 15-position protein plate used in this work. A commercial IEP kit routinely involves press-dry steps for preparing the gel for staining (TITAN Gel Immunoelectrophoresis, Helena).
26. In addition to staining for proteins in immunoprecipitates "trapped" in the electrophoretic matrix, special stains or enzyme reactions can be applied to detect the nonprotein components of the antigens and enzyme antigens themselves. During electrophoresis in agar and agarose gels, the native structure and activities of the electrophoretically separated proteins are preserved. Further, precipitating antibody binds enzymes at antigenic sites and usually leaves active sites unmasked. The open structure of immunoprecipitates thus allows characterization of ligand binding and enzyme activities

of the precipitated protein antigens. Examples include detection of lipids, calcium, iron, and esterase and oxidase activities. It is important to note, however, that azide is a potent "general" enzyme inhibitor, so that azide-containing antisera (which are common) may interfere with procedures that involve enzyme reactions. Using azide-free antisera obviously avoids this interference. The open structure of the immunoprecipitate in the agarose gel also allows additional binding of radioactively labeled antibodies to reveal specific proteins. Binding of ^{125}I-labeled IgE antibodies onto allergen-containing precipitates permitted identification of individual allergens in complex antigen extracts after autoradiography of the gel in crossed radio-IEP (CRIE) [21].

27. It was recently shown that electroimmunoprecipitated proteins can be extracted from agarose gel for identification [35]. After excising the immunoprecipitates, the precipitates were extracted from the gel and dissociated. The protein components of the extracts were analyzed using peptide mass fingerprinting following sodium dodecyl sulfate-polyacrylamide gel electrophoresis (SDS-PAGE). In addition to characterizing antibody specificity, the method can also be used to identify antigens generating unknown precipitates in crossed IEP with polyspecific antisera, including human IgG-antigen complexes electroimmunoprecipitated by secondary antibodies. This method appears to be applicable to classic IEP as well.

28. Advantages of classic IEP:
    (a) The primary advantage of classic IEP is that, in contrast to the alternative (immunotyping) method IFE, IEP is able to discern identity, partial identity, and nonidentity of the proteins under investigation [29, 30].
    (b) Classic IEP is also advantageous for assessing purity and integrity of various biochemical and pharmaceutical preparations [26, 27].

29. Disadvantages of classic IEP:
    (a) Classic IEP is very slow; it may take up to 3 days to complete a plate (18–48 h needed just for the diffusion step) *vs.* IFE with a total of ~3 h completion time.
    (b) In contrast to IFE, classic IEP lacks sensitivity to detect low concentrations of monoclonal proteins or proteins in general.
    (c) In contrast to IFE, IEP has a relatively poor resolution and often is unable to distinguish two closely migrating protein bands. Further, unlike IFE, IEP has difficulty to detect gammopathies with multiple paraproteins [36]; an increasingly common pattern due to bone marrow transplantation for lymphoid malignancies.

(d) Unlike IFE, classic IEP exhibits the "umbrella effect" where polyclonal IgG molecules mask an IgM monoclonal gammopathy [36].

(e) Although interpretation of IFE patterns can be also difficult, polyclonal increases, primarily IgG, in inflammatory conditions may be misdiagnosed by IEP as monoclonal gammopathy [36].

30. Routine IEP methods, including commercial kits for paraproteins, do not test initially for IgD and IgE heavy chains and free (unbound) κ and λ light chains. However, antibodies against these antigens are available commercially, and a pentavalent antiserum against all five heavy chains is provided in the kit for diagnostic paraprotein immunotyping (TITAN Gel Immunoelectrophoresis, Helena).

Acknowledgment

The author wishes to thank Dr. Rita Ellerbrook of the Helena Laboratories, Beaumont, TX, USA, for kindly providing illustrations to this article.

References

1. Grabar P, Williams CA Jr (1953) Méthode permettant l'étude conjugée des propriétés électrophorétiques et immunochimiques d'un mélange de proteins. Application au serum sanguine. Method permitting the combined study of the electrophoretic and the immunochemical properties of protein mixtures; application to blood serum. Biochim Biophys Acta 10:193–194. [French]
2. Martin E, Scheidegger JJ, Grabar P et al (1954) Immuno-electrophoretic study of blood serum. Bull Schweiz Akad Med Wiss 10:193–198. [French]
3. Williams CA Jr, Grabar P (1955) Immunoelectrophoretic method for the analysis of mixtures of antigenic substances. Biochim Biophys Acta 17:67–74. [French]
4. Scheidegger JJ (1955) Une mict-méthode de l'immunoelectrophorése. Int Arch Allergy Appl Immunol 7:103–110. [French]
5. Scheidegger JJ, Roulet H (1955) Practical application of the immuno-electrophoretic method; first results. Praxis 44:73–76. [French]
6. Afonso E (1964) Quantitative immunoelectrophoresis of serum proteins. Clin Chim Acta 10:114–122
7. Wilson AT (1964) Direct immunoelectrophoresis. J Immunol 92:31–34
8. Bussard A (1959) Description of a technic simultaneously combining electrophoresis and immunological precipitation in gel: electrosyneresis. Biochim Biophys Acta 34:258–260. [French]
9. Vergani C (1971) Crossover electrophoresis for the rapid detection of serum hepatitis (Australia) antigen and antibody. J Clin Pathol 24:86–87
10. Tilton RC (1978) Counterimmunoelectrophoresis in biology and medicine. CRC Crit Rev Clin Lab Sci 9:347–365
11. Ressler N (1960) Two-dimensional electrophoresis of protein antigens with an antibody containing buffer. Clin Chim Acta 5:795–800
12. Laurell CB (1965) Antigen-antibody crossed electrophoresis. Anal Biochem 10:358–361
13. Clarke HG, Freeman T (1968) Quantitative immunoelectrophoresis of human serum proteins. Clin Sci 35:403–413
14. Laurell CB (1966) Quantitative estimation of proteins by electrophoresis in agarose gel containing antibodies. Anal Biochem 15:45–52

15. Laurell CB (1972) Electroimmuno assay. Scand J Clin Lab Invest Suppl 124:21–37
16. Weeke B (1969) Quantitative immunoelectrophoresis. Rocket electrophoresis. Ugeskr Laeger 131:1419–1423. [Danish]
17. Weeke B (1973) Rocket immunoelectrophoresis. Scand J Immunol Suppl 1:37–46
18. Svendsen PJ (1973) Fused rocket immunoelectrophoresis. Scand J Immunol Suppl 1:69–70
19. Bog-Hansen TC (1973) Crossed immunoaffinoelectrophoresis. An analytical method to predict the result of affinity chromatography. Anal Biochem 56:480–488
20. Kroll J (1973) Line immunoelectrophoresis. Scand J Immunol 1:61–67
21. Weeke B, Lowenstein H (1973) Allergens identified in crossed radioimmunoelectrophoresis. Scand J Immunol Suppl 1:149–153
22. Kroll J (1973) Crossed-line immunoelectrophoresis. Scand J Immunol Suppl 1:79–81
23. Kroll J (1973) Rocket-line immunoelectrophoresis. Scand J Immunol Suppl 1:83–87
24. Ohlenschläger G, Berger I (1974) Tandem technic in Clarke, Freeman, Ressler and Laurell's 2 dimensional immunoelectrophoresis. Methods and criteria for use as a consequence of the problems of realization of reproducible separations by stability of electrical values. Z Med Labortech 15:119–126. [German]
25. Axelsen NH, Svendsen PJ (1973) Reversed rocket immunoelectrophoresis. Scand J Immunol Suppl 1:155–157
26. United States Pharmacopeial Convention (2011) Quality of biotechnological products: stability testing of biotechnological/biological products. In: The United States Pharmacopeia–National Formulary (USP 34–NF 29 through First Supplement). http://www.uspnf.com. Accessed 26 Jul 2011.
27. European Directorate for the Quality of Medicines & Health Care. (2012) Immunochemical methods. In: ep703 - European Pharmacopoeia Online; 7th Edition 73. http://online6.edqm.eu/ep703. Accessed 26 Jul 2011.
28. Nerenberg ST, Ganger C, DeMarco L (1972) Multi-antisera applicator for use in solid-support immunoelectrophoresis on cellulose acetate. Clin Chem 18:1021–1023
29. Arquembourg PC, Salvaggio JE, Bickers JN (1970) Primer of immunoelectrophoresis with interpretation of pathologic human serum patterns. Ann Arbor-Humphrey Science Publishers, Inc., Ann Arbor, MI (S. Karger, Basel {Switzerland}, Munchen, New York)
30. Williams CA (1971) Immunoelectrophoretic analysis. In: Williams CA, Chase MW (eds) Methods in Immunology and Immunochemistry, vol III. Academic Press, New York and London, pp 234–287
31. Cawley LP, Schneider D, Eberhardt L et al (1965) A simple semi-automated method of immunoelectrophoresis. Clin Chim Acta 12:105–110
32. Kostner G, Holasek A (1972) Influence of dextran and polyethylene glycol on sensitivity of two-dimensional immunoelectrophoresis and electroimmunodiffusion. Anal Biochem 46:680–683
33. Thoss K, Abendroth K (1974) The effect of polyethylene glycol and dextran on the immunoelectrophoresis (author's transl). Z Immunitatsforsch Exp Klin Immunol 147:404–412. [German]
34. Musset L, Diemert MC, Taibi F et al (1992) Characterization of cryoglobulins by immunoblotting. Clin Chem 38:798–802
35. Beyer NH, Schou C, Houen G et al (2008) Extraction and identification of electroimmunoprecipitated proteins from agarose gels. J Immunol Methods 330:24–33
36. Keren DF (1994) High-resolution electrophoresis and immunofixation. Techniques and interpretation, 2nd edn. Butterworth-Heinemann, Boston, London, Oxford, Singapore, Sydney, Toronto, Wellington.

Chapter 22

Tris-Acetate Polyacrylamide Gradient Gels for the Simultaneous Electrophoretic Analysis of Proteins of Very High and Low Molecular Mass

Monica Cubillos-Rojas, Fabiola Amair-Pinedo, Irantzu Tato, Ramon Bartrons, Francesc Ventura, and Jose Luis Rosa

Abstract

Polyacrylamide gel electrophoresis (PAGE) is one of the most powerful tools used for protein analysis. We describe the use of Tris-acetate buffer and 3–15% polyacrylamide gradient gels to simultaneously separate proteins in the mass range of 10–500 kDa. We show that this system is highly sensitive, it has good resolution and high reproducibility, and it can be used for general applications of PAGE such as Coomassie Brilliant Blue staining and immunoblotting. Moreover, we describe how to generate mini Tris-acetate polyacrylamide gels to use them in miniprotein electrophoresis systems. These economical gels are easy to generate and to manipulate and allow a rapid analysis of proteins. All these features make the Tris-acetate-PAGE system a very helpful tool for protein analysis.

Key words Electrophoresis, Giant proteins, Gradient gel, PAGE, Protein separation, Tris-acetate

1 Introduction

The sodium dodecyl sulfate polyacrylamide gel electrophoresis (SDS-PAGE) is a tool frequently used for protein analysis. Glycine-SDS-PAGE (also known as Laemmli-SDS-PAGE) and tricine-SDS-PAGE, which are based on glycine-Tris and tricine-Tris buffer systems, respectively, are the most commonly used variants to separate proteins by their size [1, 2]. Polyacrylamide gels are composed of acrylamide and the crosslinker bisacrylamide. Variations in the concentrations of acrylamide and bisacrylamide into Laemmli-SDS-PAGE and tricine-SDS-PAGE allow separating proteins in the range of 1–500 kDa. Proteins smaller than 30 kDa are well resolved using tricine-SDS-PAGE. Glycine-SDS-PAGE system is recommended for proteins larger than 30 kDa [2, 3]. The different separation characteristics of these systems are related to the pKa

Biji T. Kurien and R. Hal Scofield (eds.), *Electrophoretic Separation of Proteins: Methods and Protocols*, Methods in Molecular Biology, vol. 1855, https://doi.org/10.1007/978-1-4939-8793-1_22,

values of glycine and tricine that permit different electrophoretic mobilities and increase the resolution of proteins of a specific mass range. Thus, to separate proteins in a wide mass range, it would be necessary to use both electrophoresis systems [2]. Alternatively, it is also recommended to use a gradient gel with high percentage of acrylamide at the bottom and a lower percentage at the top to resolve giant and small proteins in the same PAGE [3].

In this chapter, we describe an electrophoresis system based on Tris-acetate gels to separate proteins in the mass range of 10–500 kDa [4]. In the last few years, several companies have commercialized Tris-acetate gels to separate proteins of different mass range. The composition of these gels is not available for costumers, and, for this reason, the researcher has no choice but to purchase precast gels and their optimized buffers. These electrophoresis systems are based on a discontinuous buffer system that results in a neutral pH environment during electrophoresis, stabilizing both proteins and gel matrix and providing a better band resolution. The relevance of the environment to which proteins are exposed was previously showed by Hachmann and Amshey using Tris-glycine gels [5]. At basic pH, there is a high possibility of protein modification, especially by alkylation of cysteine sulfhydryls, due to the cysteine pKa that ranges from 8 to 9 and because the reducing agent contained in the loading buffer does not co-migrate with the proteins. The use of near-neutral pH gel formulations, electrophoresis buffers at a pH lower than the cysteine pKa, and reducing agents (e.g., sodium bisulfite) that move into the gel ahead of the proteins to maintain a reducing environment reduces this risk significantly [5–7]. We have assessed several combinations of these parameters (concentrations, pH, acrylamide/bisacrylamide ratios, gradients) obtaining the best results using a 3–15% polyacrylamide gradient gel made with 0.2 M Tris-acetate (pH 7.0) buffer and a standard acrylamide/bisacrylamide solution (40:1) containing a 2.6% concentration of bisacrylamide, 0.042% ammonium persulfate (APS), and 0.12% *N,N,N′,N*-′-tetramethylethylenediamine (TEMED) [4].

The composition of the running buffer (50 mM tricine, 50 mM Tris, 0.1% SDS, 1.3 mM sodium bisulfite, pH 8.2) is similar to the buffer used for tricine-SDS-PAGE [2], but the antioxidant sodium bisulfite is added in order to maintain the proteins in a reduced state during the run. The sample of proteins is prepared in a sample buffer that contains lithium dodecyl sulfate (LDS): 250 mM Tris–HCl pH 8.5, 2% (w/v) LDS, 100 mM DTT, 0.4 mM EDTA, 10% (v/v) glycerol, 0.2 mM phenol red, and 0.2 mM Coomassie Brilliant Blue G. This sample buffer allows the reduction of disulfides of the proteins under mild heating conditions (10 min at 70 °C) avoiding the cleavage of Asp-Pro bonds that occurs when the sample is heated at 100 °C in Laemmli sample buffer [8]. After the run, proteins into the gel can be stained (e.g.,

Coomassie Brilliant Blue (CBB)) or transferred to a membrane [e.g., nitrocellulose or polyvinylidene fluoride (PVDF)]. To transfer the run to a PVDF membrane, we have used a transfer buffer containing 20% methanol, 25 mM bicine, 25 mM Bis-Tris, 1 mM EDTA, and 1.3 mM sodium bisulfite, pH 7.2. Under these conditions, it is possible to separate and simultaneously analyze proteins in the mass range of 10–500 kDa in a single PAGE. This system also provides high sensibility in detecting small amounts of protein: It is possible to analyze by immunoblotting proteins around 500 kDa, running samples containing only 10 μg of total protein. Interestingly, we have optimized this Tris-acetate system to use small gels (8 cm × 8 cm) in miniprotein electrophoresis systems, which are easier to manipulate and cheaper to fabricate due to reduced reagent volume requirements. Moreover the time required to run and transfer these gels is also reduced.

2 Materials

Prepare all solutions using ultrapure water (Milli-Q quality) and analytical grade reagents.

2.1 Tris-Acetate Polyacrylamide Gradient Gel Components

1. Tris-acetate (3 M, pH 7.0) gel buffer (15×): Dissolve 36.33 g Tris base in 90 mL water. Mix and adjust pH with acetic acid. Make up to 100 mL with water. Store at 4 °C.
2. LDS sample buffer (4×): 4 mL of 2.5 M Tris–HCl pH 8.5, 0.8 g LDS, 0.006 g EDTA, 5 mL of 80% glycerol, 0.75 mL of 1% Coomassie Brilliant Blue G solution (CBB), and 0.25 mL of 1% phenol red solution. Mix well and store at room temperature.
3. Acrylamide solution (40%) (37.5:1 acrylamide/bisacrylamide mix). Store at 4 °C.
4. APS (ammonium persulfate): 10% solution in water.
5. TEMED (*N,N,N′,N′*-tetramethylethylenediamine). Store at 4 °C.
6. DTT (1,4-dithiothreitol): 1 M solution in water. Aliquot and store at −20 °C. We recommend not to freeze samples again for further uses.
7. Running buffer (1×): Dissolve 8.95 g tricine, 6.06 g Tris base, 1 g dodecyl sulfate sodium salt (SDS), and 0.25 g sodium bisulfite in 1 L of water. Mix and store at 4 °C. The pH of this solution is 8.24. It is not necessary to adjust it.
8. Gradient maker (Hoefer Instruments, San Francisco, CA).
9. Stir bar and agitator.
10. Mini-PROTEAN II electrophoresis system.
11. Bicinchoninic acid (BCA) assay kit.

2.2 Immunoblotting Components

1. PVDF membranes.
2. Transfer buffer (20×): Dissolve 40.8 g bicine, 52.4 g Bis-Tris, and 3 g EDTA in 500 mL water. Store at 4 °C. For western, dilute this buffer to 1× in water with 20% methanol (e.g., for 1 L, 50 mL of 20× transfer buffer/750 mL of water/200 mL methanol and add 0.25 g of the antioxidant sodium bisulfite). The pH of this solution is 7.2. pH adjustment is not necessary.
3. Whatman chromatography papers (17 CHR).
4. Mini Trans-Blot electrophoretic transfer cell.

3 Methods

Carry out all procedures at room temperature unless otherwise specified.

3.1 Casting the Gel

1. We use the gel cassette of Mini-PROTEAN II to cast the gel (*see* Fig. 1).
2. Select and prepare the appropriate gradient solutions to separate the proteins of interest. The volumes of reagents required for a mini 3–15% polyacrylamide gradient gel (dimensions, 8 cm × 8 cm; spacers, 1.5 mm) are indicated in Table 1 (*see* **Note 1**).
3. Mix solutions before pouring into the chambers of the gradient maker. Add 4 mL of the lower polyacrylamide percentage solution (3%) into the non-outlet chamber (reservoir) (left chamber of gradient maker in Fig. 1a) and 4 mL of the higher polyacrylamide percentage solution (15%) into the outlet chamber that must be equipped with a stir bar (right chamber of gradient maker in Fig. 1a). The valves are closed at this moment. Keep the remaining volume (3 mL) of the lower percentage solution to add later (*see* **Note 2**).
4. Start pouring the mixes into the glass sandwich. Open the outlet valve. Let the higher percentage solution arrives the bottom of the glass sandwich and open immediately the other valve. In this way, solutions are mixed and form the gradient progressively while pouring. The flow must be constant. Fast flow will cause turbulence that disrupts the gradient, and slow flow rates will cause polymerization of the acrylamide before completing the pouring.
5. When solutions inside the chambers are about to consume, add 3 mL of the lower percentage solution into the reservoir chamber and wait until the total pouring of the sandwich is completed. Put the comb. Recover the volume that did not enter into the sandwich (1 mL approximately) to know when the gel has polymerized.

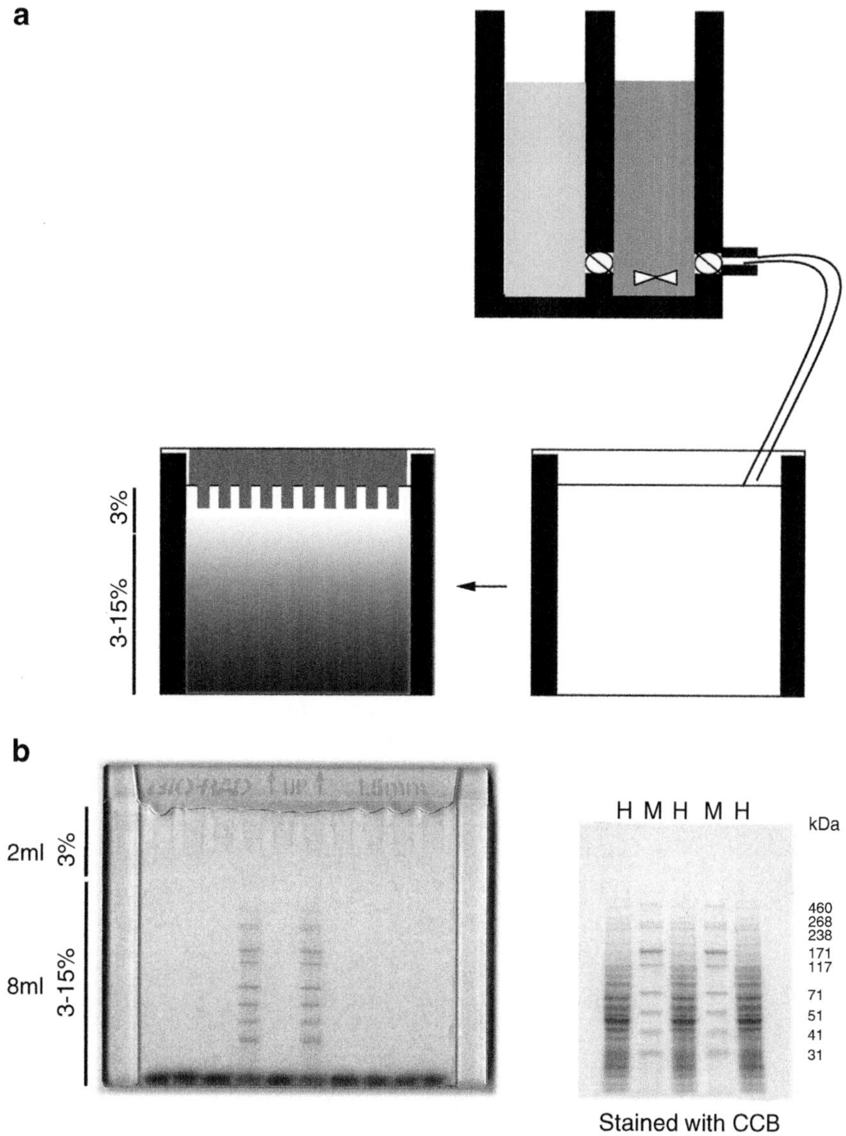

Fig. 1 The Tris-acetate polyacrylamide gel. (**a**) Schematic representation of how to make a 3–15% gradient gel with a gradient maker. The 3% and 15% polyacrylamide solutions (from a 40% acrylamide (37.5:1 acrylamide/bisacrylamide mix) stock solution) are poured in left and right compartments, respectively, of the gradient maker at the same time. Immediately after the gradient is made, a 3% solution is added before letting the gradient gel solidify between glass plates from a slab apparatus. (**b**) Example of Tris-acetate polyacrylamide gel. Gel was made in glass plates from Bio-Rad (Mini-PROTEAN II system) with 1.5 mm spacers. Volumes used are indicated (left side). Pre-stained protein standards (31, 41, 51, 71, 117, 171, 238, 268, and 460 kDa from Invitrogen) and 50 μg of protein from a lysate of HEK 293 cells were separated in a 3–15% polyacrylamide gradient gel (left side) and visualized by CBB staining (right side). Standards and lysates are indicated as M and H, respectively. Reproduced from [4] with permission from Elsevier

Table 1
Gel composition to make a 3–15% gradient gel

		3%	15%
3 M Tris-acetate pH 7.0 (15×)	(mL)	0.47	0.27
40% Acrylamide solution	(mL)	0.53	1.5
Milli-Q water	(mL)	6	2.23
Total volume	(mL)	7	4
TEMED	(μL)	8.75	5
APS 10%	(μL)	33.25	19

6. Wash the gradient maker immediately with abundant water to prevent acrylamide polymerization.
7. Leave the gel to polymerize (*see* **Note 3**).

3.2 Sample Preparation and Protein Loading

1. Lyse the samples (*see* **Note 4**).
2. Centrifuge the samples (e.g., 15,000 × *g* for 15 min at 4 °C). Collect the supernatant and determine the protein concentration by the bicinchoninic acid assay or a similar method. Add sample buffer 4× without DTT to a final concentration 1×. At this moment, you can use these samples or store them at −20 °C.
3. Add DTT (final concentration, 100 mM; e.g., 5 μL of 1 M DTT in 50 μL of total volume) when protein samples are used in PAGE (*see* **Note 5**).
4. Heat the samples for 10 min at 70 °C and load them into the wells of the gel (*see* **Note 6**).

3.3 Electrophoresis Conditions

1. Run at $\Delta V = 130$ V ($I = 115$ mA). The time run estimated for a 3–15% gradient gel is approximately 1 h and 15 min. As shown in Figs. 1b and 2, molecular weight markers ranging from 460 to 10 kDa were separated. These gels can be used for CBB staining (Fig. 1b, right) or to perform immunoblotting (Fig. 2).

3.4 Transfer Conditions

1. Submerge the fiber pads and the Whatman paper in transfer buffer. Place the gel in transfer buffer using a separate tray.
2. Activate the PVDF membrane in methanol for 2 min, wash with deionized water, and submerge in transfer buffer.
3. When the phenol red dye reaches the end of the gel, turn the power supply off. Remove the cast (glasses with the gel). Rinse the cast carefully with deionized water and equilibrate it in transfer buffer for few minutes.

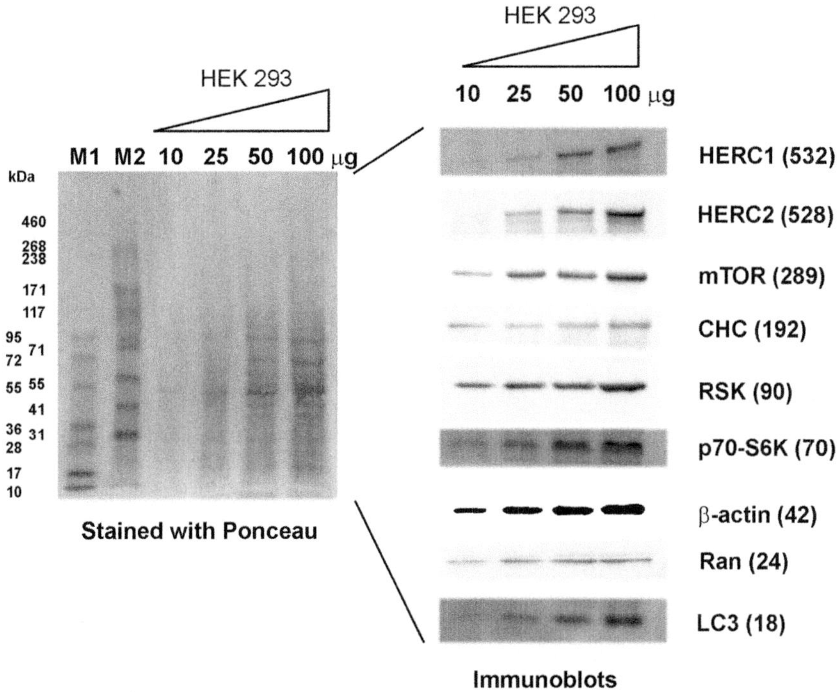

Fig. 2 Applications of the Tris-acetate polyacrylamide gel: analysis by staining and immunoblotting of proteins in the mass range of 10–500 kDa. Lysates from HEK 293 cells were loaded, and then they were run in a 3–15% polyacrylamide gradient gel, stained with Ponceau S (left side), transferred to a PVDF membrane, and analyzed by immunoblot (right side) with specific antibodies against the following proteins: HERC1 (532 kDa), HERC2 (528 kDa), mTOR (289 kDa), CHC (192 kDa), RSK (90 kDa), p70-S6K (70 kDa), β-actin (42 kDa), Ran (24 kDa), and LC3 (18 kDa). To check the system sensitivity, 10–100 μg of protein were analyzed. Two different pre-stained protein standards, M1 (10, 17, 28, 36, 55, 72, and 95 kDa from Fermentas) and M2 (31, 41, 51, 71, 117, 171, 238, 268, and 460 kDa from Invitrogen), were used (left side). Reproduced from [4] with permission from Elsevier

4. Separate the thinner glass of the gel carefully with the help of a spatula. Cover the gel with Whatman paper and separate both the gel and the paper from the cast with the help of tweezers or a spatula (*see* **Note 7**).
5. Assembly the sandwich as follows (from bottom to top, in the transfer direction): fiber pad, Whatman paper, Whatman paper with the gel, PVDF membrane, and two Whatman papers and fiber pad.
6. Slide the assembled transfer cassette into Mini Trans-Blot Electrophoretic Transfer Cell. Add the transfer buffer and connect the power supply.
7. Perform the transfer for 2 h ($I = 200$ mA) or overnight ($\Delta V = 20$ V).

4 Notes

1. APS and TEMED should be added last, immediately before pouring the gels. The APS solution should be prepared fresh to ensure good resolution of the gel.
2. It is very important to assure that there are no air bubbles between the chambers of the gradient maker before casting the gel. For air bubble removal, pour 4 mL of the lower percentage solution in the reservoir chamber, open the valve between the chambers, and allow flowing approximately 1 mL of the solution; recovering it and after that, pouring the higher percentage solution in the outlet chamber.
3. The gels can be stored at 4 °C in running buffer up to 3 days.
4. We have used CHAPS buffer to lyse the cells. This buffer contains 10 mM Tris–HCl pH 7.5, 100 mM NaCl, 0.3% CHAPS, 50 mM NaF, 0.5 mM EDTA, 1 mM sodium vanadate, 1 mM phenylmethylsulfonyl fluoride, 5 μg/mL leupeptin, 5 μg/mL aprotinin, 1 μg/mL pepstatin A, 50 mM β-glycerophosphate, 100 μg/mL benzamidine, and 1 μM E-64. After rocking during 20 min at 4 °C, lysates are centrifuged at 15,000 × g for 15 min at 4 °C. Pellets are discarded and supernatants analyzed.
5. Freeze samples in sample buffer 1×. Do not freeze the samples after thawing. Add fresh DTT just before heating the samples.
6. To eliminate traces of non-polymerized acrylamide, rinse wells with running buffer before applying samples.
7. Because the bottom of the gel is denser, it is important to start separating the glass from the bottom. Extracting the gel from glasses without the paper implies the risk of distortion of the upper part of the gel; this part is very viscous and must be handled carefully.

Acknowledgment

This study was supported by grants from Spain (BFU-2008-02084/BMC), European Union (FEDER), ISCIII (RETICRD06/0020), and Generalitat de Catalunya (2009SGR1059). MCR, FAP, and IT are supported by a doctoral fellowship from IDIBELL, Generalitat de Catalunya, and the Juan de la Cierva program, respectively.

References

1. Maizel JV (2000) SDS polyacrylamide gel electrophoresis. TIBS 25:590–592
2. Schägger H (2006) Tricine-SDS-PAGE. Nat Protoc 1:16–22
3. Casas-Terradellas E, Garcia-Gonzalo FR, Hadjebi O et al (2006) Simultaneous electrophoretic analysis of proteins of very high and low molecular weights using low-percentage acrylamide gel and a gradient SDS-PAGE gel. Electrophoresis 27:3935–3938
4. Cubillos-Rojas M, Amair-Pinedo F, Tato I et al (2010) Simultaneous electrophoretic analysis of proteins of very high and low molecular mass using Tris-acetate polyacrilamyde gels. Electrophoresis 31:1318–1321
5. Hachmann JP, Amshey JW (2005) Models of protein modification in Tris-glycine and neutral pH Bis-Tris gels during electrophoresis: effect of gel pH. Anal Biochem 342:237–245
6. Schägger H, von Jagow G (1987) Tricine-sodium dodecyl sulfate-polyacrylamide gel electrophoresis for the separation of proteins in the range from 1 to 100 kDa. Anal Biochem 166:368–379
7. Wiltfang J, Arold N, Neuhoff V (1991) A new multiphasic buffer system for sodium dodecyl sulfate-polyacrylamide gel electrophoresis of proteins and peptides with molecular masses 100,000-1000, and their detection with picomolar sensitivity. Electrophoresis 12:352–366
8. Kubo K (1995) Effect of incubation of solutions of proteins containing dodecyl sulfate on the cleavage of peptide bonds by boiling. Anal Biochem 225:351–353

Chapter 23

Diagonal Electrophoresis for the Detection of Proteins Involved in Disulfide Bonds

Ronald Saraswat and Brian McDonagh

Abstract

The formation and reduction of disulfide bonds through thiol-disulfide exchange is an important cellular mechanism for the regulation of protein activity and plays a crucial role in redox signaling. Thiol-disulfide exchange on key regulatory cysteine residues offers an efficient and reversible mechanism for regulatory control in response to alterations in the intracellular redox environment. Cysteines are one of the most rarely used amino acids in proteins, but when conserved in proteins, they usually play critical roles in protein structure, function, or regulation. Diagonal electrophoresis is a relatively simple technique to analyze the formation of protein disulfides by sequential nonreducing/reducing electrophoresis. Proteins that do not form disulfides migrate identically in both dimensions and form a diagonal after the second dimension; proteins that contained intra-chain disulfides lie above the diagonal, while proteins involved in inter-disulfides fall below the diagonal. This technique allows for the detection and subsequent identification of proteins with inter-disulfides and has been applied in combination with mass spectrometry to identify proteins involved in redox signaling cascades.

Key words Redox proteomics, Diagonal electrophoresis, Disulfide bonds, Redox relays

1 Introduction

The identification of proteins involved in disulfide bond formation in response to perturbations in the intracellular redox environment is a cornerstone of redox biology. The sequential oxidation and reduction of disulfide bonds is a key concept in redox signaling and has recently been demonstrated to be involved in signaling cascades or redox relays [1]. Cysteine (Cys) residues, although one of the most rarely used amino acids, are highly conserved in proteins when located in a regulatory position [2]. The flexibility of thiol groups is due to the wide range of oxidation states that sulfur can occupy and are often referred to as sulfur switches [3]. These thiol groups provide an important means of controlling protein function, allowing the proteome to respond to environmental, genetic, biochemical, and pathological perturbations. The formation of reactive

Biji T. Kurien and R. Hal Scofield (eds.), *Electrophoretic Separation of Proteins: Methods and Protocols*, Methods in Molecular Biology, vol. 1855, https://doi.org/10.1007/978-1-4939-8793-1_23,

oxygen species (ROS) by endogenous processes such as electron leakage from electron transport chain, enzymes such as NAD(P)H oxidase and nitric oxide synthase, or addition of prooxidants, can initiate a redox proteomic response through thiol-disulfide exchange mechanisms. Reduced thiols are initially oxidized into a sulfenic acid (–SOH) with subsequent formation of inter- or intra-disulfide bonds. Physiological concentrations of ROS can generate unique signaling cascades that play a role in a wide array of cell processes including transcription, cell proliferation, immune response, and antioxidant defense. Dysregulated ROS homeostasis or signaling responses have been reported in a wide variety of metabolic diseases including cancers [4], neurodegeneration [5], obesity [6], and diabetes [7].

As ROS are in general too reactive to have a signaling effect as a result of diffusion throughout the cell, it is now considered that they will have an effect localized to their site of generation. Sequential oxidation and reduction of redox-sensitive proteins through disulfide exchange is a potential mechanism for the generation of a redox signaling cascade. The reactivity of a protein depends on the 3D environment and accessibility of certain redox-sensitive amino acids especially Cys. Many antioxidant enzymes with catalytic Cys residues have not only evolved but are highly conserved across all kingdoms of life including isoforms of peroxiredoxins and glutathione peroxidases for the detoxification of H_2O_2. Each of these enzymes has evolved with specific degradation rates that can differ by up to an order of magnitude, and isoforms are located in specific cellular organelles, which make them ideal candidates to perform a signaling role in response to local perturbations in intracellular ROS concentrations [8]. Reversible redox modifications of the Cys or thiol group on susceptible proteins can potentially have secondary signaling roles involving the cascade of the redox signal from the site of generation to protein targets, the peroxidases are likely signaling candidates as they account for about ~90% of cellular peroxide due to their highly reactive Cys residues [9]. ROS are essential in the physiological control of cell function in biological systems and are continuously produced in living cells. Excess levels of ROS can lead to irreversible modifications on proteins such as carbonylation; however, under normal conditions, it can lead to reversible redox proteomic modifications of Cys residues that can include inter–/intra-disulfide bond formation, glutathionylation, nitrosylation, and sulfenic acid formation. Many of these reversible modifications play a regulatory or even protective role preventing further oxidative damage, and as they are reversible, protein function may be restored when normal oxidative conditions are restored. In unstressed mammalian cells, it was shown that disulfides in proteins (PSSP) account for 6% and 9.5% of protein sulfhydryls in HEK and HeLa cells, respectively, and after treatment with the thiol-specific oxidant diamide, this increased to 24% and 25%,

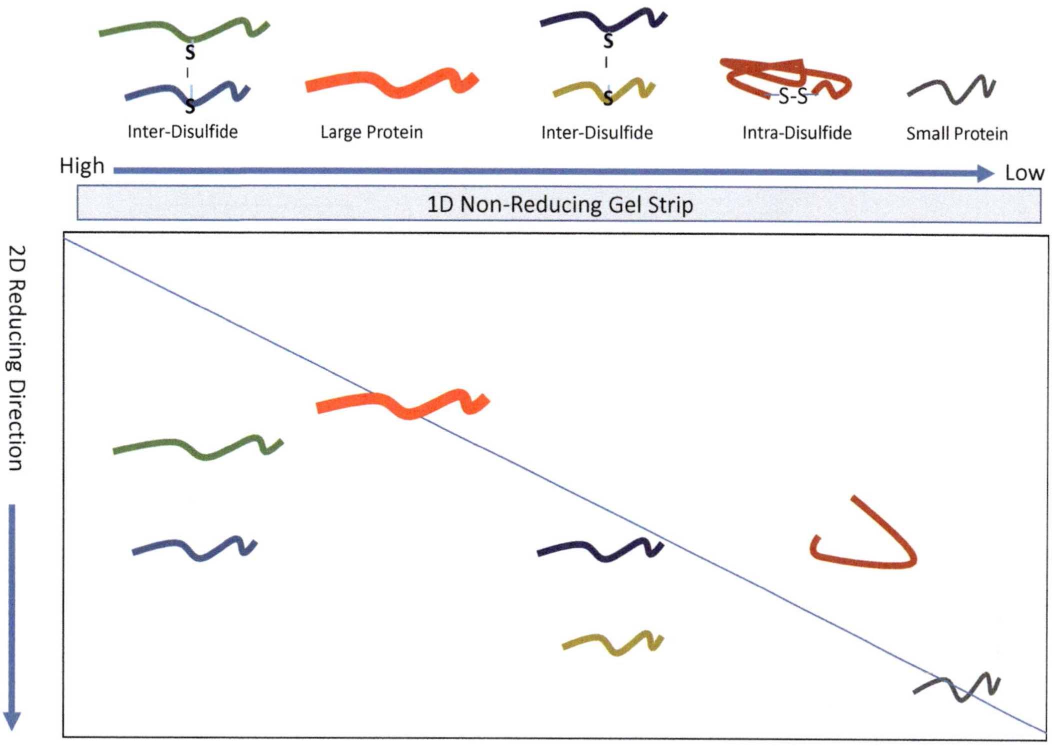

Fig. 1 Schematic of principles of diagonal electrophoresis. Protein samples are initially electrophoresed under nonreducing conditions; a gel track is excised, exposed to reducing agent, and placed orthogonally on a SDS-PAGE gel. Proteins lacking disulfides migrate forming a diagonal as they migrate identically in reducing and nonreducing conditions. Proteins with inter-chain disulfides migrate as separate polypeptides below the diagonal, while those with intra-chain disulfides migrate slower under reducing conditions and thus are above the diagonal

while the steady-state level of glutathione-protein mixed disulfides (PSSG) is less than 1% increasing to 15% after diamide treatment [10]. Therefore it would also appear that a proportion of protein thiols represent an important and significant redox buffer within the cell.

A relatively simple method developed for the analysis of both inter- and intra-chain disulfide bonds originally developed in the 1970s is the use of diagonal electrophoresis [11]. Diagonal electrophoresis involves sequential nonreducing/reducing electrophoresis, essentially manipulating the redox state of sensitive Cys residues within proteins (Fig. 1). Initially, a protein mixture is subjected to electrophoresis under nonreducing conditions in an alkylated buffer to prevent thiol-disulfide exchange. The entire lane is then excised from the one-dimensional electrophoresis gel, incubated with a reducing agent, and placed orthogonally on top of a second gel, and electrophoresis is performed again as normal under reducing conditions. After staining of the gel for proteins, proteins that do not contain disulfide bonds congregate along a diagonal since

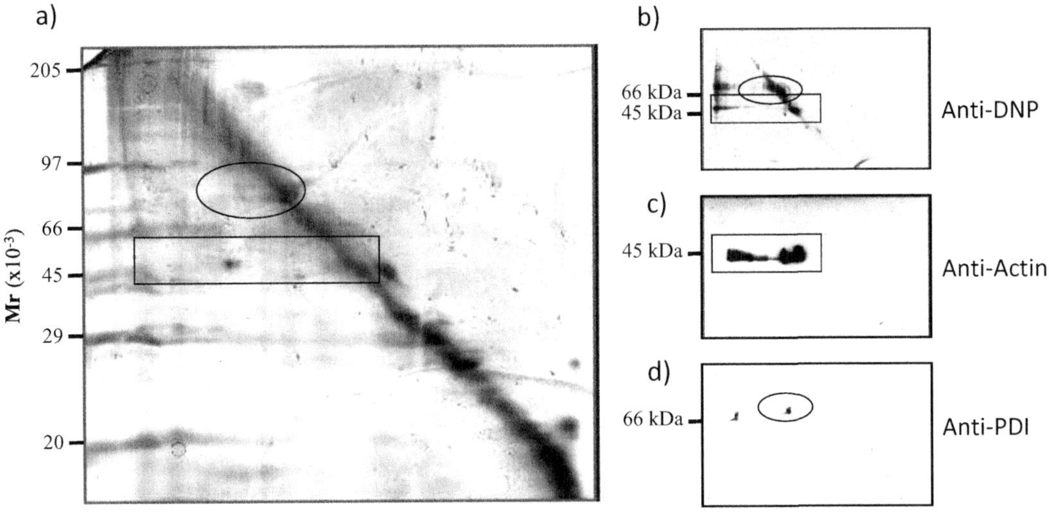

Fig. 2 (**a**) Diagonal electrophoresis of 50 μg of a derivatized protein sample with dinitrophenylhydrazine (for carbonylation). Diagonal gel was subsequently immunoblotted with various antibodies: (**b**) anti-DNP for detection of carbonylated proteins; (**c**) anti-actin, rectangular area corresponds to actin-associated proteins; and (**d**) anti-protein disulfide isomerase (PDI); the circular area corresponds to thiol exchange protein PDI

they migrate identically in both reducing and nonreducing conditions. Proteins involved in inter-chain disulfide bonds will fall below the diagonal as they have a lower apparent molecular mass under reducing conditions. Proteins that contain intra-chain disulfide bonds will migrate please above the diagonal as they have a larger apparent molecular mass. Protein spots can then be excised, tryptic digested, and identified by mass spectrometry. This technique has been used for the identification of proteins previously not known to form disulfide bonds [12, 13]. It has also been used in combination with other techniques, such as affinity purification of redox targets before separation [14], to identify proteins by immunoblotting (Fig. 2) [15].

The technique has been further expanded in a number of studies to identify interacting proteins of some of the regulatory redox proteins such as thioredoxin [14] and peroxiredoxins [1]. Introduction of redox-sensitive peroxiredoxin 2 containing an affinity tag into cells allowed the purification of interacting proteins; subsequent separation by diagonal electrophoresis allowed identification of interacting proteins by a combination of mass spectrometry and immunoblotting. In this example, STAT3 was identified as forming a redox relay that operates with peroxiredoxin 2 in the cytosolic and nuclear environments of mammalian cells [1]. This seminal article using a combination of diagonal electrophoresis and mass spectrometry identified a thiol peroxidase-transcription factor redox relay role for peroxiredoxin 2 and STAT3 in responding to intracellular H_2O_2. A previous study identified interacting partners of the thiol reductase, thioredoxin, combining affinity purification with diagonal electrophoresis and mass spectrometry, identifying 50 thioredoxin-interacting proteins including membrane proteins from the cyanobacterium, *Synechocystis* sp. 6803 [14]. Diagonal electrophoresis is a relatively simple technique that can be adapted and combined with numerous experimental procedures and molecular biological techniques, e.g. in combination with affinity purification and mass spectrometry can allow the identification of thiol interacting proteins or general investigations into the redox proteome.

2 Materials

1. An example of an alkylating buffer solution, 50 mM Tris–HCl, pH 7.5, 2 mM EDTA and 40 mM iodoacetamide (IAM) or similar alkylating agents such as N-ethylmaleimide.
2. Sample buffer (4×): 1.25 mL 0.5 M Tris–HCl, pH 6.8, 3.0 mL glycerol, 1 mL 20% SDS, and bromophenol blue.
3. 20% trichloroacetic acid (TCA).
4. Resolving acrylamide gel solution.
5. Stacking acrylamide gel solution.
6. Running buffer, 1 L containing 14.4 g glycine, 3.03 g Tris base, and 1 g SDS.
7. Equilibration buffer. (6 M urea, 0.375 M Tris–HCl, pH 8.8, 2% SDS, 20% glycerol) with either 2% dithiothreitol (DTT) or 2.5% IAM.
8. Agarose solution. 0.1% agarose in running buffer plus bromophenol blue.
9. Two sets of slab gel apparatus, e.g., ATTO and BioRad mini Protean system or alternatively slab gel apparatus with two sets of spacers of different thickness (*see* **Note 1**).
10. Power supply.
11. Rocking platform.
12. Sharp blade to excise 1D Gel.

3 Methods

3.1 Sample Preparation

1. Prepare protein extracts using an alkylating buffer solution (*see* **Note 2**).
2. Precipitate proteins in 20% TCA and resuspend in nonreducing sample buffer (4×) to give a final protein concentration of 50 μg/15 μL (*see* **Note 3**).

3.2 Electrophoresis

1. Prepare a 12% polyacrylamide resolving gel with 4% stacking gel (*see* **Note 4**).
2. Add 15 μL nonreducing sample buffer containing 50 μg of protein to a well. Use alternate wells in the nonreducing first dimension so that it will be easier to excise the lanes (*see* **Note 5**).
3. One-dimensional electrophoresis (1D E) is carried out at 60 V until dye reaches the end of the gel (*see* **Note 5**). The entire lane is excised using a sharp fine scalpel (*see* **Notes 6** and **7**).
4. Soak the entire lane for 20 min in equilibration buffer containing 2% DTT at room temperature on a rocker, followed by 20 min in equilibration buffer containing 2.5% IAM (*see* **Note 8**).
5. Rinse the gel lane in running buffer and place horizontally on a 12% resolving gel (*see* **Note 4**). Carefully slot excised lane onto slab gel using the blunt end of small spatula. It helps if the excised lane is kept moist using running buffer; gel lanes can become very fragile when dry (*see* **Note 9**).
6. A protein standard mixture can be electrophoresed alongside the diagonal gel by pipetting 5 μL of mixture onto a small cube of blotting paper (*see* **Note 10**).
7. Layer warm agarose containing bromophenol blue on top of the slab gel. Ensure there are no air bubbles between gel lane and slap gel as this will interfere with protein separation (*see* **Note 11**).
8. Carry out electrophoresis as per normal until dye front has reached the bottom of the gel. Proteins may be detected by any of the staining techniques applicable to 2DE. Proteins may also be electroblotted for immunoblot analysis. This technique may also be applied to radio- or fluorescently labeled proteins (*see* **Note 12**).

4 Notes

1. In the second dimension, it is important to use a gel with greater thickness than first dimension. For instance, in the first dimension, a gel with 0.75 or 1.0 mm spacers could be used followed by a gel of 1.5 mm thickness in the second dimension. Alternatively in larger gels, spacers with different thicknesses can be used. Also large gels can give option of running two samples from small first-dimension rigs, alongside each other [13].
2. Avoid using any reducing agents during homogenization or lysis of samples.
3. Depending on the protein concentration of the samples, precipitation may not be required but is useful if dealing with dilute samples. If using 50 μg of protein/well in the first

dimension, it is better to precipitate 200 μg of protein and resuspend in 60 μL of sample buffer overnight. This ensures complete and homogenous resuspension and also allows at least three replicates of the sample to be used. 50 μg of protein is a guideline as the precise amount will depend on the abundance of proteins of interest and size of the gel rigs. Samples can be resuspended in any sample buffer so long as there are no reducing agents present.

4. Percentages of acrylamide gels can be varied to optimize resolution for proteins of interest.
5. Electrophoresis in the first dimension is carried out slowly as if voltage and current are too high; separation of proteins in the nonreducing dimension will result in streaking of the final gel.
6. The lane width should not be exceeded as it will result in poor resolution in second dimension. Also lane edges should be kept as smooth as possible by using sharp scalpel.
7. The gel lane should not be allowed to dry during excision as it will become very brittle; use running buffer to keep moist during excision.
8. Alternative reducing and alkylating agents may be used, for instance, it is reported that NEM has a higher reactivity for reduced thiols than IAM.
9. Excised lanes can be stored at −70 °C at this stage but label the top and bottom of gel.
10. This allows easier detection of standards; however, it is not as accurate as for determining the molecular weight of proteins that have been subjected to electrophoresis in both dimensions.
11. It helps to pour agarose quickly when warm with a pipette as it fills spaces fast and prevents bubble formation.
12. A thiol-specific fluorescent reagent 5--iodoacetamidofluorescein (5-IAF) could be included after reduction with DTT. This would specifically label newly reduced thiols and offer superior sensitivity for detection of proteins by fluorescence.

Acknowledgments

This work was supported by the NUI Galway Department of Physiology and School of Medicine.

References

1. Sobotta MC, Liou W, Stocker S, Talwar D, Oehler M, Ruppert T, Scharf AN, Dick TP (2015) Peroxiredoxin-2 and STAT3 form a redox relay for H2O2 signaling. Nat Chem Biol 11:64–70
2. Marino SM, Gladyshev VN (2010) Cysteine function governs its conservation and degeneration and restricts its utilization on protein surfaces. J Mol Biol 404:902–916
3. Jones DP (2008) Radical-free biology of oxidative stress. Am J Physiol Cell Physiol 295: C849–C868
4. Mates JM, Segura JA, Alonso FJ, Marquez J (2008) Intracellular redox status and oxidative stress: implications for cell proliferation, apoptosis, and carcinogenesis. Arch Toxicol 82:273–299
5. Calabrese V, Sultana R, Scapagnini G, Guagliano E, Sapienza M, Bella R, Kanski J, Pennisi G, Mancuso C, Stella AM, Butterfield DA (2006) Nitrosative stress, cellular stress response, and thiol homeostasis in patients with Alzheimer's disease. Antioxid Redox Signal 8:1975–1986
6. O'Neill HM, Holloway GP, Steinberg GR (2013) AMPK regulation of fatty acid metabolism and mitochondrial biogenesis: implications for obesity. Mol Cell Endocrinol 366:135–151
7. Gil-del Valle L, de la CML, Toledo A, Vilaro N, Tapanes R, Otero MA (2005) Altered redox status in patients with diabetes mellitus type I. Pharmacol Res 51:375–380
8. Winterbourn CC (2014) Are free radicals involved in thiol-based redox signaling? Free Radic Biol Med 80:164–170
9. Poole LB (2014) The basics of thiols and cysteines in redox biology and chemistry. Free Radic Biol Med 80:148–157
10. Hansen RE, Roth D, Winther JR (2009) Quantifying the global cellular thiol-disulfide status. Proc Natl Acad Sci U S A 106:422–427
11. Sommer A, Traut RR (1974) Diagonal polyacrylamide-dodecyl sulfate gel electrophoresis for the identification of ribosomal proteins crosslinked with methyl-4-mercaptobutyrimidate. Proc Natl Acad Sci U S A 71:3946–3950
12. Cumming RC, Andon NL, Haynes PA, Park M, Fischer WH, Schubert D (2004) Protein disulfide bond formation in the cytoplasm during oxidative stress. J Biol Chem 279:21749–21758
13. Brennan JP, Wait R, Begum S, Bell JR, Dunn MJ, Eaton P (2004) Detection and mapping of widespread intermolecular protein disulfide formation during cardiac oxidative stress using proteomics with diagonal electrophoresis. J Biol Chem 279:41352–41360
14. Mata-Cabana A, Florencio FJ, Lindahl M (2007) Membrane proteins from the cyanobacterium Synechocystis sp. PCC 6803 interacting with thioredoxin. Proteomics 7:3953–3963
15. McDonagh B, Sheehan D (2007) Effect of oxidative stress on protein thiols in the blue mussel *Mytilus edulis*: proteomic identification of target proteins. Proteomics 7:3395–3403

Chapter 24

Identification of Proteins on Archived 2D Gels

Hiroyuki Matsumoto, Nobuaki Takemori, and Naoka Komori

Abstract

Separation technology of proteins from a complex mixture by two-dimensional gel (2D gel) was invented more than 40 years ago. With a good laboratory practice, the 2D gels are likely to be dried and stored at ambient temperature as archived record. Up until the beginning of this century, it had been difficult to identify the protein spots isolated on 2D gels. However, the advent of mass spectrometry-based proteomics protocols combined with genome information enabled us to determine the identity of a protein separated on 2D gels archived decades ago. The protocol will assist researchers to decipher molecular mechanisms involved in the system by identifying and quantifying the protein of interest from archived 2D gels.

Key words Protein identification, Archived 2D gels

1 Introduction

Identification of isolated proteins by peptide mass fingerprinting based on genome information [1, 2] has been established and being routinely used in the past decade. Quite often both one-dimensional and two-dimensional gels run on polyacrylamide slabs have been dried on a filter paper or in between cellophane films and stored at room temperature. We discovered that such archived gels and stored over two decades still serve as a source for protein identification by peptide mass fingerprinting [3]. This could serve a useful technique to confirm or even discover protein identity from dried gels stored in data archive. In our original paper [3], the in-gel digestion by TPCK–trypsin was performed by the method reported by Rosenfeld et al. [4] and adapted to mass spectrometry [5, 6]. Later we modified the protocol to optimize the efficiency of peptide mass fingerprinting [7–10]. The protocol described in this chapter is based on the recent modifications [7–10].

We routinely perform MASCOT search against the National Center for Biotechnology Information (NCBI) nonredundant database using the following parameters: (a) unlimited protein

Biji T. Kurien and R. Hal Scofield (eds.), *Electrophoretic Separation of Proteins: Methods and Protocols*, Methods in Molecular Biology, vol. 1855, https://doi.org/10.1007/978-1-4939-8793-1_24,

molecular weight and pI ranges; (b) presence of protein modifications including acrylamide modification of cysteine, methionine oxidation, protein *N*-terminus acetylation, and pyroglutamic acid; and (c) mass tolerance of ±0.25 Da. We consider the confirmation to be positive when a significant MOWSE score ($p < 0.05$) is generated.

2 Materials

1. Use sample protein spots separated by two-dimensional (2D) gel electrophoresis, stained by Coomassie blue, and dried either on a filter paper or in between cellophane papers (*see* **Notes 1** and **2**).

2.1 Stock Solutions for In-Gel Digestion

1. 50% (v/v) acetonitrile/100 mM ammonium bicarbonate (pH 8.9).
2. 50% (v/v) acetonitrile/5% (v/v) trifluoroacetic acid (TFA).
3. 0.2% (v/v) TFA solution.
4. 2,5-Dihydroxybenzoic acid (DHB) solution for MS analysis: 2% (w/v) DHB in 50% (v/v) acetonitrile/0.1% (v/v) TFA.

3 Methods

1. Soak the protein spot overnight at 4 °C in 50% (v/v) acetonitrile/100 mM ammonium bicarbonate adjusted to pH 8.9.
2. Incubate the destained gel pieces with 1 mL of acetonitrile for 30 min to dehydrate the gels.
3. After removing acetonitrile, rehydrate the gel pieces with 2 μL of trypsin solution for 30 min at 4 °C.
4. Add 50 μL of 10 mM ammonium bicarbonate and incubate at 37 °C for 5–10 h (*see* **Note 3**).
5. Transfer the solution containing tryptic digests into a new 0.65 mL polypropylene tube. To extract the rest of the tryptic digests from the gel, vigorously shake the gel in 50 μL of 50% (v/v) acetonitrile/5% (v/v) TFA for 20 min.
6. After a brief centrifugation, collect and transfer the supernatant into the new tube to mix with the original digest solution.
7. Concentrate the collected peptide solution down to near dryness using vacuum centrifuge.

3.1 Peptide Mass Fingerprinting by MALDI-TOF MS

1. Reconstitute peptide samples with 5 μL of 0.2% (v/v) TFA solution.

2. Load the peptide solution (0.5 μL) onto a MALDI-TOF sample target plate with a fresh DHB solution (0.5 μL), and dry the mixture completely at room temperature.
3. Subject the peptide samples on the MALDI sample target to peptide mass fingerprinting analysis using MALDI-TOF MS. After measuring the masses of the tryptic peptides by MALDI-TOF MS, submit the observed masses to MASCOT peptide mass fingerprinting (PMF) search program (Matrix Science, London, UK) at http://matrixscience.com.

4 Notes

1. Proteins can be identified from dried 2D gels over 20 years at room temperature by this method.
2. Typically, a small protein spot (approximately 2 mm × 2 mm) is excised.
3. During the incubation, trypsin continues to digest a protein in a gel piece, and some of the tryptic peptides come out of the gel piece.

References

1. Matsumoto H, Komori N (2000) Ocular proteomics: cataloging photoreceptor proteins by two-dimensional gel electrophoresis and mass spectrometry. Methods Enzymol 316:492–511
2. Matsumoto H, Kurono S, Matsumoto M, Komori N (2005) Mass spectrometry of biomolecules in proteomics. In: Meyers RA (ed) Encyclopedia of molecular and cell biology and molecular medicine. Wiley-VCH Verlag GmbH, Germany, pp 557–585
3. Matsumoto H, Komori N (1999) Protein identification on two-dimensional gels archived nearly two decades ago by in-gel digestion and matrix-assisted laser desorption ionization time-of-flight mass spectrometry. Anal Biochem 270:176–179
4. Rosenfeld J, Capdevielle J, Guillemot JC, Ferrara P (1992) In-gel digestion of proteins for internal sequence analysis after one- or two-dimensional gel electrophoresis. Anal Biochem 203:173–179
5. Matsumoto H, Kurien BT, Takagi Y, Kahn ES, Kinumi T, Komori N, Yamada T, Hayashi F, Isono K, Pak WL, Jackson KW, Tobin SL (1994) Phosrestin I undergoes the earliest light-induced phosphorylation by a calcium/calmodulin-dependent protein kinase in *Drosophila* photoreceptors. Neuron 12:997–1010
6. Kinumi T, Tobin SL, Matsumoto H, Jackson KW, Ohashi M (1997) The phosphorylation site and desmethionyl N-terminus of *Drosophila* phosrestin I *in vivo* determined by mass spectrometric analysis of proteins separated on two-dimensional gel electrophoresis. Eur J Mass Spectrom 3:367–378
7. Takemori N, Komori N, Matsumoto H (2006) Highly sensitive multistage mass spectrometry enables small scale analysis of protein glycosylation from two-dimensional polyacrylamide gel. Electrophoresis 27:1394–1406
8. Takemori N, Komori N, Thompson J Jr, Yamamoto M-T, Matsumoto H (2007) Novel eye-specific calmodulin methylation characterized by protein mapping in *Drosophila melanogaster*. Proteomics 7:2651–2658
9. Takemori N, Komori N, Matsumoto H (2009) Chapter 150: MS analysis of protein glycosylation. In: Walker JM (ed) The protein protocols handbook, 3rd edn. Humana Press, New York, pp 1381–1388
10. Mecklenburg K, Takemori N, Komori N, Chu B, Hardie R, Matsumoto H, O'Tousa J (2010) Retinophilin is a *NINAC*- and light-regulated phosphoprotein required for suppression of dark noise in *Drosophila* photoreceptors. J Neurosci 30(4):1238–1249

Chapter 25

Two-Dimensional Gel Electrophoresis: Vertical Isoelectric Focusing

Yaser Dorri

Abstract

Two-dimensional gel electrophoresis is one of the most powerful tools for separating proteins based on their size and charge. Two-dimensional gel electrophoresis (2-DE) is very useful to separate two proteins with identical molecular weights but different charges, which cannot be achieved with just sodium dodecyl sulfate polyacrylamide gel electrophoresis (SDS-PAGE). Here, a simpler and easier version of 2-DE is presented which is also faster than all the currently available techniques. In this modified version of 2-DE, isoelectric focusing is carried out in the first dimension using a vertical sodium dodecyl sulfate polyacrylamide gel electrophoresis (SDS-PAGE) apparatus. Following the first-dimensional IEF, each individual lane is excised from the IEF gel and after a 90° rotation, is inserted into a second-dimensional SDS-PAGE, which can be stained with Coomassie Brilliant Blue for protein analysis or immunoblotted for further analysis. This version of IEF can be run in less than 2 h compared to the overnight run required by O'Farrell's method. Difficult tube gel casting and gel extrusion as well as tube gel distortion are eliminated in our method. This method is simpler, faster, and inexpensive. Both dimensions can be done on the same SDS-PAGE apparatus, and up to ten samples can be run simultaneously using one gel.

Key words 2D gel electrophoresis, Isoelectric focusing, Sodium dodecyl sulfate polyacrylamide gel electrophoresis, SDS-PAGE, 2-DE, IEF

1 Introduction

Two-dimensional gel electrophoresis (2-DE) is a well-known and powerful technique used for separating proteins based on their size and charge. The usefulness of this method is that proteins that are separated by their molecular mass in sodium dodecyl sulfate polyacrylamide gel electrophoresis (SDS-PAGE) can be further separated based on charge. Era of electrophoresis can be reached back to the late 1800s, when Lodge introduced this technique by attaching electrodes to jelly-filled tubes and measured ionic velocities [1]. However, the results were not impressive. Despite many efforts to improve this phenomenon, a century later Laemmli introduced SDS-PAGE in the 1970s [2]. Followed by that, Margolis and

Biji T. Kurien and R. Hal Scofield (eds.), *Electrophoretic Separation of Proteins: Methods and Protocols*, Methods in Molecular Biology, vol. 1855, https://doi.org/10.1007/978-1-4939-8793-1_25,

Kenrick introduced a reliable method of 2-DE in 1969 [3], and several years later MacGillivary and Rickwood in 1974 made slight improvements to this technique [4]. In 1975, O'Farrell was able to introduce tube-based isoelectric focusing, a well-modified version of 2-DE with impressive resolution [5]. In O'Farrell's method, the IEF gel is cast within a glass tube, and the proteins of interest are electrophoresed through it. However, researchers find a number of problems with this method: (a) it is difficult to cast the gel, (b) the tube gel is difficult to extrude following an IEF run, and (c) the possibility of gel breakage during gel extrusion and the long run time which can take up to 16 h [6]. Ever since, there have been many improvements to 2-DE in terms of resolution and sample solubility; however, almost all the methods require a long run time for the first-dimensional IEF. Among the current methods of 2-DE, immobilized pH gradient (IPG), flatbed IEF, and phore isoelectric focusing system, strip gel methods are well known [6]. Sample leakage and long run time as well as high voltage requirement for running the gel are the disadvantages of these methods [7–9]. Here, a modified version of 2-DE, vertical IEF, is introduced. This method uses the same apparatus as that used in second-dimensional SDS-PAGE [6]. In this method, SDS-PAGE Mini-Protean II gel apparatus is used for the first- and second-dimensional run. Therefore, this method eliminates the use of expensive equipment required by previous methods. Faster run time and easy gel handling in this method are among the advantageous factors, compared to other 2-DE methods [6]. In this modified method, each lane of the first-dimensional IEF gel is excised, and after a short wash, it is rotated 90° and inserted onto the second-dimensional SDS-PAGE. Up to ten samples can be analyzed in one IEF run. Total run time for both dimensions in this method is less than 3 h [6]. This method is discussed in detail here.

2 Materials

Prepare all solutions with ultrapure water (prepared by purifying deionized water, to attain a sensitivity of 18 M Ω cm at 25 °C) and analytical grade reagents. Prepare and store all reagents at room temperature (unless indicated otherwise). Diligently follow all waste disposal regulations when disposing waste materials.

2.1 First-Dimensional IEF Gel Components

1. Acrylamide (30%)/Bis-acrylamide (1.5%) solution (*see* **Note 1**).
2. N,N,N,N′,N′-tetramethylethylenediamine (TEMED).
3. Riboflavin.
4. BioLyte 5–8 (Amersham Biosciences Co., NJ, USA).
5. Riboflavin (0.004%)/TEMED (0.45%) (*see* **Note 2**).

6. Ammonium persulfate: 10% solution in water (*see* **Note 3**).
7. Urea.
8. Triton X-100-20% (*see* **Note 4**).
9. Triton X-100-10% (*see* **Note 4**).
10. Acetic Acid.
11. Tris base.
12. IEF lysis buffer: 2 M thiourea, 5 M urea, 0.25% CHAPS, 0.25% Tween-20, 0.25% SB 3–10 (sulfobetaine-10), 0.25% ampholyte (isoelectric point 5–8), 10% 2-propanol, 1 mM sodium vanadate, 12.5% water-saturated isobutanol, 5% glycerol, and 0.79% ß-mercaptoethanol (*see* **Note 5**).
13. Acetic acid solution: 0.005 M, pH 5.0. Store at 4 °C.
14. Tris–HCl: 0.1 M Tris, pH 8.0. Store at 4 °C.
15. Overlay buffer: 3.0 g of urea, 2 mL of 10% Triton X-100, 0.5 mL of 40% BioLyte pH 5–8 and made to 10 mL with distilled water (*see* **Note 6**). Store at 4 °C.
16. Equilibration buffer (125 mL of 0.5 M Tris–HCl pH 6.8, 20 g of SDS, and 50 mL of ß-mercaptoethanol, made up to 1 L) (*see* **Note** 7).
17. Plastic piece (2 cm × 2 mm) [6].

2.2 SDS Polyacrylamide Gel Components

1. Acrylamide (30%)/bis-acrylamide (0.8%).
2. Resolving gel buffer: 1.5 M Tris–HCl (*see* **Note 8**). Store at 4 °C.
3. Stacking gel buffer: 0.5 M Tris–HCl, pH 6.8 (*see* **Note 9**). Store at 4 °C.
4. Ammonium persulfate: 1.5% solution in water (*see* **Note 10**).
5. SDS-PAGE running buffer: 0.025 M Tris, 0.192 M glycine, 0.1% SDS, pH 8.3 (*see* **Note 11**).
6. SDS lysis buffer (5×): 0.3 M Tris–HCl (pH 6.8), 10% SDS, 25% β-mercaptoethanol, 0.1% bromophenol blue, and 45% glycerol. Store at −20 °C.

3 Methods

3.1 First-Dimensional IEF Gel (4% IEF)

In this method, the IEF gel is cast just like a SDS-PAGE gel using a Mini-Protean II gel apparatus (used for mini SDS-PAGE gels) (*see* Fig. 1). Unlike other IEF gel methods, this method suggests an inexpensive apparatus which is cheaper than currently introduced methods. Both dimensions can be done on the same SDS-PAGE apparatus, and up to ten samples can be run simultaneously in less

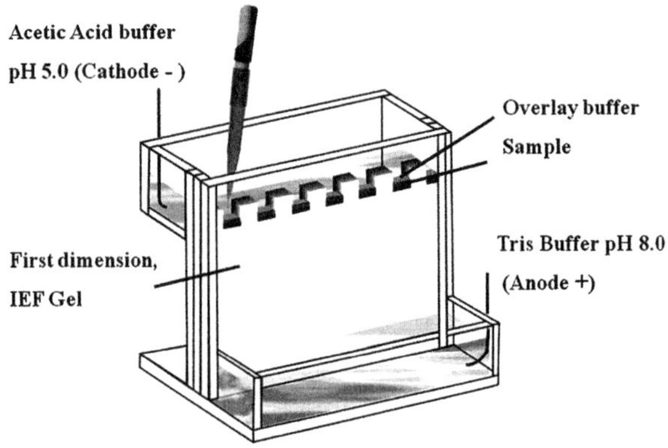

Fig. 1 Schematic representation of first-dimensional IEF gel. Mini-Protean II SDS-PAGE apparatus is used to run the first-dimensional IEF gel. A thin film of Spinkote from Beckman was coated along the sides of the apparatus to provide a tight seal and thus prevent mixing of buffers between upper and lower chambers (Adapted from ref. 6)

than 3 h, using one gel. Here, the first step of this modified version of 2-DE is introduced.

1. Set up the apparatus similar to that in SDS-PAGE (*see* Fig. 1).
2. Prepare the solution for a 4% IEF gel by adding 6.13 g of urea to 2.5 mL of deionized water and let it mix (*see* **Note 12**). Once mixed, add 1.6 mL of acrylamide (30%)/Bis-acrylamide (1.5%) solution, 1.5 mL of riboflavin/TEMED solution, 1.2 mL of 20% Triton X-100, and 0.6 mL of BioLyte 5–8. Let it mix for 10–15 min.
3. Add 200 μL APS and quickly cast the gel within a 7.25 cm × 10 cm × 1.0 mm gel cassette using 1 mm spacer (*see* **Note 13**).
4. Place a ten-well comb at the top of the gel.
5. Place the gel apparatus in front of a UV light overnight at room temperature for polymerization (*see* **Note 14**). Once polymerized, either use immediately or cover it with a wet tissue paper and wrap it thoroughly with clear plastic wrap to prevent gel from drying.
6. To run the gel, a thin film of Spinkote from Beckman was coated along the sides of the Mini-Protean II gel apparatus to provide a tight seal. This will create a greasy area along the sides and thus prevent mixing of buffers between upper and lower chambers. Then attach the polymerized IEF gel to the Mini-Protean II gel apparatus using clamps [6] (*see* Figs. 1 and 2). Then remove the comb.
7. Prepare the samples by adding 12.25 mg of urea and 34 μL of IEF lysis buffer to 5 μL of desired sample. Mice liver sample

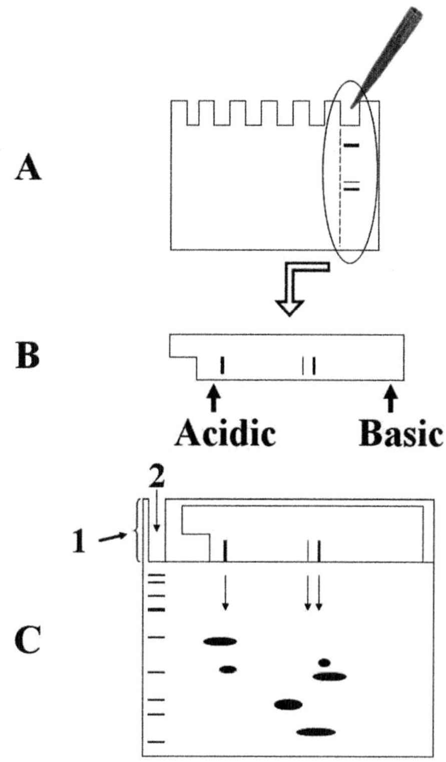

Fig. 2 Schematic representation of 2-DE: (a) each lane of the first-dimensional IEF is excised, (b) excised lane is rotated 90° counterclockwise, and (c) excised gel strip is inserted into the empty space (1) of the second-dimensional SDS-PAGE gel. Protein marker can be loaded into the well created by plastic piece (2)

(postnuclear fraction derived from 10% homogenate) is used here [6] (*see* **Note 15**).

8. Fill the top chamber with acetic acid (pH 5) buffer, including inside the wells of IEF gel.
9. Pour Tris–HCl (pH 8) in the bottom chamber of apparatus.
10. Prefocus the gel by running it at 100 V (constant voltage) for 50 min.
11. To add samples, first remove the acetic acid buffer from the upper chamber and wells using a 10 mL pipette and a 1 mL syringe (*see* **Note 16**). Place 60 μL of samples (the whole amount of each sample) in each well using a P200 pipette [6].
12. Fill the remaining space in each well with overlay buffer. Use enough overlay buffer to cover the samples and fill all the wells (including wells without samples, if any), to prevent samples from mixing with the acetic acid buffer.

13. Slowly add more acetic acid (pH 5) to the top of overlay buffer (over the wells) till the top chamber is filled. It is important to follow this step with care and slowly to prevent samples from mixing with buffer.
14. Start the running process by applying 100 V (constant voltage) for 15 min followed by 200 V (constant voltage) for an additional 10 min. Then change the setting to 20 W (constant power).
15. Run for 30 min or until a yellow horizontal line across the gel, right below the wells, is observed (*see* **Note 17**).
16. Turn off the power supply.
17. Remove the clamps and glass plate from the gel. Transfer the gel to a clean container and rinse it with water.
18. Incubate the IEF gel with equilibration buffer (125 mL or as needed) for 20 min on a shaker, in a laboratory hood.
19. Wash 2× with deionized water for 5 min.

3.2 Second-Dimensional Sodium Dodecyl Sulfate Polyacrylamide Gel Electrophoresis (SDS-PAGE)

1. Prepare a 10% SDS-PAGE resolving gel solution (solution A): Prepare the second-dimensional SDS-PAGE gel solution by mixing 2.5 mL of resolving buffer, 3.33 mL of acrylamide mixture, and 4 mL of water in a 50 mL conical flask. Degas the solution with vacuum for 10 min [6] (*see* **Note 18**).
2. Prepare another set of 10% SDS-PAGE resolving gel solution (solution B) as stated above in a separate flask. Degas the solution with vacuum for 10 min. Once finished, wrap it with plastic wrap and keep it aside (*see* **Note 18**).
3. Add 80 μL of ammonium persulfate and 20 μL of TEMED to solution A and immediately cast gel within a 7.25 cm × 10 cm × 2.0 mm gel cassette using 2 mm spacer. Allow 1.5–2.0 cm space on top, to lay the first-dimensional gel strip (*see* Figs. 3 and 4a and **Note 19**). Let the gel polymerize.
4. Excise each lane of the first-dimensional IEF vertically (*see* Figs. 2 and 3), and after a 90° counterclockwise rotation, insert it into the top of the SDS-PAGE gel, right inside the 1.5–2.0 empty space (*see* **Note 20** and Fig. 4). By doing so, the left side of the gel strip represents the acidic side, while the right side represents the basic end of the IEF gel strip (*see* Fig. 2). Keep the gel to the right side of the empty space (*see* Fig. 4).
5. Add 180 μL of ammonium persulfate and 40 μL of TEMED to solution B and immediately pour the solution into the empty space (over the IEF gel strip). This solution must fill all the remaining space around the gel strip. Before the polymerization of the top section of the gel, make a well for loading the molecular marker by inserting a 2 cm long plastic piece (2 mm thick) into the left side of the empty space (now filled

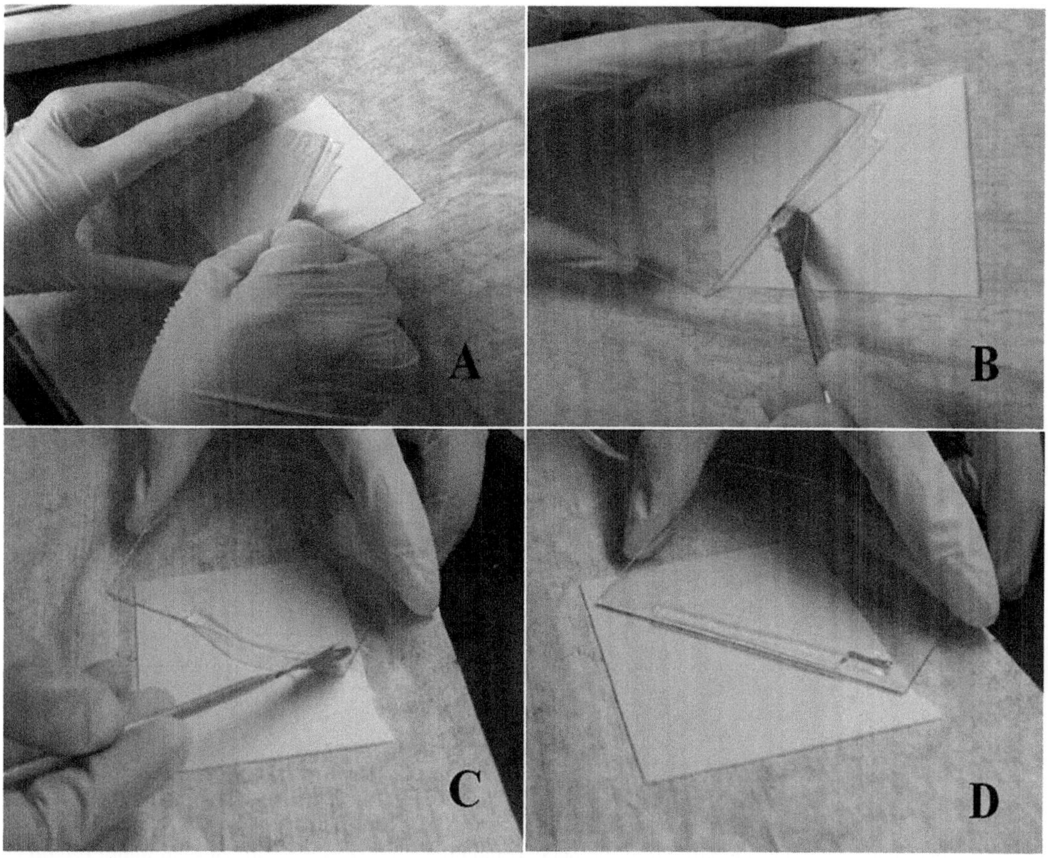

Fig. 3 Transfer of IEF gel strip onto the second-dimensional SDS-PAGE gel: (**a**) each lane of the first-dimensional IEF is excised using a blade and glass plate. (**b**) IEF gel strip is pushed over a glass plate starting from one end, using a spatula. (**c**) The other end of IEF gel strip is pushed over the glass plate. (**d**) The gel strip is completely moved on to the glass plate and is ready to be inserted on to the second-dimensional SDS-PAGE gel

with solution B), next to the gel strip (*see* Fig. 4d). Once the top portion is polymerized, remove the 2 cm plastic piece (*see* **Note 21**).

6. Once the top section of the SDS-PAGE gel is polymerized, assemble the apparatus, and fill the top and bottom chambers with SDS-PAGE running buffer.
7. Load 10–20 μL of molecular marker into the well created by plastic piece (*see* Fig. 4d).
8. Run the gel at 14 mA (constant current) until the marker passes into the bottom portion of SDS-PAGE gel (*see* **Note 22**). Then run at 30 mA (constant current) until the marker reaches the bottom of the gel.
9. Turn off power supply.

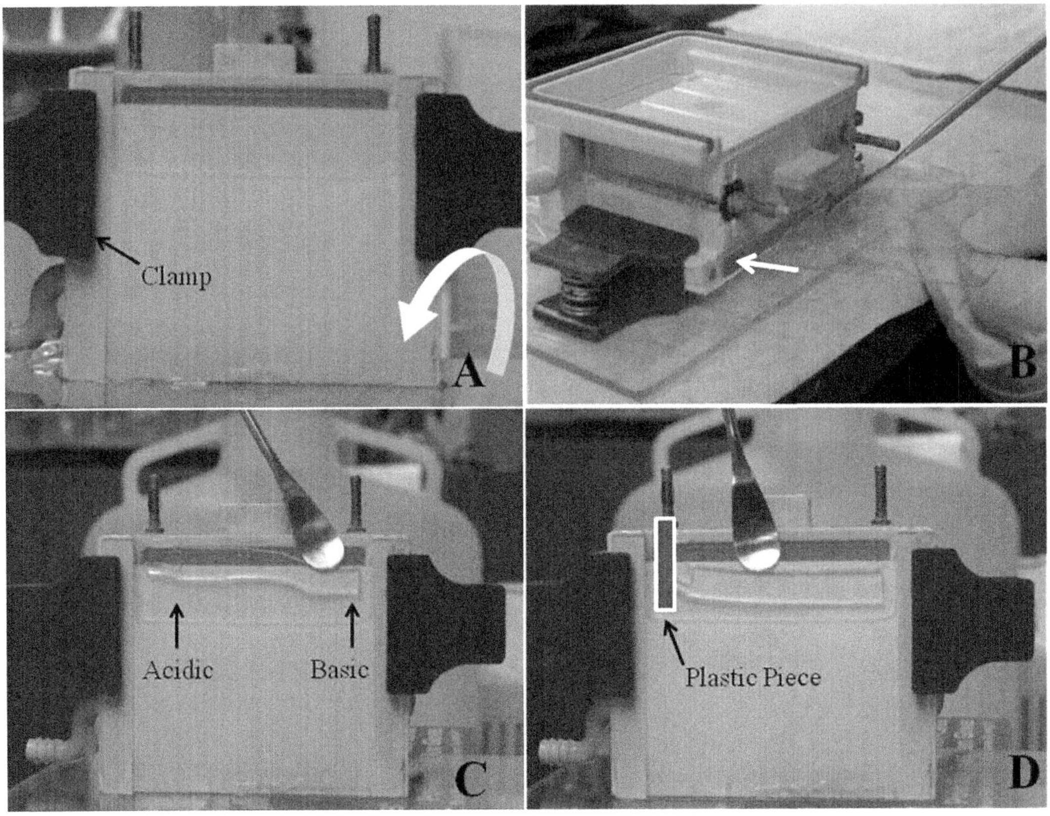

Fig. 4 The insertion of the IEF gel strip on to the second-dimensional SDS-PAGE. (**a**) Flip the Mini-Protean II SDS-PAGE apparatus to sit horizontally; (**b**) place the glass plate carrying the IEF gel strip (*see* Fig. 3d) over the glass plate of the second-dimensional SDS-PAGE apparatus. Carefully push the gel strip into the empty space. (**c**) Push the gel inside using a spatula; (**d**) place a 2 cm plastic piece to the left side of the IEF gel strip, and pour the solution B over the gel strip

10. Stain the gel using 0.05% Coomassie Brilliant Blue (CBB) for 2 h (for best result, stain overnight) [10] or use the gel for immunoblotting [11, 12].
11. Destain the gel using destaining solution and Kimwipe (*see* Fig. 5) [13].

4 Notes

1. Weigh 30 g of acrylamide and 1.5 g of Bis-acrylamide while wearing a mask, and transfer them to a 100 mL graduated cylinder containing about 50 mL of water, under a laboratory fume hood. Add a magnetic stir bar and mix it on magnetic stirrer for about 30 min. A low level of heat can speed up the mixing process. Make up to 100 mL. Wrap the bottle with aluminum foil and store it at 4 °C.

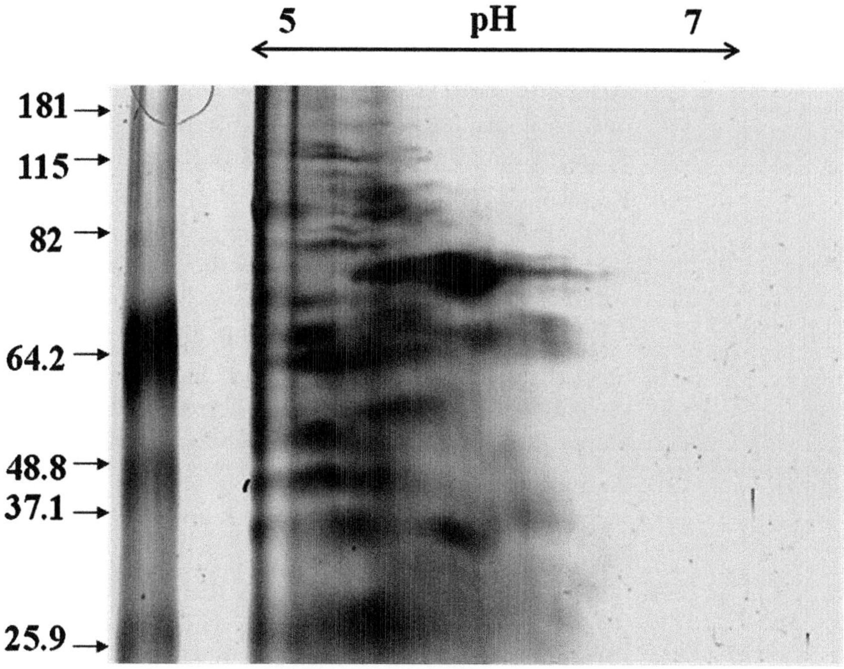

Fig. 5 CBB-stained 2-DE gel (10%): Mouse liver postnuclear fraction analyzed on 2-DE gel

2. Weigh 4 mg of riboflavin and transfer it to a 100 mL graduated cylinder containing about 50 mL of water. Add a magnetic stir bar and mix it on a magnetic stirrer for 30 min. Then add 0.45 gm of TEMED and mix for 3 min. Make up to 100 mL. Store at 4 °C, with bottle wrapped with aluminum foil.
3. Weigh 100 mg of ammonium persulfate (APS) and transfer it to 1 mL of water in a 1 mL Eppendorf tube. Mix for 1 min and keep at room temperature for immediate use. It is better to make fresh APS solution when needed.
4. Weigh 20 g of Triton X-100 and transfer it to a 100 mL graduated cylinder containing about 50 mL of water. Mix for 15 min and make up to 100 mL with water. To make 10% Triton X-100, weigh 10 mg of Triton X-100 and mix in water and make up to 100 mL with water.
5. To prepare 10 mL of IEF lysis buffer, weigh 1.52 g of thiourea, 3.0 g of urea, 25 mg of CHAPS, 25 mg of SB3–10, and 1.22 mg of sodium vanadate, and transfer them into a 10 mL graduated cylinder containing 4 mL of water. Add 25 mg of Tween-20, 25 mg of ampholyte 5–8, 1 mL of isopropanol, 1.25 mL of water-saturated buffer, 0.5 mL of glycerol, and 79.6 μL of β-ME into the solution. Add a magnetic stir bar and mix the solution on a magnetic stirrer for 30 min. Make up to 10 mL. Make 1 mL aliquots and store at 4 °C in 1 mL Eppendorf tubes.

6. To make overlay buffer, mix 2 mL of 10% Triton X-100 with 3 g of urea and 0.5 mL of BioLyte (pH 5–8). Make it to 10 mL.
7. Weigh 20 g of SDS and mix it in 500 mL of water. Add 125 mL of 0.5 M Tris–HCl (pH 6.8) and 50 mL of BME. Make up to 1 L with water. Mix for 30 min. Store at room temperature.
8. Resolving gel buffer: 1.5 M Tris–HCl, pH 8.8. Add about 100 mL of water to a 1 L graduated cylinder or a glass beaker. Weigh 181.7 g Tris and transfer to the cylinder. Add water to a volume of 900 mL. Mix and adjust pH with HCl. Make up to 1 L with water. Store at room temperature.
9. Stacking gel solution was not used in this method; however, stacking gel can be used to polymerize the IEF gel strip in the empty space above the SDS-PAGE gel [6].
10. Ammonium persulfate (1.5%): Weigh 150 mg of APS and mix in 10 mL of water. Mix for 5 min. A low level of heat can speed up the mixing process.
11. Take 100 mL of 10× native buffer (can be prepared by mixing 30.3 g of Tris base with 144 g of glycine in 500 mL of water for 30 min. Then make it to 1 L), and mix in 500 mL of water. Add 10 mL of 10% SDS. Make it to 1 L with water.
12. When making IEF gel solutions, first weigh 6.13 g of urea, and mix in 2.5 mL of water in a 10 mL conical flask for 10–15 min with medium level of heat. Add a magnetic stir bar and let it mix on magnetic stirrer. Solution should be warmed at all time; otherwise, it will form into crystals. Once urea is dissolved, add 1.5 mL of riboflavin/TEMED solution, 600 μL of BioLyte, 1.6 mL of acrylamide/bis-acrylamide, and 1.2 mL of Triton X-100. Mix in 10 min while being warmed. Add 80 μL of APS right before the use of the solution. IEF gel solution must be immediately used as it may polymerize.
13. The thickness of the first-dimensional IEF gel must be a little smaller than that of SDS-PAGE. It is because each lane of the first-dimensional IEF will be transferred on top of the second-dimensional SDS-PAGE, and since the gels expand during run time and may not fit onto SDS-PAGE apparatus on the second dimension, it would be useful to have a 1 mm thickness for the first-dimensional IEF gel and 2 mm thickness for the second dimension.
14. IEF gel was placed in front of UV light overnight as a convenience. Our results indicate 1 h UV exposure of gel is sufficient.
15. Heating the samples containing urea can speed up the mixing process.
16. Use a 10 mL pipette to remove most of the buffer solutions over the wells and then use 1 mL syringe to draw the rest of the

buffer within the wells. All the buffer solutions must be carefully removed to allow an empty space for the samples.

17. By the end of the first-dimensional run time, we observed a yellowish horizontal line, right below the wells. We noticed that if we increase the run time, proteins will run out of the gel. Therefore, this horizontal line shows the end of the first-dimensional run time.
18. For a better resolution, it is recommended to make 4–20% gradient gel. Here, a 10% gel was made as a convenience. Solution B will be used to fill the top portion of the second-dimensional SDS-PAGE, where the IEF gel strip is placed (*see* Fig. 2c). A stacking gel solution can also be used to make solution B (here, a regular SDS-PAGE resolving gel solution was used as solution B). This solution will polymerize the IEF strip on top of the SDS-PAGE gel.
19. As mentioned earlier, the thickness of the second-dimensional SDS-PAGE must be bigger than the first-dimensional IEF, so that the gel strip can be easily inserted onto the top of the SDS-PAGE (*see* Fig. 4c). Here, the thickness of the first-dimensional IEF was 1 mm, and for the second-dimensional SDS-PAGE, the thickness is 2 mm.
20. In this step, remove the glass (top plate), and let the gel remain on the back plate (*see* Fig. 3a). Rinse the gel with water, and after cleaning a glass plate with 70% ethanol, excise each lane of the first-dimensional IEF, vertically as shown in Fig. 3a, using a blade. Once each well is excised, carefully transfer each well on a glass plate using a spatula (*see* Fig. 3b). The excised IEF gel strip is flexible enough to be rotated along the sides up to 45 degrees without breaking, so start lifting one side of the IEF strip using a spatula, and place it on the glass plate. Then move to the other end of the IEF strip and move this end on the glass plate (*see* Fig. 3c). Once both ends are on the glass plate, the middle part of the IEF gel strip can be easily transferred on the glass plate by a little push. In the next step, lay the SDS-PAGE apparatus horizontally (from vertical position) so the IEF gel can easily slide into the empty space (*see* Fig. 4a). While the SDS-PAGE apparatus is in horizontal position, slowly transfer the IEF gel over the top of SDS-PAGE glass plate. Using a spatula, the IEF gel is placed over the SDS-PAGE glass plate, starting with the right side of the IEF gel (*see* Fig. 4b). Once the right side of the IEF gel strip is over the SDS-PAGE plate, push the left side of the IEF gel with spatula, and then carefully push the whole gel inside the empty space (*see* Fig. 4c). Do not force the IEF gel strip to go deep inside as the IEF gel strip will nicely fit in the empty space once the SDS-PAGE solution is poured over it. Here, the top of the

gel strip (placed in the left side of the empty space of the SDS-PAGE) represents the acidic part, and the bottom represents the basic part of the gel (*see* Fig. 4c).

21. As mentioned earlier, a 2 cm plastic piece is inserted to the left side of the SDS-PAGE empty space to create a well for the protein marker. Once the top gel portion is polymerized, slowly remove the plastic piece.
22. Since in the second-dimensional SDS-PAGE, proteins are migrating from the gel strip to the SDS-PAGE gel, it is better to run the gel, starting with a low current (14 mA). Once the protein marker moves through the SDS-PAGE gel, then change the current to 30 mA.

References

1. Lodge, O. (1886). On the migration of ions and an experimental determination of absolute ionic velocity. Rep. Brit. Assoc. Adv. Sc. 38
2. Laemmli UK (1970) Cleavage of structural proteins during the assembly of the head of bacteriophage T4. Nature 227:680–685
3. Zhen C, Jen-Fu C, Qing-Yu H (2005) Current progress in sample preparation for two-dimensional electrophoresis in proteomics. Curr Proteomics 2:259–268
4. Rabiloud T et al (2010) Two dimensional gel electrophoresis in proteomics: past, present and future. J Proteome 73:2064–2077
5. O'Farrell PH (1975) High resolution two-dimensional electrophoresis of proteins. J Biol Chem 250:4007–4021
6. Dorri Y, Kurien BT, Scofield RH (2009) A simple and faster version of two-dimensional gel electrophoresis using vertical, mini SDS-PAGE gel apparatus. Iran J Chem Chem Eng 28:51–56
7. Knisley KA, Rodkey LS (1986) Affinity immunoblotting. High resolution isoelectric focusing analysis of antibody clonotype distribution. J Immunol Methods 95:79–87
8. Zuo X, Hembach P, Echan L et al (2002) Enhanced analysis of human breast cancer proteomes using micro-scale solution isoelectrofocusing combined with high resolution 1-D and 2-D gels. J Chromatogr B Analyt Technol Biomed Life Sci 782:253–265
9. Klein J, Harding G, Klein E (2002) A new isoelectric focusing gel for two-dimensional electrophoresis constructed in microporous hollow fiber membranes. J Proteome Res 1:41–45
10. Kurien BT, Scofield RH (1998) Heat mediated quick Coomassie blue protein staining and destaining of SDS-PAGE gels. Indian J Biochem Biophys 35:385–389
11. Dorri Y, Khalili AW, Scofield RH (2009) Whole and strip nitrocellulose membrane as well as a new-line-immunoblotting of antigen using the chemiluminescence technique. Methods Mol Biol 536:417–432
12. Kurien BT, Scofield RH (1997) Multiple immunoblots after non-electrophoretic bidirectional transfer of a single SDS-PAGE gel with multiple antigens. J Immunol Methods 205:91–94
13. Dorri Y, Kurien BT (2010) Environmentally safe removal/disposal of Coomassie Brilliant Blue from gel destain and used gel stain. Anal Biochem 404:193–196

Chapter 26

A Foodomics Approach: CE-MS for Comparative Metabolomics of Colon Cancer Cells Treated with Dietary Polyphenols

Mustafa Çelebier, Clara Ibáñez, Carolina Simó, and Alejandro Cifuentes

Abstract

The potential of capillary electrophoresis-mass spectrometry (CE-MS) for metabolomics is demonstrated through the analysis of metabolites from human HT29 colon cancer cells treated and non-treated with dietary polyphenols. Prior to CE-MS analysis, four different metabolite purification strategies are investigated. Namely, the results obtained after methanol deproteinization, ultrafiltration, and two solid-phase extraction methods using C18 and polymer-based cartridges are described. These generic methods can have broad applications to analyze metabolites in a large variety of matrices and fields, including the new Foodomics area.

Key words Foodomics, Metabolomics, Colon cancer, Dietary polyphenols, CE, CE-MS

1 Introduction

One of the main goals in modern food science and nutrition is to improve our limited understanding of the roles of nutritional compounds at molecular level. Dietary polyphenols are currently receiving considerable interest for their presumed role in the prevention of various degenerative diseases such as cancer and cardiovascular diseases [1]. However, the molecular mechanisms underlying the biological activity of these compounds and their cellular targets are still unknown. Foodomics has emerged as a new discipline that studies the food and nutrition domains through the application of advanced "omics" technologies in order to improve consumer's well-being, health, and confidence [2]. The advent of "omics" techniques—interpreted broadly as a suite of high-throughput technologies for the generation, processing, and application of scientific information about the composition and functions of genome, proteome, and metabolome—has created unprecedented opportunities for increasing our understanding of how nutrients

Biji T. Kurien and R. Hal Scofield (eds.), *Electrophoretic Separation of Proteins: Methods and Protocols*, Methods in Molecular Biology, vol. 1855, https://doi.org/10.1007/978-1-4939-8793-1_26,

modulate gene and protein expression and ultimately influence cellular metabolism. In order to gain detailed insight on the molecular mechanisms governing the metabolic pathways, metabolomics is a must in the new Foodomic field. Metabolome can be defined as the full set of endogenous or exogenous low molecular weight metabolic entities of approximately <1000 Da (metabolites) and the small pathway motifs that are present in a biological system [3]. Considering the great diversity and heterogeneity of metabolites, simultaneous purification and analysis of all metabolites from a biological sample are also a challenge. Sample preparation for metabolomics will depend on the type of sample being analyzed, the subsequent method of analysis, and the metabolomic approach (targeted or nontargeted). As a result, at present no single analytical methodology or platform is available to detect, quantify, and identify all metabolites in a certain sample, although it is now generally accepted that mass spectrometry (MS)-based techniques (together with nuclear magnetic resonance) are crucial for metabolomics. However, due to the great complexity of metabolomes in biological samples, an analytical separation technique is usually coupled to MS to avoid signal suppression and subsequent loss of metabolic information. In this regard, capillary electrophoresis (CE) is particularly suited for the rapid separation of ionic and polar compounds with very high resolution using extremely small reagents and sample volumes. The coupling CE-MS is described in this work for metabolite fingerprinting [4]. Time-of-flight (TOF) mass analyzer is used in the CE-MS coupling due to its high scan speed and high mass resolution and accuracy, what perfectly fits with the narrow peaks provided by CE. In summary, four different sample preparation procedures for cytosolic metabolite extraction from human HT29 colon cancer cells are described in this work, together with a complete CE-MS approach for the study of the effect of dietary polyphenols on colorectal cancer cells' metabolome.

2 Materials

2.1 Reagents and Samples

1. Distilled water from Milli-Q system (Millipore, Bedford, MA, USA).
2. MS grade water from Scharlau (Barcelona, Spain).
3. Formic acid (98–100%) from Riedel-de Haën (Seelze, Germany).
4. Phosphate-buffered saline (PBS): 138 mM sodium chloride, 2.7 mM potassium chloride, 10 mM sodium hydrogen phosphate, pH 7.4.
5. Protease inhibitor cocktail containing 104 mM 4-(2-aminoethyl)benzenesulfonyl fluoride (AEBSF), 1.5 mM

pepstatin A, 1.4 mM E-64, 4 mM bestatin, 2 mM leupeptin, and 80 μM aprotinin.

6. Amino acids (arginine, lysine, histidine, γ-aminobutyric acid, valine, serine, leucine, threonine, glutamine, proline, aspartic acid, and tyrosine).
7. Tyramine.
8. DL-methionine sulfone.
9. 1,4-Piperazinediethanesulfonic acid (PIPES).
10. Human HT29 colon cancer cells (Universidad de Miguel Hernandez, Elche, Spain).
11. Dulbecco's Modified Eagle Medium (DMEM).
12. Rosemary polyphenolic extracts were obtained by SFE using CO_2 at 150 bar as solvent with 7% ethanol [5].

2.2 Solutions for Sample Treatment, Metabolite Extraction, and CE-ESI-TOF MS Method

1. Homogenization buffer solution for sample treatment: 10 mM Tris–HCl, 5 mM EDTA, 120 mM NaCl, pH 7.4. Dissolve 1.21 g of Trizma® base, 1.87 g of EDTA, and 3.51 g of NaCl in 95 mL of Milli-Q water, and adjust the pH to 7.4 by adding HCl. Then reach a final volume of 100 mL by adding Milli-Q water using a volumetric flask.
2. Sodium formate (5 mM) for MS calibration: Mix 5 mL of 1 M NaOH with 99.5 mL of H_2O:2-propanol (50:50 v/v) with 0.2% formic acid, and make up volume to 100 mL.
3. Sodium hydroxide (1 M): Dissolve 4 g of sodium hydroxide in 95 mL of Milli-Q water. Then reach a final volume of 100 mL using a volumetric flask.
4. Amino acids solution to corroborate metabolite identification. Dissolve amino acids in Milli-Q water at the following concentrations: Arginine (0.58 mg/mL), lysine (0.49 mg/mL), histidine (0.52 mg/mL), γ-aminobutyric acid (GABA) (0.34 mg/mL), valine (0.39 mg/mL), serine (0.35 mg/mL), leucine (0.44 mg/mL),threonine (0.40 mg/mL), glutamine (0.49 mg/mL), proline (0.38 mg/mL), aspartic acid (0.44 mg/mL), tyrosine (0.09 mg/mL).
5. 1 M formic acid, pH 1.8: Mix 3.77 mL of formic acid with 95 mL of Milli-Q water in a volumetric flask, and then reach a final volume of 100 mL.
6. Isopropanol-water (50:50 v/v): Mix 20 mL of MS grade 2-propanol with 20 mL of MS grade water.
7. Water-methanol (95:5, v/v). Mix 95 mL of Milli-Q water with 5 mL of methanol.

2.3 Instrumentation for Sample Preparation Extraction Procedures

1. Polytron® homogenizer (Capitol Scientific, Austin, TX, USA).
2. Evolute™ ABN (water-wettable polymer-based sorbent that extracts acidic, basic, and neutral analytes) cartridges (Biotage, Cardiff, Wales, UK).
3. Isolute C18 Endcapped cartridges (Biotage).
4. Amicon Ultra 3 kDa centrifugal device.
5. Savant SpeedVac SC 200.

2.4 Instrumentation for CE-ESI-TOF MS Analysis

1. High-performance capillary electrophoresis (CE) system model P/ACE 5500 (Beckman, Fullerton, CA, USA) controlled by a PC equipped with System Gold software (Beckman).
2. Syringe pump (Cole Palmer, Vernon Hills, IL, USA).
3. Uncoated fused-silica capillaries (50 μm id and 90 cm total length) (Composite Metal Services, Worcester, England).
4. Orthogonal electrospray interface model G1607A (Agilent Technologies, Palo Alto, CA, USA).
5. TOF MS instrument (micrOTOF) (Bruker Daltonics, Bremen, Germany) controlled by a PC equipped with micrOTOF control software (Bruker Daltonics).

2.5 Software for Data Processing and Analyzing

1. DataAnalysis 3.3 software (Bruker Daltonics).
2. Free available databases used for tentative identification [6–10]:

Database HMDB (http://www.hmdb.ca/).

Metlin (http://metlin.scripps.edu/).

KEGG compound (http://www.genome.jp/kegg/compound/).

KEGG Pathway (http://www.genome.jp/kegg/pathway.html).

PubChem (http://pubchem.ncbi.nlm.nih.gov/).

3 Methods

3.1 Sample Preparation

1. Grow human HT29 cells in DMEM supplemented with 5% heat-inactivated fetal calf serum, 2 mM L-glutamine, 50 U/mL of penicillin G, and 50 μg/mL of streptomycin, at 37 °C in humidified atmosphere and 5% CO_2. Treat the cell culture with 10 μM rosemary polyphenolic extract for 72 h.
2. Wash human HT29 colon cancer cells with PBS solution and then centrifuge at 1000 × *g* for 10 min to form the pellets.
3. Suspend the pellets with homogenization buffer (5.5 mL) and protease inhibitor cocktail (10 μL/10^6 cells).
4. Disrupt the cells with a polytron homogenizer (3 cycles of 30 s, in ice) and centrifuge for 14 min at 14,000 × *g* and 4 °C.

5. Discard the pellet (nuclear fraction) and take the supernatant (cytosolic fraction).
6. Centrifuge the supernatant for 1 h at 100,000 × *g* and 4 °C.
7. Store the supernatant (cytosolic fraction) at −80 °C until metabolite purification procedure is carried out. This temperature stops enzyme activity and allows samples to be stored safely without continuing metabolic activity.

3.2 Metabolite Extraction Procedures

3.2.1 Solid-Phase Extraction (SPE)

1. Use Evolute™ ABN columns and Isolute C18 Endcapped cartridges for SPE. The conditions of elution are identical for both solvents.
2. Pass 1 mL of methanol followed by 1 mL of water through the cartridges.
3. Add 350 μL of Milli-Q water to 150 μL of cytosolic fraction obtained from the cell culture.
4. Load this solution onto the column and wash with 1 mL of water-methanol (95:5, v/v).
5. Use 500 μL of methanol to elute the sample.
6. Aliquot eluted sample in 100 μL volume.
7. Vacuum dry the aliquots and store the dried extracts at −80 °C.
8. Prior to CE-MS analysis, dissolve dried extracts in 20 μL of MS grade water.

3.2.2 Protein Precipitation

1. Add 350 μL of methanol to 150 μL of cytosolic fraction obtained from the cell culture.
2. Incubate the solution at −20 °C during 2 h to reduce protein solubilization.
3. Centrifuge the suspension for 5 min at 20,000 × *g* and 4 °C.
4. Discard the pellet and collect the supernatant fraction.
5. Aliquot supernatant fraction in 100 μL volume.
6. Vacuum dry the aliquots and store the dried extracts at −80 °C.
7. Prior to CE-MS analysis, dissolve dried extracts in 20 μL of MS grade water.

3.2.3 Ultracentrifugation

1. Add 350 μL of Milli-Q water to 150 μL of cytosolic fraction obtained from the cell culture.
2. Using the Amicon Ultra 3 kDa centrifugal device, centrifuge the solution for 70 min at 14,000 × *g* and 20 °C.
3. Collect with a pipette the fraction with molecular weight lower than 3 kDa and aliquot in 100 μL volume.
4. Vacuum dry the aliquots and store the dried extracts at −80 °C.
5. Prior to CE-MS analysis, dissolve dried extracts in 20 μL of MS grade water.

3.3 CE-ESI-TOF MS Metabolic Analysis of Human HT29 Colon Cancer Cells

1. Condition all new capillaries by flushing with 0.1 M NaOH for 20 min, followed by water for 20 min and background electrolyte (BGE) for 5 min (*see* **Note 1**). Use the following BGE for CE separation (1 M formic acid, pH 1.8) (*see* **Note 2**).
2. After each run, condition the capillary with Milli-Q water for 4 min, followed by BGE for 4 min (*see* **Note 3**).
3. Set the interface between CE and TOF MS via a sheath-flow configuration, in which electrical contact at the electrospray needle tip is established via a flow of conductive sheath liquid with the following composition, isopropanol-water (50:50 v/v) (*see* **Note 4**), and delivered at a flow rate of 0.24 μL/min by a syringe pump. Use next nebulizer and drying gas conditions: 0.4 bar N_2 and 4 L/min N_2, respectively, and maintain the ESI chamber at 200 °C (*see* **Note 5**).
4. Operate the micrOTOF to acquire spectra in the m/z range of 50–600 every 90 ms. Make external and internal calibration of the TOF MS instrument (*see* **Note 6**) by introducing 5 mM sodium formate solution through the separation capillary.
5. Make sample injection at the anodic end using N_2 pressure of 0.5 psi (34.5 mbar) for 80 s (*see* **Note 7**).
6. Use +25 kV as running voltage at a constant temperature of 25 °C to carry out electrophoretic separation by using 1 M formic acid as BGE.
7. Analyze each sample in triplicate by CE-MS.
8. Depending on metabolite extraction method (*see* Subheading 3.2), four different base peak electropherograms (BPE) are obtained by CE-MS (Fig. 1).

3.4 Metabolomic Fingerprinting of Human HT29 Colon Cancer Cells Treated with Dietary Polyphenols

Follow the CE-ESI-TOF MS metabolomic approach explained in Subheading 3.3 to obtain the metabolic fingerprinting from human HT29 colon cancer cells treated and non-treated with dietary polyphenols (Fig. 2).

3.5 Data Analysis

1. Using the DataAnalysis 3.3 software, annotate the accurate m/z value, migration time, and area from each peak.
2. Depending on established m/z differences, eliminate the redundant responses from same ion, such as isotopic peaks, fragments, adducts, dimers, etc. Remove the low reliability signals such as nonpeak-like shapes or spike noise.
3. As a result of using a high mass accuracy TOF MS instrument, Generate-MolecularFormula Editor in DataAnalysis 3.3

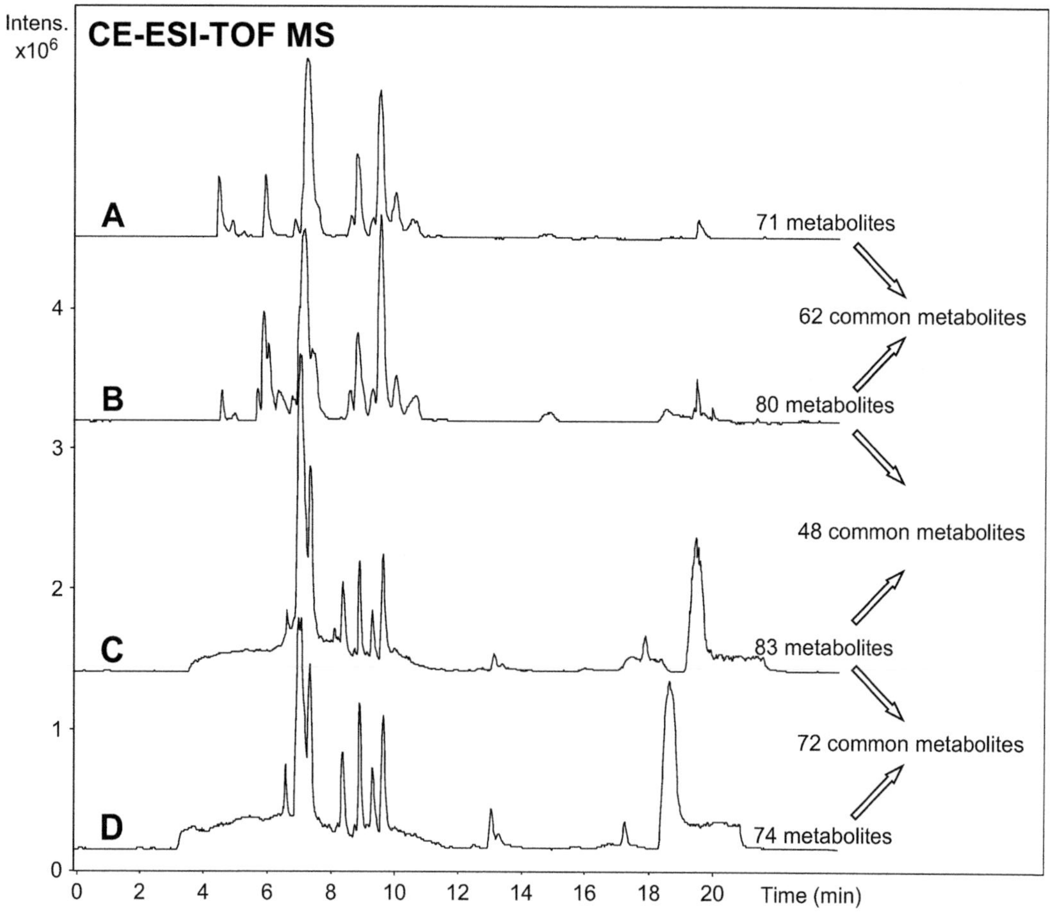

Fig. 1 CE-ESI-TOF MS base peak electropherograms (BPE) of the cytosolic fraction from HT29 colon cancer cells after four different metabolite extraction procedures. (**a**) SPE extraction using Isolute C18 Endcapped cartridge. (**b**) SPE extraction using EvoluteTM ABN cartridge. (**c**) Metabolite extraction using methanol deproteinization. (**d**) Metabolite extraction using ultrafiltration with an Amicon Ultra 3 kDa centrifugal device

software is capable of providing a list of possible elemental formulas (with errors below 5–10 ppm).

4. Enlighten standard functionalities such as minimum/maximum elemental range, electron configuration, and ring-plus double bond equivalents, and compare the theoretical and the experimental isotopic pattern (Sigma-Value™) for increased confidence in the theoretical molecular formula.
5. Use some free available databases (*see* Subheading 2.5) in order to work out tentative identification based on the obtained molecular formula.
6. Confirmation of amino acid identification is carried out by using commercial compounds (*see* Subheading 2.2).

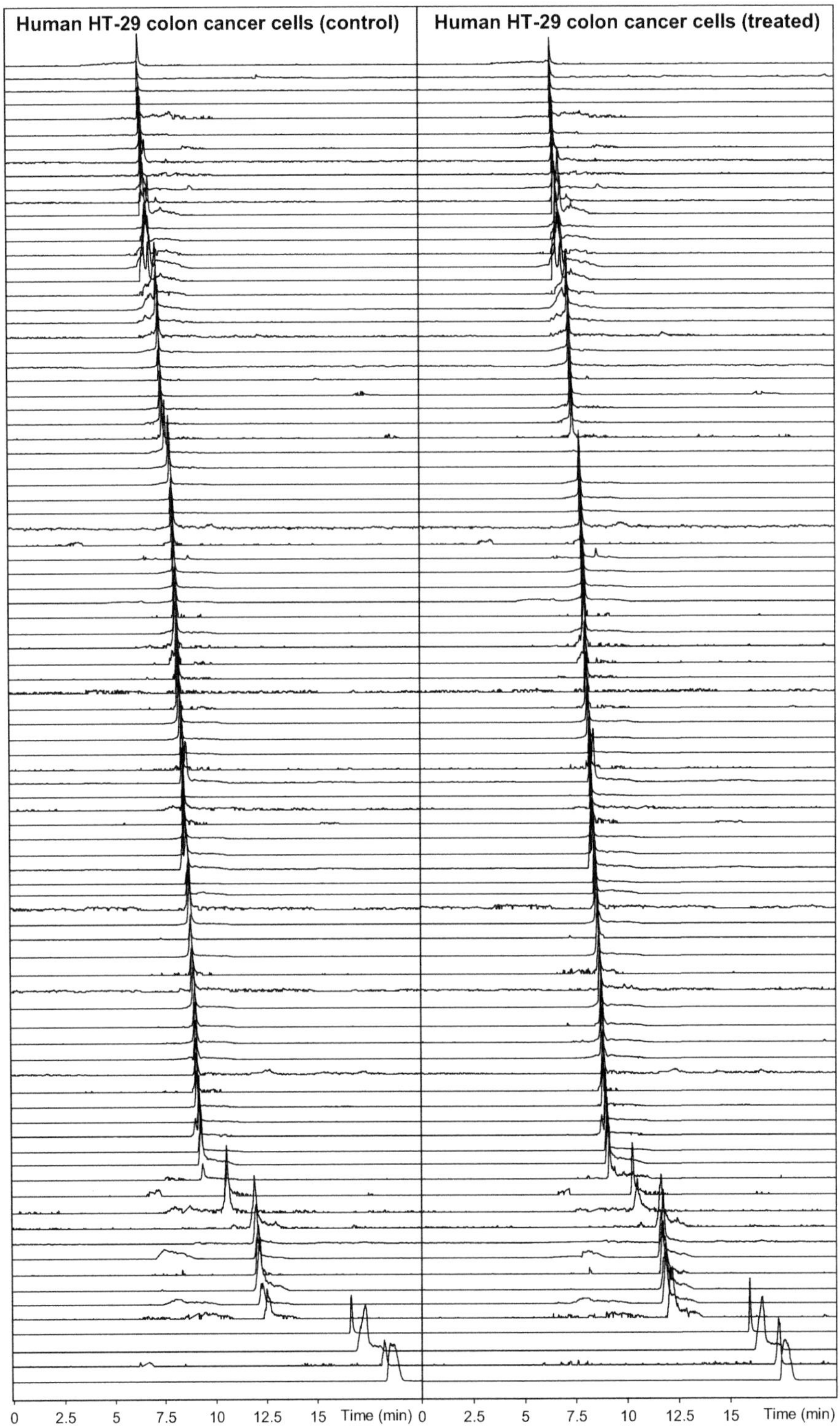

Fig. 2 CE-ESI-TOF MS extracted ion electropherograms (EIE) of the cytosolic fraction from human HT29 colon cancer cells: control (left panel) and treated with dietary polyphenols (right panel). Metabolite purification is carried out through methanol deproteinization

7. After analysis by CE-ESI-TOF MS of human HT29 colon cancer cells using different metabolite purification protocols, 71 and 80 different metabolites are detected in C18 and ABN extract, respectively. Similar extraction selectivity is obtained since 62 are common metabolites. On the other hand, 83 and 74 metabolites are detected in MeOH and ultrafiltrate extract, respectively. Similar extraction selectivity is obtained since 72 are common metabolites. Comparing ABN (80 metabolites) and methanol (83 metabolites) purification, different extraction selectivity is observed with only 48 common metabolites.
8. After data analysis of ABN and methanol extracts, a total number of 115 different metabolites with molecular masses from 90.0 to 777.2 Da are observed. Forty-four metabolites are tentatively identified and classified in several groups. Most abundant compounds are amino acids (histidine, arginine, lysine, γ-aminobutyric acid, valine, serine, leucine (or isoleucine), threonine, glutamine, proline, tyrosine, and aspartic acid) and polypeptides, although other compounds such as lipoic acid derivatives, steroid derivatives, carbohydrate derivatives, amino alcohol, amino acid phosphates, nucleosides, quaternary amines, pterins, acyl phosphates, and nucleotides are also present. Among all tentatively identified compounds, 20 of them are related to the following metabolic pathways: glycan biosynthesis and metabolism, organismal systems (excretory system), neuroactive ligand-receptor interaction, carbohydrate metabolism, lipid metabolism, nucleotide metabolism, energy metabolism, metabolism of cofactors and vitamins, environmental information processing, genetic information processing, and amino acid metabolism.
9. After CE-ESI-TOF MS analysis of metabolite extracts (obtained through methanol deproteinization), human HT29 colon cancer cells treated and non-treated with dietary polyphenols were compared. Ninety different compounds are detected with this method in less than 20 min (*see* Fig. 2). Significant differences are found for ions: 130.1602 m/z, 132.0772 m/z, 212.1167 m/z, 240.1487 m/z, 271.0557 m/z, 118.0870 m/z, 150.0586 m/z, 166.0861 m/z, 182.0802 m/z, and 307.0840 m/z. Among them, creatine, valine, tyrosine, and glutathione disulfide were identified.

4 Notes

1. Since the presence of nonvolatile components decreases the MS sensitivity, increases the background noise, and, under extreme conditions, can clog the system, volatile sheath liquids and separation BGEs must be used while working with MS.

2. A low pH should be used for the analysis in order to avoid analyte adsorption onto the inner capillary wall and confer positive charge to the analytes to improve their ESI ionization yield. Under these conditions, most amino acids, amines, nucleosides, small peptides, and, in general, basic compounds present overall positive charge, and they migrate before EOF (electroosmotic flow). Moreover, using these electrophoretic conditions allows to detect some acidic compounds (bearing some negative charge in their structure) being carried to the MS by the residual EOF.
3. In order to eliminate adsorbed compounds from the capillary wall, after each analysis, the capillary is rinsed.
4. Even some different mixtures of acetonitrile, methanol, and isopropanol with water could be used, in order to obtain the best response on sensitivity and signal stability of metabolites, 50:50 (v/v) isopropanol-water mixture is recommended as sheath liquid for the electrospray interface.
5. Even the spray is stable between 2 and 6 μL/min, it is recommended to use 4 μL/min as sheath liquid flow rate to reach the best S/N value during CE-ESI-TOF MS analysis of metabolites. Nebulizer pressure (0.2–0.7 bar) has also some effect on MS sensitivity; a nebulizer pressure of 0.4 bar is advised to obtain the best results.
6. External and internal calibration of the TOF MS instrument should be performed prior to each analysis. Masses for calibration of the TOF MS instrument are the following: 90.9766, 158.9641, 226.9515, 294.9389, 362.9263, 430.9138, 498.9012, and 566.8886 m/z.
7. Since different salt concentration of the purified samples may change the amount of injected sample, use pressure injection instead of electrokinetic injection. By applying 0.5 psi pressure for 80 s, a sample plug of 2.7 cm (3% of the capillary length) is injected.

Acknowledgments

This work was supported by AGL2008-05108-C03-01 (Ministerio de Ciencia e Innovación, Spain) and CSD2007-00063 FUN-C-FOOD projects (Programa CONSOLIDER, Ministerio de Educacion y Ciencia, Spain). MC thanks the Council of Higher Education, Turkey, for his grant. CI thanks Ministerio de Educacion y Ciencia, Spain, for her FPI grant.

References

1. Reddy L, Odhav B, Bhoola KD (2003) Natural products for cancer prevention: a global perspective. Pharmacol Ther 99:1–13
2. Cifuentes A (2009) Food analysis and foodomics. J Chromatogr A 1216:7109–7109
3. Garcia-Cañas V, Simó C, Leon C, Cifuentes A (2010) Advances in Nutrigenomics research: novel and future analytical approaches to investigate the biological activity of natural compounds and food functions. J Pharm Biomed Anal 51:290–304
4. Ramautar R, Somsen GW, de Jong GJ (2009) CE-MS in metabolomics. Electrophoresis 30:276–291
5. Herrero M, Plaza M, Cifuentes A, Ibáñez E (2010) Green processes for the extraction of bioactives from rosemary: chemical and functional characterization via ultra-performance liquid chromatography-tandem mass spectrometry and in-vitro assays. J Chromatogr A 1217:2512–2520
6. Wishart DS, Knox C, Guo AC, Eisner R, Young N, Gautam B, al e (2009) HMDB: a knowledgebase for the human metabolome. Nucleic Acids Res 37:D603–D610
7. Smith CA, O'Maille G, Want EJ, Qin C, Trauger SA, Brandon TR e a (2005) METLIN: a metabolite mass spectral database. Ther Drug Monit 27:747–751
8. Kanehisa M, Goto S, Furumichi M, Tanabe M, Hirakawa M (2010) KEGG for representation and analysis of molecular networks involving diseases and drugs. Nucleic Acids Res 38: D355–D360
9. Kanehisa M, Goto S, Hattori M, Aoki-Kinoshita KF, Itoh M, Kawashima S, al e (2006) From genomics to chemical genomics: new developments in KEGG. Nucleic Acids Res 34:D354–D357
10. Aoki-Kinoshita KF, Kanehisa M (2007) Gene annotation and pathway mapping in KEGG. Methods Mol Biol 396:71–91

Chapter 27

Characterization of New Cyclic D,L-α-Alternate Amino Acid Peptides by Capillary Electrophoresis Coupled to Electrospray Ionization Mass Spectrometry

María Dámaris Cortez-Díaz, Fanny d'Orlyé, and Anne Varenne

Abstract

The self-assembly of peptide nanotubes (PNTs) depends on the structure and chemistry of cyclic peptide (CP) monomers, impacting on their properties, which makes the choice of their monomers and their characterization a high challenge. For this purpose, we developed for the first time a capillary electrophoresis coupled to electrospray ionization mass spectrometry (CE-ESI-MS) methodology and characterized a set of eight original CP sequences of 8, 10, and 12 D,L-α-alternate amino acids with a controlled internal diameter (from 7 to 13 Å) and various properties (diameter, global surface charge, hydrophobicity). This new CE-ESI-MS methodology allows verifying the structure, the purity, as well as the stability (when stored during several months) of interesting potential precursors for PNTs that could be employed as nanoplatforms in diagnostics or pseudo sieving tools for separation purposes.

Key words Capillary electrophoresis, Electrospray ionization mass spectrometry, Cyclic peptides, Peptide nanotubes, Purity, Stability

1 Introduction

Nanotubes are becoming interesting tools for a wide range of applications in different fields such as optics, catalysis, electronics, drug delivery, and molecular transport [1]. There is a great variety in nanotube composition, such as zeolites, graphite, inorganic complexes, lipids, proteins, cyclodextrins, and peptides [1–5]. One of the most studied and employed nanotubes since the 1990s are carbon nanotubes (CNTs) [6], because of their unique structural, mechanical, and electronic properties. However, CNTs have poor solubility and size homogeneity, are quite toxic and expensive to produce, and present limited opportunities for surface covalent modifications. On the contrary, organic nanotubes allow surface functionalization, biocompatibility, and low production cost, with a reversible noncovalent bonded supramolecular

Biji T. Kurien and R. Hal Scofield (eds.), *Electrophoretic Separation of Proteins: Methods and Protocols*, Methods in Molecular Biology, vol. 1855, https://doi.org/10.1007/978-1-4939-8793-1_27,

assembly. Among them are the peptide nanotubes (PNTs) that were identified as interesting tools for various applications from biomedical to electronic ones [1, 7].

The principal advantage of PNTs is the control of the design of the CP monomers. Indeed the diameter of the PNT is controlled by the number of amino acid residues on the CP ring, achieving sub-nanometer scale in the 1–17 Å range [1, 6, 8–14], which is difficult to obtain from other structures. Furthermore, the outer surface properties of the PNTs are controlled by the amino acid composition of the sequence and by possible functionalization of the amino acid residues and/or the side chains of these amino acids [15, 16]. Finally, the internal PNT wall properties can be controlled through the incorporation of artificial amino acids with cyclohexanes and aromatic rings [1, 6, 11, 17, 18], for example. Therefore, controlling the design of the cyclic monomers allows to tailor the supramolecular architecture for novel and specific applications [6].

To characterize the structure, the morphology, the dimensions, and the optimal conditions for the self-assembly of CPs into PNTs, the cyclic monomers and supramolecular arrangements have been characterized by electron microscopy (SEM and TEM) [10, 11, 19–22], electron diffraction [23], FT-IR [8, 10, 11, 20, 23, 24] and Raman spectroscopy [20], X-ray diffraction [10, 17, 20, 25], nuclear magnetic resonance (NMR) spectroscopy [10, 11, 17, 24, 26], AFM [11, 21], optical microscopy (polarizing light and fluorescence) [21, 22], circular dichroism spectroscopy [11, 22], dynamic light scattering [11], mass spectrometry (MS) [17, 24], and molecular modeling [10, 23, 27–31]. However, there is a high need to easily verify the exact structure of the synthesized linear and cyclic peptide monomers so as to be able to tune the PNT dimensions, properties, and stability.

We developed a structural characterization of new cyclic monomers by capillary electrophoresis (CE) coupled to electrospray ionization mass spectrometry (CE-ESI-MS). Indeed, CE is a powerful separation method, which offers high resolution, efficiency, and versatility while requiring low sample and reagent consumption and short analysis time [32, 33]. The coupling to MS allowing for molecular mass determination and structural information [3, 33] provides a powerful hyphenation that combines efficiency, versatility, high resolution, and sensitivity of both techniques [34, 35]. This CE-ESI-MS methodology allows to determine, thanks to electrophoretic profiles and mass spectra, the structure and purity of cyclic peptides sequences, in an easy, rapid, and low sample consuming way.

2 Materials

Prepare all solutions using ultrapure water (produced by purifying deionized water to attain a sensitivity of 18 M Ω cm at 25 °C) and analytical grade reagents. Prepare all reagents at room temperature and store them at 4 °C. Diligently follow all waste disposal regulations when disposing waste materials.

2.1 Background Electrolyte (See Note 1)

1. 10 mM ionic strength ammonium carbonate (pH 8.0): 10.43 mM ammonia/10.06 mM ammonium carbonate (pH 8) in the separation medium. Add 10.43 mL of a 1 M ammonia solution and 10.06 mL of 1 M ammonium carbonate in 1 L of water. Degas by ultrasound for 5 min and store at 4 °C.
2. Background electrolyte preparation (*see* **Note 2**): 50/50 (v/v) 10 mM ammonium carbonate/ethanol (EtOH). Add 50–95 mL of 1 mM ionic strength ammonium carbonate (pH 8.0) to 50 mL to 5 mL EtOH, respectively.

2.2 Cyclic Peptides and Neutral Marker Solutions

1. Cyclic peptides (*see* **Note 3** and Table 1): Lyophilize and store at 4 °C.
2. Stock solutions of cyclic peptides: Dissolve at a 2–3 mM concentration in a mixture 50/50 to 95/5 (v/v) 1 mM ionic strength ammonium carbonate (pH 8.0)/EtOH (*see* **Note 4**). Store at 4 °C.
3. Neutral marker: add 0.002% dimethylformamide (DMF) and 0.5 mM of D-(+)-glucose in the background electrolyte (*see* **Note 5**).

2.3 Capillary Preparation and Storage

1. Bare-fused silica capillary: 80 cm total length, 50 μm (or 70 μm) internal diameter, 380 μm outer diameter. Condition new capillaries by successive flushes with 1 M and 0.1 M NaOH and then with water under a pressure of 93.5 kPa for 15 min each. Between runs, wash capillaries with BGE for 3 min. For storage, rinse with BGE and then water for 5 min each. Dry by air when not in use.

2.4 Sheath Liquid Composition (See Note 6)

Sheath liquid: EtOH-H_2O mixture (90:10 v/v) containing 2% v/v acetic acid. Mix 8 mL ultrapure water and 90 mL methanol of Normapure grade. Add 2 mL acetic acid in the latter solution and degas by ultrasound for 5 min before storing at 4 °C.

Table 1
Theoretical molecular mass (M) of the synthesized deprotected peptide (either linear or cyclic) sequences and the theoretical global charge at pH 8.0 (calculated by LifeProtein tool). Reproduced from [36] with permission from Elsevier

Peptide sequence	Peptide code[a]	Linear peptide theoretical M	Cyclic peptide theoretical M	Cyclic peptide theoretical charge
[-(-DAla-Asp-DAla-Asn-)$_2$-]	8aaSer1	760.3	742.3	−2
[-(-DAla-Asp-DAla-Asn-)$_3$-]	12aaSer1	1131.4	1113.4	-3
[-DAla-Asp-DAla-Asn-DAla-Trp-DAla-Asp-]	8aaSer2	813.3	831.3	−2
[-DAla-Asp-DAla-Asn-DAla-Asp-DAla-Asn-DAla-Trp -]	10aaSer2	1017.4	999.4	-2
[-DAla-Asp-DAla-Asn-DAla-Asp-DAla-Asn-DAla-Trp-DAla-Asn-]	12aaSer2	1202.5	1184.5	-2
[-DAla-Asp-DAla-Asn-DAla-His-DAla-Asn-]	8aaSer3	782.3	764.3	−1
[-DAla-Asp-DAla-Asn-DAla-His-DAla-Asn-DAla-Trp-]	10aaSer3	1039.4	1021.4	−1
[-DAla-Asp-DAla-Asn-DAla-His-DAla-Asn-DAla-Trp-DAla-Asn-]	12aaSer3	1224.5	1206.5	−1

[a]Peptide code: the number of amino acids of the sequence was used as prefix and the reference of the family sequences as suffix (Serie 1 (Ser1), Serie 2 (Ser2) and Serie 3 (Ser3))

3 Methods

Carry out all procedures at room temperature.

3.1 CE-UV Protocol

1. Analytical protocols, including capillary conditioning, sample introduction, and voltage application, were automatized and thus realized in one step (*see* **Note 7**).
2. Set the temperature in the capillary cartridge at 25 °C.
3. Fill and homogenize the capillary with the BGE by flushing for 15 min after capillary activation or 3 min in between each separation at 94.5 kPa.
4. To realize the successive injection of neutral marker and sample, apply a 5 kPa pressure for 6, 15, and 12 s for neutral marker, BGE, and sample, respectively (*see* **Note 8**).
5. Apply a voltage of 20 kV under a 5 kPa pressure to achieve the separation of the cyclic peptides (*see* **Note 9**).
6. The detection wavelengths were 200 and 214 nm.

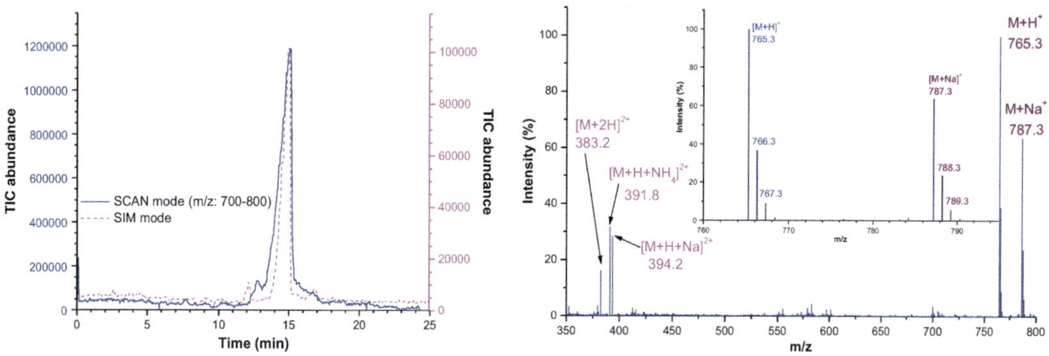

Fig. 1 CE-ESI-MS TIC electropherograms obtained in scan (plain) and SIM (dotted) (m/z value corresponding to $[M + H]^+$) modes and corresponding mass spectra for cyclic peptide of series 3 consisting in eight amino acids. The inserted figure on the left side corresponds to the isotopic mass spectra of the single-charged ion molecular mass. Reproduced from [36] with permission from Elsevier

3.2 Electrospray Interface (ESI)

1. Electrospray interface consists in a triple coaxial tube nebulizer held at ground potential.
2. Deliver the coaxial sheath liquid at a flow rate of 6 $\mu L.min^{-1}$ with an isocratic pump equipped with a splitter (1:100).
3. Set the nebulizing gas consisting in nitrogen at 100 °C under a pressure of 69 kPa (*see* **Note 10**).

3.3 Electrospray Interface Parameters

1. Connect the orthogonal ESI with a single quadrupole mass spectrometer in the positive ionization mode.
2. Use nitrogen as drying gas at 250 °C (flow rate, 6 $L.min^{-1}$).
3. Set ESI and skimmer voltage at 3000 V and 70 V, respectively.
4. Set peak width and dwell time at 0.3 min and 880 ms, respectively (*see* **Note 11**).

3.4 Data Treatment and Calibration Curve

1. Record both the overall scan mode MS signal (m/z 100–400 and m/z 400–3000) (*see* Fig. 1, left side, solid line) and the extracted ion current (EIC) signal (*see* Fig. 1, left side, dotted line) (*see* **Note 12**) for each cyclic peptide.
2. Then perform signal acquisition of the sample in the selected ion monitoring mode (SIM) at each identified and characterized m/z value (*see* Fig. 1, right side).
3. Extract the mass spectra of the cyclic peptides (*see* Fig. 1, insert on the right side) and process the numerical data using OriginPro 9.5 software. Obtain the mass spectrum data of molecular ions and adducts. Calculate the standard deviations from the replicates (*see* Tables 2 and 3).

Table 2
Theoretical (Theor.) and experimental (Exp.) mass and relative intensity of the CPs and isotopes, corresponding to protonated and adduct forms with one global charge, from CE-ESI-MS[a]. Reproduced from [36] with permission from Elsevier

Cyclic peptide	$[M+H]^+$				$[M+NH_4]^+$				$[M+Na]^+$				$[M+2Na-H]^+$			
	m/z		Intensity (%)		m/z		Intensity (%)		m/z		Intensity (%)		m/z		Intensity (%)	
	Exp.	Theor.	Exp.	Theor.	Exp.	Theor.	Exp.	Theor.	Exp.	Theor.	Exp.	Theor.	Exp.	Theor.	Exp.	Theor.
8aaSer1	743.3	743.3	100.0	100.0	760.3	760.3	100.0	100.0	765.3	765.3	100.0	100.0	787.3	787.3	100.0	100.0
	744.3	744.3	27.0	35.0	761.3	761.3	38.2	35.4	766.3	766.3	32.8	35.0	788.3	788.3	34.0	34.9
	–	–	–	–	762.3	762.3	16.9	9.0	767.3	767.3	9.3	8.8	789.3	789.3	8.8	8.8
12aaSer1	1114.4	1114.4	100.0	100.0	1131.4	1131.5	100.0	100.0	1136.3	1136.4	100.0	100.0	–	–	–	–
	1115.4	1115.4	50.7	52.0	1132.4	1132.5	47.2	52.8	1137.3	1137.4	62.0	52.0	–	–	–	–
	–	–	–	–	1133.4	1134.5	22.6	18.0	1138.3	1138.4	22.0	17.8	–	–	–	–
8aaSer2	814.3	814.4	100.0	100.0	831.4	831.4	100.0	100.0	836.3	836.3	100.0	100.0	858.3	858.3	100.0	100.0
	815.3	815.4	53.8	42.9	832.3	832.4	46.7	43.3	837.3	837.3	40.4	42.9	859.3	859.3	50.0	42.9
	816.3	816.4	15.4	11.4	833.3	833.4	15.0	11.6	838.4	838.3	10.6	11.4	–	–	–	–
10aaSer2	1000.5	1000.4	100.0	100.0	1017.5	1017.4	100.0	100.0	1022.5	1022.4	100.0	100.0	–	–	–	–
	1001.5	1001.4	52.2	51.5	1018.5	1018.4	52.4	51.8	1023.5	1023.4	57.8	51.4	–	–	–	–
	1002.5	1002.4	17.4	16.3	1019.5	1019.4	20.0	16.5	1024.5	1024.4	18.7	16.3	–	–	–	–
	–	–	–	–	1020.5	1020.5	4.8	3.9	1025.5	1025.4	4.7	3.8	–	–	–	–
12aaSer2	1185.4	1185.5	100.0	100.0	1202.5	1202.5	100.0	100.0	1207.5	1207.5	100.0	100.0	–	–	–	–
	1186.5	1186.5	60.6	60.4	1203.5	1203.5	55.4	60.8	1208.4	1208.5	53.1	60.4	–	–	–	–
	1187.5	1187.5	22.5	21.8	1204.5	1204.5	18.6	22.1	1209.4	1209.5	15.6	21.8	–	–	–	–
8aaSer3	765.3	765.3	100.0	100.0	–	–	–	–	787.3	787.3	100.0	100.0	–	–	–	–
	766.3	766.3	36.7	37.8	–	–	–	–	788.3	788.3	36.8	37.8	–	–	–	–
	767.3	767.3	8.9	9.4	–	–	–	–	789.3	789.3	8.8	9.4	–	–	–	–
10aaSer3	1022.4	1022.4	100.0	100.0	–	–	–	–	1044.4	1044.4	100.0	100.0	1066.4	1066.4	100.0	100.0
	1023.4	1023.4	53.3	54.3	–	–	–	–	1045.4	1045.4	49.3	54.3	1067.4	1067.4	60.0	54.3
	1024.4	1024.4	18.9	17.4	–	–	–	–	1046.4	1046.4	17.8	17.3	–	–	–	–
12aaSer3	1207.6	1207.5	100.0	100.0	–	–	–	–	1229.5	1229.5	100.0	100.0	1251.4	1251.5	100.0	100.0
	1208.5	1208.5	63.0	63.2	–	–	–	–	1230.6	1230.5	52.3	63.2	–	–	–	–
	1209.6	1209.5	19.6	23.2	–	–	–	–	1231.5	1231.5	16.9	23.1	–	–	–	–

[a]In all cases the double dash means that the value of m/z ratio was not observed experimentally with a significant intensity in MSD

Table 3
Theoretical (Theor.) and experimental (Exp.) mass and relative intensity of the CPs and isotopes, corresponding to protonated or adduct forms with two global charges, from CE-ESI-MS[a]. Reproduced from [36] with permission from Elsevier

Cyclic peptide	$[M+2H]^{2+}$				$[M+H+NH_4]^{2+}$				$[M+H+Na]^{2+}$				$[M+2Na]^{2+}$			
	Isotopes		Intensity (%)		Isotopes		Intensity (%)		Isotopes		Intensity (%)		Isotopes		Intensity (%)	
	Exp.	Theor.	Exp.	Theor.	Exp.	Theor.	Exp.	Theor.	Exp.	Theor.	Exp.	Theor.	Exp.	Theor.	Exp.	Theor.
8aaSer1	–	–	–	–	–	–	–	–	–	–	–	–	–	–	–	–
12aSer1	–	–	–	–	566.6	566.7	100.0	100.0	–	–	–	–	579.7	579.7	100.0	100.0
8aaSer2	–	–	–	–	–	–	–	–	–	–	–	–	–	–	–	–
10aaSer2	–	–	–	–	509.2	509.2	100.0	100.0	511.8	511.7	100.0	100.0	522.6	522.7	100.0	100.0
12aaSer2	–	–	–	–	–	–	–	–	604.7	604.7	100.0	100.0	615.2	615.2	100.0	100.0
8aaSer3	383.2	383.2	100.0	100.0	391.8	391.7	100.0	100.0	394.1	394.2	100.0	100.0	–	–	–	–
10aaSer3	511.8	511.7	100.0	100.0	520.3	520.2	100.0	100.0	522.8	522.7	100.0	100.0	–	–	–	–
12aaSer3	604.3	604.3	100.0	100.0	–	–	–	–	615.3	615.8	100.0	100.0	–	–	–	–

[a]In all cases the double dash means that the value of m/z ratio was not observed experimentally with a significant intensity in MSD

4 Notes

1. The background electrolyte must be compatible with MS. Its composition and concentration influence peak intensity, peptide charge state on the mass spectrum, and detection time. The presence of ionizable compounds in the sprayed solution causes a reduction in the MS signal intensity due to ion suppression [37–40] and background signal and decreases the mass resolution.
2. According to the cyclic peptide sequence, their solubility was modified. Therefore, a hydro-organic background electrolyte (ammonium carbonate/EtOH) was selected with different ratios so as to reach the entire solubility of each cyclic peptide. The cyclic peptides were dissolved in the BGE.
3. Height D,L-α-alternate amino acid cyclic peptides were obtained [36], whose composition, nomenclature, theoretical mass (linear and cyclic deprotected forms), as well as theoretical charge at pH 8.0 are summarized in Table 1. These novel designed cyclic peptides differ in terms of chain length within each series (8, 10, and 12 amino acids, with a theoretical diameter range from 7 to 13 Å according to [6]) and in amino acid nature so as to vary their charge density, surface group, and therefore the physicochemical characteristics of the corresponding potential PNTs. In order to promote different functionalities, the series were designed so as to provide carboxylic and amide groups for series 1; carboxylic, amide, and indole groups for series 2; and carboxylic, amide, indole, and imidazole groups for series 3. The difference in chain length and amino acid composition should allow to provide CPs of different van der Waals inner diameter and different properties for their further future use. The hydrophobicity increases, and the theoretical charge (at pH 8.0) decreases from series 1 to series 3.
4. Cyclic peptides were synthesized [36] according to the method developed by Bruce Merrifield [15, 16]. When the cyclization was finished, the cyclic peptides were solubilized in a 0.1% trifluoroacetic acid (TFA) in aqueous solution/0.08% TFA in acetonitrile (ACN) mixture at 40/60 to 50/50 (v/v), according to the peptide sequence. The solutions were filtered using a nonsterile polyvinylidene fluoride filter (0.45 μm pore size) from Whatman. They were purified by a semi-preparative HPLC, in a Phenomenex analytical column (Jupiter 10 μ, C_{18} 300 A, 250 × 10 mm, 10 μm) with 0.1% TFA in aqueous solution/0.08% TFA in ACN (gradient elution from 60/40 to 10/90 (v/v)) during 30 min, followed by a plateau to complete 40 min. The separation was performed under a

4.7 mL/min flow rate and a 115.10^5 Pa pressure and the detection by an UV-VIS detector (235 nm wavelength). The collected fractions of cyclic peptides were lyophilized in a Labconco FreeZone 2.5 Plus (Serial No. 040823971 D) lyophilizator system equipped with a hybrid pump (MTR mod. T55JXCFH-1280).

5. A mixture of two neutral compounds, DMF and glucose, was selected to be compatible with UV and MS detection, respectively.
6. The sheath liquid and ESI interface conditions were optimized to reach the best ionization-desorption yield.
7. The analysis time depends on the applied voltage, BGE nature, and peptide nature. In a basic BGE (10 mM ionic strength carbonate ammonium (8.0 pH) mixed with EtOH (50/50, v/v)), the different sequences were negatively charged, leading to a separation in positive polarity within a few minutes (3.8×10^{-4} $cm^2/V{\cdot}s$ electroosmotic mobility value). In the SIM mode, thin peaks were observed (considering the 80 cm capillary length to detector), which indicates negligible CP adsorption onto the capillary wall.
8. In these experimental conditions, the neutral markers migrate faster than the peptides. Therefore, a short zone of neutral marker is first introduced (to avoid any contact and possible interaction between marker and sample), then a short zone of BGE allows to physically separate neutral marker from sample during injection step, and finally a short zone of the sample (migrating the slowest) is introduced.
9. This method avoids the moving ion boundary phenomenon occurring when BGE is introduced in the SL. Ions from the SL enter the capillary and could shift the BGE pH [41]. This moving ionic boundary, however, could also contribute to a loss in resolution and an inversion of migration order, but the problems could be remediated by applying a gravity-induced hydrodynamic flow during mobilization [37, 42].
10. The interface corresponds to the outlet vial. The protrusion of the CE capillary from the interface needle has serious consequences on the spray performances [43–47], i.e., spray stability and ionization efficiency or yield, as it influences the mixing quality between SL and BGE. The signal intensity goes past a sharp maximum for a protruding length of 0.25 ± 0.05 mm from the SL tube (with nebulizing gas pressure set at 55 kPa and SL flow rate at 6 mL/min) [48].
11. The general configuration selected is as follows: SL helps for positive ionization of the cyclic peptides, as the positive ionization mode is more sensitive than the negative ionization one.

12. SIM/scan modes: For the cyclic peptides, the EIC (extracted ion current) signals are two to three times lower than the ones obtained in SIM mode; nevertheless, the EIC mode allows to identify any unknown compound in a single run, provided that there is no ionization suppression for compounds detected simultaneously. This online protocol thus appears as a powerful tool for fast identification of a large number of cyclic peptides, through their characterization in terms of molecular mass.

References

1. Brea RJ, Reiriz C, Granja JR (2010) Towards functional bionanomaterials based on self-assembling cyclic peptide nanotubes. Chem Soc Rev 39:1448–1456
2. Bong DT, Clark TD, Granja JR, Ghadiri MR (2001) Self-assembling organic nanotubes. Angew Chem Int Ed Engl 40:988
3. Scanlon S, Aggeli A (2008) Self-assembling peptide nanotubes. Nano Today 3:22–30
4. Gazit E (2007) Self-assembled peptide nanostructures: the design of molecular building blocks and their technological utilization. Chem Soc Rev 36:1263–1269
5. Rajagopal K, Schneider JP (2004) Self-assembling peptides and proteins for nanotechnological applications. Curr Opin Struct Biol 14:480–486
6. Chapman R, Danial M, Koh ML, Jolliffe KA, Perrier S (2012) Design and properties of functional nanotubes from the self-assembly of cyclic peptide templates. Chem Soc Rev 41:6023–6041
7. Chapman R, Jolliffe KA, Perrier S (2010) Synthesis of self-assembling cyclic peptide-polymer conjugates using click chemistry. Aust J Chem 63:1169–1172
8. Ghadiri MR, Granja JR, Buehler LK (1994) Artificial transmembrane ion channels from self-assembling peptide nanotubes. Nature 369:301–304
9. Engels M, Bashford D, Ghadiri MR (1995) Structure and dynamics of self-assembling peptide nanotubes and the channel-mediated water organization and self-diffusion. A molecular dynamics study. J Am Chem Soc 117:9151–9158
10. Ghadiri MR, Granja JR, Milligan RA, McRee DE, Khazanovich N (1993) Self-assembling organic nanotubes based on a cyclic peptide architecture. Nature 366:324–327
11. Hourani R, Zhang C, van der Weegen R, Ruiz L, Li C, Keten S, Helms BA, Xu T (2011) Processable cyclic peptide nanotubes with tunable interiors. J Am Chem Soc 133:15296–15299
12. Khazanovich N, Granja JR, McRee DE, Milligan RA, Ghadiri MR (1994) Nanoscale tubular ensembles with specified internal diameters. Design of a self-assembled nanotube with a 13-.ANG. Pore. J Am Chem Soc 116:6011–6012
13. Motesharei K, Ghadiri MR (1997) Diffusion-limited size-selective ion sensing based on SAM-supported peptide nanotubes. J Am Chem Soc 119:11306–11312
14. Motiei L, Rahimipour S, Thayer DA, Wong C-H, Ghadiri MR (2009) Antibacterial cyclic D,L-α-glycopeptides. Chem Commun (Camb) 25:3693–3695
15. AAPPTEC, L. A. a. p. p. t., Practical synthesis guide to solid phase peptide chemistry, www.aapptec.com, pp. 1–76
16. Chan W, White P (2000) Fmoc solid phase peptide synthesis: a practical approach. Oxford University Press, USA, New York
17. Amorín M, Castedo L, Granja JR (2003) New cyclic peptide assemblies with hydrophobic cavities: the structural and thermodynamic basis of a new class of peptide nanotubes. J Am Chem Soc 125:2844–2845
18. Brea RJ, Castedo L, Granja JR (2007) Large-diameter self-assembled dimers of α,γ-cyclic peptides, with the nanotubular solid-state structure of cyclo-[(L-Leu-D-MeN-γ-Acp)4-]·4CHCl2COOH. Chem Commun (Camb) 31:3267–3269
19. Montenegro J, Ghadiri MR, Granja JR (2013) Ion channel models based on self-assembling cyclic peptide nanotubes. Acc Chem Res 46:2955–2965
20. Valéry C, Pouget E, Pandit A, Verbavatz J-M, Bordes L, Boisdé I, Cherif-Cheikh R, Artzner F, Paternostre M (2008) Molecular origin of the self-assembly of lanreotide into nanotubes: a mutational approach. Biophys J 94:1782–1795

21. Mammadov R, Tekinay AB, Dana A, Guler MO (2012) Microscopic characterization of peptide nanostructures. Micron 43:69–84
22. Choi S-j, Jeong W-j, Kang S-K, Lee M, Kim E, Ryu DY, Lim Y-b (2012) Differential self-assembly behaviors of cyclic and linear peptides. Biomacromolecules 13:1991–1995
23. Hartgerink JD, Granja JR, Milligan RA, Ghadiri MR (1996) Self-assembling peptide nanotubes. J Am Chem Soc 118:43–50
24. Pérez-Alvite M, Mosquera M, Castedo L, Granja J (2011) Toward the rational design of molecular rotors ion sensors based on α, γ-cyclic peptide dimers. Amino Acids 41:621–628
25. Seebach D, Matthews JL, Meden A, Wessels T, Baerlocher C, McCusker LB (1997) Cyclo-β-peptides: Structure and tubular stacking of cyclic tetramers of 3-aminobutanoic acid as determined from powder diffraction data. Helv Chim Acta 80:173–182
26. Sun X, Lorenzi GP (1994) On the stacking of beta rings : the solution self association behavior of two partially N-methylated cyclo (hexaleucines). Helv Chim Acta 77:1520–1526
27. Rahmat F, Thamwattana N, Cox BJ (2011) Modelling peptide nanotubes for artificial ion channels. Nanotechnology 22:445707
28. Jishi RA, Flores RM, Valderrama M, Lou L, Bragin J (1998) Equilibrium geometry and properties of cyclo[$(Gly\text{-}d\text{-}Ala)_4$] and {Cyclo [$(Gly\text{-}d\text{-}Ala)_4$]}$_2$ from density functional theory. J Phys Chem A 102:9858–9862
29. Lewis JP, Pawley NH, Sankey OF (1997) Theoretical investigation of the cycle peptide system cyclo(D-Ala-Glu-D-Ala-Gln)m=i-4. J Phys Chem B 101:10576–10583
30. García-Fandiño R, Granja JR, D'Abramo M, Orozco M (2009) Theoretical characterization of the dynamical behavior and transport properties of alpha,gamma-peptide nanotubes in solution. J Am Chem Soc 131:15678–15686
31. Hartgerink JD, Clark TD, Ghadiri MR (1998) Peptide nanotubes and beyond. Chemistry 4:1367–1372
32. Medina-Casanellas S, Domínguez-Vega E, Benavente F, Sanz-Nebot V, Somsen GW (2014) Low-picomolar analysis of peptides by on-line coupling of fritless solid-phase extraction to sheathless capillary electrophoresis-mass spectrometry. de Jong, G. J. J Chromatogr A 1328:1–6
33. Suntornsuk L (2010) Recent advances of capillary electrophoresis in pharmaceutical analysis. Anal Bioanal Chem 398:29–52
34. Olivares LA, Nguyen NT, Yonker CR, Smith RD (1987) On-line mass spectrometric detection for capillary zone electrophoresis. Anal Chem 59:1230–1232
35. Smith RD, Barinaga CJ, Udseth HR (1988) Improved electrospray ionization interface for capillary zone electrophoresis-mass spectrometry. Anal Chem 60:1948–1952
36. Cortez-Diaz MD, d'Orlyé F, Gutierrez-Granados S, de Leon-Rodriguez LM, Varenne A (2016) Design, synthesis, and characterization of new cyclic d,l-α-alternate amino acid peptides by capillary electrophoresis coupled to electrospray ionization mass spectrometry. Anal Biochem 502:8–15
37. Jensen PK, Pasa-Tolic L, Peden KK et al (2000) Mass Spectrometric detection for capillary isoelectric focusing separations of complex protein mixtures. Electrophoresis 21:1372–1380
38. Minarik M, Foret F, Karger BL (2000) Fraction collection in micropreparative capillary zone electrophoresis and capillary isoelectric focusing. Electrophoresis 21:247–254
39. Foret F, Muller O, Thoren J et al (1996) Analysis of protein fractions by micropreparative capillary isoelectric focusing and matrix-assisted laser desorption time-of-flight mass spectrometry. J Chromatogr A 716:157–166
40. Chartogne A, Gaspari M, Jespersen S et al (2002) On-target fraction collection for the off-line coupling of capillary isoelectric focusing with matrix-assisted laser desorption/ionization mass spectrometry. Rapid Commun Mass Spectrom 16:201–207
41. Yang L, Lee CS, Hofstadler SA et al (1998) Capillary isoelectric focusing-electrospray ionization fourier transform ion cyclotron resonance mass spectrometry for protein characterization. Anal Chem 70:3235–3241
42. Zhang C-X, Xiang F, Pasa-Tolic L et al (2000) Stepwise mobilization of focused proteins in capillary isoelectric focusing mass spectrometry. Anal Chem 72:1462–1468
43. Ohnesorge J, Sänger-van de Griend C, Wätzig H (2005) Quantification in capillary electrophoresis-mass spectrometry: Long- and short-term variance components and their compensation using internal standards. Electrophoresis 26:2360–2375
44. Geiser L, Rudaz S, Veuthey JL (2003) Validation of capillary electrophoresis-mass spectrometry methods for the analysis of a pharmaceutical formulation. Electrophoresis 24:3049–3056
45. Nunez O, Moyano E, Galceran MT (2002) Capillary electrophoresis-mass spectrometry for the analysis of quaternary ammonium herbicides. J Chromatogr A 974:243–255

46. Sanz-Nebot V, Benavente F, Balaguer E et al (2003) Capillary electrophoresis coupled to time of flight-mass spectrometry of therapeutic peptide hormones. Electrophoresis 24:883–891
47. Nilsson SL, Bylund D, Jörnten-Karlsson M et al (2004) A chemometric study of active parameters and their interaction effects in a nebulized sheath-liquid electrospray interface for capillary electrophoresis-mass spectrometry. Electrophoresis 25:2100–2107
48. Mokaddem M, Gareil P, Belgaied JE et al (2008) A new insight into suction and dilution effects in capillary electrophoresis coupled to mass spectrometry *via* an electrospray ionization interface. Part I-Suction effect. Electrophoresis 29:1957–1964

Chapter 28

"Microchip Electrophoresis," with Respect to "Profiling of Aβ Peptides in the Cerebrospinal Fluid of Patients with Alzheimer's Disease"

Mohamad Reza Mohamadi, Romain Verpillot, Myriam Taverna, Markus Otto, and Jean-Louis Viovy

Abstract

Aggregation of beta-amyloid peptides especially Aβ1-42 in amyloid plaques is one of the major neuropathological events in Alzheimer's disease. This event is normally accompanied by a relative reduction of the concentration of Aβ1-42 in the cerebrospinal fluid (CSF) of patient developing the signs of Alzheimer's disease. Here, we describe methods for isolation and for microchip gel electrophoresis of Aβ peptides in polydimethylsiloxane (PDMS) microfluidic chip. The method was applied to compare the relative concentration of Aβ1-42 with other Aβ peptides, for example, Aβ 1-40 in CSF. In order to increase the sensitivity of detection, Aβ peptides in the CSF samples were first captured and concentrated using magnetic beads coated with specific anti-Aβ antibodies.

Key words Microchip electrophoresis, Beta-amyloid, Cerebrospinal fluid (CSF), Alzheimer's disease (AD)

1 Introduction

The growing number of Alzheimer's patients, associated with the aging of the population, raises a strong demand for efficient treatment and also for reliable methods for diagnosis of AD in the early stages. It is commonly accepted that AD has a very long (>10 years) asymptomatic evolution period during which irreversible neurological damages are progressively happening. Although there is no short-term hope of treatments able to reverse the existing damages, there are more and more hopes to find neuroprotective treatments able to retard the disease's progression. Due to the slow evolution of the disease, even a moderately efficient treatment can have a strong impact in quality of life and survival, if diagnosis can be made early enough. Currently diagnosis of AD is mainly based on clinical and neurological symptoms. However, at early stages

Biji T. Kurien and R. Hal Scofield (eds.), *Electrophoretic Separation of Proteins: Methods and Protocols*,
Methods in Molecular Biology, vol. 1855, https://doi.org/10.1007/978-1-4939-8793-1_28,

neurological symptoms are less clear, and distinguishing AD from other forms of dementia or even normal aging is challenging. Therefore, there are increasing demands for biomarker screening and developing methods for the detection of AD-specific markers at early stages.

The most important and widely accepted biomarkers for AD are beta-amyloid peptides. Aβ peptides are formed by proteolytic cleavage of a transmembrane protein called amyloid precursor protein (APP) [1–4]. Cleavage at different sites is possible, and several Aβ peptides are found in measurable quantities in biological fluids of healthy individuals. The ratios of concentrations of these different peptides are relatively stable in the CSF of healthy individuals [5]. During the progression of AD, Aβ peptides, especially Aβ1-42, are accumulated in amyloid plaques. Therefore, a selective reduction of Aβ1-42 in CSF from AD patients was reported in several studies, while the total level of Aβ peptides was found to be in the normal range [6].

Currently, conventional slab gel followed by western blotting is the principal method for quantitation of Aβ peptides in CSF samples [5]. However, slab gel electrophoresis is time-consuming, difficult to be automated, and hard to be used for screening the large number of samples that is required for biomarker studies. Recent advances of microchip electrophoresis (MCE) demonstrated the potential of MCE as a reliable analytical method for the profiling of proteins and peptides in biological samples [6]. As compared to conventional electrophoresis methods, MCE involves significant advantages, such as low sample consumption and a strong potential for automation and integration with different on-chip sample preparations [7, 8].

Here, we introduce a gel-based MCE method for the analysis of fluorescently labeled Aβ peptides in human CSF. Due to the complexity and variability of real CSF sample, and to the low concentration of Aβ peptides as compared to other highly abundant peptides and proteins in CSF, a sample preparation step prior to the MCE analysis is necessary. We used an immunocapture method based on capturing of Aβ peptides using magnetic beads coated with anti-Aβ peptides [9–13]. Since the publishing of the first version of the current chapter, we have completed the integration of the immunocapture step with microfluidic electrophoresis [10, 11]. In the integrated microfluidic device, first, immunocapture of Aβ peptides is done on a column of self-assembled magnetic beads coated with antibodies specific to Aβ peptides, then the captured Aβ peptides are eluted and preconcentrated on a nanoporous membrane made of photopolymerized hydrogel, and finally the preconcentrated sample was sent to an electrophoresis channel with fluorescent detection (Fig. 1). Using this method, we were able to process submilliliter sample volume in a semiautomated way. By integrating the nano-membrane, total assay time was reduced,

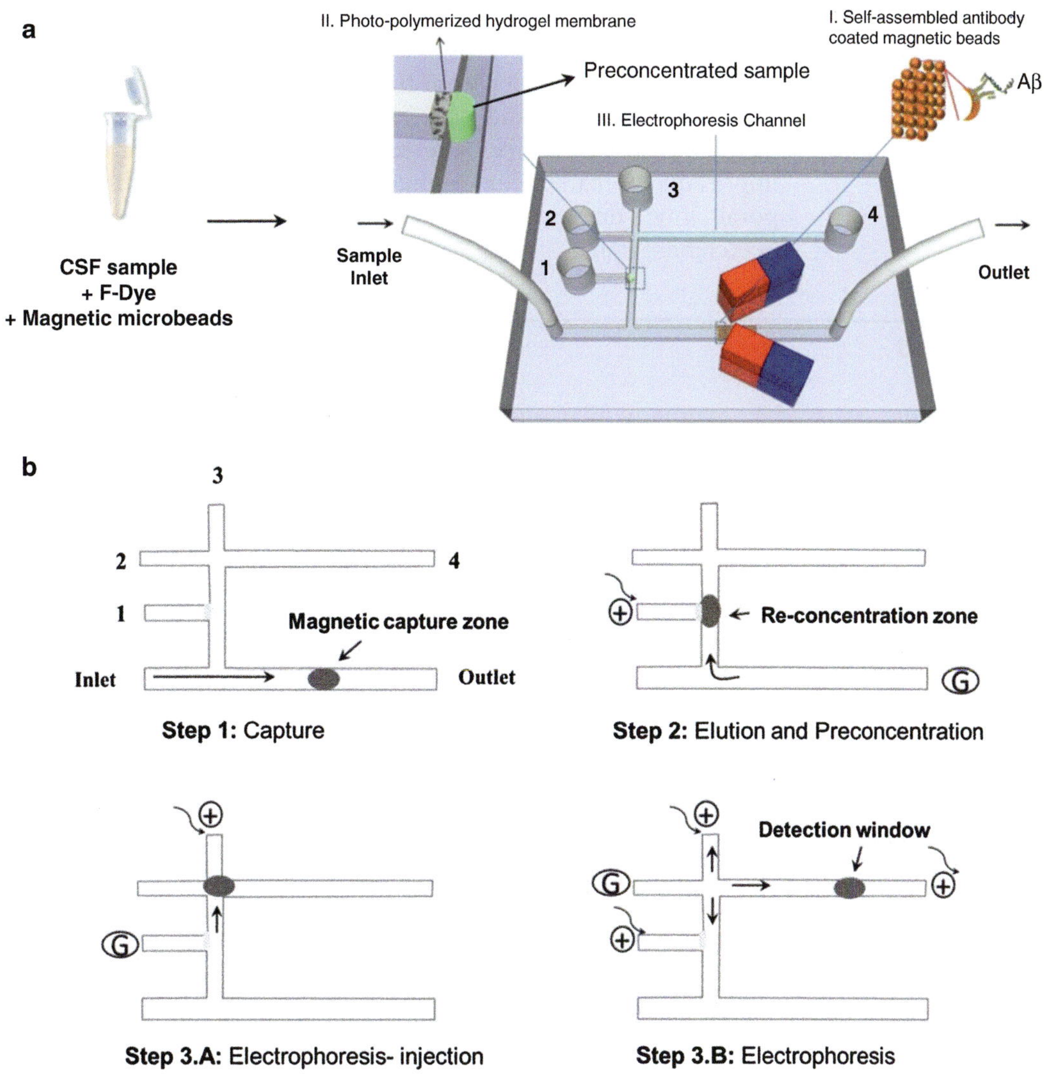

Fig. 1 Schematic design of the integrated chip. (**a**) CSF sample was labeled with the fluorescent dye off-chip, mixed and incubated with magnetic microbeads, and then injected into the integrated microchip. Schematic layout of the integrated microfluidic device is explained. The integrated chip includes three compartments: (I) a plug of self-assembled magnetic beads coated with antibody inside the microchannel for immunocapture of the target peptide (Aβ), (II) a photopolymerized hydrogel plug for preconcentration of the eluted sample, and (III) a MCE unit coupled with fluorescence detection (reservoirs 1–4 were used for applying voltage during preconcentration and electrophoresis steps). (**b**) In step 1, using a syringe pump, sample was injected to the chip from the inlet, and the target peptide was captured on the antibody-coated magnetic beads. In step 2, by applying voltage, target was eluted from the beads and concentrated on the photopolymerized membrane. In step 3, the concentrated target was injected to the electrophoresis channel (From ref 10 with permission Copyright © 2015 AIP Publishing LLC)

and also off-chip reconcentration or buffer exchange steps were not needed. The device was applied to the capture and analysis of differently truncated peptides Aβ (1–37, 1–39, 1–40, and 1–42) and was able to detect as low as 25 ng of synthetic Aβ peptides spiked in undiluted cerebrospinal fluid (CSF). The device was also tested with CSF samples from healthy donors. CSF samples were fluorescently labeled and premixed with the magnetic beads and injected into the device. The results indicated that Aβ1-40, an important biomarker for distinguishing patients with frontotemporal lobe dementia from controls and AD patients, was detectable [10].

2 Materials

2.1 Standard Peptides, Fluorescent Dye, and Antibodies

1. Synthetic Aβ peptides (Aβ1-37, Aβ1-38, Aβ1-39, Aβ1-40, and Aβ1-42) were purchased from AnaSpec (Fremont, CA, USA). We followed the protocols proposed by the producer for dissolving and aliquoting the original batch of standard Aβ peptides (*see* **Note 1**). Typically Aβ peptides should be dissolved in alkaline buffer or ddH_2O (depends on the type) at concentration of 1 mg/mL. 5–10 μL aliquots can be kept in −20 °C for a few months or −80 °C for long time. Fluoprobe-488 NHS ester was purchased from Interchim (Montlucon, France). Fluoprobe-488 NHS is a powder and should be prepared in dimethyl sulfoxide (DMSO) at concentration of 10 mg/mL. Prepare 5–10 μL aliquots of the fluorescent dye and keep it in dark and in −20 °C for few months.
2. Monoclonal antibody anti-Aβ 6E10 was purchased from Covance (Covance, Emeryville, CA).

2.2 RIPA Buffer

1. Prepare 50 mM Tris–HCl pH 7.2 including 150 mM NaCl, 1% Triton X-100, 0.1% SDS, 1 mM sodium orthovanadate (Na_3VO_4), 1 mM phenylmethanesulfonyl fluoride (PMSF), 1 mM ethylenediaminetetraacetic acid (EDTA), and 1 mM protease inhibitor cocktail (*see* **Note 2**).

2.3 Borate Buffer 2× for Fluorescent Labeling

1. Prepare 80 mM borate in ddH_2O and adjust the pH at 10.5 using NaOH.

2.4 Material for Preparation of SU-8 Master

1. Silicon wafer, 2 in. (5.1 cm) diameter circular (Neyco, France), SU-8 2050 resin (Microchem Corp, Newton MA, USA) (*see* **Note 3**). Microchem's SU-8 developer (Microchem Corp, Newton MA, USA), UV exposure machine, with wavelength range of 360–400 nm (speed plate 650, MC10, Marcheti) (*see* **Note 4**), photomask, and isopropyl alcohol.

2.5 Material for Preparation of Polydimethylsiloxane (PDMS Mold)

1. PDMS kit containing PDMS elastomer and curing agent (Sylgard184, Dow Corning, Wiesbaden, Germany). Spin coater machine, glass microscope slides (75 mm × 25 mm × 1 mm thick).
2. A plasma oxidizer (PDC-32G plasma cleaner, Harrick Scientific, Ossining, NY) is needed.

2.6 Material for Microchip Capillary Electrophoresis

1. Methylcellulose (MC-400) was purchased from Sigma (Sigma Chemical Co., St. Louis, MO, USA).
2. High-voltage programmable power supply: We used A HVS448 3000 V power supply (LabSmith, Livermore, CA, USA) controlled by a personal computer, platinum wire electrodes, and electric connectors. An inverted fluorescence microscope to detect in real time fluorescently labeled peptides: we used Olympus IX71, equipped with a 10 × 0.3 NA objective lens (Olympus, Tokyo, Japan). A 100-W mercury arc lamp for illumination. A CCD camera connected to the microscope: we used a Nikon digital DS-Qi1 camera; an image grabbing software or an image grabbing/processing software: we used NIS Elements by Nikon) which allows online intensity measurement of the recorded images by CCD.

3 Methods

Carry out all procedures at room temperature unless otherwise specified.

3.1 Preparation of Photomask

1. Design the microchip using designing software such as AutoCAD or QCad. A typical design for MCE is shown in Fig. 2a. Set the channel width at 100 μm by choosing the line width in your design. Set the distances from the intersection of injection and separation microchannels to the buffer reservoir (BR), the sample reservoir (SR), and sample waste reservoir (SW) at 5 mm and to the buffer waste reservoir (BW) at 35 mm. Print the designed microchip onto a transparency film using high-resolution printer (at least 3000 dpi or higher). Our photomasks were printed by SELBA (Versoix, Switzerland); other local retailer may be available for high-resolution printing (*see* **Note 5**).

3.2 Preparation of Master for PDMS Microchip

1. Microchips were fabricated by rapid prototyping and PDMS technology [14], starting with a master composed of a positive relief of SU-8 resin on a silicon wafer made by soft lithography (Fig. 2b) (*see* **Note 6**). Generally photolithography consisted of several steps as mentioned below:

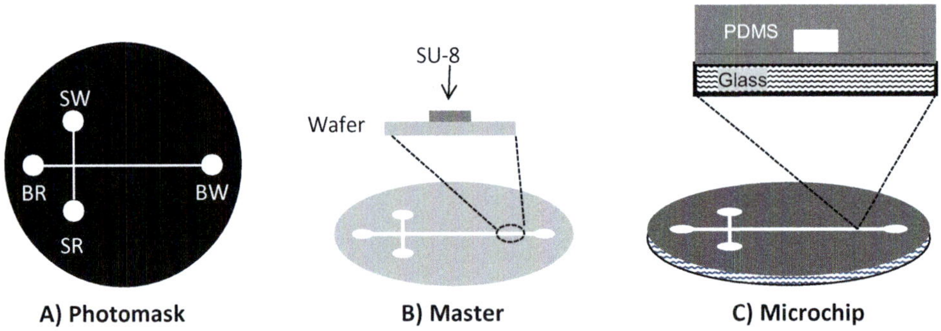

Fig. 2 Schematic depicting three steps for preparing microchip. Figure is not to scale. Part **a** shows a typical microchip design layout with a on the photolithography mask; the indications on the figure are BR, buffer reservoir; BW, buffer waste reservoir; SR, sample reservoir; and SW, sample waste reservoir. Part **b** silicon wafer on which SU-8 resin is photopolymerized. Part **c** shows the prepared microchip. If you follow the same protocol given in this chapter, channel cross section will be 100 μm (width) by 50 μm (depth)

3.2.1 Wafer Pretreatment and Spin Coating of Initiator

1. To obtain maximum process reliability, substrates should be clean and dry prior to applying the SU-8 resist. Start with a solvent cleaning, or a rinse with dilute acid, followed by a deionized water rinse. Where applicable, substrates should be subjected to a piranha etch/clean (sulfuric acid and hydrogen peroxide) (*see* **Notes** 7 and **8**).
2. To dehydrate the surface, bake at 120 °C for at least 5 min on a hot plate. Treat the surface of the wafer using oxygen plasma for 1 min. Put the wafer on spin coater and pour about 1 mL of MCC Primer 80/20 (HMDS) and spin coat for 20 s at 500 RPM and acceleration of 300 rpm/s (*see* **Note 9**).

3.2.2 Su-8 Resin Coating on the Silicon Wafer by Spin Coating

1. Spin coating is a step for spreading photoresist on the surface of wafer evenly. It ensures the consistency in the depth of microchannels of the device. In addition, the thickness and as a result the depth of the microchannel depend on the spinning time and speed. Pour about 1 mL of the photoresist resin per inch of wafer in the center of the wafer and let it spread naturally. Put the wafer at the center of the spinner and first spin the wafer at low speed of 500 rpm for 10 s with acceleration of 100 rpm/s and then speed up at 2000 rpm for 30 s with an accelerating ramp of 300 rpm/s (*see* **Note 10**).

3.2.3 Soft Baking or Pre-Exposure Baking of the Coated Wafer

1. This step is used to evaporate the solvent and densify the photoresist film on the surface of the wafer. Place the coated wafer in an oven or hot plate set at 65 °C for 3 min and then in an oven or hot plate set at 95 °C for 6 min (*see* **Note 11**).

3.2.4 Ultraviolet Exposure

1. The photomask controls the UV exposure on the photoresist-covered wafer allowing the transfer of microchannel structures onto the wafer. Place the negative photomask on the coated

wafer and fix it with paper tapes. Program the UV exposure machine to 30 s exposure time. Start the UV exposure (*see* **Note 12**).

3.2.5 Postexposure Bake

1. Postexposure baking of exposed wafer permits the exposed photoresist to cross-link, thereby allowing it to be more resistant to the developer in the next step. Place the coated wafer in an oven or hot plate at 65 °C for 1 min and then at 95 °C for 7 min. Let the wafer cool to room temperature.

3.2.6 Developing of Exposed Wafer

1. The unexposed photoresist will be removed by SU-8 developer, leaving the cross-linked photoresist which forms inverted microchannels in the wafer. After the developing step, the imprinted microchannels will be obviously visible. Submerge the exposed wafer completely in an undiluted SU-8 developer in a beaker. Shake the beaker vigorously in the developer for at least 1 min or until the unexposed photoresist, as can be signified by white spots on the wafer, is completely washed off (*see* **Note 13**). Rinse the wafer thoroughly with isopropyl alcohol in another clean beaker. If still the white spots can be observed on the surface of the wafer, repeat the developing step and again clean the surface of wafer with fresh running isopropanol. Dry the wafer using N_2 gas or air stream.

3.3 PDMS Replication of Si Wafer Mold

1. After fabrication of wafer mold, the PDMS replica can be easily replicated by a simple molding procedure.
2. Mix well the PDMS elastomer and curing agent in the ratio of 10:1 (V/V) in a single-use plastic container.
3. Place the silicon wafer mold in a plastic petri dish and pour the mixture in the petri dish to cover completely the surface of the wafer.
4. Place the uncovered petri dish in a vacuum degasser for 15 min to remove the bubbles from the PDMS mixture. When there is no any other bubble in the PDMS mixture, take it out. Cover the petri dish and label your name and date of the experiment on it.
5. Leave the petri dish in an oven at 67 °C for 2 h to cure the PDMS mixture. Take the PDMS mold out and wait for reaching to the room temperature. Peel off the PDMS microchip from the wafer mold with a sharp scalpel. Make reservoirs on the PDMS microchip with a 3 or 5 mm circular punch (*see* **Note 14**).

3.4 Preparation of Glass Slide Coated with PDMS

1. Rinse the glass slide with acetone and then after with ethanol. Dry the glass slide with N_2 gas or air stream. Put the glass slide in the oxygen plasma source. Set the plasma machine for 1 min. Put the glass slide on the spin coater and pour 2 mL prepared PDMS elastomer mixture on the glass slide. Set the spin coater at 1000 rpm for 20 s and with acceleration of 300 rpm/s (*see* **Note 15**).

3.5 Adhering PDMS Microchip to PDMS-Coated Glass Slide Using Oxygen Plasma Bonding

1. The PDMS has self-adhesive property, hence allowing reversible sealing with most materials such as glass and polymers. The sealing will be improved if the surfaces of both the PDMS and sealing substrate are free of particles. If there is particle on the PDMS surface, wash the surface with ethanol. Dry the microchip by N_2 gas. In order to achieve more robust chips, we recommend inducing permanent bonding, thanks to plasma treatment: Place the microchip and PDMS-coated glass slide in the oxygen plasma source. Set the plasma machine for 1 min. Bring out the glass slide and the PDMS microchip. Locate the PDMS microchip on the PDMS-coated glass and press naturally (*see* **Note 16**). Keep the microchip in 67 °C oven for at least 1 h before use. Store your prepared chip in dust-free box such as single-use petri dishes before use (Fig. 2c).

3.6 Surface Coating and Microchip Electrophoresis

1. To control the electroosmotic flow in the microchannel, the surface of the microchannel should be coated. Inject a solution of 0.2% of poly(dimethylacrylamide-co-allyl glycidyl ether), (PDMA-AGE) in ddH_2O into the microchannels. Keep the chip in the dark for at least 30 min and then rinse out the excess polymer by extensive ddH_2O washing (*see* **Note 17**).

3.7 Preparation of Methylcellulose (MC) 2%

1. Prepare 2% MC in borate 40 mM pH 9.5. Prepare 25 mL hot water (about 40–45 °C). Add 1 g MC to the hot water and gently mix it using a glass rod. Use the glass rod to gently press the big agglomerated MC formed in the hot water (*see* **Note 18**). Add 25 mL 80 mM borate buffer pH 9.5 plus 0.01% Tween-20 which has been kept in refrigerator for at least 1 hour before use. Mix the buffer and polymer and keep it at 4 °C for few more hours until the polymer solution is completely clear [15–17].

3.8 Preparation and Labeling of Standard Aβ Peptides

1. Take an already prepared aliquot of Aβ, add equal volume of borate buffer pH 10.5 (2×) [18].
2. Add a given volume of fluorescent dye to the peptide solution. If your peptide is prepared in NH_4OH aqueous solution, add equal volume of borate buffer pH 10.5 (1×) to your peptide solution.
3. Lyophilize the sample and dissolve the powder in desire volume of borate buffer solution (pH 10.5 (1×) (*see* **Note 19**). Typically a tenfold excess (molar ratio) of the fluorescent dye should be used for labeling (*see* **Note 20**). Incubate the mixture at 4 °C for 1 h in dark (*see* **Note 21**).

3.9 Preparation of CSF Samples

1. The CSF samples analyzed in this study were obtained from patients attending the outpatient memory clinic in Ulm (*see* **Note 22**). Collection and analysis of the CSF sample were

approved by the Ethics Committee in Ulm. All samples were anonymized before transmission to our laboratory (*see* **Note 23**). CSF was taken, aliquoted within 2 h, and stored at −80 °C [19].

2. CSF can be directly used for the labeling reaction. Add 10 μL of Fluoroprobe NHS-488 (10 mg/mL in DMSO) to 250 μL of human CSF.
3. Incubate the mixture at 4 °C for 1 h in dark. The labeled CSF was then used for immunocapture of Aβ peptides (*see* **Note 24**).

3.10 Preparation of Magnetic Beads

1. Add 80 μL of monoclonal antibody anti-Aβ (1 mg/mL) to 2 mL of magnetic microparticles M280 Dynabeads (10 mg/mL, 6.7 × 108 beads/mL) coated with sheep anti-mouse IgG. The activated beads can be stored in 4 °C for few months.

3.11 Affinity Immunocapture of Labeled Aβ and CSF

1. Take 25 μL of the activated magnetic beads in 1 mL tube and add 1 mL RIPA buffer to the beads.
2. Let the beads rotate for 5–10 min in the RIPA buffer.
3. Use a magnetic particle separator to keep the beads and remove the buffer.
4. Add 250 μL RIPA buffer to 250 μL labeled CSF sample or to the 25 μL of labeled standard Aβ peptides.
5. Add the mixture to the magnetic beads.
6. Let the mixture to rotate at 4 °C and in dark overnight.
7. Use a magnetic separator to separate the beads from the mixture.
8. Wash the beads by adding 1 mL RIPA and rotate for 10 min.
9. Use a magnetic particle separator to separate the beads.
10. Repeat the washing steps three times.
11. Wash the beads with 1 mL ddH$_2$O following the same procedure for washing the beads with RIPA buffer.
12. Add 200 μL 0.16% NH$_4$OH in ddH$_2$O to the beads and use sonication (*see* **Note 25**) for 10 min at room temperature.
13. Use a magnetic particle separator to separate the beads.
14. Use a Speedvac to completely dry the eluted solution and resolve the dried eluted peptides in electrophoresis buffer (borate 40 mM, pH 9.5).
15. Keep the sample for MCE analysis in dark and at 4 °C before use.

3.12 Microchip Capillary Electrophoresis

1. Place 10–15 μL of MC 2% in sample waste reservoir. Slight pressure using a syringe is necessary to introduce the MC solution into the microchannel. Make sure that MC solution has filled all the channels and reached all reservoirs.

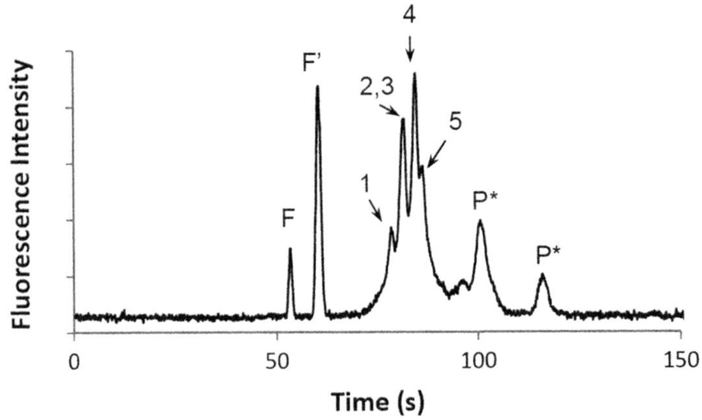

Fig. 3 Electropherogram from the separation of a mixture of five Aβ peptides labeled with Fluoprobe-488 (Aβ1-37(1), Aβ1-38 (2), Aβ1-39 (3), Aβ1-40 (4), and Aβ1- 42 (5)) after immunocapture (Reprinted with permission from ANALYTICAL CHEMISTRY 2010, 82, 7611–7617, Copyright 2010 American Chemical Society)

2. Use microscope to make sure there is no air bubble trapped in the channels. Remove the excess of MC solution in buffer waste reservoir.
3. Place 15 μL electrophoresis buffer (borate 40 mM pH 9.5) in all the reservoirs. Fix the platinum wire electrodes in the reservoirs (*see* **Note 26**).
4. Replace the buffer in sample reservoir with 15 μL of sample when ready to run the voltage program. Run the recording software before running the voltage program. The voltage program includes two step, injection and separation.
5. To inject the sample, apply +550 V in SW reservoir while keeping other reservoirs at ground for 60 s.
6. Switch to separation step by applying +250 V in SW and SR reservoirs and + 950 V in BW while keeping BR at ground.
7. Measure the total fluorescent intensity at 5 mm to the BW in the separation microchannel and plot the measured intensity against recording time to achieve electropherograms (Figs. 3 and 4).

4 Notes

1. Instead of using alkaline buffer which is normally recommended for dissolving Aβ peptides, you may also dissolve and aliquot each Aβ peptide (Aβ1-37, Aβ1-38, Aβ1-39, Aβ1-40, and Aβ1-42) in aqueous solution of 0.16% ammonium hydroxide (NH_4OH); this will help to limit the aggregation of peptides during their storage. Note that NH_4OH is an active compound and may interfere with fluorescent labeling, thus

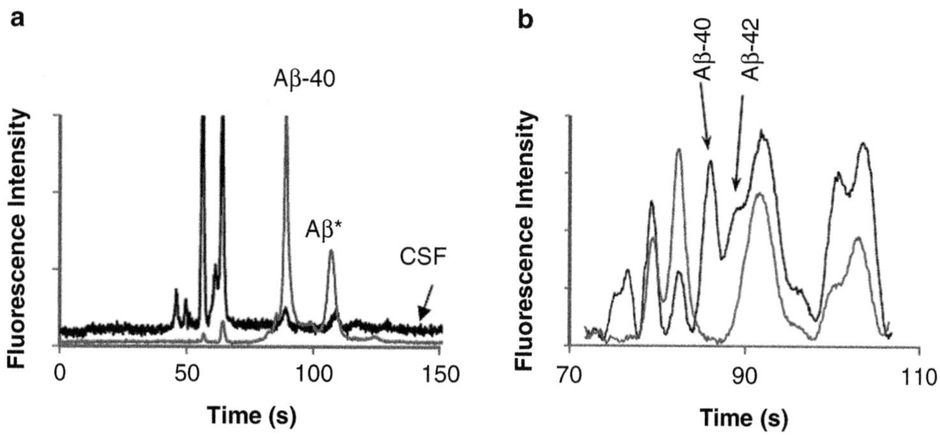

Fig. 4 (**a**) Electropherogram of labeled CSF. CSF (250 μL) was labeled with Fluoprobe-488, and the Aβ peptides in the labeled CSF were isolated using magnetic beads coated with anti-Aβ. An electropherogram of standard Aβ1-40 labeled with Fluoprobe-488 is superimposed to identify the peak from Aβ1-40 in CSF. Aβ* corresponds to the peak for aggregated or modified peptides. (**b**) Electropherogram of Aβ peptides immunocaptured from 250 μL of CSF which were labeled after immunocapture. An electropherogram (dashed electropherogram) of a blank control showing the peaks related with the fluorescent dye is superimposed in this electropherogram. The separation window of Aβ1-40 and Aβ1-42 is out of the zone for strong peaks related with the fluorescent dye (Reprinted with permission from ANALYTICAL CHEMISTRY 2010, 2010, 82, 7611–7617, Copyright 2010 American Chemical Society)

an additional lyophilization step to remove all traces of NH_4OH prior to the labeling step is necessary.

2. The buffer data are mentioned in molarity. In general, the buffer volume needed for microchip electrophoresis is very low, so the buffers can be prepared in 50 mL volumes. All the buffers should be filtered by 0.2 μm filters and kept in refrigerator before use.
3. Beware that SU-8 resin are light and temperature sensitive.
4. Beware that chronic UV exposure may lead to skin cancer and eye cataract. Wear UV glasses and leave the room during the UV exposure.
5. For printing smaller structure in the photomask, higher resolutions are recommended.
6. Soft lithography normally requires clean room facilities, a clean room (a room away from dusts) for doing photolithography techniques, and a constant room temperature for consistency in the coating by SU-8 resin. Instead of using clean room facility, photolithography can be done under a hood with laminar clean air; however, there is higher risk of defects.
7. These steps are crucial if glass slides are used as substrate for lithography, but they can be avoided when using commercial silicon wafer prepared for photolithography.

8. Safety note: Piranha is a very strong oxidizing agent and reacts violently with organic materials or solvents, so it must be handled with extreme precaution under chemical hood. Before using piranha, ask your institution about the procedure for piranha waste disposal in your institution.
9. Adding MCC primer 80/20 (HMDS) step is recommended to improve adherence of SU-8 resin to the silicon wafer.
10. During the spin coating process step, a buildup of photoresist may occur on the edge of the substrate. In order to remove this edge, a small stream of isopropanol using a syringe at the edge of the wafer can be used. This procedure will help the photomask to be placed into close contact with the wafer, resulting in improved resolution and aspect ratio.
11. Let the wafer cool to room temperature. If you see wrinkles on the SU-8 film, repeat the heating procedure until wrinkles disappear.
12. UV exposer can be done using an automated aligner, which is normally part of clean room facilities for lithography. Caution: UV exposure is dangerous for the eyes and potentially may cause skin cancer; use UV protection glasses and protect your skin when using UV.
13. Long developing step may detach the polymerized structure too and should be avoided. Caution: developer is toxic and should be used under chemical hood.
14. Normal biopsy punches can be used, and special punches for cutting PDMS with wide range of diameters are also commercially available.
15. Pay attention that the upper side of the glass when it is put in the plasma container should be used for PDMS coating.
16. Don't overpress the PDMS microchip as the channels may be attached to the surface. Be careful to put the surfaces together which are exposed to the plasma.
17. We used a homemade PDMA-AGE synthesized copolymer kindly provided by Marcella Chiari. The protocol for synthesis of PDMA-AGE is published (*see* ref. 20) and thus excluded from this chapter.
18. Adding cold buffer should be done in 5–10 min for preparation of MC solution in water. During this time keep mixing the MC solution using the glass stick.
19. Aβ peptide mixture can be prepared and lyophilized prior to labeling this way. The labeling conditions for all the peptides in the mixture are similar. Lyophilized peptide samples can be stored several days before performing the labeling reaction.

20. The ratio of added fluorescent dye depends on concentration of your peptides. If you are preparing peptides with concentration lower than 10 μg/mL, higher ratios (50-fold) of fluorescent dye are recommended.
21. Labeled peptides could be used in the same day if it is kept in the refrigerator and in dark.
22. Obtaining CSF from patients should be done by specialist within protocols approved by clinical ethical committe and processed in specialized laboratories (e.g., neurochemical labs).
23. Follow the procedure given by your institution, following national and international legislation, for using biological samples from patients.
24. Fluorescent labeling can also be done after immunocapture and on the eluted peptides.
25. Normal ultrasonic bath used in your laboratory with average power of 100 W can be used.
26. Be aware of electric shock when working with high-voltage power supply, use appropriate tapes to fix the electrode, and avoid any direct contact with the electrode during electrophoresis.

Acknowledgments

This work was supported by European grants LSHB CT 06 037953 (NeuroTAS) and FP7-NMP-09-246513 (NADINE).

References

1. Haass C, Schlossmacher MG, Hung AY et al (1992) Amyloid β-peptide is produced by cultured cells during normal metabolism. Nature 359:322–325
2. Hardy J, Selkoe DJ (2002) The amyloid hypothesis of Alzheimer's disease: progress and problems on the road to therapeutics. Science 297:353–356
3. Klein WL, Krafft GA, Finch CE (2001) Targeting small Aβ oligomers: the solution to an Alzheimer's disease conundrum? Trends Neurosci 24:219–224
4. Lambert MP, Barlow AK, Chromy BA et al (1998) Diffusible, nonfibrillar ligands derived from Aβ1–42 are potent central nervous system neurotoxins. Proc Natl Acad Sci U S A 95:6448–6453
5. Wiltfang J, Esselmann H, Bibl M et al (2002) Highly conserved and disease-specific patterns of carboxyterminally truncated Aβ peptides 1–37/38/39 in addition to 1–40/42 in Alzheimer's disease and in patients with chronic neuroinflammation. J Neurochem 81:481–496
6. Lewczuk P, Wiltfang J (2008) Neurochemical dementia diagnostics: state of the art and research perspectives. Proteomics 8:1292–1301
7. Jakeway SC, de Mello AJ, Russell EL (2000) Miniaturized total analysis systems for biological analysis. Fresenius J Anal Chem 366:525–539
8. Chovan T, Guttman A (2002) Microfabricated devices in biotechnology and biochemical processing. Trends Biotechnol 20:116–122
9. Mohamadi MR, Svobodova Z, Verpillot R et al (2010) Microchip electrophoresis profiling of Aβ peptides in the cerebrospinal fluid of patients with Alzheimer's disease. Anal Chem 82:7611–7617

10. Mohamadi MR, Svobodova Z, Bilkova Z et al (2015) An integrated microfluidic chip for immunocapture, preconcentration and separation of β-amyloid peptides. Biomicrofluidics 9:054117
11. Svobodova Z, Mohamadi MR, Jankovicova B et al (2012) An integrated microfluidic chip for immunocapture, preconcentration and separation of β-amyloid peptides. Biomicrofluidics 6:024126
12. Mai TD, Pereiro I, Hiraoui M et al (2015) Magneto-immunocapture with on-bead fluorescent labeling of amyloid-β peptides: towards a microfluidized-bed-based operation. Analyst 140:5891–5900
13. Mai TD, Ferraro D, Aboud N et al (2017) Single-step immunoassays and microfluidic droplet operation: towards a versatile approach for detection of amyloid-β peptide-based biomarkers of Alzheimer's disease. Sensors Actuators B 255:2126–2135
14. Xia Y, Whitesides GM (1998) Soft lithography. Angew Chem Int Ed Engl 37:550–575
15. Mohamadi MR, Mahmoudian L, Kaji N et al (2007) Dynamic coating using methylcellulose and polysorbate 20 for nondenaturing electrophoresis of proteins on plastic microchips. Electrophoresis 28:830–836
16. Mesbah K, Verpillot V, Chiari M et al (2014) Neutral polymers as coatings for high resolution electrophoretic separation of Ab peptides on glass microchips. Analyst 139:6547–6555
17. Mohamadi MR, Kaji N, Tokeshi M et al (2007) Online Preconcentration by transient Isotachophoresis in linear polymer on a poly(methyl methacrylate) microchip for separation of human serum albumin immunoassay mixtures. Anal Chem 79:3667–3672
18. Verpillot R, Esselmann H, Mohamadi MR et al (2011) Analysis of amyloid-β peptides in cerebrospinal fluid samples by capillary electrophoresis coupled with LIF detection. Anal Chem 83:1696–1703
19. Jesse S, Brettschneider J, Süssmuth SD et al (2011) Summary of cerebrospinal fluid routine parameters in neurodegenerative diseases. J Neurol 258:1034–1041
20. Chiari M, Cretich M, Horvath J (2000) New adsorbed coatings for capillary electrophoresis. Electrophoresis 21:909–916

Chapter 29

Measuring Low-Picomolar Apparent Binding Affinities by Minigel Electrophoretic Mobility Shift

Karen A. Lewis, Sarah E. Altschuler, and Deborah S. Wuttke

Abstract

Measuring protein/DNA interactions that have apparent binding affinity constants in the low-picomolar range presents a unique experimental challenge. To probe the sequence specificity of telomere binding proteins, our laboratory has developed an electrophoretic mobility shift assay protocol that allows for the routine measurement of $K_{D,app}$ values in the 1–20 pM range. Here, we describe the protocol and highlight the particular considerations that should be made to successfully and reproducibly measure high-affinity interactions between proteins and single-stranded DNA.

Key words Native gel, EMSA, DNA binding protein, Binding assay, High-affinity binding

1 Introduction

Several DNA-binding proteins, including the lactose repressor LacI and some telomere end-binding proteins, interact with their cognate ligands with low to sub-picomolar binding affinities [1, 2]. These interactions are difficult to measure experimentally, primarily due to the extremely small signal that is generated under the concentrations needed to measure K_D's and the level of accuracy of the concentration of protein. Recent advances in microscale thermophoresis and analytical ultracentrifugation have enabled these equilibrium-based methods to measure high-affinity interactions under certain circumstances ($K_{D,app} < 50$ pM) [3, 4]. Because such tight interactions usually have relatively slow off-rates, the apparent binding constants for these systems can also be measured with non-equilibrium methods like native electrophoretic mobility shift assays (EMSAs) or double-filter binding assays. As very little specialized equipment is needed for an EMSA, this method is accessible to a wide range of laboratories. The literature contains excellent guidance and instruction in optimizing EMSAs for protein-nucleic acid assays [5]. In this protocol, we focus on the

Biji T. Kurien and R. Hal Scofield (eds.), *Electrophoretic Separation of Proteins: Methods and Protocols*, Methods in Molecular Biology, vol. 1855, https://doi.org/10.1007/978-1-4939-8793-1_29,

experimental conditions that are required to measure low- to sub-picomolar binding interactions using EMSAs.

A system with a $K_{D,app}$ in the low pM range requires the use of femtomolar concentrations of ligand, which serves two purposes. First, it allows the assumption that $[P]_{free} = [P]_{total}$, which simplifies the equation used to fit the data and solve for the $K_{D,app}$ [6]. Second, it ensures that the experiment includes conditions in which there is fully free ligand [6]. With some advanced planning in reagent preparation and attention to experimental protocol, it is possible to robustly detect such small amounts of signal using radiolabeled oligonucleotide ligands, allowing for the reproducible and accurate measurement of apparent binding affinities [6].

Here, we describe our laboratory protocol for measuring the binding affinity of telomere end-binding proteins for their cognate ssDNA ligands, which have apparent dissociation constants ($K_{D,app}$) of 2–20 pM [7–9]. The most critical step in being able to reliably and consistently measure these tight binding events is to use the freshest radioisotope possible. The binding assay is performed with the hottest radioisotope available, such that the [γ-^{32}P]-ATP has a specific activity of >5800 Ci/mmol. As the protein concentration needs to be known accurately, it is essential to use tips and tubes made from a low-retention plastic polymer to avoid non-specific adsorption of the very low concentrations of protein.

2 Materials

2.1 Gel Electrophoresis

1. Bio-Rad mini-PROTEAN vertical electrophoresis cell or equivalent.
2. Intact gel combs, 15-well, 1.5 mm thick.
3. Casting stand.
4. Glass plates (short and tall, 1.5 mm spacer).
5. Mini-Protean multi-casting chamber (optional).
6. 40% polyacrylamide solution (29:1 acrylamide/bisacrylamide).
7. Tetramethylethylenediamine (TEMED).
8. 10% w/v ammonium persulfate (APS).
9. 10× Tris-borate-EDTA, pH 8.4 TBE: 890 mM Tris, 890 mM boric acid, 20 mM EDTA.
10. Gel dryer.
11. 3 mm Whatman paper.
12. Plastic wrap.
13. 0.5 mil (12.7 μm) plastic sheets.
14. Impact sealer.

15. Solvent-resistant laboratory marker or laboratory marking pen.
16. Stir bar.
17. Stir plate.

2.2 DNA Labeling Reaction

1. Nuclease-free water.
2. Oligonucleotide.
3. Illustra MicroSpin G-25 columns.
4. γ-^{32}P-ATP (PerkinElmer, freshest lot possible).
5. T4 polynucleotide ligase.
6. 10× T4 reaction buffer.

2.3 Binding Reaction

1. Appropriate binding buffer.
2. Purified recombinant protein.
3. Low-retention 96-well plates.
4. Low-retention filter tips.
5. 12-channel multichannel pipettor.
6. Disposable buffer reservoir.

2.4 Detection and Analysis

1. Phosphor screen.
2. Phosphor imager.
3. Gel quantitation software (ImageQuant, ImageJ, or similar).
4. Curve fitting software capable of multivariate fitting (SigmaPlot, Kaleidagraph, or similar).

3 Methods

Carry out all procedures on ice unless otherwise specified or as appropriate for the protein/DNA system of interest.

3.1 Preparation of Native Polyacrylamide Gels

The thicker gel is important for these experiments to increase the signal for tight binding systems; in a 15-well, 1.5 mm minigel, each well can hold ~25–27 μL.

1. If needed, prepare 10× TBE (*see* **Notes 1** and **2**).
2. Combine 10× TBE, acrylamide, glycerol, and water to create a "gel stock solution." The recipe for one 1.5 mm 5% polyacrylamide gel with 5% glycerol is:

Reagent	Volume	Final
10× TBE	1.0 mL	1×
40% acrylamide (29:1)	1.25 mL	5%
100% glycerol	0.5 mL	5%

Milli-Q H_2O	7.25 mL
10% APS	50 μL
TEMED	30 μL

3. Mix well but not too vigorously, being mindful to not over-aerate the solution. This solution can be stored at 4 °C in the dark for up to 4 months (*see* **Note 3**).
4. Prepare the glass gel plates for hand-casting (*see* **Notes 4** and **5**).
5. Measure out 10 mL of gel stock per gel to be poured. Add the appropriate amounts of 10% APS and TEMED and mix gently (*see* **Note 6**).
6. Gently insert the combs into the plates, being careful to avoid introducing air bubbles (*see* **Note** 7).
7. Allow gels to polymerize for 15–30 min.
8. While binding reactions are incubating (*see* Subheading 3.3, **step 9**), set up the gels in the electrophoresis apparatus. It is especially important to set the gels up early if the gels are at room temperature. As this protocol does not use loading dye, it is critical to outline and number each well on the large (spacer) plate to help keep track of each well during loading. We recommend using a solvent-resistant marker like the laboratory marker or laboratory marking pen, as the ink in these pens is more resistant to being rinsed off of the plate by the running buffer.
9. Fill the inner chamber with prechilled running buffer (1× TBE + 5% glycerol). Fill the outer chamber about 1/3 full with the same cold buffer (*see* **Note 8**).
10. Place a stir bar in the outer chamber, and place the assembled apparatus in the cold room or refrigerator (*see* **Note 8**).

3.2 Radiolabeling of DNA Oligonucleotide

1. Reconstitute the lyophilized DNA oligonucleotide to approximately 100 μM in nuclease-free water.
2. Determine concentration of the oligonucleotide by absorption spectroscopy, using the appropriate molar extinction coefficient at 260 nm.
3. Dilute to 25 μM with nuclease-free water.

4. On the day of radioisotope delivery (i.e., when the radioisotope is >5800 Ci/mmol), carry out the end-labeling reaction per the manufacturer's instructions. We use T4 polynucleotide kinase (New England BioLabs), performing the ligation in a 10 μL reaction volume (*see* **Notes 9** and **10**).
5. Heat inactivates the DNA ligase by incubating the reaction at 95 °C for 20 min.
6. During the heat inactivation incubation, prepare the G-25 column.
7. Briefly vortex to resuspend the resin.
8. Snap off the bottom tab and slightly loosen the cap.
9. Insert into 2 mL collection tube.
10. Centrifuge at 6000 × *g* for 10 s at room temperature. We find it most time-efficient to use a personal-size benchtop centrifuge (e.g., Spectrafuge mini).
11. Discard flowthrough.
12. Transfer G-25 column to a 1.5 mL microcentrifuge tube.
13. Add 40 μL prechilled nuclease-free water to the heat-inactivated labeling reaction, for a total volume of 50 μL.
14. Apply the entire 50 μL reaction to G25 column, and again centrifuge at 6000 × *g* for 10 s at room temperature.
15. Retain flowthrough as the radiolabeled DNA oligonucleotide. Hold on ice for the preparation of the binding reaction as described below, and then store remainder at −20 °C in appropriate storage conditions for radioactivity (*see* **Note 11**).

3.3 Binding Reaction

1. Determine protein concentration by A_{280} and appropriate molar extinction coefficient.
2. Prepare the binding buffer using 10× salts stock, BSA, and glycerol. The composition of the binding buffer is critical for accurately and reproducibly measuring binding constants. Parameters that should be considered include pH, buffering agent, salt concentration, and concentration of divalent cations (*see* **Note 12**).
3. Using low-retention filter tips, prepare serial dilutions of protein in binding buffer to 2× final concentrations in microcentrifuge tubes (*see* **Notes 13** and **14**).
4. Transfer 15 μL of each protein dilution to a 96-well plate on ice.
5. Dilute radiolabeled oligonucleotide to 2× final concentration in a final volume of at least 500 μL, so that the solution is of sufficient volume to fill the base of a disposable buffer reservoir (*see* **Note 15**).

6. Transfer the diluted radiolabeled oligonucleotide to a disposable buffer reservoir.
7. With a multichannel pipettor and low-retention filter tips, add 15 μL of radiolabeled oligonucleotide to each of the wells that contain the 2× protein solutions.
8. Cover the plate with a piece of parafilm that is cut to the same size as the plate. Ensure a solid seal on each well by pressing firmly in each well with the pads of your fingers.
9. Incubate on ice. Initial experiments should be carried out at several incubation times to determine the appropriate incubation time to ensure that the binding reaction has reached equilibrium.
10. During incubation, prepare the electrophoresis apparatus as described in Subheading 3.1, **step 8.**

3.4 Native Gel Electrophoresis

1. Load the binding reactions directly into the wells of the prechilled gel. Up to ~27 μL of a reaction containing 15% glycerol can be loaded into a well of a 15-well, 1.5 mm gel that is prepared in running buffer containing 5% glycerol.
2. Note that this protocol does not use a loading dye, which would dilute the amount of radiolabeled DNA signal.
3. To make it easier to load samples and ensure that they are loaded in the correct order, we use a laboratory marker to mark the back of the large spacer plate. We outline the base of each well and then number the wells below the marks.
4. To make sure the samples sink to the bottom of the well, we take advantage of the difference in refractive index between the running buffer and the sample solution. Schlieren waves will be visible as the binding reaction (containing 15% glycerol) is dispensed into the running buffer (which contains only 5% glycerol).
5. For higher throughput, we found that multiple wells can be loaded simultaneously using a multichannel pipettor. The micropipettor and type of gel-loading tips should be optimized before proceeding with radiolabeled oligonucleotide. We use a 12-channel multipipettor that does not have O-rings on the tip holders. After samples are aspirated into three tips, the user gently squeezes the outer two tips into the correct geometry for the wells. One should practice extreme caution when handling the sample-containing tips to make sure that the vacuum seal is not broken (i.e., no sample is lost) and to avoid radioactive contamination of gloves, instruments, and surroundings.
6. Place ice packs in the outer chambers of the gel chamber (*see* **Note 16**).

7. Fill the outer gel chambers with running buffer.
8. Electrophorese the gels at 200 V for 20–30 min. The short run time helps with exchange problems and shorter off-rates, and the high voltage allows for the short run time. The ice packs keep the gels cool at this high voltage.
9. If longer run times are required to improve the separation of free and bound oligonucleotide, consider exchanging the ice packs if your protein is sensitive to temperature.
10. When optimizing run times, be aware that the unbound radiolabeled oligonucleotide may run off the end of the gel and contaminate the running buffer and gel box. The free oligonucleotide can be retained in the gel with a 5–10 mm "plug" of 30% polyacrylamide at the bottom of the gel. For reference, in our lab a 12-nt ssDNA will run off a 5% gel that is prepared as described above in ~45 min (*see* **Note 17**).
11. Promptly disassemble completed gels, and place on two layers of 3 mm Whatman filter paper. Check all disassembled buffers and equipment for radioactivity contamination.
12. Place plastic wrap over the gels and dry on a gel dryer for ~25 min per gel.
13. Expose the phosphor screen for two nights when using ≤5 pM final concentration of radiolabeled oligonucleotide. We find that exposing the screen for longer times (i.e., three nights) artificially inflates the measured $K_{D,app}$ values due to increased background signal (*see* **Note 18**).

3.5 Quantitation and $K_{D,app}$ Calculation

The quantitation method is outlined briefly here. Additional details, including how to identify the appropriate background correction, have been published elsewhere [6].

1. Use gel imaging software (ImageQuant, ImageJ, or similar) to quantitate the free and bound DNA bands (*see* **Note 19**).
2. Evaluate the sum of counts (bound + free) across all lanes. Total counts per lane should be approximately the same. If the sum of total counts is not equal across lanes, then it means that some of the labeled ligand was not detected. There are several causes for this, including uneven sample loading or rapid dissociation of the bound complex during electrophoresis or aggregation in the well. This needs to be addressed, as any subsequent curve fitting will have considerable error.
3. Calculate the fraction of ligand bound as the ratio of bound ligand over total ligand (i.e., the sum of bound and free ligand).
4. Plot fraction bound as function of total protein concentration.
5. Fit the data to the simplified form of the binding isotherm to solve for $K_{D,app}$:

$$\frac{[\mathrm{PL}]}{L_\mathrm{T}} = s\left(\frac{P_\mathrm{T}}{P_\mathrm{T} + K_{\mathrm{D,app}}}\right) + O \quad (1)$$

where $\frac{[\mathrm{PL}]}{L_\mathrm{T}}$ is the fraction of ligand bound; $[P]_\mathrm{T}$ is total protein concentration; s is a scalar offset for binding saturation; O is a background offset; and $K_{\mathrm{D,app}}$ is the apparent dissociation constant [10].

4 Notes

1. Be attentive when making both the TBE stock and gel running buffer solutions. Any discrepancies in gels or running buffer could affect the reproducibility of results.
2. The boric acid in 10× TBE may precipitate during storage. This can be handled in one of two ways: (1) prepare 5× TBE and adjust all volumes accordingly or (2) use 10× TBE to make ~500 mL of gel stock and use the remaining 10× TBE to make gel running buffer (1× TBE + 5% glycerol).
3. For higher-throughput analyses (5–10 gels per week), we recommend preparing at least 500 mL of gel stock. It can be stored in an opaque or amber bottle at 4 °C for up to 4 months.
4. The thickness of the 1.5 mm spacers can be challenging when pouring in a side-by-side caster. Heavily used gaskets are often pockmarked, as a result from a poor seal with the plates that can leak air bubbles into the gel. We recommend reserving a set of four new gaskets just for pouring native gels; alternatively, one could use gel tape. In general, we avoid this problem by using the multi-caster to prepare 10–12 gels at a time.
5. If more than eight gels will be needed within 7–10 days, the Bio-Rad Mini-PROTEAN III Multi-Caster can easily hold up to 12 sets of 1.5 mm plates.
   (a) The dead volume of the multi-caster requires about 20 mL more of gel solution than if the gels are poured individually. For example, for 12 gels, the gel solution recipe would be a total volume of 140 mL.
   (b) Before pouring the gel solution, loosely place nine to ten of the combs above the gels. When the multi-caster has been filled with gel solution, these combs can be pressed into place quickly and the remaining combs inserted.
   (c) Store gels in the fridge with the combs left in place, wrapped in a wet paper towel and sealed in a plastic bag for up to a month.
6. When pouring more than two gels at a time (either in side-by-side casters or using the multi-caster), the amounts of APS and

TEMED should be reduced so that there is sufficient time to insert all the combs before the gel starts polymerizing. When pouring 12 gels, we typically use the amounts of 10% APS and TEMED for eight gels.

7. Be very careful when handling the combs. Combs with broken outer sealing tabs prevent proper polymerization of the wells with native gel solution. Store the combs away from the ones for general lab use, and treat them with the utmost care.

8. Because the native gels are run at a high voltage (200 V), it is important to carry out electrophoresis in a cold environment to dissipate the heat that is generated from the high voltage. We recommend running the gels in an ambient temperature of 4 °C with prechilled gels and prechilled running buffer, as well as ice packs that were frozen at −70 °C.

9. For detection of sub-picomolar apparent $K_{D,app}$ S, it is *critical* that the specific activity of the [γ-^{32}P]-ATP be as high as possible. We have had the most success using [γ-^{32}P]-ATP that has a specific activity of 6000 Ci/mmol at the time of synthesis and calibration. The order for the [γ-^{32}P]-ATP should be placed with the manufacturer so that the freshest lot ships on the day of synthesis and calibration for next-day delivery. A full set of replicate binding assays are then completed within 4 days, at which time the [γ-^{32}P] will have decayed to a specific activity of 5590 Ci/mmol when the phosphor screen is imaged. When using <1 pM of labeled oligonucleotide, it was difficult to reliably measure the signal when the specific activity of the ^{32}P was less than 5400 Ci/mmol at the time of imaging.

   We purchase [γ-^{32}P]-ATP radionuclides from PerkinElmer with a specific activity of 6000 Ci/mmol at a concentration of 10 mCi/mL (catalog numbers: in a lead-lined container, NEG002Z: in a lead-free container, BLU002Z). Fresh lot schedules can be accessed at their website (http://www.perkinelmer.com/lab-solutions/resources/docs/CAL_2017_Radiochemical_Fresh_Lot_Schedule.pdf).

10. Take appropriate precautions for working with the highly radioactive material during the labeling reaction, as recommended by your institutional safety protocols. For example, our laboratory members wear both a torso badge and a ring badge on the non-dominant hand to thoroughly monitor exposure.

11. Determining the concentration of radiolabeled DNA is challenging, as it would potentially contaminate cuvettes and instrumentation with radioactivity. Therefore, we make the conservative assumption that the G-25 spin column recovers

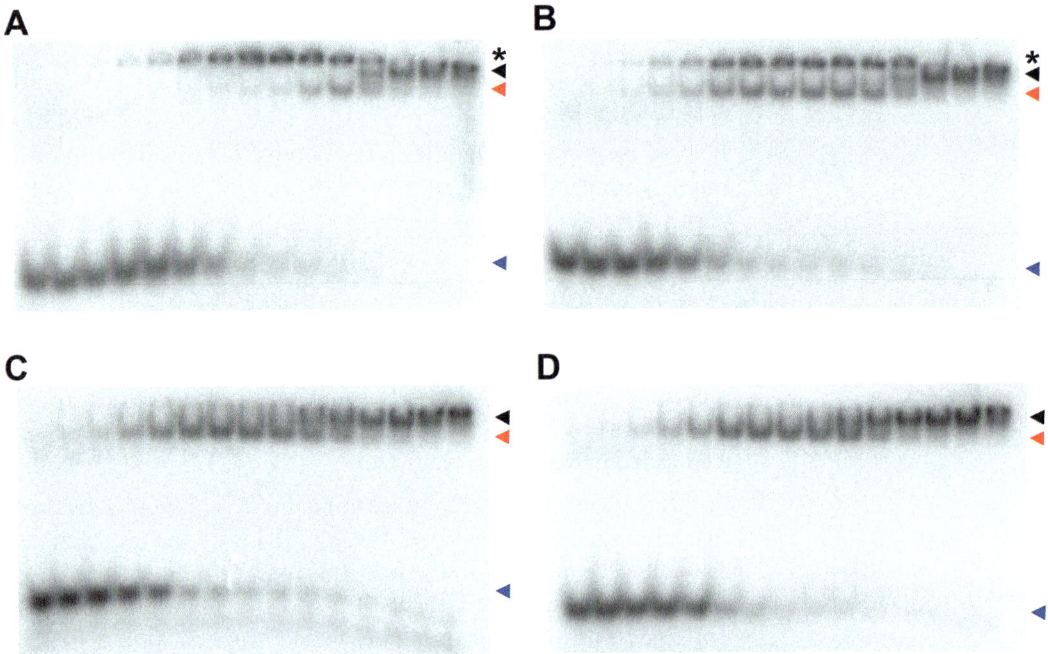

Fig. 1 Binding buffer affects protein-DNA complex formation and stability. *S. cerevisiae* Cdc13 was incubated with 1 pM oligonucleotide (Tel11, 5′-GTGTGGGTGTG) in one of four buffers at pH 7.5: (**a**) HEPES, (**b**) tricine, (**c**) sodium phosphate, or (**d**) Tris–HCl. Each binding reaction contained 50 mM buffering agent, pH 7.5, 75 mM KCl, 1 mM DTT, 1 mg/mL BSA, and 15% glycerol. The binding reactions were separated in 6.7% polyacrylamide 1 × TBE gels containing 5% glycerol in 1 × TBE running buffer containing 5% glycerol. All gels were run at 200 V for 35 min at 4 °C. Red and black arrows indicate the monomer and dimer species, respectively; blue arrows indicate free ssDNA; asterisks indicate well-shifted ssDNA. *Adapted with permission from Figures 1 and 2* [6], *under Creative Commons Attribution NonCommercial 4.0 International License (CC BY-NC 4.0)*

80% of the total DNA for oligonucleotides up to 20 nucleotides in length.

12. The binding buffer composition should be carefully considered and optimized for each system, because the components in the binding buffer may interact with the complex being studied and alter the complex formation in many ways [5, 6, 11]. For example, protein/DNA complexes may aggregate and be retained in the well (a "well shift"), rather than separating in the polyacrylamide matrix. A well shift prevents the quantitation of the EMSA, as one cannot determine whether the ligand has been specifically bound or is simply aggregating with the protein. The choice of salt concentration and/or buffering agent can minimize the formation of well shifts (Fig. 1).

    Additionally, we recommend carrying out binding reactions at the same pH as the running buffer, if possible. This ensures that there is as little change as possible as the reactions run in the gel. For conventional TBE, that pH is 8.4. Similarly, the addition of 10% more glycerol than the running buffer to the sample facilitates gel loading. We include glycerol in our binding reactions so that the binding equilibrium is not disturbed by the addition of glycerol immediately prior to loading the gels.

13. The serial dilution of protein should cover at least two orders of magnitude of concentration. The choice of protein concentrations will require optimization. Concentrations should be evenly distributed across the dilution, and the majority of the samples should produce fraction bound values between 0.1 and 0.9. Because the hyperbolic curve fit is dominated by the fraction bound values at high protein concentrations, an increased confidence in the fits can be obtained by using protein concentrations that span three orders of magnitude [6, 11] (Fig. 2). For example, for a protein with an apparent binding affinity of 10 pM, we would prepare the following set of 15 protein solutions at 2× final concentration: 0.002, 0.004, 0.02, 0.04, 0.2, 0.4, 2, 4, 20, 40, 200, 400, 2000, 4000, and 20,000 pM.

14. Low-retention tips are critical for accurate final protein concentrations in the binding reaction. We recommend the low-retention plastic consumables manufactured by Eppendorf and USA scientific, because these brands use a low-retention plastic polymer, rather than a coating.

15. When considering the amount of radiolabeled ligand to use, the final concentration in the binding reaction must be large enough to detect the signal. Using the conservative assumption described in **Note 11**, we determined that ~15 attomoles of radiolabeled DNA is the lower limit for reliable detection. This translates to ~0.5 pM DNA in a 30 μL binding reaction. Additionally, if one wants to use the simplified form of the binding equation, care should be taken to make sure that the final DNA concentration is sufficiently below the K_d of the interaction being measured to allow for the assumption that $[P]_{free} = [P]_{total}$.

16. To prepare custom-sized ice packs, cut heavy plastic (approx. 0.5 mil thick) in dimensions that will fit in the outer chamber of the gel box. Seal three sides with impact sealer and fill with deionized H_2O. Seal the fourth side while minimizing air bubbles. Freeze flat overnight at <70 °C. Between uses, rinse with deionized H_2O and allow to thoroughly dry before refreezing.

17. When pouring a 30% "plug" to retain free oligonucleotide, the plug layer should be poured and allowed to polymerize before pouring the separating gel. Gels with a plug cannot be poured ahead of time and stored in the fridge, as over time the high acrylamide content of the plug pushes the plates apart and destroys the gel.

18. Phosphor screens should be thoroughly cleaned with phosphor screen cleaner. Screens should be regularly checked for background defects and residual signal by exposing the screen in an

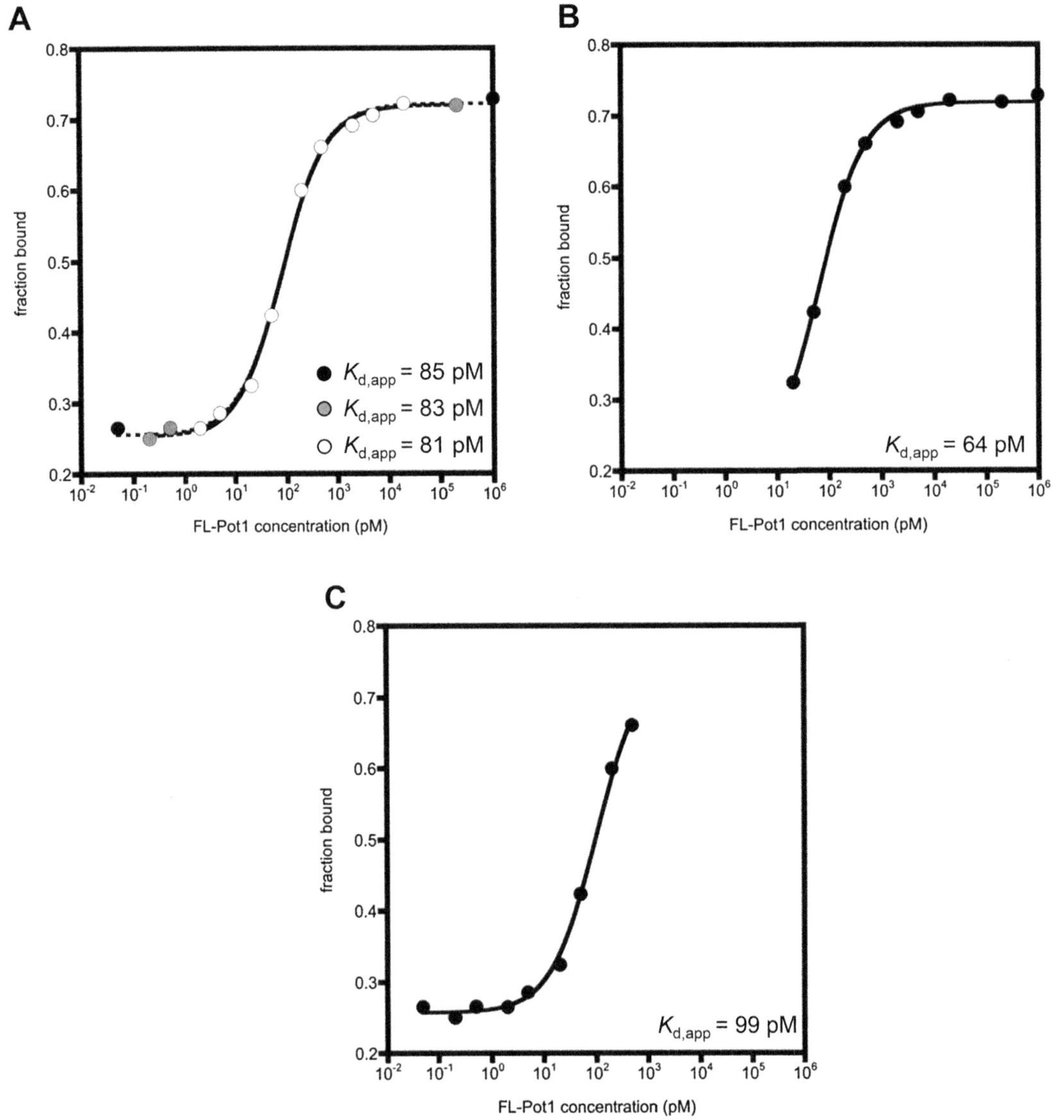

Fig. 2 Selection of protein concentrations affects the robustness of the fitted binding constant. Full-length *S. pombe* Pot1 ("FL-Pot1") was incubated with 1.5 pM oligonucleotide (T3A complemer, 5'-GGATACGGTTACGGT) in binding buffer (20 mM Tris, pH 8.4, 50 mM NaCl, 5 mM DTT, 15% glycerol, 1 mg/mL BSA) before separation in 6.7% polyacrylamide 1 × TBE gels with 5% glycerol using 1 × TBE running buffer with 5% glycerol. All gels were run at 200 V for 20 min at 4 °C. The data were quantitated and fit using three approaches to demonstrate the utility of choosing a wide range of protein concentrations. (**a**) A range of protein concentrations that is two orders of magnitude above and below the $K_{D,app}$ is needed to obtain a reliable fitted value (open circles). An additional order of magnitude above and below the binding constant (gray circles) increases the fitted value by 2.5%, while a protein range of four orders of magnitude above and below (black circles) increases the fitted value by 5%. (**b**) Without a baseline of unbound ligand, the fitted value for the $K_{D,app}$ is artificially low. (**c**) In contrast, lack of binding saturation artificially inflates the fitted $K_{D,app}$ value. Data sets shown in (**b**) and (**c**) are subsets of the data shown in (**a**)

empty, clean cassette for two nights and then scanning on a phosphorimager.

19. The quantitation of the free and bound counts can be carried out multiple ways, depending on the stability of the complex during electrophoresis [6]. Briefly, when protein/DNA complexes were initially bound but dissociated during the short gel run, there may be a smear of signal throughout the lane. Care should be taken to optimize binding and gel conditions to promote complex stability and minimize smearing, including adjusting the composition of the binding buffer and/or the acrylamide concentrations in the gel. However, a small amount of smearing can still be accommodated in quantitation by defining the "bound" counts as the full length of the lane above the free oligonucleotide migration distance. If there are tight, discrete bands for both free and bound DNA, then smaller boxes can be quantitated.

Acknowledgments

Funding for this work was provided by the National Institutes of Health – National Institute for General Medical Sciences (R01GM059414 to D.S.W., F32GM093528 and R15GM119096 to K.A.L., and T32GM065103 to S.E.A.) and the National Science Foundation (MCB-0617956 to D.S.W.). The content is solely the responsibility of the authors and does not necessarily represent the official views of the National Institutes of Health or the National Science Foundation.

References

1. Dickey TH, Altschuler SE, Wuttke DS (2013) Single-stranded DNA-binding proteins: multiple domains for multiple functions. Structure. 21:1074–1084. https://doi.org/10.1016/j.str.2013.05.013
2. Dong F, Spott S, Zimmerman O et al (1999) Dimerisation mutants of lac repressor. I. A monomeric mutant, L251A, that binds lac operator DNA as a dimer. J Mol Biol. 290:653–666. https://doi.org/10.1006/jmbi.1999.2902
3. Zhao H, Mayer ML, Schuck P (2014) Analysis of protein interactions with Picomolar binding affinity by fluorescence-detected sedimentation velocity. Anal Chem 86:3181–3187. https://doi.org/10.1021/ac500093m
4. Jerabek-Willemsen M, André T, Wanner R et al (2014) MicroScale Thermophoresis: interaction analysis and beyond. J Mol Struct 1077:101–113. https://doi.org/10.1016/j.molstruc.2014.03.009
5. Hellman LM, Fried MG (2007) Electrophoretic mobility shift assay (EMSA) for detecting protein–nucleic acid interactions. Nat Protoc 2:1849–1861. https://doi.org/10.1038/nprot.2007.249
6. Altschuler SE, Lewis KA, Wuttke DS (2013) Practical strategies for the evaluation of high-affinity protein/nucleic acid interactions. J Nucleic Acids Investig 4:19–28. https://doi.org/10.4081/jnai.2013.e3
7. Altschuler SE, Croy JE, Wuttke DS (2012) A small molecule inhibitor of Pot1 binding to Telomeric DNA. Biochemistry 51:7833–7845. https://doi.org/10.1021/bi300365k
8. Lewis KA, Pfaff DA, Earley JN et al (2014) The tenacious recognition of yeast telomere

sequence by Cdc13 is fully exerted by a single OB-fold domain. Nucleic Acids Res 42:475–484. https://doi.org/10.1093/nar/gkt843

9. Dickey TH, McKercher MA, Wuttke DS (2013) Nonspecific recognition is achieved in Pot1pC through the use of multiple binding modes. Structure 21:121–132. https://doi.org/10.1016/j.str.2012.10.015

10. Anderson EM, Halsey WA, Wuttke DS (2003) Site-directed mutagenesis reveals the thermodynamic requirements for single-stranded DNA recognition by the telomere-binding protein Cdc13. Biochemistry 42:3751–3758. https://doi.org/10.1021/bi027047c

11. Goodrich JA, Kugel JF (2007) Binding and kinetics for molecular biologists. Cold Spring Harbor Laboratory Press, Cold Spring Harbor, NY

Chapter 30

Identification of Proteins Interacting with Single Nucleotide Polymorphisms (SNPs) by DNA Pull-Down Assay

Bhupinder Singh and Swapan K. Nath

Abstract

Single nucleotide polymorphisms (SNPs) are the most abundantly found common form of DNA variation in the human genome. Many genetic association studies have proved that some of these SNPs are involved in regulating several cellular/physiological processes ranging from gene regulation to disease development. Analysis of the protein complex that binds to these SNPs is a crucial step in studying the mechanisms by which gene expressions are regulated in *cis*- or *trans*-acting manner. Commonly used techniques to determine DNA-protein interaction, such as electrophoretic mobility shift assay (EMSA), have limited value for simultaneously analyzing a large number of proteins in the complex. Furthermore, this assay is tedious and time-consuming and often requires radiolabeled probe as well as extensive optimization. Here, we describe a pull-down assay before performing the EMSA, which helps in the detection of differentially-bound protein(s) in an allele-specific manner. The assay is easy to perform and does not require radiolabeling of DNA probes. Biotinylated DNA probe bound to streptavidin beads can be complexed with protein (s) from cell nuclear lysate, nonspecific proteins were washed out, and only protein(s) having high affinity to SNP-specific DNA were detected on SDS-PAGE and identified by mass spectrometry.

Key words SNP, EMSA, Pull-down assay, SDS-PAGE, GWAS, Mass spectrometry

1 Introduction

The central goal of human genetics is to identify specific DNA variants that contribute significantly to population variation in each trait or phenotype. The Encyclopedia of DNA Elements (ENCODE) project consortium data demonstrated that ~80% of the human genome is biochemically active [1]. SNPs (single nucleotide polymorphisms) are the most commonly occurring genetic variations in the human genome and account for more than 80% of all variations in the human genome [2]. A SNP is a DNA point variation at a single base pair position with frequency of more than 1% in a population making it different from mutation which is rare with frequency of less than 1% [3]. It has been estimated that the human DNA sequence carries a SNP in every 500–1000

Biji T. Kurien and R. Hal Scofield (eds.), *Electrophoretic Separation of Proteins: Methods and Protocols*,
Methods in Molecular Biology, vol. 1855, https://doi.org/10.1007/978-1-4939-8793-1_30,

nucleotides (The International SNP Map Working Group, 2001) [4]. Over the years, SNPs are shown to have potential to modulate the transcriptional regulation and hence change in gene expression. They have also proved to be extremely informative in disease gene mapping and tracing evolutionary histories of populations [5]. After the completion of sequencing of human genome project, many follow-up association studies have aimed to pinpoint the predisposing SNPs and explore the functional role of these SNPs.

SNPs are distributed throughout the human genome; however, the associated variants identified by the recent genome-wide association studies (GWAS) demonstrated that about 88% of these variants fall outside the protein coding regions, especially the intronic or intergenic regions [6, 7]. Some of those SNPs present in the regulatory (i.e., promoter/enhancer) regions and noncoding RNAs have been shown to produce insightful effects on the transcription of its target genes, and the expression of a substantial proportion of genes is influenced by polymorphism in regulatory elements [5, 8]. The physical locations of these SNPs emphasize the regulatory role of these SNPs, and the challenge is to detect the actual causal variants and define the functional consequence of these regulatory SNPs.

Transcriptional regulation of gene expression is controlled through the binding of sequence-specific DNA binding proteins (i.e., transcription factors) to the regulatory regions of genes. The presence of these proteins/factors is dependent upon the cell type being examined and the stimulus to which the cell are exposed. Knowledge of the transcription factors present during any given time can be important in generating a more thorough understanding of how a cell or tissue responds to its environment. Therefore, identification of these transcription factors or regulatory protein (s) required for the expression of a specific gene can provide a better understanding of the molecular mechanisms involved and might suggest new therapies that specifically target an individual gene or set of genes.

The electrophoretic mobility shift assay (EMSA), also known as gel retardation or gel shift assay, is widely used for detection of sequence-specific DNA binding proteins [9, 10]. This assay can be used to determine, in both a qualitative and quantitative manner (densitometry), if a particular transcription factor is present within the nuclei of the cells and tissue of interest or to identify an unknown DNA binding protein which may control the expression of a gene of interest. The assay is based upon the ability of a transcription factor to bind in a sequence-specific manner to a radiolabeled oligonucleotide probe and retard its migration through a polyacrylamide gel. Nevertheless, this technique has its limitations; it requires a radiolabeled probe and results in nonspecific protein bands in SDS-PAGE, which makes it difficult to quantitate sequence-specific interacting protein. To overcome this difficulty, we herewith describe the DNA pull-down assay which helps to wash out nonspecific proteins, thereby allowing the proteins possessing strong binding affinity to DNA, which can be detected by the EMSA. These proteins can later be identified by mass spectrometry. The assay is relatively simple and does not require radiolabeled probes. Further, this technique can detect the allele-specific interaction of protein(s) in wild type versus mutated genotype of SNP. Furthermore, this technique also showed enrichment of low abundant targets as well as isolation of intact complex.

2 Materials

Prepare all solutions using ultrapure water (prepared by purifying deionized water, to attain a sensitivity of 18 MΩ-cm at 25 °C) and analytical grade reagents. Prepare and store all reagents at room temperature (unless indicated otherwise). Diligently follow all waste disposal regulations when disposing waste materials. We do not add sodium azide to reagents.

2.1 Nuclear Extract Isolation

1. Phosphate-buffered saline (PBS), pH 7.4.
2. 0.5 M sodium fluoride (NaF), store at 4 °C.
3. 100 mM phenylmethylsulfonyl fluoride (PMSF) solution in isopropanol, store at −20 °C.
4. Dithiothreitol (DTT), store at −20 °C.
5. Leupeptin, store at −20 °C.
6. 1.25 M β-Glycerophosphate disodium salt, store at 4 °C.
7. 1 M sodium vanadate, store at −20 °C.
8. 1 M potassium chloride (KCl), store at room temperature (RT).
9. 1 M HEPES, stored at 4 °C.
10. 1 M magnesium chloride hexahydrate ($MgCl_2$), stored at RT.
11. 1 M sucrose, stored at RT.
12. 10% Igepal CA-630 (NP-40), stored at RT.
13. 1 M sodium chloride (NaCl), stored at RT.
14. 0.5 M EDTA, stored at RT.
15. PBS buffer containing inhibitors (PBSI): 0.5 mM PMSF, 25 mM β-glycerophosphate, 10 mM NaF, stored at 4 °C.
16. Buffer A: 10 mM HEPES, pH 7.9, 1.5 mM $MgCl_2$, 10 mM KCl, 300 mM sucrose, 0.5% NP-40, stored at 4 °C.

17. Buffer B: 20 mM HEPES, pH 7.9, 1.5 mM $MgCl_2$, 420 mM NaCl, 0.2 mM EDTA, 2.5% glycerol, stored at 4 °C.
18. Buffer D: 20 mM HEPES, pH 7.9, 100 mM KCl, 0.2 mM EDTA, 8% glycerol, stored at 4 °C.

2.2 Protein Assay

1. BCA Protein Assay reagent kit: Reagent A containing sodium carbonate, sodium bicarbonate, bicinchoninic acid, and sodium tartrate in 0.1 M sodium hydroxide, and Reagent B containing 4% cupric acid.
2. Albumin standard 2 mg/mL.
3. 96-well plate.
4. Microplate spectrophotometer, benchmark plus.
5. Microplate manager version 5.2.

2.3 Pull-Down Procedure

1. Dynabeads M-280 streptavidin-coated beads.
2. Magnetic separation tube rack.
3. 2× B/W buffer [10 mM Tris-HCl (ph 7.5), 1 mM EDTA, 2 M NaCl].
4. 5× EMSA buffer (100 mM HEPES (pH 7.2), 160 mM KCl, 0.5 mM EDTA (pH 8) 50% glycerol, 350 ng/μL BSA, 40 ng/μ L poly dI-dC, 12.5 mM DDT).
5. Phosphate-buffered saline (PBS), pH 7.4.
6. Rocking platform.
7. Microcentrifuge.
8. Wash buffer (1× EMSA buffer plus 500 mM NaCl).
9. Biotinylated (5′-biotin TEG) labeled sense strand DNA (wild type and mutated). Non-labeled oligo for antisense DNA (wild type and mutated) (Integrated DNA Technologies, Coralville, IA) (*see* **Note 1**).

2.4 SDS-PAGE

1. 2× Laemmli sample buffer: 950 μL of Laemmli sample buffer mixed with 50 μL of 2-mercaptoethanol.
2. 4–15% gradient polyacrylamide gel in Tris–HCl.
3. Running buffer (10×): 250 mM Tris, 1.92 M glycine, 1% SDS. Prepare 1× running buffer by diluting 100 mL of 10× buffer with 900 mL deionized water.

3 Methods

3.1 Nuclear Extract Isolation

1. All procedures are done on ice or in a controlled temperature room at 4 °C.
2. Remove medium from cultured cells and wash them once with cold PBS. Add PBSI buffer (1 mL/10-cm dish), and harvest

cells with a rubber scraper. Cells are collected into a 50-mL conical tube, which is centrifuged at 550 × *g* for 5 min (*see* **Note 2**).

3. Remove supernatant. Transfer the pellet to a 1.5-mL microcentrifuge tube, and centrifuge at 1500 × *g* for 30 s.
4. To buffers A, B, and D, add the following inhibitors: 0.5 mM PMSF, 1 mM Na_3VO_4, 0.5 mM DTT, 1 μg/mL leupeptin, 25 mM β-glycerophosphate, and 10 mM NaF.
5. Remove supernatant and resuspend the pellet in two package cell volume of buffer A with inhibitors. Keep on ice for 10 min. Vortex briefly, and centrifuge at 2600 × *g* for 30 s.
6. Remove supernatant, and resuspend the pellet in 2/3 package cell volume of buffer B with inhibitors.
7. Sonicate the mixture for 5 s. Centrifuge at 10,400 × *g* for 5 min (*see* **Note 3**).
8. Dilute the supernatant isovolumetrically with buffer D with inhibitors. Aliquot and store at −80 °C (*see* **Note 4**).

3.2 Protein Assay

1. Prepare standard albumin concentrations at 50, 100, 250, 500, 750, 1000, 1500, and 2000 μg/mL.
2. Pipet 10 μL of standards and samples each in duplicate onto a 96-well plate.
3. Mix 50 parts of buffer A with 1 part of buffer B.
4. Add 200 μL of mixed buffer A and B to standards and samples.
5. Incubate plate at 37 °C for 30 min.
6. Read the protein content at 562 nm on a microplate spectrophotometer.

3.3 Pull-Down Procedure

1. Dissolve single-stranded DNA to make duplex. Resuspend the oligonucleotide to 100 μM in water, and mix both single strands in a 1:1 ratio. For preparation of duplex, use 62 μL sense oligonucleotide and 62 μL antisense oligonucleotide, and add 376 μL water. Anneal this mixture by heating to 95 °C and then cooling slowly to room temp. Store this at −20 °C (*see* **Note 5**).
2. Resuspend the Dynabeads™ in the vial (i.e., vortex for >30 s or tilt and rotate for 5 min). Use 40 μL of beads for each reaction. Wash the beads with buffer on magnetic rack. Remember to include a beads only control (*see* **Note 6**).
3. Resuspend the beads according to the direction in 2× B/W buffer in twice the original volume.
4. Divide the beads into different tubes as required.

5. Add annealed oligonucleotide to the beads. 25 μL of annealed oligonucleotide for a total of 8 μg of DNA. Add water to bring the volume up to dilute the 2× B/w buffer to 1×.(for 80 μL of beads in 2× B/W buffer, add 25 μL of annealed DNA and 55 μL of water for a total of 160 μL).
6. Incubate the mixture at RT, rotating, for 15 min.
7. Wash the beads 3× using the magnetic stand to separate the beads each time. Add 8 μL of 1 μg/μL poly dI-dC per 1 mL of this wash buffer to make the first wash after the binding reaction.
8. After the final wash, resuspend the beads in 1× B/W buffer without NaCl, and transfer the beads with bound DNA into a fresh tube.
9. Separate again and add your binding reaction cocktail.
10. Make up a binding reaction cocktail, and keep on ice. 1× binding reaction (500 μL total): 100 μL 5× EMSA buffer and 100 μg of nuclear lysate water to 500 μL.
11. After the final wash with 1× B/W buffer (no salt) from above, add 500 μL of binding cocktail to the beads.
12. Incubate at RT, rotating, for 20 min.
13. Pull down the beads and wash 3× with 300 μL of wash buffer, using buffer spiked with dI-dC for the first wash.
14. With the final wash, transfer the beads to a new tube (this prevents carrying over proteins stuck to the plastic) (*see* **Note 7**).
15. Separate the beads, remove the wash buffer, and add 25 μL of 1× Laemmli SDS loading buffer, with B-ME.
16. Boil the beads for 5 min, spin down, separate, and load supernatant onto a gel (*see* **Note 8**).

3.4 SDS-PAGE

1. Remove 4–15% gradient "ready to use" gel from plastic, remove the comb, and cut the tape along the black line across the entire gel. Pull out the gel (*see* **Note 9**).
2. Set up the electrophoresis unit.
3. Place the gel onto the mini-gel holder cassette of the electrophoresis unit.
4. Load the 10-well gel with 5 μL of standards or 25 μL of the supernatant protein samples.
5. Add 1× running buffer to the electrophoresis cell.
6. Cover the electrophoresis cell with the cell lid.
7. Connect the electrophoresis unit to a power supply. Set the voltage at 150 V.

8. Run the gel at 150 V until the marker dye has run off the edge of the gel.
9. Remove gel and rinse with 1× transfer buffer precooled in ice (4 °C).
10. Set up the electrophoretic transfer cell.
11. Soak fiber pad, filter paper, and nitrocellulose membrane in 1× transfer buffer before placing them in the gel cassette.
12. Place nitrocellulose membrane between gel and anode. Inspect to make sure that no air bubble is trapped between gel and membrane.
13. Connect the electrophoretic transfer unit to a power supply.
14. Place the unit on ice or at 4 °C.
15. Run the gel at 90 V for 1 h.
16. Stain the gel with Coomassie brilliant blue R-250 (*see* **Note 10**).
17. Protein bands which show significant binding difference between wild-type and mutated DNA used for mass spectrometry analysis.

4 Notes

1. We design 30–40-base-long (can be up to 500 bp) oligonucleotides keeping the SNP position in center. The sequence is sent to Integrated DNA Technologies (IDT). 5'-biotinylated sense and antisense single-stranded oligonucleotides are synthesized, and the sequence is verified by IDT.
2. A 50-mL conical tube is used to ensure a maximal yield of cells.
3. In our experience, a 5-s sonication is sufficient to break the cells. Sonication for a longer period may disrupt the nucleus.
4. Nuclear extracts so prepared do not have detectable cytosolic proteins.
5. For DNA 4 μg of biotinylated double-stranded oligonucleotides are used for each assay; we mix 4 μg of sense with 4 μg of antisense oligonucleotides to generate 4 μg of 5′-biotinylated double-stranded oligonucleotide in each experiment.
6. Add an equal volume of washing buffer, or at least 1 mL, and mix (vortex for 5 s, or keep on a roller for at least 5 min). Place the tube on a magnet for 1 min and discard the supernatant. Remove the tube from the magnet, and resuspend the washed Dynabeads™ in the same volume of washing buffer as the initial volume of Dynabeads™ taken from the vial.

7. The incubation is important to centrifuge the complex at a low speed. A higher speed of centrifugation may disrupt the complex.
8. The step causes dissociation of transcription factors from the complex.
9. We have found the Bio-Rad mini-gel to be convenient. However, mini-gels from other manufacturers may be used.
10. Please *do not* stain the gels using boiling or microwave to speed up the staining or destaining process. Please *never* use overhead projector foils for gel scanning as this prevents the gel from being used for MS experiments.

Acknowledgments

This work was supported by the National Institutes of Health (NIH) grants AR060366, MD007909, and AI107176.

References

1. Consortium EP (2012) An integrated encyclopedia of DNA elements in the human genome. Nature 489(7414):57–74
2. Sherry ST et al (2001) dbSNP: the NCBI database of genetic variation. Nucleic Acids Res 29 (1):308–311
3. Brookes AJ (1999) The essence of SNPs. Gene 234(2):177–186
4. Sachidanandam R et al (2001) A map of human genome sequence variation containing 1.42 million single nucleotide polymorphisms. Nature 409(6822):928–933
5. Shastry BS (2002) SNP alleles in human disease and evolution. J Hum Genet 47(11):561–566
6. Manolio TA et al (2009) Finding the missing heritability of complex diseases. Nature 461 (7265):747–753
7. Hindorff LA et al (2009) Potential etiologic and functional implications of genome-wide association loci for human diseases and traits. Proc Natl Acad Sci U S A 106(23):9362–9367
8. Hoogendoorn B et al (2003) Functional analysis of human promoter polymorphisms. Hum Mol Genet 12(18):2249–2254
9. Fried M, Crothers DM (1981) Equilibria and kinetics of lac repressor-operator interactions by polyacrylamide gel electrophoresis. Nucleic Acids Res 9(23):6505–6525
10. Garner MM, Revzin A (1981) A gel electrophoresis method for quantifying the binding of proteins to specific DNA regions: application to components of the *Escherichia coli* lactose operon regulatory system. Nucleic Acids Res 9(13):3047–3060

Chapter 31

Horizontal Agarose Gel Mobility Shift Assay for Protein-RNA Complexes

Jennifer A. Ream, L. Kevin Lewis, and Karen A. Lewis

Abstract

Recent advances in agarose gel electrophoresis protocols established conditions for the high-resolution separation of DNA and RNA using higher voltages combined with short run times. We subsequently developed a protocol for using these conditions to measure the binding affinity of a protein for an RNA ligand on an agarose gel. This native gel mobility shift assay is highly accessible, using common molecular biology reagents found in most laboratories. Here, we describe the protocol for carrying out native agarose gel electrophoresis to characterize the binding affinity of a protein for an RNA ligand. The electrophoresis time is less than 10 min, which minimizes the dissociation of protein and ligand. We have used the p19 siRNA binding protein and its cognate dsRNA ligand to demonstrate strategies for identifying optimal conditions to measure apparent binding constants using this agarose gel shift system.

Key words Gel shift, Mobility shift assay, EMSA, RNA binding assay, Native agarose gel

1 Introduction

Native gel electrophoretic mobility shift assays (EMSAs) are widely used to measure the association between biomolecules, such as proteins and nucleic acid oligonucleotides [1]. While this method is inherently non-equilibrium, it requires very little specialized equipment. As a result, an EMSA can be a highly accessible approach for evaluating interactions between biological macromolecules [2]. Conventional EMSA protocols use vertical polyacrylamide gels, which can involve handling of neurotoxic unpolymerized acrylamide solution [3].

Building on previous advances in horizontal agarose gel electrophoresis [4, 5], we have established a method for separating protein-RNA complexes on high-percentage agarose gels in less than 10 min [6]. The short run time is especially useful for EMSAs, as it reduces the amount of time that could allow the bound complex to dissociate. The oligonucleotide ligand is

Biji T. Kurien and R. Hal Scofield (eds.), *Electrophoretic Separation of Proteins: Methods and Protocols*, Methods in Molecular Biology, vol. 1855, https://doi.org/10.1007/978-1-4939-8793-1_31,

detected using SYBR Gold, a low-toxicity stain that can detect nanogram quantities of nucleic acid. This novel system was used to determine the apparent binding constant for the p19 siRNA-binding protein and its cognate dsRNA ligand. The binding affinity was found to be comparable to the apparent binding constant obtained from a conventional polyacrylamide gel assay performed concurrently [6].

Because this method requires only standard molecular biology reagents and equipment, it is affordable and accessible to a wide range of laboratories. Here, we describe our protocol for rapid native agarose gel electrophoresis to measure quantitatively the association of protein and RNA.

2 Materials

2.1 General Reagents

1. Double-deionized water for buffers.
2. Autoclaved water for reactions (*see* **Note 1**).

2.2 Gel Electrophoresis

1. Horizontal agarose gel apparatus system with a gel-pouring tray and comb.
2. Gel-sealing tape (3/4 in. wide).
3. Agarose.
4. 0.5× TB buffer: 45 mM Tris base, 45 mM boric acid, pH 8.2–8.4 [4].
5. 10,000× SYBR Gold.
6. Containers for staining and destaining (*see* **Note 2**).
7. Gel imaging system.

2.3 Binding Reaction Components

1. Purified protein of interest.
2. Purified RNA oligonucleotides of interest.
3. RNA dilution buffer: 5 mM Tris–HCl, pH 8.0 at room temperature.
4. Binding buffer (*see* **Note 3**).
5. 30% glycerol.

2.4 Quantitation

1. Gel imaging system (Alpha Innotech Imager, BioRad XRS+ ChemiDoc system, or similar).
2. Imaging software capable of quantitation (ImageQuant, ImageJ, ImageLab, or similar).
3. Software capable of nonlinear multivariate curve fitting (SigmaPlot, Kaleidagraph, or similar).

3 Methods

3.1 Annealing Complementary Single-Stranded RNA Oligonucleotides

1. Prepare equimolar mixtures of single-stranded RNA oligonucleotides in the buffered solution of choice. For the reactions described below, each single-stranded RNA was resuspended in nuclease-free water to a stock concentration of ~140 μM. Annealing reactions contained 3.8 μM of each RNA oligonucleotide in RNA dilution buffer.
2. Gently mix the solution.
3. Incubate at 90 °C in a dry heating block for 4 min.
4. Remove tubes to room temperature and allow to equilibrate for 30 min.
5. Confirm annealing with a 3.5% agarose gel (*see* Subheadings 3.2 and 3.4).
6. Store annealed RNA in small aliquots at −20 °C and avoid freeze-thaw cycles. Under these conditions, we found that the annealed dsRNA was stable for several months at −20 °C.

3.2 Preparing a High-Percentage Agarose Gel

1. Seal the ends of an agarose gel tray with gel-sealing tape, and place it on a level surface in a 60 °C incubator for 2–5 min to prewarm the plastic.
2. Meanwhile, prepare a solution of 2.5–3.0% agarose dissolved in 0.5× TB buffer (*see* **Notes 4** and **5**).
3. With the gel tray still in the incubator, pour the dissolved agarose solution into the tray. Place the comb and incubate at 60 °C for 2 min (*see* **Note 6**).
4. Remove the gel tray from the incubator, and allow the gel to solidify at room temperature on a level bench for at least 20 min.

3.3 Binding Reaction

1. Set up the binding reactions in a total volume of 12–14 μL (*see* **Note** 7).
2. Allow reactions to equilibrate under appropriate conditions of time and temperature. Note that several incubation times may need to be evaluated to confirm that the binding reaction has reached equilibrium. For p19, we incubated at room temperature for 1 h.
3. If glycerol was not included in the binding buffer, add 30% glycerol to each reaction to bring to a final concentration of 5% (v/v) glycerol in preparation for gel loading.

3.4 Running the Gel

1. Remove the gel tape and comb, and place the gel tray in the gel rig, making sure the gel is completely submerged in fresh 0.5× TB buffer. The level of the buffer should be no more than 4–5 mm above the surface of the gel [4] (*see* **Note 8**).

2. Load ~13 μL of binding reaction + glycerol (*see* **Note 9**).
3. Set the power supply to 20–30 V/cm (for a 14-cm-long gel that translates to 300–420 V). Electrophorese for 6–10 min (*see* **Note 10**).

3.5 Imaging the Gel

1. Staining: place the gel in a dish, and cover it with 2× SYBR Gold, freshly prepared in 0.5× TB buffer, diluted from the 10,000× SYBR Gold stock solution. Shake vigorously for 40 min at room temperature (*see* **Notes 11** and **12**).
2. Destaining: Decant the staining solution, and replace it with enough 0.5× TB buffer to cover the gel. Shake vigorously for 15 min at room temperature (*see* **Note 12**).
3. Image under UV light in an electronic gel imaging system.

3.6 Quantitation and $K_{D,app}$ Calculation

1. Use gel imaging software (ImageQuant, ImageJ, or similar) to quantitate the free and bound DNA bands (*see* **Note 13**).
2. Evaluate the sum of counts (bound + free) across all lanes. Total counts per lane should be approximately the same, because all of the RNA ligand should be detectable in either partition (bound or free).
3. Calculate fraction of ligand bound = (bound/(bound + free)).
4. Plot fraction bound as function of total protein concentration.
5. Fit the data with the quadratic form of the binding isotherm to solve for $K_{D,app}$ (*see* **Note 14**):

$$\frac{[\mathrm{PL}]}{L_{\mathrm{T}}} = \left(\frac{K_{\mathrm{D,app}}[P]_{\mathrm{T}} + [L]_{\mathrm{T}} - \sqrt{\left(K_{\mathrm{D,app}}[P]_{\mathrm{T}} + [L]_{\mathrm{T}}\right)^2 - 4[P]_{\mathrm{T}}[L]_{\mathrm{T}}}}{2[L]_{\mathrm{T}}}\right) \quad (1)$$

where $[P]_T$ is the total protein concentration, $[L]_T$ is the total ligand concentration, [PL] is the fraction of protein and ligand in complex with one another, and $K_{D,app}$ is the apparent dissociation constant.

4 Notes

1. Water should be double-deionized and sterilized and ideally nuclease-free; however, it does not need to be DEPC-treated. For reaction buffers, we use commercial nuclease-free water (e.g., from integrated DNA technologies).
2. Be sure to use a staining container that is appropriate for the staining agent; for example, SYBR nucleic acid stains should not be used in glass containers (per manufacturer's recommendation).

3. The binding buffer contents and conditions should be optimized for the system of interest, including buffering agent, salt concentration, glycerol content, and incubation temperature [6, 7]. For p19 siRNA binding protein, we used the 1× p19 siRNA binding buffer provided by the manufacturer (20 mM Tris–HCl, 100 mM NaCl, 1 mM EDTA, 1 mM TCEP, 0.02% Tween-20, pH 7.0 at 25 °C) and incubated for 1 h at room temperature (~22–25 °C).
4. Agarose content should be optimized for the protein/RNA system of interest. We evaluated a range of concentrations from 2.5 to 3.5% agarose. The ~50 mL solution is prepared by microwaving on full power for 45 s, mixing by swirling the flask, and then microwaving on full power in 8 s increments until fully dissolved. We found that 2.5% and 3.0% agarose produced less background fluorescence, which was compatible with SYBR gold staining.
5. For a gel with dimensions of 11 cm × 14 cm, we prepare gel solutions of 45–55 mL volume. This combination of volume and surface area produces gels that are 6–8 mm thick, which we found to be optimal to reduce background fluorescent signal from the agarose (Fig. 1).
6. The high agarose content of the gel solution will cause it to solidify quickly if poured at room temperature. By pouring the gel at higher temperatures (60 °C), it prevents distortions in the gel matrix caused by uneven cooling, which would compromise the quality of the electrophoresis.
7. Ligand concentration should be optimized for the protein and detection systems being used. For high percentage agarose gels that are 6–8 mm thick and stained with SYBR gold, we have found that the minimum amount of dsRNA needed for robust signal over background is ~20 ng.
8. While 0.5× TB running buffer can be reused for electrophoresis, we found that repeated reuse (more than two

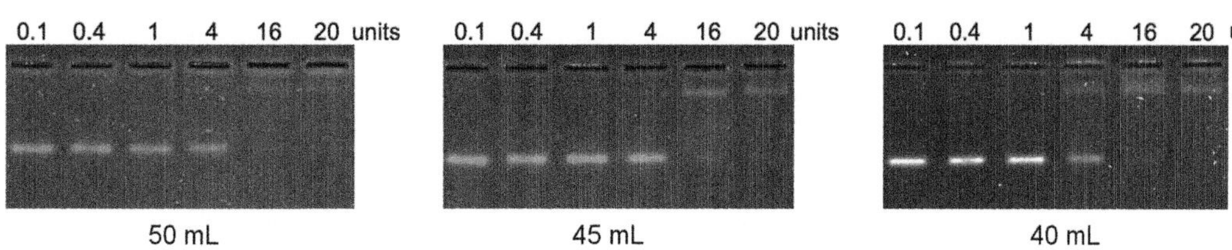

Fig. 1 Reducing the volume and thickness of the gel improves band signal. 2.5% agarose gel solutions were prepared in volumes of 50, 45, and 40 mL and used to analyze identical binding reactions. All three gels were stained with SYBR Gold and destained in 0.5× TB. The background fluorescence of the SYBR Gold-stained agarose was significantly reduced in the thinner gels. (Data reproduced with permission from Ream et al. [6], Figure 1)

electrophoresis runs) resulted in heavily smeared and uninterpretable bands.

9. To maximize signal, the volume of sample that is loaded onto the gel should be as large as the well can contain. For example, our 10-well gel of dimensions 11 cm × 14 cm × 0.8 cm uses combs that are 1 mm thick, allowing for an approximate well volume of ~ 12–13 μL. If thicker gels or bigger combs are used, more samples can be loaded. Such adjustments may allow for a lower concentration of RNA ligand to be used in the initial binding reaction, since a larger quantity can be loaded into the well.
10. Gel running time and voltage should be optimized for the system of interest. Higher voltages may allow for faster run times and good band separation but can cause blurry or streaky bands if set too high. We have found that 20–30 V/cm produces the best results in our system (Fig. 2).
11. For a gel with dimensions of 11 cm × 14 cm × ~0.8 cm, we used 50 mL of staining solution in a container that was ~15% larger than the gel.
12. Destaining took place in a larger container and used 100–200 mL of destaining solution. SYBR gold has a strong pH dependence, and we found that preparing either the staining or destaining solutions with deionized water resulted in poor signal (Fig. 3). In particular, we observed a significant reduction in band sharpness and intensity when compared to staining and destaining in 0.5× TB. Longer incubations for either staining or destaining (more than 40 min and 15 min, respectively) are not recommended, as they will result in fuzzy and/or undetectable bands due to RNA diffusion within the agarose matrix (Fig. 3).

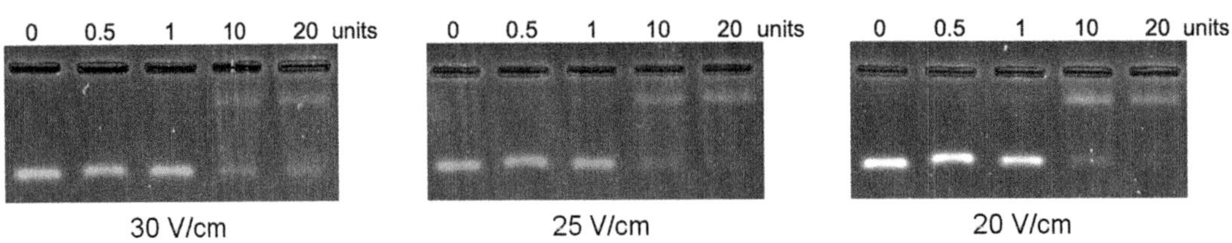

Fig. 2 Electrophoresis voltage must be optimized for each protein/RNA system. Identical binding reactions of 0, 0.5, 1, 10, and 20 units of p19 protein were separated on identical 2.5% agarose gels at three different voltages: 30, 25, and 20 V/cm, where cm refers to the length of the agarose gel. While higher voltages may allow for reduced run times, we observed that the higher voltages reduced the stability of the protein-RNA complex. The 30 V/cm voltage in particular shows a considerable amount of free RNA at the two highest protein concentrations, while the 20 V/cm separation shows robust RNA-protein complex and very little free RNA. (Data reproduced with permission from Ream et al. [6], Figure 2)

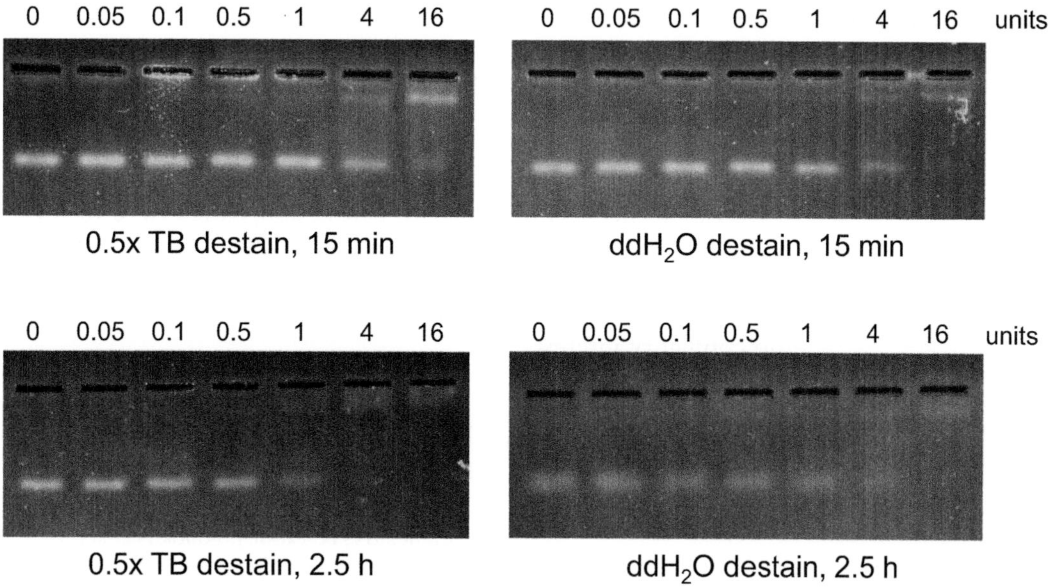

Fig. 3 Destaining protocol affects the quality of band resolution. *Top panels*: The appropriate destaining buffer was identified by comparing 15-min destain incubations in either 0.5× TB or ddH_2O. The intensity and resolution of bands were superior when gels were destained in 0.5× TB. *Bottom panels*: Destaining too long will negatively affect band quality. Identical gels were incubated in either 0.5× TB or ddH_2O destaining solution for 2 h. Regardless of destain solution, the resulting bands were fuzzy and low-intensity, preventing quantitation or interpretation

13. There are multiple approaches to defining bands for quantitation. Because of the short run times of these experiments, the complex is less likely to dissociate during electrophoresis. The sharp, discrete bands can then be easily defined. However, if the complex dissociates during the run, there may be a smear of SYBR gold signal in the lane (*see* Fig. 2, left and center panels). Consequently, the entire smear should be included in the definition of "bound" oligonucleotide, using a box that covers the entire lane above the "free" oligonucleotide migration distance.
14. Because the amount of RNA ligand needed for detection by SYBR gold is likely to be greater than the apparent K_D of the interaction, the binding reaction conditions prevent the common assumption that $[P]_{free} \approx [P]_{total}$. As a result, the simplified form of the binding isotherm cannot be used [7]. However, fluorescently or radiolabeled oligonucleotides could be used in place of SYBR gold staining for more sensitive detection of the ligand. By using smaller amounts of ligand, this assumption could then be made, and fitting could be carried out with the simplified form of the binding isotherm [7].

Acknowledgments

This work was supported by the National Science Foundation (NSF 1407736) through the Houston Louis-Stokes Alliance for Minority Participation Scholars Program (J.A.R., scholar) and the National Institutes of Health – National Institute for General Medical Sciences (GM099049 to L.K.L. and GM119096 to K.A. L.). The content is solely the responsibility of the authors and does not necessarily represent the official views of the National Institutes of Health or the National Science Foundation. All authors declare no conflict of interest.

References

1. Lane D, Prentki P, Chandler M (1992) Use of gel retardation to analyze protein-nucleic acid interactions. Microbiol Rev 56:509
2. Helwa R, Hoheisel JD (2010) Analysis of DNA–protein interactions: from nitrocellulose filter binding assays to microarray studies. Anal Bioanal Chem 398:2551–2561. https://doi.org/10.1007/s00216-010-4096-7
3. Hellman LM, Fried MG (2007) Electrophoretic mobility shift assay (EMSA) for detecting protein–nucleic acid interactions. Nat Protoc 2:1849–1861. https://doi.org/10.1038/nprot.2007.249
4. Sanderson BA, Araki N, Lilley JL et al (2014) Modification of gel architecture and TBE/TAE buffer composition to minimize heating during agarose gel electrophoresis. Anal Biochem 454:44–52. https://doi.org/10.1016/j.ab.2014.03.003
5. Rodriguez BV, Pescador J, Pollok N et al (2015) Impact of size, secondary structure, and counterions on the binding of small ribonucleic acids to layered double hydroxide nanoparticles. Biointerphases 10:041007. https://doi.org/10.1116/1.4936393
6. Ream JA, Lewis LK, Lewis KA (2016) Rapid agarose gel electrophoretic mobility shift assay for quantitating protein: RNA interactions. Anal Biochem. https://doi.org/10.1016/j.ab.2016.07.027
7. Altschuler SE, Lewis KA, Wuttke DS (2013) Practical strategies for the evaluation of high-affinity protein/nucleic acid interactions. J Nucleic Acids Investig 4:19–28. https://doi.org/10.4081/jnai.2013.e3

Chapter 32

Applications of Immobilized Metal Affinity Electrophoresis

Bao-Shiang Lee, Lasanthi P. Jayathilaka, Jin-Sheng Huang, and Shalini Gupta

Abstract

Immobilized metal affinity electrophoresis (IMAEP) is a technique in which metal ions are integrated in a localized polyacrylamide gel section and which possess an insignificant electrophoretic migration. IMAEP has been implemented into both native and sodium dodecyl sulfate (SDS) one-dimensional (1-D) and two-dimensional (2-D) polyacrylamide gel electrophoresis (PAGE) systems. The gel piece containing metal ions is overlaid on top of an individual well in 1-D IMAEP, and gel strip is overlaid on top of the two-second dimensional polyacrylamide gel in 2-D IMAEP, respectively. This method is used to capture/enrich metal ion binding proteins. Due to the preferential binding between iron (III), manganese (II), or aluminum (III) ions and the phosphate group, these metal ions are used to capture/enrich phosphoproteins from a mixture of proteins. Data shows that SDS not only does not interfere with the extraction of phosphoproteins but also help unmasking phosphate groups by unfolding the phosphoproteins to facilitate metal ion-phosphate binding while supplying the protein with negative charges. In addition, IMAEP together with avidin functional affinity electrophoresis (AFAEP) and antibody affinity electrophoresis (AAEP) have been used to demonstrate the separation of proteins by their functionalities. AFAEP is used here to capture biotinylated proteins using avidin and AAEP is used here to capture protein G using IgG.

Key words Immobilized metal affinity electrophoresis (IMAEP), Immobilized metal affinity column (IMAC), Avidin functional affinity electrophoresis (AFAEP), Antibody affinity electrophoresis (AAEP), Dimension of functionality, Phosphoproteins, SDS-PAGE, Immobilized pH gradient (IPG), Pro-Q diamond phosphoprotein gel stain, β-Casein, Mass spectrometry

1 Introduction

An innovative immobilized metal affinity electrophoresis (IMAEP) technique by embedding metal ions in a polyacrylamide gel section of either one-dimension (1-D) or two-dimensional (2-D) polyacrylamide gel electrophoresis (PAGE) systems for the extraction of phosphoproteins from a protein mixture has been reported by us [1–4]. This method with immobilized Fe^{3+} metal ion has been used by other researchers to assess the autokinase activity of wildtype and mutants of human αA-crystallin [5]. The underlying mechanism of

Biji T. Kurien and R. Hal Scofield (eds.), *Electrophoretic Separation of Proteins: Methods and Protocols*, Methods in Molecular Biology, vol. 1855, https://doi.org/10.1007/978-1-4939-8793-1_32,

IMAEP is the same as in immobilized metal affinity column (IMAC) which is the most widely used technique for the enrichment of phosphoproteins or phosphopeptides and relies on the electrostatic interaction between the negatively charged phosphate group and positively charged metal ions like Fe(III), Mn(II), Al(III), etc. [6–11]. In IMAEP, the metal ion is embedded in the gel matrix by adding the metal ion into acrylamide gel solution before polymerization with ammonium persulfate. We have experimented with a wide variety of metal ions [1], including iron (III) chloride, iron (II) chloride, nickel (II) chloride, manganese (II) chloride, magnesium (II) chloride, silver chloride, tin (II) chloride, scandium (III) oxide, aluminum (III) chloride, vanadium (III) chloride, gallium (III) nitrate, cupric sulfate, chromium (III) chloride, zirconium (IV) hydrogen phosphate, sodium chloride, potassium chloride, titanium (III) acetate, calcium (II) chloride, and titanium (IV) oxide nanopowder at various concentrations and volumes. A protein mixture of nonphospho- and phosphoproteins and highly complex protein cell extracts is used to demonstrate the capture of phosphoproteins using IMAEP. Results show that electrophoretic migration of phosphoproteins ceases upon encounter with metal ions like Al^{3+}, Ti^{3+}, Fe^{3+}, Fe^{2+}, and Mn^{2+} whereas non-phosphorylated proteins migrate normally [1–4]. SDS is incorporated into IMAEP in our subsequent study [2], in which protein samples are heated with 2% w/v SDS for 15 min at 85 °C. Results show that SDS not only has no detrimental effect on the extraction of phosphoproteins of IMAEP but also unmasks phosphate groups by unfolding the phosphoproteins to facilitate metal ion-phosphate interaction. To extend the usefulness of IMAEP method, we have implemented the IMAEP technique into the high-resolution two-dimensional (2-D) SDS-PAGE [12–15] which allows for better separation, identification, and quantification of larger number of proteins [3, 4]. In 2-D SDS-PAGE, proteins are separated by isoelectric focusing (IEF) using immobilized pH gradient (IPG) gel strip in the first dimension, which distinguishes proteins according to their isoelectric points (*pI*s), followed by SDS-PAGE in a second dimension, which resolves proteins according to their molecular weights. In 2-D IMAEP [3], a rectangular gel strip containing metal ions is overlaid on top of the two-dimensional polyacrylamide gel to extract phosphoproteins. These captured phosphoproteins are readily identifiable using standard biochemical or proteomic techniques such as western blot and in-gel trypsin digestion/mass spectrometry. One of the advantages of 2-D IMAEP over IMAC is that it not only enriches the phosphoproteins but also separates the phosphoproteins based on their *pI*s. To further expand the possible applications of IMAEP, it has been incorporated into a so-called dimension of functionality, separation of proteins by PAGE using their functionalities [16, 17], which beside IMAEP also contains two additional gel sections: avidin functional affinity electrophoresis (AFAEP) and antibody affinity electrophoresis (AAEP). AFAEP is instrumental in the capture of biotinylated proteins/peptides [17–22], and AAEP is used to capture specific antigens [23, 24]. A protein mixture of β-casein, protein G, and biotinylated BSA is used to demonstrate the separation/capture of the IMAEP-AFAEP-AAEP gel. Results show that electrophoretic migration of phosphorylated protein β-casein ceases upon encounter with IMAEP section, migration of biotinylated BSA ceases upon encounter with AFAEP section, and migration of protein G ceases upon encounter with AAEP section.

2 Materials

Prepare all solutions in 18 MΩ water with clean containers and use ultrapure reagents. Solutions and reagents are stored at room temperature unless mentioned otherwise. Always keep safety considerations in mind.

2.1 1-D Native Immobilized Metal Affinity Electrophoresis

1. 30% stock monomer acrylamide gel solution. Store at 4 °C.
2. 4× resolving gel buffer: 1.5 M Tris–HCl, pH 8.8. Store at 4 °C.
3. Native gel running buffer: 25 mM Tris–HCl, 192 mM glycine, pH 8.3. Store at 4 °C.
4. 10% (w/v) Ammonium persulfate (APS) solution in water (prepare fresh).
5. *N*,*N*,*N*′,*N*′-Tetramethylethylenediamine (TEMED).
6. 2× native sample buffer: 125 mM Tris–HCl, pH 8.8, 20% glycerol, 0.005% bromophenol blue. Store at 4 °C.
7. 50 × 8 × 1 mm plastic strips (homemade).
8. Metal ion solution: 1 M iron (III) chloride in water, 0.2 M iron (III) chloride in 2-methyltetrahydrofuran, 1 M manganese (II) chloride in water, 1 M cupric sulfate in water, 1 M aluminum (III) chloride in water, 1 M titanium (III) acetate in water. Store at 4 °C.
9. 100 × 100 × 1 mm empty gel cassette (Invitrogen, Carlsbad, CA).
10. Power supply: PowerEase 500 programmable power supply.

2.2 SDS 1-D Immobilized Metal Affinity Electrophoresis

1. 10% Sodium dodecyl sulfate (SDS (*see* **Note 1**).
2. SDS gel running buffer: 25 mM Tris, 192 mM glycine, 0.1% SDS, pH 8.3. Store at 4 °C.
3. 2× SDS sample buffer: 125 mM Tris–HCl, pH 6.8, 20% v/v glycerol, 0.2 M DTT, 4% SDS, 0.005% bromophenol blue. Store at 4 °C.

4. Gradient maker (Hoefer, Holliston, MA).
5. Light acrylamide gel solution: 7.5% (w/v) acrylamide gel solution, containing resolving gel buffer, APS, and TEMED (prepare fresh).
6. Heavy acrylamide gel solution: 15% (w/v) acrylamide gel solution, containing resolving gel buffer, APS, and TEMED (prepare fresh).
7. MCF-7 human breast carcinoma cells (American Type Culture Collection, Rockville, MD).
8. Lysis buffer: 50 mM HEPES, pH 7.5, 150 mM NaCl, 10% glycerol, 1% Triton X-100, 1 mM EGTA, 1.5 mM $MgCl_2$, 100 mM serine/threonine phosphatase inhibitor sodium fluoride (NaF), 1 mM tyrosine phosphatase inhibitor sodium orthovanadate (Na_3VO_4).

2.3 2-D Immobilized Metal Affinity Electrophoresis

1. Rehydration buffer: 8 M urea, 2 M thiourea, 4% CHAPS, 2 mM EDTA, pH 8.0, 250 mM DTT, 2 mM tributyl phosphine, 0.5% ampholytes, and 0.005% bromophenol blue (prepare fresh).
2. Precast 7 cm immobilized pH gradient (IPG) pH 3–10 dry gel strip (GE Bioscience, Piscataway, NJ, USA). Store at −20 °C.
3. PROTEAN IEF cell (Bio-Rad Laboratories, Hercules, CA).
4. Equilibrium solution I: 50 mM Tris–HCl (pH 8.8), 6 M urea, 30% glycerol, 2% SDS, 0.005% bromophenol blue, and DTT (1% w/v) (prepare fresh).
5. Equilibration buffer II: 50 mM Tris–HCl (pH 8.8), 6 M urea, 30% glycerol, 2% SDS, 0.005% bromophenol blue, iodoacetamide (2.5% w/v) (prepare fresh).
6. Pro-Q diamond phosphoprotein gel stain (Molecular Probes, Eugene, OR).
7. PageRuler™ Prestained Protein Ladder (Fermenta, Burlington, Ontario, Canada).
8. FX fluorescent imager (Bio-Rad).

2.4 Dimension of Functionality

1. Avidin. Store at −20 °C.
2. β-Casein. Store at −20 °C.
3. Biotinylated BSA. Store at −20 °C.
4. Bovine IgG. Store at −20 °C.
5. Metal ion solution: 1 M manganese (II) chloride in water. Store at 4 °C.
6. Protein G. Store at −20 °C.

3 Methods

Always manipulate gels and solutions with gloved hands. Perform all procedures at room temperature unless indicated otherwise. Strictly follow all safety regulations.

3.1 1-D Native Immobilized Metal Affinity Electrophoresis

1. Prepare the native 7.5% acrylamide gel solution by mixing 7.5 mL of 30% stock monomer acrylamide gel solution, 7.5 mL 4× resolving gel buffer, 14.8 mL deionized water, and 10 μL TEMED (*see* **Note 2**). Add 190 μL 10% APS aqueous solution to the gel solution immediately before casting the gel.
2. Each sample well (18 × 8 × 1 mm) is made by inserting one plastic strip at a correct depth in a 100 × 100 × 1 mm empty gel cassette, and 8 mL of the gel solution is poured into the cassette to cast the gel (*see* **Note 3**).
3. Following gel formation and removal of the plastic strips, 75–200 μL of the native 7.5% acrylamide gel solution, containing APS, TEMED, and 1–4 μL of 1 M metal ion solution is added to the wells and allowed to polymerize (*see* **Note 4**).
4. The gel is run using native gel running buffer at ~3 W with a current of ~22 mA and a voltage of ~125 V for ~1.5 h with PowerEase 500 programmable power supply. Fe^{3+}, Mn^{2+}, and Cu^{2+} IMAEPs are run without loading the protein sample before and after electrophoresis (Fig. 1a) (*see* **Note 5**). In addition, Al^{3+}, Ti^{3+}, Fe^{3+}, Mn^{2+}, and Cu^{2+} IMAEPs are run with human albumin, carbonic anhydrase, dephosphorylated a-casein, human hemoglobin, a-casein, b-casein, and phosvitin (Fig. 1b–d) (*see* **Note 6**). Gels are stained with Coomassie blue G-250 or Zn^{2}—reverse staining and documented using a scanner or a digital camera and a light box.
5. The metal-protein complex is excised from the IMAEP gel, and the captured phosphoproteins are identified by standard proteomic techniques using in-gel trypsin digestion and mass spectrometry (Fig. 2).
6. To evaluate the concentration effect of the metal ions, IMAEPs are run with different amount of incorporation of metal ions from 0.1 to 4 μL (Fig. 3).
7. To assess the sample loading capacity, IMAEPs are run with different amount of a protein from 1 to 200 μg (Fig. 4).

3.2 1-D SDS Immobilized Metal Affinity Electrophoresis

1. The procedure of performing 1-D SDS IMAEP is the same as in 1-D native IMAEP except that 0.1% SDS is added to the native acrylamide gel solution and/or native gel running buffer. In addition, 4% SDS is added to the 2× native sample

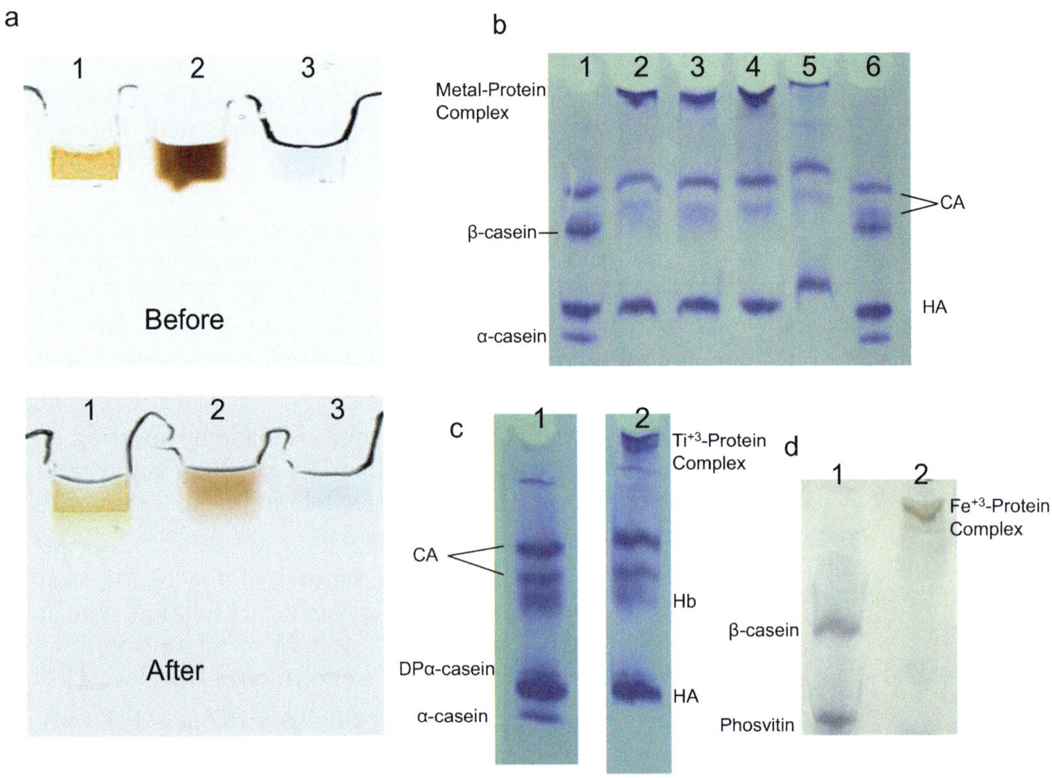

Fig. 1 (**a**) IMAEP without protein sample of a 7.5% native polyacrylamide gel before and after gel run with 1.5 μL of 1 M Fe^{3+} (Lane 1), Mn^{2+} (Lane 2), or Cu^{2+} (Lane 3) incorporation. (**b**) IMAEP of human albumin (HA), carbonic anhydrase (CA), α-casein, and β-casein with no metal (Lane 1), 1.5 μL of 1 M Al^{3+} (Lane 2), Ti^{3+} (Lane 3), Fe^{3+} (Lane 4), Mn^{2+} (Lane 5), or Cu^{2+} (Lane 6) incorporation. (**c**) IMAEP of hemoglobin (Hb), HA, CA, α-casein, and dephosphorylated α-casein (DP α-casein) with Ti^{3+} (Lane 2) or no metal incorporation (Lane 1). (**d**) IMAEP of phosvitin and β-casein with ferric ion (Lane 2) or no metal incorporation (Lane 1). Zn^{2+}-reverse staining was used to visualize phosvitin. All other proteins were visualized with Coomassie blue G-250 staining. Proteins (10 μg each) are dissolved in native sample buffer. Electrophoresis was done at ~3 W, 125 V, and 22 mA (Reproduced by permission of the Association of Biomolecular Research Facilities)

buffer, and protein sample is heated at 80 °C for 15 min (*see* **Note 7**).

2. MCF-7 cell is maintained in Modified Eagle's Medium supplemented with 10% fetal bovine serum and incubated at 37 °C with 5% CO_2 until 100% confluent. Harvested cells are lysed with lysis buffer with 10% proteinase inhibitor cocktail. Debris is removed by centrifugation at 14,000 × *g* for 20 min at 4 °C. The lysates with protein concentration 17 mg/mL (Bradford protein assay) were stored at −80 °C until used.
3. The 1-D native and SDS IMAEP is run at ~3 W with a current of ~22 mA and a voltage of ~125 V for ~1.5 h (Fig. 5) (*see* **Note 8**). Gels are stained with Coomassie blue G-250 and documented using a scanner or a digital camera and a light box.

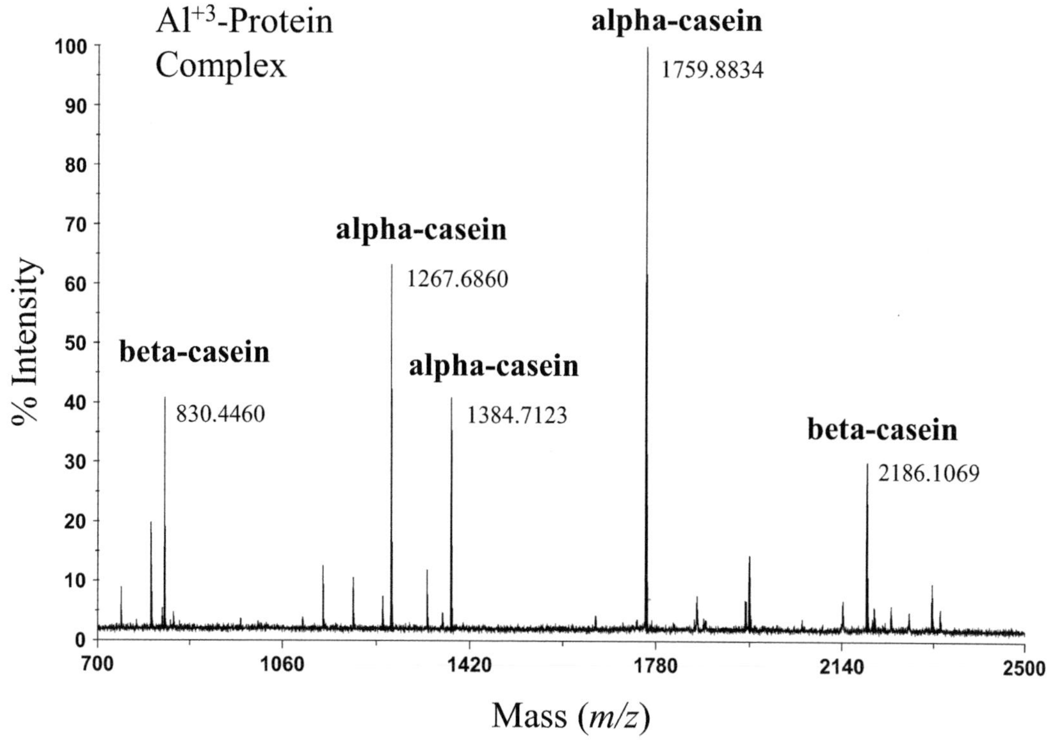

Fig. 2 Matrix-assisted laser desorption/ionization time-of-flight (MALDI-TOF) mass spectrometry of the in-gel trypsin digestion of the Al^{3+}-protein complex in Fig. 1b, Lane 2. External mass calibration was done using Bradykinin fragments 1–7 at *m/z* 757.3997, angiotensin II (human) at *m/z* 1046.5423, $P_{14}R$ (synthetic peptide) at *m/z* 1533.8582, ACTH fragment 18–39 (human) at *m/z* 2465.1989, and insulin oxidized B (bovine) at *m/z* 3494.6513. Internal mass calibration was done using trypsin auto-digestion peaks at *m/z* 842.5100 and 2211.1046 (Reproduced by permission of the Association of Biomolecular Research Facilities)

4. The metal–protein complex is excised from the IMAEP gel and the captured phosphoproteins are identified by standard proteomic techniques using in-gel trypsin digestion and mass spectrometry (Fig. 6) (*see* **Note 9**).

3.3 2-D Immobilized Metal Affinity Electrophoresis

1. A protein mixture (10 g each protein) of amyloglucosidase from *Aspergillus niger* (*pI* 3.6, MW 100 kDa), β-casein from bovine milk (*pI* 4.6, MW 24 kDa), human albumin (*pI* 5.2, MW 66 kDa), carbonic anhydrase from bovine erythrocytes (*pI* 6.6, MW 30 kDa), and lysozyme from chicken egg white (*pI* 9.3, MW 14 kDa) is dissolved in 120 μL rehydration buffer. This solution is used to rehydrate a 7 cm IPG pH 3–10 IPG dry gel strip overnight.
2. The IPG gel strip is run with the PROTEAN IEF cell for a total of 50,000 Vh.
3. After the IEF run, the IPG gel is equilibrated with the equilibrium solution I for 20 min, followed by equilibrating with

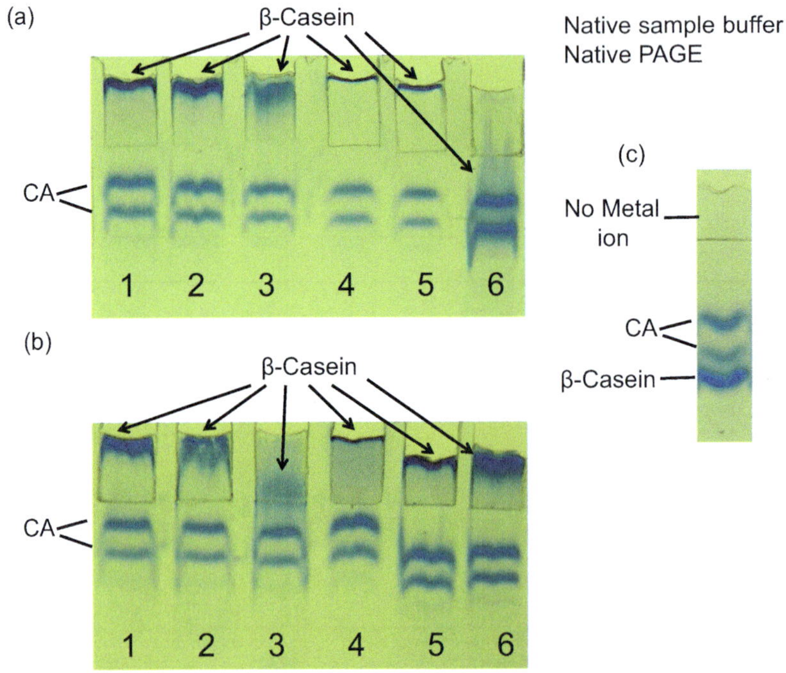

Fig. 3 (**a**) Native sample buffer and native IMAEP gel of carbonic anhydrase (CA) and β-casein with 1.5 μL of 1 M Fe^{3+} (H_2O, Lane 1), 1.0 μL of 1 M Fe^{3+} (H_2O, Lane 2), 0.5 μL of 1 M Fe^{3+} (H_2O, Lane 3), 1.5 μL of 1 M Mn^{2+} (Lane 4), 1.0 μL of 1 M Mn^{2+} (Lane 5), 0.5 μL of 1 M Mn^{2+} (Lane 6) incorporation. (**b**) Native sample buffer and native IMAEP gel of carbonic anhydrase (CA) and β-casein with 1.5 μL of 1 M Al^{3+} (Lane 1), 1.0 μL of 1 M Al^{3+} (Lane 2), 0.5 μL of 1 M Al^{3+} (Lane 3), 1 μL of 0.2 M Fe^{3+} (MeTHF, Lane 4), 0.5 μL of 0.2 M Fe^{3+} (MeTHF, Lane 5), and 0.25 μL of 0.2 M Fe^{3+} (MeTHF, Lane 6) incorporation. All proteins were visualized with Coomassie blue G-250 staining. 10 μg each protein is used. Electrophoresis was done at ~3 W, 125 V, and 22 mA

equilibrium solution II for another 20 min. The IPG gel is then overlaid on top of a second dimension 7.5–15% gradient IMAEP gel.

4. The second dimension 7.5–15% gradient IMAEP gel is generated according to the following procedure: native 7.5% and 15% acrylamide gel solutions (3.9 mL) are poured into light and heavy solution reservoir chambers of a gradient maker and are used to fill an empty gel cassette. A 2-D sample comb is inserted on top of the gel (*see* **Note 10**).
5. Following gel formation and removal of the 2-D sample comb, one larger well of 70 × 8 × 1 mm for the IMAEP gel strip and IPG strip and one smaller well of 2 × 8 × 1 mm for the protein marker are created. 525 μL native 7.5% acrylamide gel solution with APS, TEMED, and 7 μL

   1 M Fe^{3+} solution is added to the larger well and allowed to polymerize.

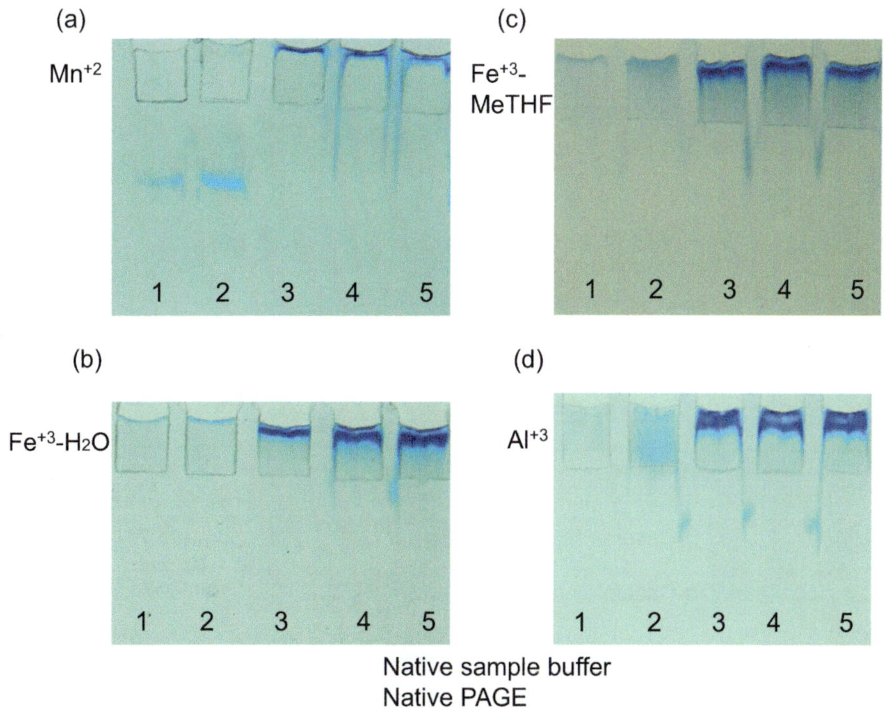

Fig. 4 (**a**) Native sample buffer and native IMAEP gel of 5 μg β-casein with 0.25 μL of 1 M Mn^{2+} (Lane 1), 20 μg β-casein with 0.25 μL of 1 M Mn^{2+} (Lane 2), 50 μg β-casein with 1 μL of 1 M Mn^{2+} (Lane 3), 100 μg β-casein with 1 μL of 1 M Mn^{2+} (Lane 4), or 200 μg β-casein with 1 μL of 1 M Mn^{2+} (Lane 5) incorporation. (**b**) Native sample buffer and native IMAEP gel of 1 μg β-casein with 0.5 μL of 1 M Fe^{3+} (H_2O, Lane 1), 5 μg β-casein with 0.5 μL of 1 M Fe^{3+} (H_2O, Lane 2), 50 μg β-casein with 1 μL of 1 M Fe^{3+} (H_2O, Lane 3), 100 μg β-casein with 1 μL of 1 M Fe^{3+} (H_2O, Lane 4), or 200 μg β-casein with 1 μL of 1 M Fe^{3+} (H_2O, Lane 5) incorporation. (**c**) Native sample buffer and native IMAEP gel of 1 μg β-casein with 0.1 μL of 1 M Fe^{3+} (MeTHF, Lane 1), 5 μg β-casein with 0.1 μL of 1 M Fe^{3+} (MeTHF, Lane 2), 50 μg β-casein with 0.5 μL of 1 M Fe^{3+} (MeTHF, Lane 3), 100 μg β-casein with 0.5 μL of 1 M Fe^{3+} (MeTHF, Lane 4), or 200 μg β-casein with 0.5 μL of 1 M Fe^{3+} (MeTHF, Lane 5) incorporation. (**d**) Native sample buffer and native IMAEP gel of 1 μg β-casein with 0.5 μL of 1 M Al^{3+} (Lane 1), 5 μg β-casein with 0.5 μL of 1 M Al^{3+} (Lane 2), 50 μg β-casein with 1.5 μL of 1 M Al^{3+} (Lane 3), 100 μg β-casein with 1.5 μL of 1 M Al^{3+} (Lane 4), or 200 μg β-casein with 1.5 μL of 1 M Al^{3+} (Lane 5) incorporation. All proteins were visualized with coomassie blue G-250 staining. 10 μg each protein is used. Electrophoresis was done at ~3 W, 125 V, and 22 mA

6. The first dimension of IPG strip gel is overlaid on top of the IMEAP gel, and then the second dimension of the 2-D IMAEP is run at ~3 W with a current of ~22 mA and a voltage of ~125 V for ~1.5 h (Fig. 7). Gels are stained with Coomassie blue G-250 or Pro-Q diamond phosphoprotein gel stain and documented using a scanner or a FX fluorescent imager.

3.4 Dimension of Functionality

1. Prepare the native 7.5% acrylamide gel solution by mixing 7.5 mL of 30% stock monomer acrylamide gel solution, 7.5 mL 4× resolving gel buffer, 14.8 mL deionized water,

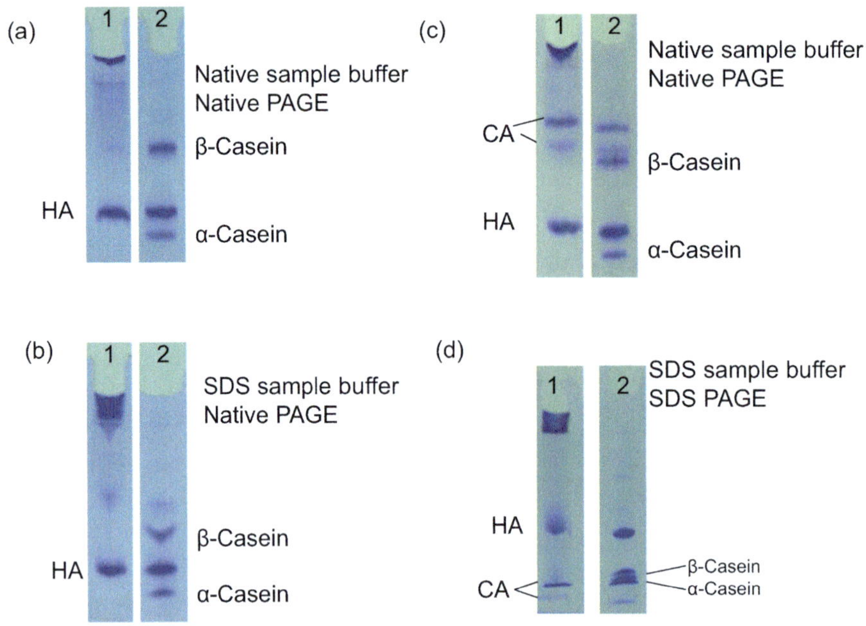

Fig. 5 (**a**) Native sample buffer and native IMAEP gel of human albumin (HA), α-casein, and β-casein with 2 μL of 1 M Fe^{3+} (Lane 1) or no metal incorporation (Lane 2). (**b**) SDS sample buffer and native IMAEP gel of human albumin (HA), α-casein, and β-casein with 2 μL of 1 M Fe^{3+} (Lane 1) or no metal incorporation (Lane 2). (**c**) Native sample buffer and native IMAEP gel of human albumin (HA), carbonic anhydrase (CA), α-casein, and β-casein with 2 μL of 1 M Fe^{3+} (Lane 1) or no metal incorporation (Lane 2). (**d**) SDS sample buffer and SDS IMAEP gel of human albumin (HA), carbonic anhydrase (CA), α-casein, and β-casein with 2 μL of 1 M Fe^{3+} (Lane 1) or no metal incorporation (Lane 2). All proteins were visualized with Coomassie blue G-250 staining. 10 μg each protein is used. Electrophoresis was done at ~3 W, 125 V, and 22 mA (Reprinted with permission from Wiley-VCH)

and 10 μL TEMED. Add 190 μL 10% APS aqueous solution to the gel solution immediately before casting the gel.

2. Sample wells (size from 18 × 8 × 1 to 25 × 8 × 1 mm) are made by inserting plastic strips at correct depths in a 100 × 100 × 1 mm empty gel cassette and 8 mL of the gel solution is poured into the cassette to cast the gel.
3. Following gel formation and removal of the plastic strips, 75–200 μL of the native 7.5% acrylamide gel solution containing APS, TEMED, and 1 μL of 1 M metal ion solution (IMAEP), 250 μg of avidin (AFAEP), or 200 μg bovine IgG (AAPE) is added to the wells and allowed to polymerize.
4. The gel is run using native gel running buffer at ~3 W with a current of ~22 mA and a voltage of ~125 V for ~1.5 h with PowerEase 500 programmable power supply. The IMAEP, AFAEP, and AAEP are run with a mixture of β-casein, biotinylated BSA, and protein G (10 μg each) (Fig. 8). Gels are stained with coomassie blue G-250 staining and documented using a scanner or a digital camera and a light box.

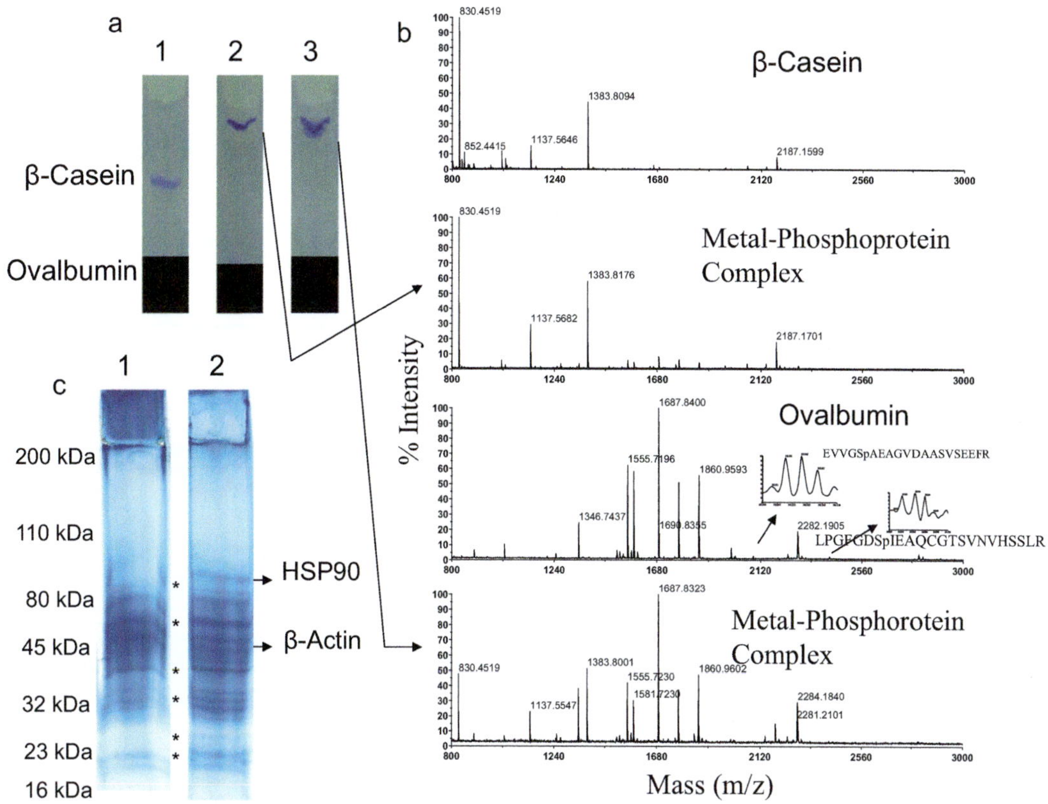

Fig. 6 (**a**) Native sample buffer and native IMAEP gel of ovalbumin and β-casein with no metal (Lane 1) or 2 μL of 1 M Fe^{3+} incoporation (Lane 2) and SDS sample buffer and native IMAEP gel of with 2 μL 1 M Fe^{3+} (Lane 3). (**b**) Matrix-assisted laser desorption/ionization time-of-flight (MALDI-TOF) mass spectrometry of the in-gel trypsin digestion of β-casein, the Fe^{3+}–phosphoprotein complex in Lane 2, ovalbumin, the Fe^{3+}–phosphoprotein complex in Lane 3. External mass calibration was done using Bradykinin fragments 1–7 at *m/z* 757.3997, angiotensin II (human) at *m/z* 1046.5423, P[14]R (synthetic peptide) at *m/z* 1533.8582, ACTH fragment 18–39 (human) at *m/z* 2465.1989, and insulin oxidized B (bovine) at *m/z* 3494.6513. Internal mass calibration was done using trypsin auto-digestion peaks at *m/z* 842.5100 and 2211.1046. (**c**) SDS sample buffer and SDS IMAEP gradient gel (7.5–15%) of total protein extracts of MCF-7 human breast carcinoma cells with 2 μL of 1 M Fe^{3+} (Lane 1) or no metal incorporation (Lane 2). All proteins were visualized with Coomassie blue G-250 staining. 100 μg of total protein is loaded onto each lane. Electrophoresis was done at ~3W, 125 V, and 22 mA for 2 h. Asterisks indicate possible phosphoproteins. HSP90 stands for heat shock protein 90 (Reprinted with permission from Wiley-VCH)

4 Notes

1. 10% SDS can be prepared by dissolving 10 g of ultrahigh pure SDS with gentle warming. Precipitation will form if the solution is at lower temperature. SDS undergoes hydrolysis at elevated temperatures causing decomposition of alkyl sulfates into fatty alcohols and sodium sulfate.

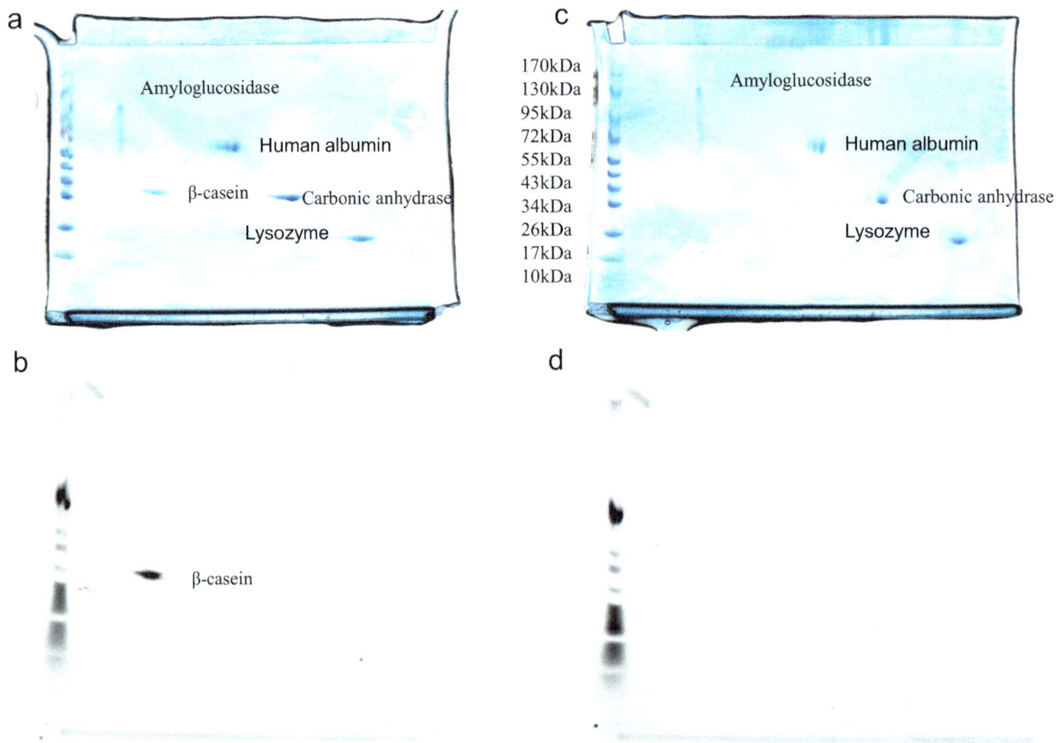

Fig. 7 Two-dimensional polyacrylamide gel electrophoresis of a protein mixture of amyloglucosidase, β-casein, human albumin, carbonic anhydrase, and lysozyme. (**a** and **b**) Normal 2D-PAGE (**c** and **d**) Fe^{3+} 2-D IMAEP. A 7.5% to 15% gradient native Tris–glycine polyacrylamide gel with SDS Tris–glycine gel running buffer was used. All proteins were visualized with Coomassie blue G-250 staining for (**a** and **c**). All proteins were visualized with Pro-Q diamond phosphoprotein staining for (**b** and **d**). Proteins (10 μg each) are dissolved in 2-D sample buffer. Electrophoresis was done at ~3 W, 125 V, and 22 mA. PageRuler™ Prestained Protein Ladder (Fermenta, Burlington, Ontario, Canada) was used to estimate the molecular masses of the proteins (Reproduced by permission of the Association of Biomolecular Research Facilities)

2. The gel solution is degassed under vacuum in a 50 mL BD Falcon conical tube sealed with a rubber septum.
3. Scotch tape is used to hold the plastic strips during the gel-forming process.
4. The amount of the metal ion should be adjusted to optimize the IMAEP performance. Polymerization time for the gel can be controlled by changing the amount of the 10% APS. Pipetting up and down the acrylamide gel solution produces a more homogenized IMAEP gel.
5. The colors of the metal ions including yellow (Fe^{3+}), burgundy (Mn^{2+}), and blue (Cu^{2+}) stay fixed in the gel through the electrophoresis confirming that the metal ions have been immobilized.

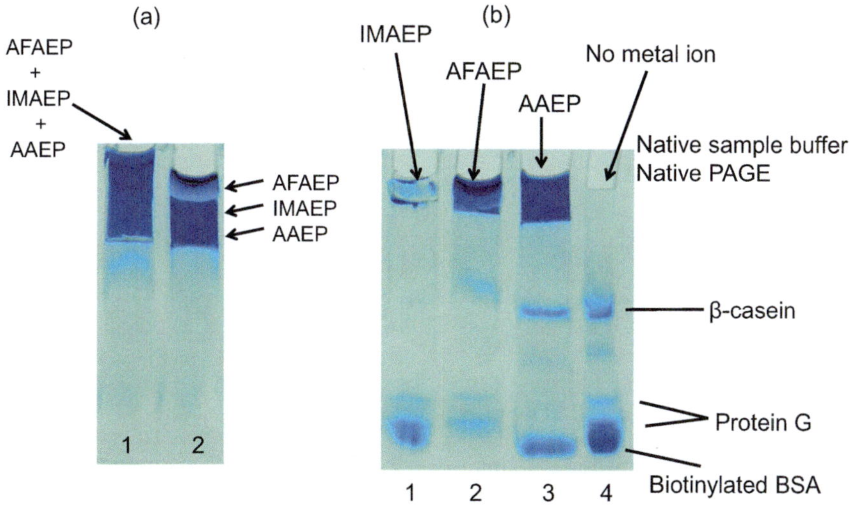

Fig. 8 IMAEP, AFAEP, and AAPE of a protein mixture of biotinylated BSA, β-casein, and protein G. (**a**) IMAEP + AFAEP + AAEP of a 7.5% native polyacrylamide gel run with mixture of 1 μL of 1 M Mn^{2+}, 250 μg of avidin, and 200 μg bovine IgG (Lane 1), 250 μg of avidin (top), 1 μL of 1 M Mn^{2+}(middle), or 200 μg bovine IgG (bottom) (Lane 2) incorporation. (**b**) IMAEP + AFAEP + AAEP of a 7.5% native polyacrylamide gel run with 1 μL of 1 M Mn^{2+}(Lane 1), 250 μg of avidin (Lane 2), and 200 μg bovine IgG (Lane 3), nothing (Lane 4) incorporation. All other proteins were visualized with Coomassie blue G-250 staining. Proteins (10 μg each) are dissolved in native sample buffer. Electrophoresis was done at ~3 W, 125 V, and 22 mA

6. Results show that among the many metal ions tested here, only Al^{3+}, Ti^{3+}, Fe^{3+}, Fe^{2+}, and Mn^{2+} are able to catch the phosphoproteins through metal–phosphate ion-pair interaction.
7. SDS treatment imparts negative charges to proteins and unfolds the protein.
8. Results demonstrate that there is no detrimental effect on the metal–phosphoprotein interaction by treating the sample with SDS because there is no unbound phosphoprotein observed in the gel, sharper protein bands are observed, and the phosphate groups are exposed by unfolding the protein for metal–phosphate interaction.
9. Phosphoproteins, like heat shack protein 90 (HSP90) and phosphorylated β-actin which had been observed previously in several different types of cancer cells [25–27], are clearly retained by the immobilized Fe^{3+} ion with MCF-7 protein extract. Protein extracts from HEK293 human embryonal kidney cells, *Saccharomyces cerevisiae*, Chinese hamster ovary cells, and mouse brain all show similar results with either Fe^{3+} or Al^{3+}.
10. We simply reused the comb that comes with precast 2-D gel from Invitrogen.

Acknowledgment

We thank the support of the Research Resources Center at the University of Illinois at Chicago.

References

1. Lee BS, Jayathilaka GDL, Huang JS, Gupta S (2008) Immobilized metal affinity electrophoresis: a novel method of capturing phosphoproteins by electrophoresis. J Biomol Tech 19:106–108
2. Lee BS, Jayathilaka GDL, Huang JS, Decresce D, Borgia JA, Zhou X, Gupta S (2008) Modification of the immobilized metal affinity electrophoresis using sodium dodecyl sulfate polyacrylamide gel electrophoresis. Electrophoresis 29:3160–3163
3. Gupta S, Jayathilaka GDL, Huang JS, Lee BS (2010) Two-dimensional immobilized metal affinity electrophoresis for capturing a phosphoprotein. J Biomol Tech 16:160–162
4. Lee BS, Lateef SS, Jayathilaka GD, Huang JS, Gupta S (2012) One-dimensional and two-dimensional immobilized metal affinity electrophoresis. Methods Mol Biol 869:275–285
5. Khoshamana K, Yousefia R, Tamaddonb AM, Abolmaalib SS, Oryanc A, Moosavi-Movahedid AA, Kurganov BI (2017) The impact of different mutations at Arg54 on structure, chaperone-like activity and oligomerization state of human αA-crystallin: The pathomechanism underlying congenital cataract-causing mutations R54L, R54P and R54C. Biochim Biophys Acta 1865:604–618
6. Anderson L, Porath J (1986) Isolation of phosphoproteins by immobilized metal (Fe^{+3}) affinity chromatography. Anal Biochem 154:250–254
7. Witze ES, Old WM, Resing KA, Ahn GN (2007) Mapping protein post-translational modifications with mass spectrometry. Nat Methods 4:798–806
8. Gaberc-Porekar V, Menart V (2001) Perspectives of immobilized-metal affinity chromatography. J Biochem Biophys Methods 49:335–360
9. Neville DC, Rozanas CR, Price EM, Gruis DB, Verkman AS, Townsend RR (1997) Evidence for phosphorylation of serine 753 in CFTR using a novel metal-ion affinity resin and matrix-assisted laser desorption mass spectrometry. Protein Sci 6:2436–2445
10. Zachariou M (2004) Immobilized metal ion affinity chromatography. In: Aguilar M (ed) HPLC of Peptides and Proteins. Humana Press Inc., New Jersey, p 89
11. Biswas S, Sarkar A, Misra R (2017) Iron affinity gel and gallium immobilized metal affinity chromatographic technique for phosphopeptide enrichment: a comparative study. Biotechnol Biotechnol Equip 31:639–646. https://doi.org/10.1080/13102818.2017.1293492
12. O'Farrell PH (1975) High resolution two-dimensional electrophoresis of proteins. J Biol Chem 250:4007–4021
13. Bjellqvist B, Ek K, Righetti PG, Gianazza E, Gorg R, Westermeier R, Postel W (1982) Isoelectric focusing in immoblized pH gradients: principle, methodology and applications. J Biochem Biophys Methods 6:317–339
14. Hames BD (1998) Electrophoresis of Protein. Oxford University Press, Oxford
15. Issaq HJ, Veenstra TD (2008) Two-dimensional polyacrylamide gel electrophoresis (2D-PAGE): advances and perspectives. BioTechniques 44:697–700
16. Lee BS, Gupta S, Krisnanchettier S, Lateef SS (2004) Catching and separating protein ligands by functional affinity electrophoresis. Anal Biochem 334:106–110
17. Lee BS, Krisnanchettier S, Lateef SS, Gupta S (2006) New methods of affinity electrophoresis. Curr Anal Chem 2:243–251
18. Lee BS, Gupta S, Krisnanchettier S, Lateef SS (2004) Countercurrent affinity electrophoresis of biotinylated proteins. Anal Biochem 330:178–180
19. Lee BS, Krisnanchettier S, Lateef SS, Gupta S (2005) Mass spectrometric detection of biotinylated peptides captured by avidin functional affinity electrophoresis. Rapid Commun Mass Spectrom 19:886–892
20. Lee BS, Gupta S, Krisnanchettier S, Lateef SS (2005) Capturing SDS treated biotinylated protein and peptide by avidin functional affinity electrophoresis with or without SDS in the gel running buffer. Anal Biochem 336:312–315
21. Lee BS, Krisnanchettier S, Lateef SS, Lateef SS, Gupta S (2013) Isotope-coded affinity tagging technique using avidin functional affinity electrophoresis: An alternative to an avidin affinity column. J Chi Chem Soc 53:745–750

22. Lee BS, Krisnanchettier S, Lateef SS, Gupta S (2007) Capturing biotinylated proteins and peptides by avidin affinity electrophoresis. Methods Mol Biol 418:51–61
23. Lee BS, Gupta S, Krisnanchettier S, Lateef SS (2004) Catching protein antigen by antibody affinity electrophoresis. Electrophoresis 25:3331–3335
24. Lee BS, Krisnanchettier S, Lateef SS, Gupta S (2005) Capturing sodium dodecyl sulfate treated protein antigen by antibody affinity electrophoresis. Electrophoresis 26:511–513
25. Rush J, Moritz A, Lee KA, Gao A, Goss VL, Speck EJ (2005) Immunoaffinity profiling of tyrosine phosphorylation in cancer cells. Nat Biotechnol 23:94–101
26. Giorgianni F, Zhao Y, Desiderio DM, Beranova-Giorgianni S (2007) Toward a global characterization of the phosphoproteome in prostate cancer cell: identification of phosphoproteins in the Lncap cell line. Electrophoresis 28:2027–2034
27. Wang S, Zheng Y, Yu Y, Xia L, Chen G, Yang Y, Wang L (2008) Phosphorylation of beta-actin by protein kinase C-delta in camptothecin analog-induced leukemic cell apoptosis. Acta Pharmacol Sin 29:135–142

Chapter 33

Isoelectric Focusing in Agarose Gel for Detection of Oligoclonal Bands in Cerebrospinal and Other Biological Fluids

Gyorgy Csako

Abstract

Isoelectric focusing (IEF) coupled with immunodetection (immunofixation or immunoblotting) has become the leading technique for the detection and study of oligoclonal bands (OCBs) in cerebrospinal fluid (CSF) and also is increasingly used in other body fluids such as the tear and serum. Limited commercial availability of precast agarose IEF gels for research and a need for customization prompted reporting a detailed general protocol for the preparation and casting of agarose IEF gel along with sample, control, and isoelectric point marker preparation and carrying out the focusing itself for CSF OCBs. However, the method is readily adaptable to the use of other body fluid specimens and, possibly, research specimens such as culture fluids as well.

Key words Agarose gel, Cerebrospinal fluid, Control specimen, Gel casting, Immunoglobulins, Isoelectrofocusing, Immunodetection, Oligoclonal bands, pI marker, Tear

1 Introduction

Detection of clonally produced immunoglobulins (Igs) in the blood and various body fluids has become a critical diagnostic test in clinical practice and also is of major research interest. One of the most important applications is in the area of neurological diseases. Intrathecal Ig production is detected by the presence of oligoclonal Igs in the CSF and oligoclonal band (OCB) testing for IgG in the CSF is now the primary laboratory test for the diagnosis and follow-up of multiple sclerosis [1–5]. As a noninvasive alternative for specimen collection, the tear has been recently proposed to replace CSF for OCB testing [6, 7].

Based on charge differences mostly due to amino acid sequence variations, clonal production of Igs traditionally has been detected by the presence of distinct bands of highly restricted mobility on zone electrophoresis using agar gel, cellulose acetate membrane,

Biji T. Kurien and R. Hal Scofield (eds.), *Electrophoretic Separation of Proteins: Methods and Protocols*, Methods in Molecular Biology, vol. 1855, https://doi.org/10.1007/978-1-4939-8793-1_33,

agarose or polyacrylamide gels [3]. Posttranslational modifications such as glycosylation may also alter the charge on an Ig molecule. Consequently, Igs originating from the same clone of cells and sharing identical primary structure can migrate in different and limited regions of a gel upon isoelectric focusing (IEF). IEF is electrophoresis in a pH gradient, which concentrates ("focuses") proteins at their isoelectric points (pIs) and allows them to be separated on the basis of charge differences potentially as small as a pI of 0.01. The resultant pattern reflects the microheterogeneity, also called limited heterogeneity, of proteins such as Igs. Although alternative methods such as high-resolution agarose gel electrophoresis [8, 9], capillary electrophoresis [10] and microLC-ESI-Q-TOF mass spectrometry [11] have been reported for OCBs, IEF in combination with highly sensitive protein stain/immunodetection remains the most widely used laboratory method for assessing intrathecal Ig synthesis. A variety of IEF methods using polyacrylamide or agarose gels with or without immunodetection for human CSF OCB testing have been developed over the past three decades. Agarose IEF methods combined with immunodetection (immunofixation or immunoblotting) for increasing detection sensitivity and providing high specificity have been "standardized" and now are proposed as the "gold standard" for detecting IgG OCBs in the human CSF in diagnostic laboratories [2]. There are semiautomated commercial versions of this kind of IEF methods, such the Hydragel 3/9 Isofocusing with immunofixation (HYDRASYS and HYDRASYS-2, Sebia Inc., Norcross, GA, USA), IgG IEF with immunoblotting (SPIFE® 3000, Helena Laboratories, Beaumont, TX) and CSF Isoelectrofocusing with immunoblotting (Interlab G26, Interlab Sri, Rome, Italy) which have been approved for clinical diagnostic testing by the US Federal Drug Administration (FDA). A polyacrylamide IEF method coupled with immunodetection (CSF Analysis kit for PhastGel™, GE Healthcare, Piscataway, NJ, and SERVA Electrophoresis GmbH, Heidelberg, Germany) also is available commercially. While detection and study of OCB patterns with IEF kits involve paired serum and CSF specimens [1–5], there is a continued need to have manual versions that are amenable to modifications for laboratory research with these and other type of fluids. The following is the description of agarose IEF for the study of OCBs in CSF. However, this method is readily adaptable and modifiable to the study of OCBs in other body fluids and in select research specimens such as culture fluids.

2 Materials

All solutions are prepared with "distilled water" (refers here to ultrapure water: deionized water further purified to obtain a resistance of 18 MOhm at 25 °C) and high-grade analytical reagents at room temperature (15–25 °C). Reagents and other materials are stored at room temperature, unless stated otherwise. Unless obtained commercially, all reagents used in these methods are prepared and stored without adding sodium azide. Follow all waste disposal regulations for discarding reagents and disposables. As a safe laboratory practice to minimize or eliminate exposure to hazardous chemicals and infectious agents, always wear clean gloves. If the gels are to be further processed for silver staining, make sure that the gloves are also powder-free when handling the gel. Wash gloves with soap and water and pat dry.

2.1 Components of Agarose IEF Gel Solution

1. Distilled water.
2. IEF Agarose (GE Healthcare [formerly Amersham Biosciences], Piscataway, NJ, USA).
3. Sorbitol.
4. Carrier ampholytes (a mixture of amphoteric substances): Pharmalyte broad range pH 3–10 (GE Healthcare). Store refrigerated at 4–8 °C (*see* **Note 1**).

2.2 Preparation of Agarose IEF Gel Solution

Prepare an ~1% agarose solution freshly for casting one 125 × 250 × 0.5 mm agarose IEF gel (pH 3–10) essentially according to a procedure reported previously (Instructions 52-1536-00 AF (GE Healthcare Bio-Sciences AB, Uppsala, Sweden, 2006).

1. Add 0.2 g dry IEF agarose powder and 2.4 g sorbitol to 18 mL of distilled water.
2. Heat the mixture to boiling in a water bath or by using an electric heater equipped with a magnetic stirrer.
3. Once the solids are dissolved (after ~10 min), cool the solution to 60–70 °C before adding 1.3 mL of the carrier ampholytes (Pharmalyte broad range pH 3–10).
4. Mix thoroughly, keep the solution heated at the same temperature, and proceed immediately to cast the gel as in Subheading 2.5 (*see* **Note 2**).

2.3 Component and Accessories for Agarose IEF Gel Casting

1. GelBondR film for agarose, 124 × 258 mm (GE Healthcare).
2. Glass plate, 125 × 260 × 3 mm—support for GelBondR in gel casting (GE Healthcare).
3. Glass plate, 125 × 240 mm, 0.5 mm—parallel spacer for gel casting (GE Healthcare).
4. Plastic syringe, 20 mL with short flexible tubing attached. Alternatively, use a plastic pipette of appropriate caliber.
5. FlexiClamps (GE Healthcare).
6. Roller (GE Healthcare).

2.4 Assembly of Mold for Agarose IEF Gel Casting

Assemble the mold for gel casting essentially according to the instructions in the Multiphor II Electrophoresis System User Manual (Code No. 18-1103-43, Amersham Biosciences [now GE Healthcare], Uppsala, Sweden, 1998).

1. Clean the glass plates with a mild detergent, and then rinse them with distilled water and dry.
2. Pipet 2–3 mL of distilled water (*see* **Note 3**) onto the 3-mm-thick glass plate and center the GelBond® film with its hydrophilic side up over the plate (*see* **Note 4**).
3. Cover the film with a sheet of blotting paper.
4. Starting longitudinally at one end of the glass plate, begin pressuring the blotting paper with the film beneath it evenly with the roller to eliminate air bubbles and to seal the film to the plate with a minimum of interfacing water. Remove the paper and remove any excess water with a tissue.
5. Place the glass plate with spacer on top of the film, leaving 10 mm free at both ends.
6. Hold the plates securely together with one FlexiClamp on each of the long sides.
7. Place the assembled mold in a heating cabinet at 70 °C for 10 min.

2.5 Casting of Agarose IEF Gel (See Note 5)

Cast the gel essentially according to the instructions in the Multiphor II Electrophoresis System User Manual (Code No. 18-1103-43, Amersham Biosciences [now GE Healthcare], Uppsala, Sweden, 1998) (*see* **Note 6**).

1. Take the prewarmed mold out of the heating cabinet.
2. Slowly fill the syringe or pipette with the molten agarose solution. Avoid getting air bubbles into the syringe or pipette.
3. Hold the mold by one hand at a 20–45 degree angle with support plate positioned lower and inject or pipet by the other hand the agarose solution at a moderate, steady rate between the glass plate with spacer and the GelBond film at the elevated end. The gel solution will fill the space by capillary action.
4. As the mold fills, gradually lower the angle to decrease the flow of the gel solution.
5. When completely filled, position the mold horizontally and allow the agarose solution to gel at room temperature for 15 min.
6. Before opening the mold, carefully remove the excess (10 mm wide) gel from both "open" ends with a scalpel.
7. Remove the two FlexiClamps and position the mold on one long side.

8. Grasp the GelBondR film and glass plate with spacer with one hand and the 3 mm glass plate with the othe and pull them gently apart (*see* **Note 7**).
9. First remove the 3 mm glass plate, and then the glass plate with spacer from the gel.
10. Place the agarose gel supported by the GelBondR film in a humidified chamber at 4 °C for at least 1 h (*see* **Note 8**).
11. The gel is now ready for use. Alternatively, store the gel in a humidified chamber at 4 °C for up to 3 days.

2.6 Instrument and Accessories for Agarose IEF Gel Electrophoresis

1. Horizontal IEF system (*see* **Note 9**): Multiphor® II Flatbed Electrophoresis System, including electrophoresis unit Multiphor® II, thermostatic circulator MultiTempr® III, and EPS 3501 Power Supply ([Pharmacia, then Amersham Biosciences] now GE Healthcare).
2. IEF electrode strips (GE Healthcare).
3. EPH/IEF sample application foil (agarose), 24 samples, 2–4 μL (GE Healthcare) (*see* **Note 10**).
4. Sample application syringe, 1–5 μL, with stoppers (e.g., Hamilton microliter syringe from Thermo Fisher Scientific).
5. Filter paper: EPH electrode wicks, paper 104 × 253 mm (GE Healthcare) or Whatman blotting paper.
6. Paper towels (*see* **Note 11**).
7. Anolyte (anode [+] solution): 0.05 M sulfuric acid. Follow safety precautions!
8. Catholyte (cathode [−] solution): 1 M sodium hydroxide. Follow safety precautions!

2.7 Specimens

1. Collect CSF and serum specimens with standard medical procedures. Store at 2–8 °C and use within 48 h freeze at −20 °C for up to 3 months or at −70 °C for up to 12 months (*see* **Note 12**).
2. Use neat or unadulterated CSF (*see* **Note 13**).
3. Determine the IgG concentrations in both CSF and serum (*see* **Note 14**).
4. Prepare samples for agarose IEF by diluting both CSF and serum specimens with physiologic saline to the same concentration range deemed appropriate for the detection sensitivity of the method to be used (e.g., 1–2 mg total IgG/dL might be appropriate for an immunofixation-based method) (*see* **Note 15**).

2.8 Controls

1. OCB-positive and OCB-negative specimens. Store small aliquots (10–20 μL) frozen in 80 μL sealed capillary tubes at −20 °C for up to 3 months, at −70–80 °C for up to a year,

and avoid or minimize repeat freezing–thawing (≤3 cycles!) (*see* **Note 16**).

2. Bring frozen aliquots of the controls to room temperature (preferably by briefly [~5 min] placing them in a 37 °C water bath), mix well with brief (3–5 s) vortexing after thawing, and then dilute the same way as the unknown patient samples.
3. Suggest using an OCB-positive CSF in combination with an OCB-negative serum in each gel.

2.9 pI Markers

1. Broad range pI marker: GE Healthcare Broad pI kit, pH 3–10, a lyophilized mixture of stable, salt-free, highly purified proteins (Thermo Fisher Scientific Inc.). Store refrigerated at 4–8 °C (*see* **Note 17**).
2. Reconstitute lyophilized pI marker preparation according to the manufacturer's instructions. Store frozen in small aliquots (10–20 μL) in 80 μL sealed capillary tubes at −20 °C for 3 months or, preferably, at −70 °C for up to 12 months.

3 Methods

Carry out all procedures at room temperature unless otherwise specified. For general guidance and safety issues consult instrument's operating manual.

3.1 IEF System Set Up

1. Connect the electrophoresis unit MultiphorR II to the thermostatic circulator MultiTemp® III.
2. Check the cooling platform with a leveler and, if necessary, adjust the level.
3. Thoroughly clean the cooling surface with distilled water and towel dry.
4. Set the temperature to 10 °C and begin equilibration of the electrophoresis chamber at least 15 min before starting IEF.

3.2 Gel Placement

1. Bring the agarose IEF gel to room temperature (10–20 min). Place the gel on a clean, level surface (clean filter paper or clean plastic tray). Do not lay the gel on the bench top!
2. Lightly blot the gel with one filter paper/gel blotter to remove the surface moisture. After 10–30 s gently peel off the paper from the gel surface and also blot excess water from around the perimeter of the gel.
3. Pipette 1–2 mL of distilled water onto the center of the cooling platform and spread it with a gloved finger.
4. Place the gel with the plastic film facing down in the middle of the cooling surface in the electrophoresis chamber using the

screened template on the platform to center the gel. While holding the (opposite) diagonal corners, apply the gel to the platform by bending it in a "U shape." This minimizes or avoids getting air bubbles under the gel.

5. If bubbles form, with the edge of a thin spatula, gently lift the edge of the gel, and then lay the gel back down on the platform to remove the bubble.
6. Blot excess moisture with paper towel from the base of the electrophoresis chamber (around the gel) being careful to not disturb the gel (*see* **Note 18**).

3.3 Electrode Strip Placement

1. Soak one electrode strip into 3 mL anolyte (0.05 M sulfuric acid).
2. Blot the strip for ~1 min on a filter paper to remove excess liquid.
3. Place the blotted strip to the anodic side of the gel.
4. Soak a second strip into 3 mL catholyte solution (1 M sodium hydroxide) and repeat **step 2**.
5. Place this strip 100 mm away from the anodic strip on the gel. The strips should be well aligned (parallel) and completely inside the gel on the respective anodic and cathodic gel side.
6. Make sure that the electrode strips have good, even contact with the gel.

3.4 Sample Application

1. Position the sample application foil (also called application "template" or "mask") across the gel such a way that the samples can be loaded at 20 mm from the anodic strip on the gel. Make sure that no air bubbles become trapped between the template and the gel (*see* **Note 19**).
2. With the syringe, apply 2–4 μL (volume dependent on sensitivity of the OCB detection method) of the prediluted controls and paired CSF and serum samples (side by side!), and if applicable, the pI marker over each slot, without touching the gel surface.
3. Allow samples to absorb into the gel (usually within 3–5 min).
4. Gently peel off the template when the last sample has been completely absorbed and the sample sites look dry (max. 20 min).

3.5 Running IEF

1. Place the electrode holder on the electrophoresis unit in the shallow depressions.
2. Carefully center the EPH/IEF electrodes over the strips and connect electrical leads.
3. Lower carefully, so that the electrodes rest on and make good contact with the electrode strips.

4. Place the safety cover over the unit and connect the leads into the power supply.
5. Run IEF at 10 W for 30 min and then at 20 W until focusing is completed after 1500 volt hrs. Limit voltage to 1250 V. Total run time is ~90 min at 10 °C.

3.6 Ending IEF

1. Upon completion of IEF (~90 min), turn the power supply off (*see* **Note 20**).
2. Disconnect the Multiphor II^{R} unit from the power supply.
3. Remove the safety lid from the unit.
4. Carefully remove the electrode holder.
5. Remove the gel from the electrophoresis chamber.
6. Gently pull the electrode strips from the gel and discard them.
7. Process the gel immediately as intended for direct protein staining (first placing it in a fixative) or for immunodetection by methods described elsewhere (*see* **Note 21**).
8. Remove the electrodes from the electrode holder.
9. Rinse the electrodes with distilled water to remove the strong acidic and basic solutions.
10. Air dry or carefully dry with paper towel the electrodes for storage.
11. Check the platinum wire for possible damage.

3.7 Interpretation of IEF (See Notes 17 and 22)

1. Visually assess the IEF patterns.
2. First, determine the reliabilityof the run by confirming expected performance of the control samples (e.g., "OCB-positive" CSF and "OCB-negative" serum).
3. If applicable, check the performance of the pI markers (Fig. 1).
4. In case of CSF OCB testing, evaluate the presence, strength, location and number of distinct bands in the side-by-side paired unknown CSF and serum samples (Fig. 2).
5. Report results according to one of five IgG OCB patterns in CSF for the evaluation of multiple sclerosis [2]. This classification is from the 2005 consensus report of all participants in an international panel of multiple sclerosis and CSF diagnostic method experts. The five classic patterns are as follows:

   Type 1, no bands in CSF and the serum sample (normal CSF);

   Type 2, IgG OCB's in CSF, not in the serum sample (intrathecal IgG synthesis found in multiple sclerosis);

   Type 3, OCB's in CSF (like Type 2) and additional identical OCB's in CSF and the serum sample (like Type 4) (still indicative of intrathecal IgG synthesis; occurs in multiple

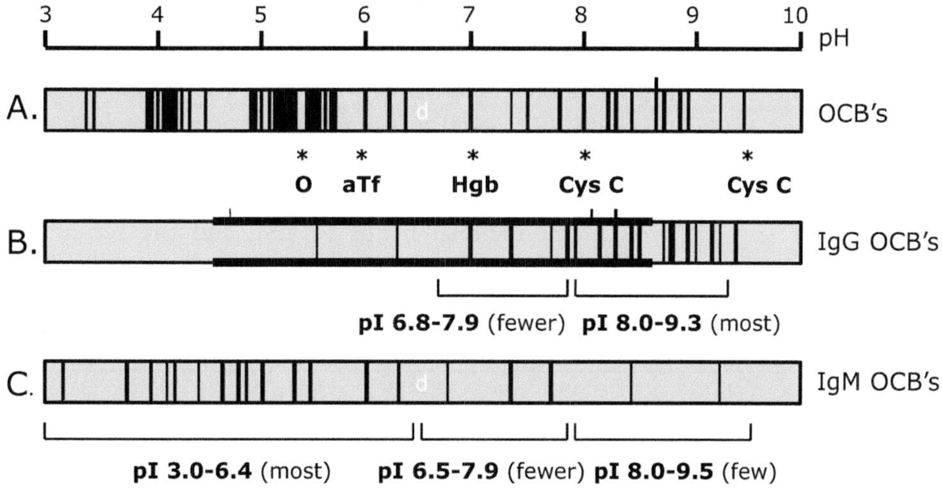

Fig. 1 Scheme for the location and relative frequency of common oligoclonal bands (OCBs) and for the location of common non-Ig OCBs in IEF patterns of CSF. Based largely on data reported in [3, 9, 12–14]. (**a**) Direct protein staining: Asterisks show location of O, origo (application point) and common non-Ig bands such as aTf, asialotransferrin; Hgb, hemoglobin (occurring in hemolyzed specimens) and Cys C, cystatin C ("γ-trace protein"). Note that two abundant CSF proteins (albumin, pI $\sim$4.8 and prealbumin, pI $\sim$4.7) are located near the application point but are not identified in the pattern. (**b**) Immunodetection for IgG: Heavy horizontal lines indicate location of polyclonal IgG (pI range 4.7–8.6). (**c**) Immunodetection for IgM

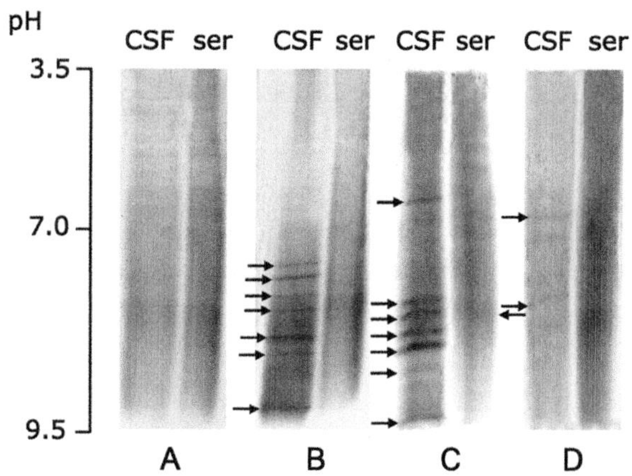

Fig. 2 Paired CSF and serum (ser) samples used as controls (**a–d**) in the survey, as analyzed by one center whose reported band numbers were the nearest to the median values in each control. IEF was in agarose gel and staining of nitrocellulose paper after affinity-mediated protein blotting was IgG-specific. *Arrows* indicate OCBs. (Franciotta D, Lolli F. Interlaboratory reproducibility of isoelectric focusing in oligoclonal band detection. Clinical Chemistry Aug 2007;53(8):1557–1558. DOI: 10.1373/clinchem.2007.089052 (Reproduced with permission from the American Association for Clinical Chemistry)

sclerosis and brain inflammation in systemic disease, e.g., sarcoidosis);

Type 4, identical OCB's in CSF and the serum sample (systemic not intrathecal immune reaction such as seen in Guillain-Barrè syndrome, when OCB's passively transferred in the CSF through a leaky blood–CSF barrier); and

Type 5, monoclonal bands in CSF and the serum sample (found in myeloma or monoclonal gammopathy of undetermined significance, MGUS).

4 Notes

1. For higher resolution of IgG OCB's with IEF, the pH gradient can be expanded in the basic region with Pharmalyte broad range pH 8–10.5 (GE Healthcare). In turn, using ampholytes in the range of pH 3–8 improves the resolution for IgM OCB's. (Fig. 1).
2. Agarose solution remains liquid until the temperature drops to about 40 °C.
3. As alternatives, 0.1% solution of a non-ionic detergent such as Triton X-100® and ethanol/water (95/5 vol.) has been suggested with no clear advantage over water.
4. The GelBond® Film, a transparent, flexible polyester sheet, has a specially treated hydrophilic side that firmly adheres to agarose gels as they form. This property can be tested by observing the behavior of a drop of water on both sides of the film: the water will "spread" on the hydrophobic surface, whereas it will "bead" on the hydrophobic side. Since GelBond® is nonporous, it cannot be used with electroblotting or capillary blotting applications (Document #18652-1007-03, Lonza Rockland, Inc., Rockland, ME, 2007).
5. Precast Agarose IEF gel, supported on GelBond®, Lonza Rockland) along with necessary accessories is available commercially and adaptable to OCB testing: (Lonza Rockland) IsoGel IEF plate, pH 3–10, 125 × 100 mm (Thermo Fisher Scientific Inc. or Westburg BV, Leusden, The Netherlands). Using the Multiphor® II system, the IEF running conditions reported for protein separation with this gel (2.5 μL sample/site applied, gel prefocused at 1 W for 10 min, then focused at 2000 V [max.], 25 mA [max.], and 25 W [max.] for 60 min at 10 °C) would be applicable to CSF OCB testing as well.
6. If commercial precast agarose IEF gel is not preferred, gel casting mold is not available, and/or GelBond® sizes other than those fitting the commercially available mold are to be used, horizontal agarose IEF gels can be prepared using an

open casting procedure outlined previously (Lonza GelBond® Film protocol sheet for Casting Agarose Gels on GelBond Film, Document #18652-1007-03, Lonza Rockland). However, this technique is optimal only for thicker (0.5–1.0 mm) gels and carries the risk of possible nonuniformity and evaporative loss. In order to assure a uniform thickness, gels must be cast on a level surface. Briefly, first adhere the GelBond^R film (cut 1–2 mm smaller than the glass plate) to a clean glass plate with its hydrophilic side up as described under Materials (*see* Subheading 2.4, **items 1–4**). As with mold casting, prewarm the glass plate with GelBond® film attached prior to gel casting. Quickly spread the molten agarose solution containing the ampholytes over the prewarmed film on top of the glass plate. Make sure that the agarose solution will not overflow the edge of the GelBond® film and run under the film. Cool the gel at room temperature for 15 min and, once it has set, refrigerate it in a humidified chamber for at least 1 h. The gel is now ready for use. Alternatively, the gel can be stored in a humidified chamber at 4 °C for up to 3 days (*see* **Note 8**).

7. Agarose gel poorly adheres to glass.
8. Either commercially available humidity chambers (e.g., humidity chamber 125 × 260 mm, GE Healthcare) or a tightly closed plastic container with wet towel at the bottom and a leveled glass plate placed over short glass rods (or plastic pipettes) can be used.
9. IEF commonly is run in horizontal gels for the following reasons:
   (a) Since IEF separates only according to the charge, soft gels with large pore sizes are used to allow for easy migration. These soft gels may slide down between vertical glass plates.
   (b) IEF requires exact temperature control. A horizontal ceramics cooling plate connected to a thermostatic circulator, or a peltier cooling plate is optimal for this purpose.
   (c) Horizontal equipments meet best the safety precautions for the high voltages needed to obtain sharply focused IEF bands.
   (d) A horizontal gel with an open surface allows sample application to a defined location within the pH gradient to minimize or avoid aggregation and precipitation of some proteins.
10. As an alternative to using sample application foil, IEF sample application pieces (GE Healthcare) can be used. However, these pieces require much larger sample volumes. They are

soaked with 15–20 μL of sample solution before placing them on the gel surface for 30–45 min.

11. Use only white paper towels because colored paper towels may exude a dye into the gel and electrode strips.
12. CSF and serum specimens were claimed to be stable for 7 days at "ambient" temperature, 14 days "refrigerated," 3.5 months frozen at −20 °C, and for 3 freeze–thaw cycles [8].
13. Because of the high sensitivity of currently available methods for OCB testing, concentrating CSF is now considered obsolete [2]. Further, concentrating CSF leads to artifacts, including protein losses that differ from sample to sample and may reach as much as 50% for IgG [3].
14. Standardized commercial methods currently are available for serum and CSF IgG measurements (e.g., immunonephelometry by Siemens BN-II).
15. Reliable detection of CSF OCBs requires standardization so that identical amounts of a patient's CSF IgG and serum IgG are applied side by side on the gel. In most cases, the CSF IgG concentration requires little adjustment for IgG OCB testing, and only the serum requires dilution. Rarely, the CSF IgG concentration may be too low, and in such cases, despite its recognized shortcomings, concentrating CSF specimens is necessary to attain the detection sensitivity of the particular method used for OCBs. Illustrating the rarity of these cases, in a large study, 98.2% (388/395) of the CSF specimens received for OCB testing had an IgG concentration of >1 mg/dL, and the lowest IgG concentration was 0.7 mg/dL [8].
16. Controls help to determine the reliability of any given IEF run. For proper interpretation of the results, recognized control samples (e.g., an OCB-positive CSF and an OCB-negative serum) should be included in each run. While commercial kit methods for diagnostic IEF testing with immunodetection for CSF IgG OCBs include appropriate control specimens, such specimens are not readily obtainable outside of kit users. If commercially not available, collect and save aliquots of single or pooled CSF and serum specimens that have been previously analyzed and confirmed for the presence or absence of OCB for use as positive and negative OCB controls.
17. Although routine use of pI markers is not required for diagnostic CSF OCB testing, these markers may be useful (Figs. 1 and 2) to:
    (a) Confirm the analytical performance of the IEF method.
    (b) Study the properties of the OCBs for research purposes.
    (c) Recognize non-Ig bands in the absence of immunodetection.

18. This step is important to avoid fanning.
19. Unlike conventional electrophoresis, the point of application of the sample on the gel is not as critical in IEF. This is because whatever the starting point may have been, the proteins in the sample will concentrate in the pH regions equal to their isoelectric point. Nevertheless, the site of sample application still can be optimized for "best"performance (*see* **Note 9d**).
20. Because IEF is an end-point method, the stop time is less important than in conventional electrophoresis. Proteins actually become more focused as time progresses beyond the "stop time."
21. Immunodetection methods for OCBs are advantages over traditional direct protein staining methods both in terms of higher sensitivity and greater specificity [2, 3, 13]. Since the overwhelming majority (>99%) of all OCBs are known to be of the IgG type, diagnostic assays most commonly test for IgG OCBs. However, IEF tests followed by immunodetection using monospecific antibodies to IgG subclasses, IgM, IgA, kappa and lambda chains also have been developed for CSF OCBs [3, 13]. Furthermore, the utility of testing for specific antibodies among OCBs by using IEF coupled with "reverse" immunodetection (i.e., with use of antigens) has also been well established. Examples of this approach include detection of antiviral [15, 16] and paraneoplastic antibodies [17, 18] in serum and/or CSF.
22. Additional comments on evaluation of IgG OCB patterns in CSF
    (a) CSF analysis methods vary substantially and experts in multiple sclerosis and CSF diagnostic techniques repeatedly addressed the need for standardization. Obviously, the lack of reproducibility could lead to false-negative/false-positive results in critical CSF samples, i.e., samples with few and weak bands. In a comprehensive survey [19], 20 experienced Italian centers were asked to blindly analyze freshly collected paired CSF and serum samples from four patients (controls A–D, Fig. 1 - Fig. 2 in current chapter) with clinically isolated syndrome, a disorder that converts into multiple sclerosis in ~50% of cases. All participants used IEF with immunoblotting for IgG. IEF was performed with agarose/polyacrylamide gels from the following suppliers: Helena ($n = 9$), homemade ($n = 4$), Pharmacia ($n = 3$), Amersham ($n = 2$), Sebia ($n = 1$), and Cambrex ($n = 1$). The main finding of the survey was the unacceptably large interlaboratory variation not only in OCB numbering but also in qualitative reporting of the OCB pattern and in differentiating OCB-positive from

OCB-negative samples. The minimum and maximum (median) band numbers in control samples for the 20 centers were as follows: A [0–6, (0), serum; 0–6, (0) CSF], B [0–15, [3]; 3–26, [13]], C [0–8 (0); 5–20, [9]], D [0–5, (0); 0–7, [2]] (kappa statistic for inter-observer agreement was not significant for each control). Finding of great variation in OCB identification in control D, with 65% of survey responses indicating OCB-positive and 35% OCB-negative results, was particularly disturbing. These observations indicate that, despite major efforts, CSF analysis for OCB detection remains poorly standardized even in experienced laboratories.

(b) Recently, a "refinement" based on differentiating between two "distinctive" IgG OCB patterns in the CSF was reported to improve the diagnostic utility for multiple sclerosis [20]. The first, termed "delta" pattern, was characterized by prominent, discrete, well-defined, usually multiple (>4) bands, whereas the second, termed "theta" pattern, was more subtle, with fewer (usually <4), faint and weaker bands. According to the authors, only the delta pattern was sensitive and specific for multiple sclerosis, whereas the theta pattern was never found in these patients. Hence, the diagnostic significance of IgG OCBs in CSF for multiple sclerosis and clinically isolated syndromes might be improved by restricting interpretation to the delta banding pattern alone.

(c) Because of frequently low contrast and artifacts, the interpretation of complex electrophoretic profiles can be both difficult and subjective. In order to ease analysis and reduce subjectivity, recently a semiautomated image analysis of gel electrophoretic patterns for OCB detection in CSF has been proposed [21]. In this method, color images are sequentially converted to grayscale, bands are realigned, artifacts are removed, and then 2D images are converted to a signal, before detecting, thresholding and editing peaks to optimize profiles. Comparing the results to those obtained by an expert biologist, the variability seemed to be similar to inter-expert variability reported in the literature.

References

1. Polman CH, Reingold SC, Edan G et al (2005) Diagnostic criteria for multiple sclerosis: 2005 revisions to the "McDonald Criteria". Ann Neurol 58:840–846
2. Freedman MS, Thompson EJ, Deisenhammer F et al (2005) Recommended standard of cerebrospinal fluid analysis in the diagnosis of multiple sclerosis: a consensus statement. Arch Neurol 62:865–870
3. Link H, Huang Y-M (2006) Oligoclonal bands in multiple sclerosis cerebrospinal fluid: An update on methodology and clinical usefulness. J Neuroimmunol 180:17–28

4. Regeniter A, Kuhle J, Mehling M et al (2009) A modern approach to CSF analysis: pathophysiology, clinical application, proof of concept and laboratory reporting. Clin Neurol Neurosurg 111:313–318
5. Schäffler N, Köpke S, Winkler L et al (2011) Accuracy of diagnostic tests in multiple sclerosis—a systematic review. Acta Neurol Scand 124:151–164
6. Devos D, Forzy G, de Seze J et al (2001) Silver stained isoelectrophoresis of tears and cerebrospinal fluid in multiple sclerosis. J Neurol 248:672–675
7. Calais G, Forzy G, Crinquette C et al (2010) Tear analysis in clinically isolated syndrome as new multiple sclerosis criterion. Mult Scler 16:87–92
8. Fortini AS, Sanders EL, Weinshenker BG et al (2003) Cerebrospinal fluid oligoclonal bands in the diagnosis of multiple sclerosis. Isoelectric focusing with IgG immunoblotting compared with high-resolution agarose gel electrophoresis and cerebrospinal fluid IgG index. Am J Clin Pathol 120:672–675
9. Nakashima I, Fujihara K, Sato S et al (2005) Oligoclonal IgG bands in Japanese patients with multiple sclerosis. A comparative study between isoelectric focusing with IgG immunofixation and high-resolution agarose gel electrophoresis. J Neuroimmunol 159:133–136
10. Sanders E, Katzmann JA, Clark R et al (1999) Development of capillary electrophore.is as an alternative to high resolution agarose electrophoresis for the diagnosis of multiple sclerosis. Clin Chem Lab Med 37:37–45
11. Barnidge DR, Kohlhagen MC, Zheng S et al (2015) Monitoring oligoclonal immunoglobulins in cerebral spinal fluid using microLC-ESI-Q-TOF mass spectrometry. J Neuroimmunol 288:123–126
12. Laurenzi MA, Link H (1979) Characterisation of the mobility on isoelectric focusing of individual proteins in CSF and serum by immunofixation. J Neurol Neurosurg Psychiatry 42:368–372
13. Mehta PD (1991) Diagnostic usefulness of cerebrospinal fluid in multiple sclerosis. Crit Rev Clin Lab Sci 28:233–251
14. Sharief MK, Thompson EJ (1992) Distribution of cerebrospinal fluid oligoclonal IgM bands in neurological diseases: a comparison between agarose electrophoresis and isoelectric focusing. J Neurol Sci 109:83–87
15. Mehta PD, Patrick BA, Thormar H et al (1982) Oligoclonal IgG bands with and without measles antibody activity in sera of patients with subacute sclerosing panencephalitis (SSPE). J Immunol 129:1983–1985
16. Virtanen JO, Wohler J, Fenton K et al (2014) Oligoclonal bands in multiple sclerosis reactive against two herpesviruses and association with magnetic resonance imaging findings. Mult Scler 20:27–34
17. Storstein A, Monstad SE, Honnorat J et al (2004) Paraneoplastic antibodies detected by isoelectric focusing of cerebrospinal fluid and serum. Neuroimmunol 155:150–154
18. Stich O, Rauer S (2007) Antigen-specific oligoclonal bands in cerebrospinal fluid and serum from patients with anti-amphiphysin- and anti-CV2/CRMP5 associated paraneoplastic neurological syndromes. Eur J Neurol 14:650–653
19. Franciotta D, Lolli F (2007) Interlaboratory reproducibility of isoelectric focusing in oligoclonal band detection. [Letter]. Clin Chem 53:1557–1558
20. Lin MW, Suan D, Lenton K et al (2012) Differentiating patterns of oligoclonal banding in the cerebrospinal fluid improves diagnostic utility for multiple sclerosis. Pathology 44:248–250
21. Boudet S, Peyrodie L, Wang Z et al (2016) Semi-automated image analysis of gel electrophoresis of cerebrospinal fluid for oligoclonal band detection. Conf Proc IEEE Eng Med Biol Soc 2016:744–747

Chapter 34

A Combined Free-Flow Electrophoresis and DIGE Approach to Compare Proteins in Complex Biological Samples

Kim Y. C. Fung, Chris Cursaro, Tanya Lewanowitsch, Leah Cosgrove, and Peter Hoffmann

Abstract

Free-flow electrophoresis has been applied in numerous studies as a protein separation technique due to its multiple advantages such as fast and efficient sample recovery, high resolving power, high reproducibility and wide applicability to protein classes. As a stand-alone platform, however, its utility in comparative proteomic analysis is limited as protein samples must be run sequentially rather than simultaneously which introduces inherent variability when attempting to perform quantitative analysis. Here we describe an approach combining fluorescent CyDye technology (DIGE) with free-flow electrophoresis to simultaneously separate and identify differentially expressed proteins in a model cell system.

Key words Free-flow electrophoresis, FFE, DIGE, Proteomics

1 Introduction

Separation of proteins in biological matrices is a major challenge in proteomics and has resulted in the development of numerous technologies primarily involving either electrophoresis or chromatography as the mode of separation. Recent advances in free-flow electrophoresis (FFE) have drawn attention to its potential in the broader proteomics community. FFE is a liquid-based separation method amenable to a wide variety of protein classes, and as a result it is used successfully as a preparative or analytical pre-fractionation tool in proteomics [1–8]. The continuous nature of the sample loading allows for relatively high volumes of sample to be analyzed which is important for detecting proteins present at low abundance. Separation occurs across an established, well-defined pH gradient and can be performed under native or denaturing conditions. Because the process occurs in aqueous media, sample recovery is highly efficient and potentially quantitative. The versatility of FFE is also highlighted by its compatibility with many commonly used

Biji T. Kurien and R. Hal Scofield (eds.), *Electrophoretic Separation of Proteins: Methods and Protocols*, Methods in Molecular Biology, vol. 1855, https://doi.org/10.1007/978-1-4939-8793-1_34,

protein analytical methods such as 1D or 2D gel electrophoresis, chromatography, enzymatic assays, immunoassays, and mass spectrometry.

In its standard format, FFE is not optimal for comparative proteomic studies as protein samples must be run sequentially rather than simultaneously. If an additional separation method is used (e.g., HPLC or 1-DE), the inherent variability associated with comparing multiple images or traces must be taken into consideration when performing comparative or (semi-) quantitative analysis. Here we describe an approach where fluorescent labeling of proteins is performed prior to separation by FFE and 1-DE to allow direct comparison of two protein samples and also facilitates visualization of intact proteins and their biologically relevant modified forms (e.g., isoforms, variants, post translational modifications). This methodology can be readily applied to measure differential protein expression between any treatment groups.

2 Materials

All solutions are prepared using ultrapure water with a resistance of ≥18 MΩ/cm at 25 °C and analytical grade reagents. Use only the highest quality urea and thiourea that have low conductivity. Wear appropriate personal protection when handling potentially harmful reagents and always observe local waste disposal regulations when disposing off waste.

2.1 Cell Lysates and Cydye Labelling Reagents

1. Obtain cell lysate from HT29 cells and HT29 cells treated with butyrate for 48 h as described in [9].
2. Rehydration buffer for solubilization of protein pellet (*see* **Note 1**): 7 M urea, 2 M thiourea, 2% CHAPS, 30 mM Tris. Weigh 10.5 g urea, 3.8 g thiourea, 0.5 g CHAPS and 0.09 g Tris and transfer to 50 mL polypropylene tube (e.g., Falcon tube). Add 25 mL of water and mix gently to dissolve.
3. Protein assay kit (e.g., Bradford, bicinchoninic acid, or EZQ).
4. CyDye DIGE Fluor minimal dyes: Cy2, Cy3 and Cy5 (GE Healthcare, Piscataway, NJ, USA). Prepare stock solution by addition of equal volumes of freshly opened, anhydrous *N*, *N*-dimethylformamide (DMF, >99%) (*see* **Note 2**). For example, for each 25 nmol CyDye vial add 25 μL DMF (final concentration 1 nmol/μL). Aliquot and store at −80 °C wrapped in aluminum foil. These dyes are light sensitive and exposure to light should be minimized.
5. Quenching solution: 10 mM lysine. Weigh 0.015 g L-lysine, transfer to 15 mL polypropylene tube, and add 10 mL ultrapure water. Aliquots can be stored at −80 °C.

2.2 Free Flow Electrophoresis Components

2.2.1 Apparatus

1. FFE apparatus [e.g., Prometheus™ FFE apparatus (BD Biosciences, Bedford, MA, USA)] (*see* Fig. 1 for a schematic diagram of a FFE apparatus).
2. 96 well microtiter plates for initial performance test (laminar flow) and pI test.

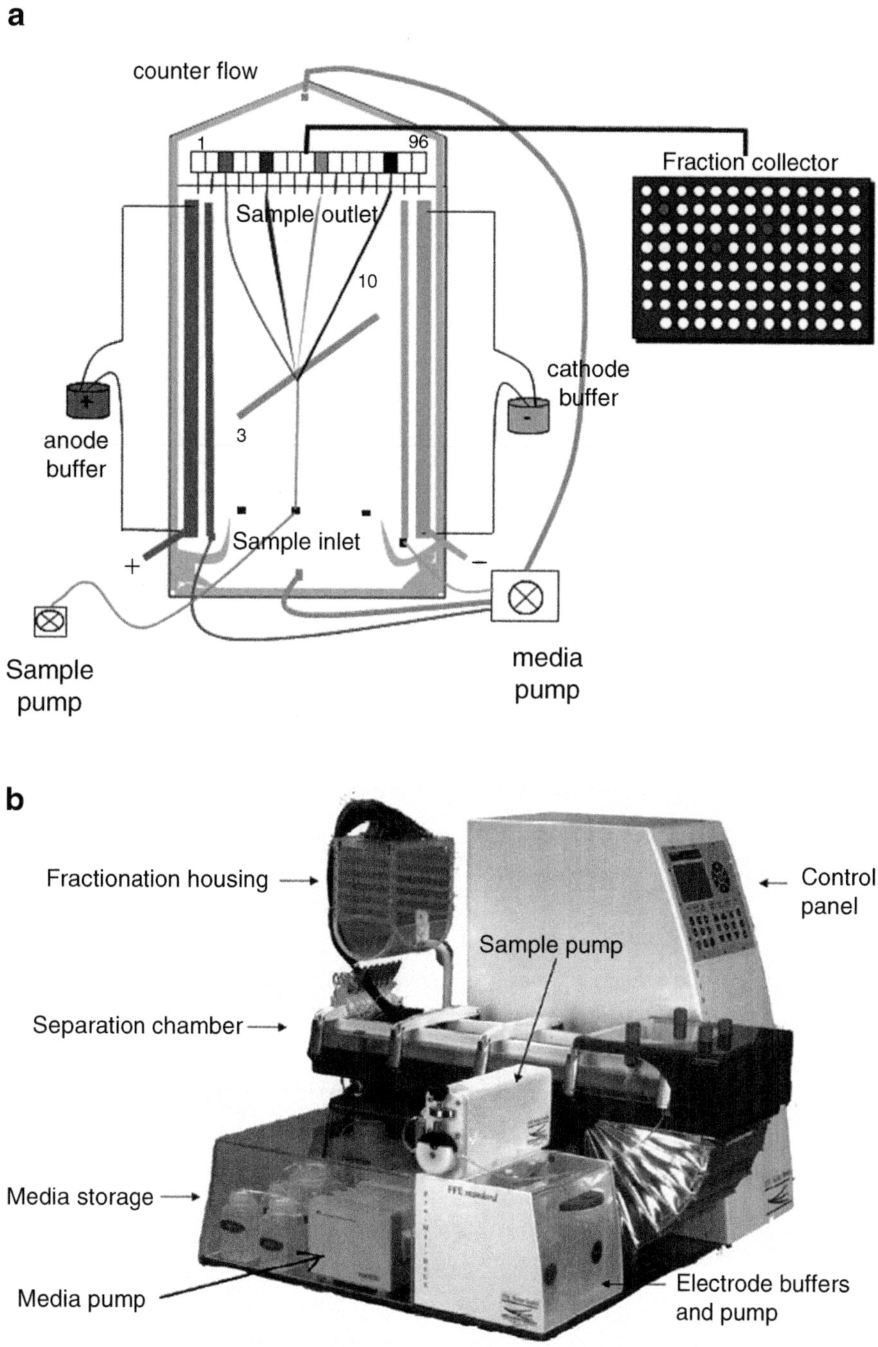

Fig. 1 (**a**) Schematic diagram of FFE apparatus. (**b**) BD Biosciences ProMetHeus™FFE apparatus

3. 96 × 2.2 mL polypropylene deep-well plates for collection of FFE fractions.
4. pH meter with micro pH probe (e.g., Thermo™ Ross micro pH probe; Thermo Fisher Scientific, Waltham, MA, USA).
5. Conductivity meter.
6. Micro plate reader (capable of absorbance readings at 280 nm, 410 nm, 510 nm).

2.2.2 Reagents

Weigh the viscous liquids such as HPMC, glycerol, and ProLytes as this is more accurate than volumetric measurement. For each medium, mix thoroughly in a warm (~37 °C) water bath, or sonicating water bath if available, until all reagents are dissolved. For best results, the separation media should be freshly made and used within 2 or 3 days.

1. Hydroxyl propyl methyl cellulose stock solution (HPMC; Sigma-Aldrich, St Louis, MO, USA): 0.8% (w/w) in ultrapure water. Prepare by slowly adding 8 g HPMC to 1 L of ultrapure water in a glass beaker with constant agitation, e.g., using a magnetic stirrer. This may take several hours, so it is best left overnight. Decant into a 1 L Schott bottle and store in the dark at 4 °C.
2. 2-(4-Sulfophenylazol)1,8-dihydroxy-3,6-napthalene disulfonic acid (SPADNS; BD Biosciences): 1% (v/v) in ultrapure water (or the appropriate Separation Medium as required).
3. Carrier ampholytes to establish pH gradient: ProLytes 1, 2 and 3 (BD Biosciences).
4. Anode stabilization medium: 12.5% (w/w) glycerol, 0.05% (w/w) HPMC, 100 mM H_2SO_4, 6 M urea, 1.7 M thiourea (*see* **Note 3**).
5. Separation Medium 1: 12.5% (w/w) glycerol, 0.05% (w/w) HPMC, 14.5% (w/w) ProLyte 1, 6 M urea, 1.7 M thiourea (*see* **Note 4**).
6. Separation Medium 2: 12.5% (w/w) glycerol, 0.05% (w/w) HPMC, 23.2% (w/w) ProLyte 2, 6 M urea, 1.7 M thiourea (*see* **Note 5**).
7. Separation Medium 3: 12.5% (w/w) glycerol, 0.05% (w/w) HPMC, 19.4% (w/w) ProLyte 3, 6 M urea, 1.7 M thiourea (*see* **Note 6**).
8. Cathode stabilization medium: 12.5% (w/w) glycerol, 0.05% (w/w) HPMC, 14.5% (w/w) 100 mM NaOH, 6 M urea, 1.7 M thiourea (*see* **Note** 7).
9. Counterflow Medium: 12.5% (w/w) glycerol, 0.05% (w/w) HPMC, 6 M urea, 1.7 M thiourea (*see* **Note 8**).
10. Anode electrolyte: 100 mM H_2SO_4 (at least 300 mL).

11. Cathode electrolyte: 100 mM NaOH (at least 300 mL).
12. pI markers: diluted 1/20 with Separation Medium 2.
13. (Optional) Standard protein mixture in water: 0.25 mg/mL stock solutions of myoglobin, cytochrome c and bovine serum albumin (BSA) with bromophenol blue (0.01%) (Bromophenol blue will bind to BSA and stain BSA blue). Combine 100 μL of each and dilute to a final volume of 1 mL with Separation Medium 2.

2.3 SDS Polyacrylamide Gel Components

2.3.1 Apparatus

1. Precast polyacrylamide gels including appropriate buffer system, sample loading buffer, and electrophoresis apparatus (*see* **Note 9**).
2. Gel imaging system (e.g., Ettan DIGE Imager or Typhoon imager capable of producing and detecting excitation/emission wavelengths for CyDyes (Cy2, 488/520 nm; Cy3, 532/580 nm; Cy 5, 633/670 nm).
3. Image analysis software which enables densitometry analysis (e.g., ImageQuant TL; GE Healthcare).

2.3.2 Reagents

1. Centrifugal molecular weight filters (e.g., Millipore Ultrafree 10 kDa or Vivaspin).
2. Standard molecular weight markers (ECL Plex Fluorescent Rainbow Markers).
3. Gel fixing solution: 40% ethanol, 10% acetic acid in ultrapure water.
4. Coomassie Brilliant Blue G-250 (CBB) dye stock: 5% (w/v) in ultrapure water. Store at room temperature.
5. Coomassie Brilliant Blue G-250 working solution: For 100 mL, dissolve 8 g ammonium sulfate in 77 mL ultrapure water. Add 1.9 mL 85% phosphoric acid, 20 mL methanol, and 1.6 mL of the 5% CBB dye stock.

3 Methods

3.1 Labelling of Protein Lysate with CyDyes

1. Resolubilize protein pellet in rehydration buffer. Keep volumes to a minimum to obtain a final protein concentration of approx 8–10 mg/mL. Perform protein assay to determine concentration of each sample. Transfer 650 μg of untreated (Sample 1) and treated (Sample 2) HT29 cells to clean individual Eppendorf tubes. Prepare the Internal Standard by combining 325 μg of each sample in a separate Eppendorf tube.
2. To Sample 1, add 500 pmol of Cy3 dye (equivalent to 0.5 μL of 1 nmol/μL stock). To Sample 2, add 500 pmol of Cy5 dye. To

the Internal Standard, add 500 pmol of Cy2 dye. Incubate samples on ice for 30 min in the dark.

3. Quench labeling reaction with 10 mM L-lysine (1 μL).
4. Combine labeled protein samples and dilute with FFE Separation Medium 2 to a final concentration of approx 1.5 mg/mL. Store in the dark at 4 °C (*see* **Notes 10** and **11**).

3.2 Free Flow Electrophoretic Separation of Samples

Clean and assemble the FFE apparatus as described in the manufacturer's operating manual. Set the temperature of the separation chamber to 10 °C. Fill the system with ultrapure water ensuring all air bubbles are removed and leaks are not present. Ensure the 96-fraction collection tubes are dispensing at an equal rate.

3.2.1 Performance Test for Laminar Flow

This test is performed to ensure that the laminar flow is stable (i.e., the separation media is being delivered at a constant flow into the separation chamber and that turbulence does not exist) and is essential to ensure reproducibility of separations.

1. Move the separation platform into the horizontal position. Set the media pump speed to 80 mL/h. (or use specifications provided by manufacturer). Allow the system to equilibrate (10–15 min).
2. Introduce SPADNS via inlet tubes 2, 4, and 6. (Do not introduce SPADNS close to the electrodes as this will lead to contamination of electrode membranes.) Allow the SPADNS to flow through the separation chamber and check that the dye lines are straight and well defined (sharp edges and of a deep red color).
3. When the SPADNS reaches the end of the separation chamber, begin collecting into a 96-well microtiter plate (approx 100 μL fractions). The microtiter plate can be visually inspected as the dye should elute as well-defined fractions spanning 12–15 wells (*see* Fig. 2a). (The SPADNS fractions should appear red.) Alternatively, the absorbance of the microtitre plate can be read at 510 nm to quantify the SPADNS in the wells.
4. When the results are satisfactory, flush the system with ultrapure water to remove SPADNS (at least 20 min). If the results are unsatisfactory, flush the system with water, and if necessary disassemble the apparatus and clean the system.

3.2.2 Establish the pH Gradient and Perform the pI Test

These methods have been optimized to establish a linear pH gradient from pH 3 to 9. This can be customized to form a pH gradient suitable for your application.

1. Introduce the separation, stabilization, and counterflow media via the appropriate inlets. Allow the solutions to flow through the system ensuring air bubbles and leaks are not present.

a

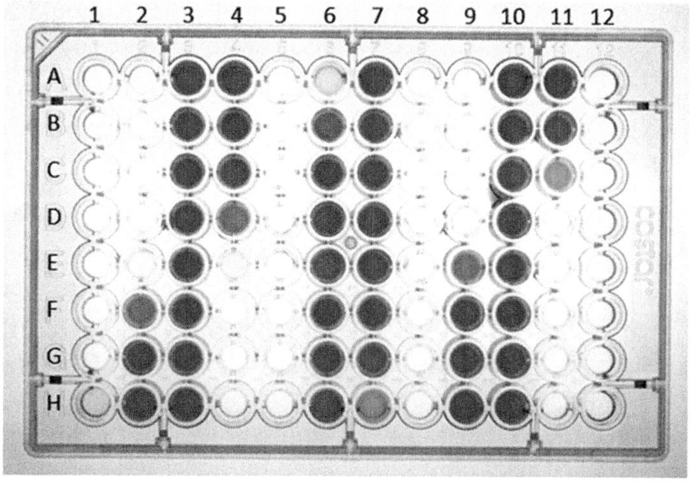

b

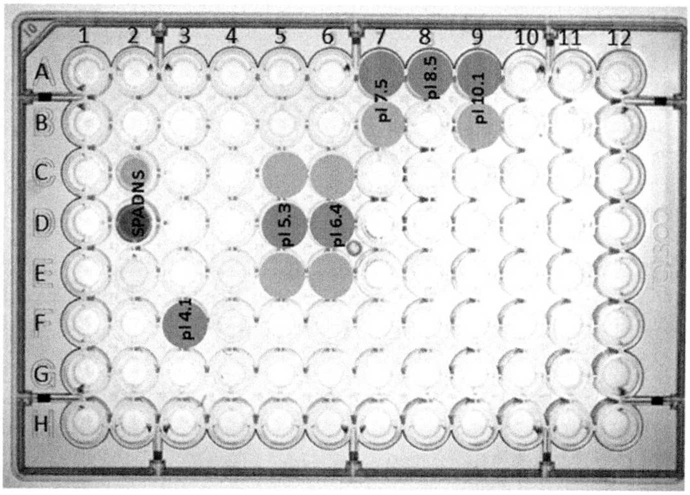

Fig. 2 (**a**) Performance test showing the expected elution pattern of the SPADNS dye. The leftmost fraction from the FFE chamber (anodic side) is delivered to well A1, with the next fraction delivered to B1, etc., down the column and then proceeding to A2. The H12 fraction is collected from the tube closest to the cathodic side of the chamber. (**b**) pI test showing the expected pattern of the markers and SPADNS

2. Introduce the anode and cathode electrolytes, and switch on the electrode pump.
3. Adjust the voltage, current, and power settings to 1000 V, 50 mA, and 60 W, respectively (or use settings recommended by the manufacturer). Set the media flow rate (80 mL/h or as recommended by manufacturer) and allow the system to equilibrate.
4. Introduce the pI markers at a rate of 2 mL/h (or use settings recommended by the manufacturer). Allow the colored markers to flow through the separation chamber. A sharp red band

(SPADNS) should define the anodic boundary of the separation field. The pI markers should appear as six sharply defined yellow bands within the separation chamber.

5. Collect fractions into a 96-well microtiter plate. The microtiter plate can be visually inspected (*see* Fig. 2b). The SPADNS should appear in 1–2 fractions, and each individual pI marker should span <4 fractions. Alternatively, read the absorbance of the microtiter plate at 410 nm (pI markers) and 510 nm (SPADNS). Measure the pH of each well containing pI marker to ensure each has eluted at its appropriate isoelectric point.
6. If the result of this test is acceptable, disconnect the sample tubing from the FFE apparatus and then flush with ultrapure water. Reconnect the tubing and flush with media with the sample pump in the reverse direction.

3.2.3 FFE Separation of CyDye-Labelled Protein Sample

Analysis of the standard protein mixture (**steps 1–4**) is optional and can be omitted by experienced users. Proceed to **step 5** for analysis of CyDye-labelled protein sample.

1. Introduce a small amount of SPADNS (1% w/w in Separation Medium 2) to help define the beginning of the protein sample. SPADNS is used to help visualize the sample limits (analogous to the dye front in SDS-PAGE). Introduce the standard protein mixture and follow this with a small amount of diluted SPADNS.
2. Begin collecting samples into 96-deep-well plates a few minutes after the initial SPADNS introduced reaches the end of the separation chamber. Collect a small volume (approx 100 μL) into a 96-well microtiter plate to measure the pH of each fraction.
3. When the end of sample limits is reached (defined by SPADNS), stop sample collection and turn off the voltage. Flush the sample pump and chamber with ultrapure water (at least 20 min).
4. Measure the pH of each well of the 96-well microtiter plate to determine if proteins have eluted at their isoelectric point. Absorbance readings at 280 nm will identify the wells where the proteins have eluted.
5. Proceed with the CyDye-labeled protein sample using the same process as detailed above, i.e., introduce a small amount of diluted SPADNS (1% w/w in Separation Medium 2), followed by labeled protein sample and then followed by diluted SPADNS. Collect fractions into a 96 × 2.2 mL polypropylene deep well plate and also collect a small volume (approx 100 μL) into a standard 96-well microtiter plate to measure the pH of each fraction. Store fractions in the dark (e.g., wrapped in aluminum foil) at 4 °C (short-term overnight storage) or −80 °C (longer-term storage).

6. This process can be repeated to separate multiple protein samples. Ensure sample chamber and pump are flushed with water between each sample and the separation chamber does not contain any visible precipitation.
7. Once the experiment has been completed, stop the voltage, flush the system with ultrapure water, and follow manufacturer protocols for system shutdown.

3.3 SDS Polyacrylamide Gel Electrophoresis

3.3.1 Sample Concentration

The volume of each fraction is approx 500 μL (or greater depending on the volume collected). A small volume from each fraction can be analyzed directly by 1D PAGE (*see* Subheading 3.3.2), or alternatively, each fraction can be concentrated to increase sensitivity of downstream mass spectrometry identification. Two potential methods for sample concentration are detailed below:

1. Concentration of each sample using centrifugal filters (e.g., 5 or 10 kDa cutoff) (*see* **Note 12**).
   (a) Apply each sample to individual centrifugal filters and process as described by the manufacturer ensuring centrifugation limit (*g* force or rpm) is not exceeded.
   (b) Transfer the retentate to a clean Eppendorf tube ready for SDS-PAGE.
2. Protein precipitation using acetone (*see* **Note 13**).
   (a) Transfer each fraction to 15 mL polypropylene tubes, and add 5–10 volumes of ice-cold acetone to each. Incubate at 4 °C overnight in the dark (e.g., wrapped in aluminum foil).
   (b) Spin each sample at 4 °C for 15 min using the maximum g force recommended by the manufacturer.
   (c) Decant supernatant, and rinse the pellet three times with 100 μL ice-cold acetone, and gently disperse the pellet each time to remove residual buffers.

3.3.2 SDS PAGE, Gel Scanning and Image Analysis

1. Resuspend fractions to be analyzed in sample loading buffer, using a clean Eppendorf tube for each sample.
2. Assemble and clean the electrophoresis apparatus as per manufacturer operating instructions. Prepare the precast gel cassettes and gel running buffers according to manufacturer instructions.
3. Load the molecular weight marker into the outer most well of the gel, and load each fraction to be analzsed into the remaining individual wells.
4. Perform electrophoresis using the conditions recommended by the manufacturer.
5. Following electrophoresis, carefully pry the gel cassette open and transfer gel to clean plastic tub. Rinse gently with ultrapure water (two to three times) and remove from gel scanner.

6. Scan the gel at the appropriate settings for CyDyes (Cy2: 488/520 nm; Cy3: 532/580 nm; Cy 5: 633/670 nm).
7. Remove the gel to a clean plastic container, rinse gently with ultrapure water and apply enough Coomassie fixing solution to just cover the gel. Cover the container with plastic wrap, and allow the gel to incubate for 1 h while gently agitating (e.g., using an orbital shaker). Decant the fixing solution, and add enough CBB working solution to just cover the gel. Cover container with plastic wrap and leave to stain overnight while gently agitating (*see* **Note 14**).
8. Gently decant Coomassie stain into waste container and gently rinse gel several times with ultrapure water. Gels can be stored in ultrapure water at 4 °C for a maximum of 2–3 days.
9. Perform image analysis of gels to determine the protein bands that are differentially expressed as a result of treatment. To calculate the fold change in protein expression, potential bands of interest must first be normalized to the Internal Standard sample (*see* Fig. 3). Protein bands of interest can be excised and in-gel digestion performed, and the protein can then be identified by mass spectrometry.

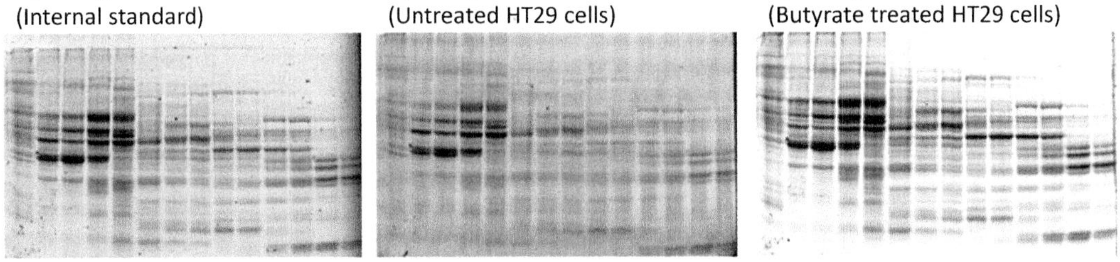

$$\text{Normalised expression for Untreated band} = \frac{\text{Untreated (Cy3)}}{\text{Internal standard (Cy2)}}$$

$$\text{Normalised expression for Treated band} = \frac{\text{Treated (Cy5)}}{\text{Internal standard (Cy2)}}$$

$$\text{Fold change} = \frac{\text{Normalised expression for Treated band}}{\text{Normalised expression for Untreated band}}$$

Fig. 3 1D gel image of selected FFE fractions scanned at the appropriate settings for each respective CyDye. The formula to calculate the fold change in protein expression of a band of interest using densitometry is shown below

4 Notes

1. Protein pellets can be resolubilized in an alternate buffer system, but the compatibility of each buffer reagent with the CyDye labeling procedure should be determined by consulting the product insert.
2. *N*,*N*-Dimethylformamide is a hazardous, potentially toxic chemical and should be used in a chemical fume hood. Avoid inhalation and contact with the skin. Refer to MSDS and product insert for safety information regarding handling, storage, and disposal.
3. To make 80 g (~70 mL) of Anode Stabilization Medium, weigh 25.20 g urea and 9.05 g thiourea and add to a 100 mL wide-neck Schott bottle or similar, containing 23.25 mL of ultrapure water and 7 mL of 1 M H_2SO_4. Place the bottle on a balance, and using pipettes, add glycerol (10 g) and 0.8% HPMC stock solution (5 g). Do not heat excessively as this decomposes the urea, and store the reagents in a cool, dark place until use. The pH and conductivity of the solution should be ~1.7 and ~6.6 mS, respectively.
4. To make 80 g (~70 mL) of Separation Medium 1, weigh 25.20 g urea and 9.05 g thiourea, and add to a 100 mL Schott bottle containing 19.2 mL of ultrapure water. Place the bottle on a balance, and using pipettes, add ProLyte 1 (11.6 g), glycerol (10 g), and 0.8% HPMC stock solution (5 g). The pH and conductivity of the solution should be ~4.5 and ~260 μS, respectively.
5. To make 160 g (~140 mL) of Separation Medium 2, weigh 50.45 g urea and 18.11 g thiourea, and add to a 250 mL Schott bottle containing 24.4 mL of ultrapure water. Place the bottle on a balance and using pipettes, add ProLyte 2 (37.12 g), glycerol (20 g), and 0.8% HPMC stock solution (10 g). The pH and conductivity of the solution should be ~7.4 and ~430 μS, respectively.
6. To make 80 g (~70 mL) of Separation Medium 3, weigh 25.20 g urea and 9.05 g thiourea, and add to a 100 mL Schott bottle containing 15.3 mL of ultrapure water. Place the bottle on a balance, and using pipettes, add ProLyte 3 (15.52 g), glycerol (10 g), and 0.8% HPMC stock solution (5 g). The pH and conductivity of the solution should be ~10.2 and ~350 μS, respectively.
7. To make 160 g (~140 mL) of Cathode Stabilization Medium, weigh 50.45 g urea and 18.11 g thiourea, and add to a 250 mL Schott bottle containing 47.5 mL of ultrapure water and 14 mL of 1 M NaOH. Place the bottle on a balance, and

using pipettes, add glycerol (20 g) and 0.8% HPMC stock solution (10 g). The pH and conductivity of the solution should be ~14.0 and ~5.4 mS, respectively.

8. To make 500 g (~435 mL) of Counterflow Medium, weigh 156.75 g urea and 56.29 g thiourea and add to a 500 mL Schott bottle containing 193.2 mL ultrapure water. Place the bottle on a balance, and add glycerol (62.5 g) and 0.8% HPMC stock solution (31.25 g). The pH and conductivity of the solution should be ~9.8 and ~50 µS, respectively.
9. Alternatively, polyacrylamide gels can be hand cast. Low fluorescent glass plates (GE Healthcare, catalogue #80-6442-14) can be used to minimize gel handling during the scanning procedure and minimize potential for contamination (e.g., contamination with keratin which can interfere with mass spectrometry identification).
10. Ensure the total salt content of the sample does not exceed 25–50 mM as high salt concentrations will interfere with the FFE separation process. If required, reduce the salt concentration by diluting further in Separation Medium 2. Remove any precipitate material by centrifugation.
11. We found that final protein concentrations over 2 mg/mL resulted in precipitation occurring in the separation chamber. Proteins may precipitate when they reach their isoelectric point. If precipitation occurs, dilute the protein sample further in Separation Medium 2.
12. Concentration of samples using centrifugal filters is limited by the presence of glycerol and HPMC in the sample. These are highly viscous solutions and do not easily pass through the filters, i.e., they will also be concentrated along with protein. Excess glycerol and HPMC may also affect resolution of protein bands by 1D PAGE.
13. Alternate protein precipitation methods can be used, but care must be taken to ensure the integrity of the CyDye label is maintained. For example, the low pH associated with TCA precipitation methods may reverse the amide linkage between the NHS ester reactive group of the CyDye and the epsilon amino group of lysine in proteins.
14. Alternate staining methods can be used, e.g., silver staining, to visualize less abundant proteins. If silver staining the gels, use a mass spectrometry compatible method to reduce interference with downstream analysis.

References

1. Hannig K (1982) New aspects in preparative and analytical continuous free-flow cell electrophoresis. Electrophoresis 3:235–243
2. Hannig K, Heidrich H (1990) Free-flow electrophoresis. GIT Verlag, Darmstadt
3. Hoffmann P, Ji H, Moritz RL et al (2001) Continuous free-flow electrophoresis separation of cytosolic proteins from the human colon carcinoma cell line LIM 1215: a non two-dimensional gel electrophoresis-based proteome analysis strategy. Proteomics 1:807–818
4. Islinger M, Eckerskorn C, Volkl A (2010) Free-flow electrophoresis in the proteomic era: a technique in flux. Electrophoresis 31:1754–1763
5. Krivankova L, Bocek P (1998) Continuous free-flow electrophoresis. Electrophoresis 19:1064–1074
6. Weber G, Bocek P (1996) Optimized continuous flow electrophoresis. Electrophoresis 17:1906–1910
7. Weber G, Islinger M, Weber P et al (2004) Efficient separation and analysis of peroxisomal membrane proteins using free-flow isoelectric focusing. Electrophoresis 25:1735–1747
8. Weber PJ, Weber G, Eckerskorn C (2004) Isolation of organelles and prefractionation of protein extracts using free-flow electrophoresis. Curr Protoc Protein Sci Chapter 22:Unit 22.5
9. Fung KY, Cursaro C, Lewanowitsch T et al (2011) A combined free-flow electrophoresis and DIGE approach to identify proteins regulated by butyrate in HT29 cells. Proteomics 11:964–971

Chapter 35

SDS-PAGE for ^{35}S Immunoprecipitation and Immunoprecipitation Western Blotting

Edward P. Trieu and Ira N. Targoff

Abstract

This report discusses recent methods of sample preparation and gel electrophoresis for ^{35}S immunoprecipitation (IP) and IP western blotting. In both methods, IP is used to obtain purified proteins, and sodium dodecyl sulfate polyacrylamide gel electrophoresis is used to separate the proteins on a gel. In ^{35}S IP, the proteins are radiolabeled and visualized on film by fluorography; in IP blotting, proteins are transferred onto nitrocellulose paper, and antibodies are used to detect specific proteins. A similar IP and SDS-PAGE method can be used for both procedures, but IP blotting has the potential advantages of improvement in sensitivity for low-abundance proteins and enhanced specificity for identification of proteins from a mixture. Some of the technical adaptations discussed here to facilitate IP blotting and avoid loss of beads or purified proteins may also be useful for ^{35}S IP.

Key words Immunoprecipitation, ^{35}S immunoprecipitation, Immunoblot (Western blot), Autoantibody, Autoantigen, Dimethyl pimelimidate, Isotachophoresis, SDS-PAGE

1 Introduction

Sodium dodecyl sulfate polyacrylamide gel electrophoresis (SDS-PAGE) of proteins is routinely used in modern laboratories to separate mixtures of proteins into individual bands before applying other methods such as fluorography and Western blotting for analysis of the proteins. In 1964, Ornstein designed the polyacrylamide gel with a discontinuous buffer system [1] based on the 1897 isotachophoresis theory of Kohlrausch [2]. When he observed proteins as a blue band stacked into a thin band from an isoelectric focusing gel electrophoresis, he knew that pH played an important role in the stacking mechanism and explored the theory of Kohlrausch. Laemmli improved on this system in 1970 by using SDS and adjusted pH [3]. This system of SDS-PAGE continues to be widely used to fractionate different kinds of proteins and analyze different kinds of protein mixtures. For example, SDS-PAGE is

Biji T. Kurien and R. Hal Scofield (eds.), *Electrophoretic Separation of Proteins: Methods and Protocols*, Methods in Molecular Biology, vol. 1855, https://doi.org/10.1007/978-1-4939-8793-1_35,

used for determining the molecular weight of proteins and for the analysis of proteins isolated by immunoprecipitation from cell extracts. For ^{35}S IP, the cells are labeled by growing in appropriate media with reduced amount of methionine, after the addition of [^{35}S]-methionine. The cells incorporate [^{35}S]-methionine to produce ^{35}S-labeled proteins and are used as antigen source for IP. ^{35}S IP sensitivity can be adjusted by changing the amount of methionine used. Beginning with the labeling method of Pollard in 1994, methionine was used at 1 mg/L in the labeling medium [4]. However, it was found that this level did not provide strong enough labeling to detect some of the proteins that were present in the cell in low copy numbers. Therefore, the level of cold methionine in the labeling medium was reduced to 0.5 mg/L (*see* Fig. 1). This decrease in cold methionine resulted in an increase in the signal of ^{35}S in the gel. However, later, the sensitivity level was raised to a higher level by using mainly [^{35}S]-methionine in the labeling medium for pulse-chase labeling of the proteins [5].

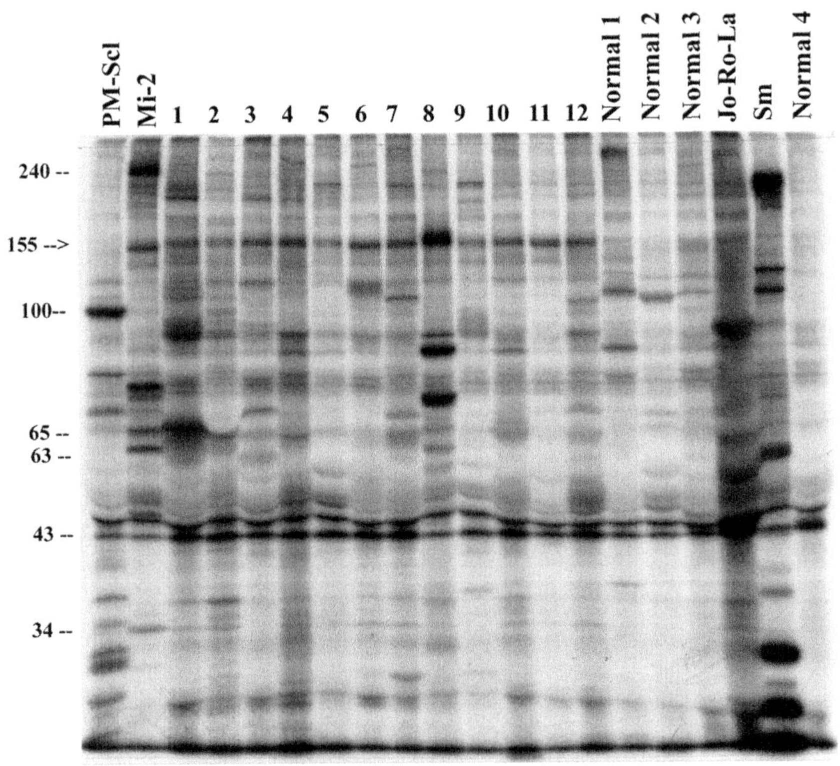

Fig. 1 SDS-PAGE fluorography of ^{35}S IP with 0.5 mg/L of methionine in the labeling medium. Before proceeding to the Western blot IP, 12 anti-p155 autoantibody-positive sera (lanes 1–12) were screened with an IP of ^{35}S-labeled cell extract. Lanes 1–12 show sera that precipitated a protein similar to that precipitated by the p155 prototype sera shown in lane 6; however, several sera show additional coexistent auto antibodies. The control sera used in the other lanes are Mi-2, PM-Scl, Jo-Ro-La, and Sm. Normal is serum from a normal subject. Molecular weight markers in kDa are shown on the left side of the gel [6]

SDS-PAGE has been used in clinical testing as part of ^{35}S IP and IP blotting for identification of autoantibodies in patient sera [6]. For example, ^{35}S IP has long been used to detect and identify autoantibodies of significance for autoimmune myositis [7]. ^{35}S IP more recently revealed additional autoantibodies, anti-p155/140 and anti-MJ-140, which were more difficult to conclusively identify by simple SDS-PAGE of immunoprecipitates. IP blotting was employed to confirm results when these antibodies were suspected after ^{35}S IP testing. Initially, the prototype serum was used for IP, followed by separation by SDS-PAGE and transfer, after which the test sera were used for blotting to confirm identity with prototype serum. However, some test sera with ^{35}S IP results suggesting anti-p155/140 were negative for any bands by IP-blot, even when the test serum itself was used for both the IP and blotting steps. This suggested exclusive reaction with conformational epitopes. When such IP-positive, blot-negative sera were tested in a reverse manner, using test serum for the IP step and prototype serum for the blotting step, the expected bands were detected [6, 7]. This method has also been successfully applied to detection of other autoantibodies.

The overall procedure has four sections: (1) sample preparation, (2) gel preparation to separate proteins from the samples, (3) gel electrophoresis of protein samples, and (4) fluorography and Western blotting.

2 Materials

2.1 Sample Preparation

2.1.1 ^{35}S Immunoprecipitation

1. Protein A-Sepharose-CL-4B (GE Healthcare, Sweden).
2. Immunoprecipitation (IP) buffer: 10 mM Tris, pH 8.0, 0.5 M NaCl, 0.1% NP-40, or Igepal CA-630 (Sigma-Aldrich, St. Louis, MO, USA). Store at 4 °C.
3. Human or animal sera. Remember to use exposure control plan to protect both personnel and environment from blood-borne pathogens.
4. Bovine serum albumin (BSA) blocking solution: 4% BSA in 10 mM Tris, pH 8.0, 0.5 M NaCl, 0.01% NP-40, or Igepal CA-630. Store at −70 °C in a small aliquot of 5–10 mL. A small amount of detergent in the solution aids in even coating of solution inside the tube for blocking.
5. HeLa or K562 cell lines obtained from the American Type Culture Collection.
6. Complete medium: MEM Eagle with Earle's Salts (Sigma-Aldrich) supplemented with 10% bovine calf serum, glutamine, and penicillin/streptomycin. Store at 4 °C.

7. Labeling medium: MEM Eagle (modified) with Earle's Salts without L-glutamine, L-methionine. Store at 4 °C.
8. Complete labeling medium: above medium with 0.5 mg/L of methionine.
9. Dialyzed fetal bovine serum (Sigma): Store in small aliquots, 5–10 mL at −70 °C not more than 6 months.
10. L-Glutamine–penicillin–streptomycin solution, 100× (Sigma). Store at −20 °C.
11. Methionine, cell-labeling grade, L-[^{35}S] (Perkin Elmer, Waltham, MA, USA).
12. Radiolabeling box: A box used to contain small flasks of cells for radiolabeling with the top open and was made up of 1/2-in.-thick Plexiglas. The Plexiglas shields the other cell culture flasks that are in the same incubator from the β-rays. A typical dimension for the outside of the box is 6 × 9 × 5½ in.
13. Cell culture flasks, 25 cm^2.
14. L-Methionine: 1 mg/mL of methionine, sterile filter and store at 4 °C.
15. Activated carbon-loaded paper, size 8 ×10 in. Use to absorb the ^{35}S waste product from cell metabolism.
16. Lysis buffer for ^{35}S-IP: 1% NP-40 or Igepal CA-630 in IP buffer supplemented with 100× protease inhibitor cocktail. Prepare immediately before use.
17. Whatman Vacu-Guard Filter (Fisher Scientific, Dallas, TX, USA). Put on the vacuum line to avoid contamination of ^{35}S to the vacuum system.
18. Transfer pipette, capillary tip.
19. SDS sample buffer, 1×: 10 mM Tris, pH 6.8, 10% glycerol, 2% SDS, 0.0025% bromophenol blue, 1% mercaptoethanol. Store at −20 °C in 5–10 mL aliquots.
20. Hamilton syringe, 50 μL.

2.1.2 Immunoprecipitation Blotting

1. Micro Bio-Spin Chromatography Columns, caps, and tip enclosures (Bio-Rad, Hercules, CA, USA) or 96-well filter plates with silicone mat closure that seal both the top of the plate and the bottom drain ports (Arcticwhite LLC, Bethlehem, PA, USA). The alternative 96-well filter plate is Nunc 96-well Filter Plates with PET frit (Fisher Scientific).
2. Costar, V-well-shaped untreated, clear polystyrene 96-well plate with silicone plate seal.
3. Antibody-binding buffer (ABB): 0.2 M triethanolamine, pH 8.9, 0.5 M NaCl. Store at 4 °C.

4. Cross-linking solution: 30 mg/mL of dimethyl pimelimidate (DMP) in ABB. Prepare immediately before use. Store DMP at 4 °C.
5. Quenching buffer: 0.1 M Tris, pH 8.5, 0.5 M NaCl. Store at −20 °C.
6. Unbound antibody elution solution: 1 M acetic acid, 0.15 M NaCl. Store at room temperature (RT).
7. Lysis buffer for IP blotting: 0.3% NP-40 or Igepal CA-630 in IP buffer supplemented with 100× protease inhibitor cocktail.
8. Micro incubator shaker: Model E36 Micro Incubator Shaker (Bionexus Inc., Oakland, CA, USA) holds two 96-well plates or titer plate shaker (LabLine Instruments, Inc., Melrose Park, ILL, USA) holds four 96-well plates.
9. Sonicator (ultrasonic processor).
10. SDS sample buffer, 4×: 40 mM Tris, pH 6.8, 40% glycerol, 8% SDS, 0.01% bromophenol blue, 4% mercaptoethanol. Store at −20 °C in 5–10 mL aliquots.
11. Protein-free (PBS-T) blocking buffer.
12. Filter paper (Whatman 3MM) or Pellon # 806, reusable. Pellon # 806 Stitch-N-Tear—white has the same thickness with filter paper. Wet Pellon sheets with 70% ethanol, rinse with distilled water, and store in a zip bag with the protein transfer buffer for use each time.

2.2 Gel Preparation

1. Protean II xi gel apparatus from Bio-Rad, 16 cm glass plates, 25-well combs, and spacers for each of 0.75 and 1.5 mm gels.
2. Polyacrylamide solution, 30%: Polyacrylamide is not stable once in solution. Prepare immediately or a few days before use. Store at 4 °C in a dark bottle or cover with aluminum foil. Polyacrylamide is a neurotoxin when it is a monomer. Therefore polymerize excess solution before disposal.
3. Resolving buffer, 4×: 1.5 M Tris–HCl, pH 8.8. Store in 50 mL aliquots at −20 °C.
4. Stacking buffer, 4×: 0.5 M Tris–HCl, pH 6.8. Store in 50 mL aliquots at −20 °C.
5. Water-saturated butanol: Add 10 mL of water into a 50 mL tube and add 40 mL of butanol. Cap the tube, shake well, and let sit to separate; use the water-saturated butanol in the top layer. Store at RT.
6. Ammonium persulfate solution, 10%: Store at −20 °C in 2 mL aliquots. Reuse the remaining solution if re-frozen at −20 °C.
7. TEMED.

2.3 Gel Electrophoresis of Protein Samples

1. A small power supply with a current capacity of 500 mA or higher.
2. Refrigerated circulator.
3. Running buffer, 10×: 250 mM Tris, 1.92 M glycine, 1.0% SDS. Store at RT.
4. Molecular weight markers for western blot gel: Fisher BioReagents EZ-Run Rec Protein Ladder.
5. Prestained molecular weight markers for ^{35}S-gel: Precision Plus Protein all Blue Standard.

2.4 Fluorography and Western Blotting

2.4.1 Fluorography

1. Gel drier with vacuum: Savant SG210D SpeedGel.
2. X-Ray film: Hyperfilm MP or ECL, size 8 × 10 in.
3. X-Ray cassette for fluorography, size 10 × 12 in. or 14 × 17 in.
4. X-Ray developing machine and a dark room.
5. Whatman filter paper, grade 3, size 46 × 57 cm.
6. 0.5 M sodium salicylate: Sodium salicylate is light sensitive when in solution. Store the solution in a dark bottle at RT. Over time, the solution turns dark yellow and blue and should be discarded.
7. A roller or a plastic spreader: Plastic spreader is a piece of flexible plastic for spreading soft things such as glue onto a surface.

2.4.2 Western Blotting

1. Nitrocellulose paper.
2. Trans-Blot cell.
3. Protein transfer buffer: 25 mM Tris, 192 mM glycine, 20% methanol, 20 mM sodium hydroxide, 0.005% SDS (pH 8.3). Store at 4 °C.
4. Rocking shaker: Labquake Shaker. When shaking, the diluted serum solution becomes a wave over the membrane and thus can use less volume for the incubation.

3 Methods

3.1 Sample Preparation

First, the antibody of interest from animal or human serum binds to the protein A-Sepharose beads during incubation of serum with protein A-Sepharose beads. The wash buffer removes unbound antibodies and other serum proteins. Then, the antibodies on the protein A-Sepharose bind their specific antigen during incubation with a homogenate or cell extract. The immunoprecipitate forms on the Sepharose beads from the reaction of the antibody with the antigen. Then, the wash buffer is used to wash away unbound proteins from the beads. Following the wash, SDS is used to separate the antigen from the beads. At this step, for the ^{35}S IP discussed here, SDS sample buffer releases both the antigen and the antibody from the beads; for the IP-blot procedure, SDS sample buffer releases only the antigen from the beads because the antibody is cross-linked to protein A-Sepharose beads.

3.1.1 ^{35}S Immunoprecipitation

1. *Prepare antibody-protein A-Sepharose.* Label on the cap of 0.5 mL microcentrifuge tubes a number for each serum sample to be used and add into each tube the following: 400 μL of IP buffer, 10 μL of protein A-Sepharose CL-4B 1:1 with IP buffer, and 10–15 μL of serum.
2. Cap and place tubes into a plastic freezer box that has dividers for 50 tubes to keep the tubes organized and in the same orientation. Let the tubes agitate on the rotator for 2 h to overnight at 4 °C.
3. Centrifuge briefly (approximately 30 s) at 5000 × *g* and aspirate the buffer to above the bead level about 10 μL. Add 400 μL of blocking solution into each tube and place on rotator overnight at 4 °C. BSA binding to the beads is necessary to reduce the background from ^{35}S-labeled proteins.
4. Wash the beads thrice with IP buffer, and store the beads in a refrigerator until the ^{35}S-labeled protein is ready.
5. *Labeling of HeLa cells for ^{35}S immunoprecipitation:* Grow HeLa cells to near confluence with a little space between the cells in a 25 cm^2 flask with Complete Medium (*see* **Note 1**).
6. To prepare 50 mL of complete labeling medium with 0.5 mg/L of methionine, 10% DFCS, and penicillin/streptomycin in labeling medium, add into a 50 mL tube the following: 44.5 mL of labeling medium, 25 μL of 1 mg/mL of methionine solution, 500 μL of L-glutamine-penicillin-streptomycin solution, and 5 mL of dialyzed fetal calf serum. Warm up to 37 °C just before use (*see* **Note 2**).
7. Wash the flask of cells thrice with 2–2.5 mL of complete labeling medium. The washes remove the Complete Medium that has 15 mg/L of methionine.
8. Now cells are ready for labeling. Each flask is usually labeled at the "medium" sensitivity level as described in Table 1, and that is assumed in the steps below (*see* **Note 3**).
9. After pipetting out the remaining wash medium, add into the flask 1.7 mL of the Complete Labeling Medium and 500 μCi of [^{35}S]-methionine with a filtered pipette tip. Place the flask into the radiolabeling box and cover with activated carbon-loaded paper, and place the box into a cell culture incubator (37 °C, 5% regulated CO_2 level and humidified).

Table 1
Sensitivity levels for ^{35}S-labeling

^{35}S labeling level	Weak	Medium	Strong	Unit in
Cold methionine	1	0.5	0	mg/L
Volume of media used per flask of cells	2.5	1.7	1.3	mL
^{35}S-Methionine, 0.5 mCi/flask	25–30	25–30	25–30	μL
Specific activity of ^{35}S per mL	200	294	385	μCi/mL
Artifacts or background	Weak	Low	High	–

10. Culture the cells with [^{35}S]-methionine overnight for 16–20 h.
11. *Prepare ^{35}S-labeled cell extract:* In the morning, remove the box containing the [^{35}S]-methionine flask of cells from the incubator and place on an absorbent paper. Remove the used ^{35}S-medium from the flask with a capillary tip transfer pipette or with an aspirator connected to a vacuum flask for collection of all ^{35}S radioisotope waste.
12. Wash the cells thrice with 2–2.5 mL of ice-cold PBS per flask.
13. After removing the wash buffer, use the cells immediately or store for a few days at −70 °C, but not longer than 7 days because ^{35}S decays, changes its property, and comes off from the protein.
14. Add into the flask 0.5 mL of ice-cold lysis buffer. Turn the flask up and down on each side to distribute the lysis buffer, and wait for about 5 min or until the cells come off the flask surface. The detergent from the lysis buffer dissolves the cell membrane, and the high salt in the buffer disrupts the chromatin structure. Therefore no viscosity from the DNA.
15. Transfer 0.5 mL of ^{35}S-labeled cell extract into a 2 mL centrifuge tube with a capillary tip transfer pipette. Rinse the flask surface for the remaining cell extract and cell debris with another 500 μL of lysis buffer. Combine the lysis buffer wash with the cell extract in the 2 mL centrifuge tube. Use one tube of 2 mL per flask of cells. Cap the tubes and vortex to release more antigens from the cell debris.
16. Centrifuge for 5 min at 12,000 × *g* at RT or preferably at 4 °C.
17. Transfer the ^{35}S-labeled cell extract from each tube with 1 mL filtered pipette tip into a 15 mL conical tube and dispose off the cell debris pellet. Bring the volume of ^{35}S-labeled cell extract from 1 to 4 mL with 2 mL of BSA blocking solution and 1 mL of IP buffer. Each flask of cells will thus provide enough labeled cell extract for ten samples (400 μL of ^{35}S-labeled cell extract per sample). Be sure to measure the volume of the cell extract

with a little bit over so that all tubes will get the same volume of cell extract.

18. Cap the conical tube and invert to mix. Keep the ^{35}S-labeled cell extract on ice.
19. *Immunoprecipitation:* Add 400 μL of ^{35}S-labeled cell extract into each tube with antibody-protein A-Sepharose (*see* Subheading 3.1.1, **steps 1–4**), cap, and place on the rotator to rotate for 2.5–3 h at 4 °C.
20. Centrifuge the tubes to bring the beads down to bottom of the tube. Aspirate the ^{35}S-labeled cell extract into a flask for ^{35}S radioisotope waste container, and wash the beads 4× with 500 μL of IPP buffer. Discard the ^{35}S-labeled cell extract and the wash buffer from the flask into a jug for ^{35}S radioisotope waste, and log the usage of radioisotope.
21. Wash 2× with 5 mM Tris–HCl, pH 6.8 (1/100 dilution from 4× stacking buffer).
22. Aspirate the Tris buffer to gel level with a 200 μL pipette tip or with a Hamilton syringe. Add 30 μL of 1× SDS sample buffer into each tube, cap, and vortex. Store the samples at −70 °C (*see* **Note 4** for troubleshooting).

3.1.2 Immunoprecipitation Blotting

1. *Prepare the antibody cross-links to protein A-Sepharose:* Add into each well of a 96-well plate with V-shaped well bottom 20 μL of 1:1 protein A-Sepharose CL-4B beads to IP buffer and 20 μL of undiluted serum and 150 μL of IP buffer. Put a silicone-sealing mat on the 96-well plate to seal the wells. Place the plate on a rotator, and let it rotate on the rotator for 2 h to overnight at 4 °C. Do not use the Micro Bio-Spin columns in this step because the porous support void volume isolates some of the antibody and protein A-Sepharose from the binding.
2. The next day, use a 200 μL wide bore pipette tip for each well; transfer the beads from each well of the 96-well plate to each Micro Bio-Spin column. Place the columns on vacuum manifold or tall tubes to collect the flow-through [7].
3. Rinse the wells thrice with 200 μL of PBS to collect the remaining beads.
4. Wash the beads thrice on the Micro Bio-Spin columns with 300 μL of PBS, and blot the bottom of the columns on a piece of paper towel or a piece of sponge to remove the excess water in the bottom of the columns.
5. Plug the bottom of the columns with tip enclosures, and push each column onto a well of a deep-well microplate. Micro Bio-Spin columns fit well onto a 96 × 2 mL/square well plate, and the micro incubator shaker holds the deep-well microplate for the shaking. Add into each Micro Bio-Spin

column 100 μL of cross-link solution. Place on the micro incubator shaker for 15 min at low speed. Do not place the columns on a rotator unless the columns are full with liquid because the beads will stick onto the side of the columns or onto the caps and may be lost.

6. When time is up after 15 min, immediately add 200 μL of quenching solution into the columns, and continue to shake for 10 min. The quenching solution neutralizes the remaining binding sites while the beads are still in the suspension to avoid cross-linking between the beads when they settle down to the bottom of the column.
7. Unplug the bottom of the columns, and place the columns on vacuum manifold or tall tubes. Wash twice with 300 μL of PBS.
8. Add 200 μL of unbound antibody elution solution into the columns, and immediately wash the beads thrice with 300 μL of PBS. Blot the bottom of the columns on a piece of paper towel or a piece of sponge to remove the water in the bottom of the columns that may contain some residual acetic acid.
9. Plug the bottom of the columns, add about 200 μL of IP buffer or PBS, and put on the cap cover. Keep the columns in the refrigerator overnight or over the weekend.
10. *Prepare the K562 cell extract:* Plan ahead for the amount of cell extract to be used. For each column with 10 μL of beads, 100–150 μL of cell extract is needed. K562 cell extract can be prepared in a large amount for several columns, and extra extract can be frozen at −70 °C with 10% glycerol for later use.
11. Add 10 μL of 100× protease inhibitor cocktail to a 15 mL conical tube containing 1 mL of K562 frozen cell pellet, and let the cells thaw in a beaker containing tap water at RT. Vortexing the conical tube a few times will help to thaw the cells quicker.
12. Add 5 mL of lysis buffer for IP blotting containing 0.3% NP-40 and 1× protease inhibitor cocktail into the conical tube with thawed K562 cell pellet. Cap the tube and mix.
13. Place the conical tube in a beaker containing ice and water, and sonicate the cells (power of 30 W, six times with 20 s sonication and 10 s rest between each sonication).
14. Transfer the cell extract into a centrifuge tube or microcentrifuge tubes and centrifuge at 10,000 × *g* for 20 min at 4 °C. Pour the cell extract into a new tube and discard the cell debris pellet.
15. Take the columns out from the refrigerator and unplug each column to drain IP buffer or PBS. Plug each column, add 100–150 μL of K562 cell extract, and put on caps. Push the columns onto the wells of a 96 deep-well plate that is on the

micro incubator shaker, and incubate them with shaking for 3 h at 4 °C.

16. When incubation is complete, take the columns out from the 96 deep-well plate. Unplug each column to drain the cell extract into a well of the 96 deep-well plate. Wash the beads 8–10× with 300 μL of ice-cold IP buffer.
17. Wash each column twice with 300 μL of ice cold 5 mM Tris, pH 6.8 (from 1/100 dilution of the stacking buffer).
18. Shake off the water from each column into the sink, and hit lightly on the bottom of the column onto a piece of sponge or a piece of paper towel to absorb some of the excess water that may still be inside the column (*see* **Note 5**).
19. When no more water issues out from the bottom and the outside of columns are dried, put each column into a 1.5 mL microcentrifuge tube.
20. Add 7 μL of 4× SDS sample buffer into each column. The porous support of the column and the beads could hold up to about 20–25 μL of the water. Centrifuge briefly. Take the same sample buffer from each 1.5 mL microcentrifuge tube, and put it back into the same column to more completely elute the antigen from the column. Centrifuge again, about 30 s to 1 min at 5000 × *g*. Label each 1.5 mL tube with the antigen name and cap the tubes. There is about 30–35 μL of sample in each tube.
21. Store the samples at −70 °C until gel electrophoresis.

3.2 Gel Casting

The proteins from the samples are separated on a polyacrylamide gel using the discontinuous buffer system of Laemmli [3].

3.2.1 Prepare the Gel-Casting Chamber

1. Read the manual of the manufacture for instructions to set up the gel apparatus. In short, to set up the gel-casting chamber for one gel with 16 cm glass plates for the Bio-Rad Protean II xi, place the following on a clean surface in this order: longer glass plate, one two spacers on each side, and the 16 cm glass plate on top of the two spacers. For ^{35}S IP, use 0.75 mm spacers to avoid cracking of the ^{35}S gel while drying, but for Western blot IP, use 1.5 mm spacers.
2. While holding the glass plate-spacer sandwich upright on a flat surface such as a glass plate, pull two spacers both side outward a little, and then push the left side clamp to the left side of the glass plates. Do the same to the right side of the glass plates with right side clamp. Make sure both glass plates and the spacers are even inside of both clamps by pushing them together and the spacer ends are flat on the bottom of the glass plates by pushing the top of spacers down. Now, while holding one clamp steady, tie up the clamp of the right side of

the glass plate and then the left side of the glass plates. Check the bottom of the two glass plates. If they are not smooth to touch, repeat the setup.

3. Attach the gel-casting chamber (glass plate-spacer sandwich) onto the base of the gel-casting stand, and insert the comb and mark on the short glass plate 1–1.5 cm from the bottom of the comb with a Sharpie marker for the resolving gel level.

3.2.2 SDS Polyacrylamide Gel Casting

1. The volume of gel solution for ^{35}S IP samples with Protean II xi gel apparatus is 17 mL, while the volume of gel solution for IP blotting samples is 34 mL.
2. Add into each 125 mL side-arm vacuum flask the resolving gel solution and the stacking gel solution according to Table 2, but leave out 10% ammonium persulfate solution (10% AP) until the last step to polymerize the gel because it can polymerize the gel solution without TEMED if the solution is left overnight at 4 °C. Use double distilled water for all solutions.
3. Degas the resolving gel solution for at least 5 min at RT.
4. Add 0.34 mL of 10% AP to a flask that contains 34 mL of resolving gel solution, swirl lightly to mix. While holding the gel-casting chamber at about a 45° angle to the bench top, pour the solution into the gel-casting chamber and repeat if there are two gels. Fill the solution to the mark of the resolving gel level.
5. Apply water-saturated butanol across the top, and let the solution sit for about 5–10 min to polymerize.
6. Prepare stacking gel solution and degas at RT for at least 1 min.

Table 2
Gel solution for 8.5% SDS-PAGE [8]

Gel solutions	Resolving		Stacking	
Final percent use	8.5%		3.9%	
Final volume use	34.00 mL		10.00 mL	
0.50 M Tris pH 6.8	0		2.50 mL	[0.125 M]
1.5 M Tris pH 8.8	8.50 mL	[0.375 M]	0	
30% polyacrylamide	9.63 mL		1.30 mL	
Water	14.82 mL		6.00 mL	
5% SDS	0.68 mL	0.1% w/v	0.20 mL	0.1% w/v
10% AP	0.34 mL	1.0% w/v	0.05 mL	0.5% w/v
TEMED	24 μL	0.07% v/v	10 μL	0.1% v/v

7. When the resolving gel solution polymerizes, pour off the butanol from the top of the gel into the sink, and use double distilled water to flush out the remaining butanol with a squirt bottle. Lay the gel-casting chamber on one side to let the water come out, and blot dry the top of the glass plates with a paper towel. To remove the water inside of the upper gel chamber, lay the gel horizontally, and insert into the upper gel chamber two pieces of Kimwipe folded over onto a piece of transparent film with the same width of the gel chamber. Do not let the paper touch the gel.
8. Add 0.05 mL of 10% AP to a flask that contains 10 mL of stacking gel solution, swirl to mix, and while holding the gel-casting chamber at about 45° angle to the bench top, pour the stacking gel solution into the gel-casting chamber and insert the comb between the glass plates. If there are bubbles on the comb, pull the comb up and push down hard to remove bubbles.
9. Place a piece of plastic or Saran wrap on top of the upper gel, and place a metal clamp on top to reduce space between the comb and the glass plates. To avoid drying of the stacking gel, put the whole gel with casting stand into a plastic bag before pouring of the stacking gel. Add a piece of wet sponge or a piece of wet paper towel onto the casting stand, and close the end of the plastic bag. Leave the gel overnight or over the weekend at RT.

3.3 Gel Electrophoresis of Protein Samples

3.3.1 Gel Electrophoresis of Protein Samples

1. Pull the comb slowly out from the gel (water can help to pull the comb out). If the gel fingers are not straight, straighten them with a Hamilton syringe needle. Mark the bottom of each well with a Sharpie marker, or use the water to adhere a template with pre-marks of the bottom of the wells on a strip of transparency to the glass plate.
2. Attach the gel with the clamps to the cooling unit of the gel apparatus to form the upper buffer reservoir. If running only one gel, block the other side of the cooling unit similarly with a piece of plastic to replace the glass plates.
3. Fill up the lower reservoir tank with 2–2.5 L of running buffer. Fill up the upper reservoir with 250–300 mL of running buffer. After the gel is assembled with the cooling unit on the bench top, wait for a few minutes to check for any leak; if not, put the gel assembled with the cooling unit and with the running buffer on top into the lower tank buffer.
4. Take the samples out from the −70 °C freezer along with the molecular weight markers. Thaw the samples with warm water.
5. To denature the samples, heat at 95 °C for 5 min or microwave for 90 s [9].

6. Use a Hamilton syringe to load ^{35}S-IP samples and pre-stained molecular weight markers. For western blot IP, use 200 μL narrow tips. Wash the Hamilton syringe thrice with distilled water to get rid of the remaining sample and the beads. The beads can clog up the syringe if the plunger is pulled out rapidly.
7. When the loading is finished, turn on the cooling unit, put on the lid of gel apparatus, and connect the leads from the lid to a power supply. Turn on the power supply and set the current for gel electrophoresis. For one gel (0.75 mm thickness) of ^{35}S IP, use 13 mA; for one gel (1.5 mm thickness) of western blot IP, use 26 mA. If running 0.75 and 1.5 mm gels together in a single-gel apparatus or in separate gel apparatus, use 39 mA—a combined current for two gels.
8. Turn up the current when the blue tracking dye enters the resolving gel. For ^{35}S-gel, use 15 mA; for western blot IP gel, use 30 mA.
9. Let the gel run until the blue dye front reaches near the bottom about 1–0.5 cm. It takes for about 5½–6 h for ^{35}S-gel and for Western blot gel.
10. Turn off the power supply and cooling unit and disassemble the gel apparatus. Take the gel with the cooling unit out from the lower tank, and place it into a large tray. Release the gel from the cooling unit, and remove the clamps from the glass plates. The gel is ready for the next step.

3.4 Fluorography and Western Blotting

When the gel electrophoresis is complete, remove the gel from the gel apparatus and the glass plates and process according to either ^{35}S IP or western blot IP. For ^{35}S IP, the gel is soaked in sodium salicylate, a fluor that absorbs the radiation from ^{35}S and emits light through the gel [10]. Dry the gel in a gel dryer after it absorbs the sodium salicylate. The gel side of dried gel is exposed to X-ray film in a cassette to record the image of ^{35}S-labeled proteins in the gel. For western blot IP, the proteins from the gel are electrophoretically transferred onto a membrane according to western blot protocol [7].

3.4.1 Fluorography

1. Separate the glass plates with a gel wedge. Remove the stacking gel with a long blade.
2. Drop the gel into a shallow tray containing 100 mL of 0.5 M sodium salicylate and place on a shaker at low speed for 30 min.
3. Place a piece of Saran wrap onto a large piece of plastic or cardboard. Put on gloves to avoid contact with sodium salicylate solution. Pull the gel up with fingers or a gel clip from Bio-Rad, and place the gel on the Saran wrap. Hold the gel

steady with the fingers since it is slippery. However, the gel can be held more securely with a gel clip.

4. Cut a piece of Whatman filter paper #3, about 1–2 in. larger than the gel all around. Spray 70% ethanol onto the filter paper with a squirt bottle to make it damp and wrinkles start to show up. Use a paper towel to remove the excess ethanol and the paper become flat. Place the damp filter paper on the gel.
5. Use a roller or a plastic spreader over the filter paper to drive the excess sodium salicylate solution out from the gel surface. Dried sodium salicylate on the gel surface quenches ^{35}S signal.
6. Place the porous support of the gel dryer on top, and turn the whole assembly over. The Saran wrap side should be on the top and the paper side is on the porous support of the gel dryer. If the Saran wrap is on the bottom, the gel will not dry.
7. Remove the large piece of plastic or cardboard on top, and position the gel with the porous support onto the gel dryer. Cover the gel with the silicon cover and turn on the power. Let the gel dryer run at 80 °C with vacuum for 1–2 h. Do not take the gel out before it is dried because the gel can break immediately into pieces. If it is dried, by barely touching with finger, no bump is from the gel but from the filter paper.
8. When the drying is done, take the gel out from the gel dryer. Trim the excess paper around the gel to fit into the film cassette. Use an autoradiography marking pen or a radioactive pen to mark the molecular weight markers on the gel or the orientation of the gel (*see* **Note 6**).
9. In the dark room, position a film inside the cassette (without intensifier screen) and place the gel side, with no Saran wrap, on the film and close the cassette [11]. If the cassette has an intensifier screen that cannot be removed, then place a piece of paper on the screen and use that side for the film (*see* **Note 7**). Develop the film for a few days in a −70 °C freezer.

3.4.2 Western Blotting

1. Set up protein transfer for western blotting. The procedure has been described previously [7]. In brief, set up the protein transfer onto a nitrocellulose paper under water to avoid bubbles: fill into a tray the protein transfer buffer and put in a piece of Whatman filter paper #3 and a piece of nitrocellulose paper larger than the gel. Push them down to the bottom of the tray. Drop the gel into the tray, and orient the gel to the middle of the nitrocellulose paper. Place a second filter paper into the water and align with the gel. Take the gel sandwich out from the tray and place on a glass plate with the nitrocellulose paper side on top. Roll out the buffer between the gel and the nitrocellulose paper. Put the first Scotch-Brite pad on the positive charge side of the cassette. Transfer the gel sandwich onto

the first Scotch-Brite pad with the membrane side on the bottom. Put the second Scotch-Brite pad on top of gel sandwich and close the cassette. Put the cassette into the protein transfer tank with the nitrocellulose paper side toward red lead. Fill up the tank with protein transfer buffer, and transfer the proteins at 200 mA overnight at 4 °C in a refrigerator (*see* **Note 8**).

2. The next day, take the membrane out from the tank, stain with Fast Green and air-dry for at least 15 min, cut the membrane into strips, and block with 5% nonfat dry milk in PBS-T for 1–2 h at 37 °C.
3. Antibody development with 1/100 diluted serum in PBS-T with 1% nonfat dry milk or with protein-free (PBS-T) blocking buffer (*see* **Note 9**) is on a rocking shaker overnight at 4 °C. The membrane sizes can vary. Hence, place the membrane into a plastic container or a boat made out from aluminum foil, if there is no other way. If turning aluminum foil into a boat, it is a perfect tray for membrane strips because it can vary in size and uses less liquid; however, be aware that leaking may occur because of erosion.
4. Continue with conjugate incubation and development with substrate. After substrate development, rinse the membrane strips thrice with distilled water to remove the substrate, and then stop the reaction with 70% ethanol (*see* **Note 10**). Remove the water and pour into the tray 70% ethanol, enough to cover the membranes. Ethanol dehydrates the membrane so that the adhesive glue can bind to the wet membrane. Thus, while the membrane strips are still wet with ethanol, align them together with face downward on a piece of transparency. Secure the strips together by putting on the back of them with a piece of large size packing tape. Alternatively, apply spray glue on a transparency and apply this to the back of the membrane strips. Air-dry and document the membrane (Fig. 2).

4 Notes

1. Each 25 cm^2 flask of cells contains about two million cells when fully confluent. To prepare the flask with cells to use on the same day, use about 1.5–1.8 million cells to seed into the flask. It takes about 5 h for the cells to settle down and attach. HeLa cells attach to the flask surface to grow. Therefore, it is much easier for the washing of the cells. However, when cells in suspension are used, use half of the normal cell density to label the cells [4]. Be sure the cells are healthy and alive before using for IP.

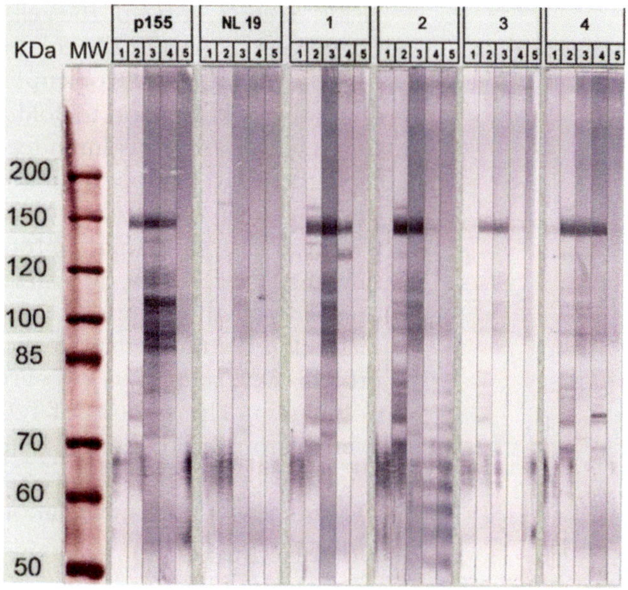

Fig. 2 IP blotting after screening with ^{35}S IP. Lanes 1–4: Sera subjected to Western blot IP. The immunoprecipitates from lanes 1–4 were screened for p155. Lane 1 is positive-control sera for p155. Lane 2 is normal serum. The p155 positive sera are on lanes 1, 2, 3, and 4. Each lane was cut into five strips and developed with five different serums: (1) normal rabbit, (2) anti-p155 rabbit, (3) anti-p155 reference sera, (4) patient sera, and (5) normal

2. There is a practice of pre-gassing for ^{35}S-labeling medium. Depending on the depth of media, it takes from 1 to 4 h for the oxygen to diffuse into the media and reach the cells. The cells need the oxygen to grow. Thus, it would help them by preincubating some of the labeling medium in a flask in the CO_2 incubator at least a few hours before use.
3. If the ^{35}S signal for the protein is too weak, the "strong" labeling level can be used; however, remember that the background level would also increase. More often, the IP test fails because of insufficient protein. The low-sensitivity level is commonly used to enhance the weak signal of proteins by reducing the background because when the background is high, it is really difficult to see the protein bands among the artifacts.
4. Problems may arise with ^{35}S IP because it is a very sensitive system. First, the whole gel may turn dark suddenly for no apparent reason. Four possible causes include: (a) the beads from a bad lot of protein A-Sepharose bind to ^{35}S-labeled proteins; (b) denatured BSA in the blocking step binds to proteins nonspecifically (to test for this, remove the blocking step); (c) cell debris is incompletely removed from the ^{35}S-labeled cell extract; and (d) low pH from the buffer denatures ^{35}S-labeled proteins which can bind nonspecifically to the

beads. Second, some of the cell material cannot pass through the porous support of the column when using column for the IP. To overcome this problem from the washing step after IP, dilute the cell extract to tenfold, and let the beads settle down to the bottom of the column for 20–30 min, and then remove the buffer from the top. Aspirate the cell extract or wash buffer down to near the beads level and repeat the same for four washes. Third, signal may be too weak so that it takes a longer time than expected for the film exposure. This problem may be caused by methionine in the dialyzed FCS. Do not store the dialyzed FCS more than 6 months. Fourth, there may be protein overloading and loss of resolution in the area of immunoglobulin. The amount of IgG may be too high for other proteins to migrate into it [12]. Try reducing the amount of protein A-Sepharose beads or reconstitute the beads again with the buffer with no centrifugation to sediment the beads.

5. Do not centrifuge the columns at high speed to remove the excess water inside the columns before elution of the purified protein because it will remove the purified protein. Once the beads are dried by high-speed centrifugation, there is only a trace or a little bit of purified protein remaining with the beads. Thus, the protein can come off from the beads. The alternative of shaking off the water by hand while keeping the protein inside the beads is by centrifugation at the lowest speed of the centrifuge such as 300–400 rpm. Try a few columns and if the filter and the beads are still wet, it is safe to use that speed of the centrifuge. The shaking off of the water by hand has a disadvantage in that the volume of the sample sometimes is too much for the well of the gel to hold, and it can lead to crossover of the samples to the other wells. However, removing the excess water from the column by centrifugation gives a smaller sample volume that the well of the gel can hold, and the sample volume is mostly the same from one to another.
6. To make a radioactive pen, take a Sharpie permanent marker with ultrafine point, pull the tip out, and add 2–3 μL of [^{35}S]-methionine into the pen with 25 μL of ethanol and reinsert the pen tip. The pen will be ready to use in a few days.
7. Intensifying screens work well only for highly energetic penetrative emission of ^{32}P by absorbing the energy that extend far away from the film back onto the screen and emit out the blue light. The intensifying screens do nothing for less energetic emission from ^{3}H, ^{14}C, or ^{35}S but may decrease the resolution. Therefore, for fluorography with ^{3}H, ^{14}C, or ^{35}S, do not use an intensifier screen [11]. ^{35}S emits β rays 30 cm into air but only 0.25 mm into water or plastic, a very short distance and require a close contact of the gel side to the film [9]. A piece of aluminum foil can stop ^{35}S β rays completely. If there is more

than one gel in the cassette, put a piece of aluminum foil between the gels.

8. Protein transfer buffer usually lost the activity after a few use. Sodium hydroxide is the main source of conductivity; therefore if there is a drop in the antibody signal, add more of 5.0 M sodium hydroxide solution to the protein transfer buffer to regain its activity.
9. Sometimes, 1% nonfat dry milk in PBS-T is not a good solution to use for antibody development because it usually forms curd and becomes frozen while in incubation at 4 °C. Therefore, a blocking solution developed by Pierce, using no protein for blocking of the membrane, is a perfect choice to use because it never become frozen at 4 °C. Therefore, protein-free (PBS-T) blocking buffer is good for the primary and the secondary antibody dilution; however, it is not good for the blocking of the membrane due to its large molecular weight. The molecule of this blocking solution is so big that it hinders the binding site of the proteins, so the antibody cannot fully access them. As a result, the signals from the antibodies are greatly reduced if blocking the membrane with protein-free (PBS-T) blocking buffer compared to nonfat dry milk as a membrane blocking solution.
10. How to stop the membrane from color development (bands become darker) even if it is in 70% ethanol due to long time of handling for mounting: The best way is to dry the membrane right after it reaches the stopping point. Denaturing the enzyme by drying is more effective than using hot water at 95 °C. Since the hot water can cause shrinking of the membrane like it does to some plastic.

References

1. Ornstein L (1964) Disc electrophoresis: I. Background and theory. Ann N Y Acad Sci 121:321–349
2. Kohlrausch F (1897) Ueber Concentrations-Verschiebungen durch Electrolyse im Inneren von Lösungen und Lösungsgemischen. Ann J Phys Chem 62:209–239
3. Laemmli UK (1970) Cleavage of structural proteins during the assembly of the head of bacteriophage T4. Nature 227:680–685
4. Pollard JW (1994) The in vivo isotopic labeling of proteins for polyacrylamide gel electrophoresis. Methods Mol Biol 32:67–72
5. Paulsson KM, Jevon M, Wang JW et al (2006) The double lysine motif of tapasin is a retrieval signal for retention of unstable MHC class I molecules in the endoplasmic reticulum. J Immunol 176:7482–7488
6. Targoff IN, Mamyrova G, Trieu EP et al (2006) A novel autoantibody to a 155-kd protein is associated with dermatomyositis. Arthritis Rheum 54:3682–3689
7. Trieu EP, Gross JK, Targoff IN (2009) Immunoprecipitation—western blot for protein of low abundance. In: Kurien BT, Scofield RH (eds) Protein blotting and detection: methods and protocols, Methods in molecular biology, vol 536. Humana Press, New York, pp 259–275
8. Hames BD (ed) (1998) Gel electrophoresis of proteins: a practical approach, The practical approach series, 3rd edn. Oxford University Press, Oxford

9. Horscroft NJ, Roy P (1997) Thermal denaturation of proteins for SDS-PAGE analysis by microwave irradiation. Biotechniques 22:224–226
10. Bonner WM (1987) Autoradiograms: ^{35}S and ^{32}P. Methods Enzymol 152:55–61
11. Laskey RA (1994) Radiometric methods for detection in blots. In: Dunbar BS (ed) Protein blotting: a practical approach. In: Rickwood D, Hames BD (series eds) The practical approach series, Gel Electrophoresis of Proteins - A Practical Approach Second Edition Edited by B.D. HAMES and D. RICKWOOD - The Practical Approach Series. pp 121–132.
12. Kessler SW (1981) Use of protein A-bearing staphylococci for the immunoprecipitation and isolation of antigens from cells. Methods Enzymol 73:442–459

Chapter 36

A Multichannel Gel Electrophoresis and Continuous Fraction Collection Apparatus for High-Throughput Protein Separation and Characterization

Ming Dong, Megan Choi, Mark D. Biggin, and Jian Jin

Abstract

We developed a multichannel gel electrophoresis system that continuously collects fractions as protein bands migrate to the bottom of gel columns. The device uses several short linear gel columns, each of a different percent acrylamide, to achieve a separation power similar to that of a long gradient gel. A "counter-free-flow" elution technique allows continuous and simultaneous fraction collection from multiple channels at low cost. Using the system with SDS-PAGE, 300 μg samples of protein can be separated and eluted into 48–96 fractions over a mass range of 10–150 kDa in 2.5 h. Each eluted protein can be recovered at 50% efficiency or higher in ~500 μL. The system can also be used for native gel electrophoresis, but protein aggregation limits the loading capacity to about 50 μg per channel and reduces resolution. This system has the potential to be coupled with mass spectrometry to achieve high-throughput protein identification.

Key words Continuous elution electrophoresis, Counter-free-flow, High throughput, Multichannel electrophoresis

1 Introduction

Proteomic characterization of complex protein samples by mass spectrometry requires pre-fractionation at both the protein level and peptide level. PAGE is a widely used method to fractionate proteins due to its high resolution. For example, SDS-PAGE is often used to process tagged pull-down samples [1]. One limitation of all such gel-based methods is that it is laborious and time-consuming to extract protein out of a gel. Some commercially available systems can elute protein from the bottom of an electrophoretic gel column. These systems, however, suffer from high sample dilution and low throughput.

In order to separate and collect protein efficiently over a broad mass/size range with high throughput, we have developed a

Biji T. Kurien and R. Hal Scofield (eds.), *Electrophoretic Separation of Proteins: Methods and Protocols*, Methods in Molecular Biology, vol. 1855, https://doi.org/10.1007/978-1-4939-8793-1_36,

Table 1
Gel concentration and fractionation range

	Gel concentration		MW range (kDa)
SDS single mode	12%		<25
	10%		25–100
	8%		>100
Native single mode	4%		>400
	6%		150–350
	8%		45–180
	Top gel	**Lower gel**	
SDS multimode	10%	12%	10–100
	8%	12%	10–150
	7%	10%	25–150
Native multimode	5%	8%	45–400

multichannel electrophoresis and continuous fraction collection apparatus [2]. This system can be operated in two different modes. In the "single" mode, we use a single-piece, long polyacrylamide gel, and the proteins are separated and eluted directly into the fraction collector. This mode can achieve effective separation over a finite mass range only, however the range being chiefly determined by the concentration of polyacrylamide used (Table 1). In "multi" mode, we use two or three short linear gel segments of different polyacrylamide concentrations (Fig. 1). We first stack these gels one on top of one another, with the lowest concentration gels uppermost (Fig. 1b). The sample is loaded on the very top. After an initial period of electrophoresis when migrating faster, low molecular weight protein bands have traveled into the lower gel segments, and the upper gel segments are removed from the stack and placed on to separate elution devices. Electrophoresis is then resumed simultaneously on each gel segment and protein bands are captured in the liquid phase using the fraction collector (Fig. 1c). This multimode provides an overall separation power similar to that of a traditional long gradient gel for proteins over a broad mass range and allows slow-migrating, larger molecular weight protein bands to be eluted in a realistic time.

A prototype 16-channel apparatus (Fig. 2) has been constructed and tested. It includes multiple gel boxes (each containing four channels of gel column), four elution units, and two fraction collectors. Gel boxes were constructed for easy and quick "docking" and "undocking" of one box on the top of another. Each gel box also contains a cathode electrode and electrophoresis running

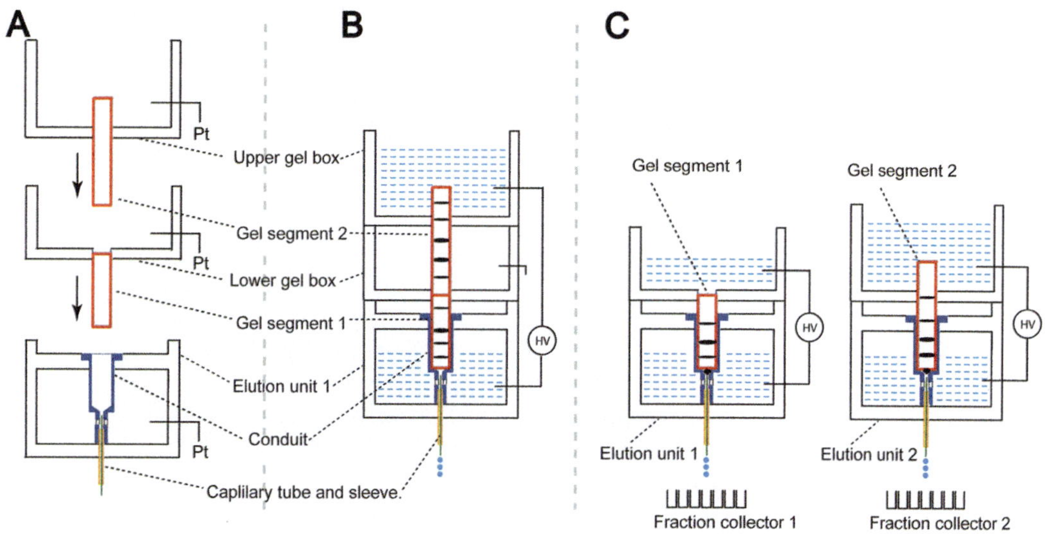

Fig. 1 System design and operation scheme for counter-free-flow electrophoresis. (**a**) Upper and lower gel boxes and an elution unit. Each gel box houses four gel columns (only one gel column is shown) and contains a cathode electrode and electrophoresis running buffer container. The upper gel (Gel segment 2) has a lower polyacrylamide concentration, while the lower column gel (Gel segment 1) has a higher polyacrylamide concentration. (**b**) The two gel boxes and the elution unit are stacked to form an approximation of a gradient gel. Gel segment 1 is inserted into a conduit in elution unit 1, and Gel segment 2 is stacked on top of this unit. Then, an initial phase of electrophoresis is used to separate proteins from a complex mixture loaded at the top of Gel segment 2. High-mobility protein bands migrate into Gel segment 1, whereas lower mobility proteins remain in Gel segment 2. (**c**) After separating (unstacking) the two gel columns, the buffer container associated with Gel segment 1 is filled, whereas Gel segment 2 is inserted into a conduit into elution unit 2 (*see* text and Fig. 2). Finally, electrophoresis is resumed on both gel columns to further separate and elute the protein bands captured in each gel segment. To stack more than two gel segments in a separation, additional gel boxes designed like Gel segment 1 can be included because they are constructed to allow "self-stacking"

buffer container. Gels are poured and polymerized inside each column (made out of a glass tube). Each elution unit contains an anode electrode and another running buffer container. The key feature that allows continuous and effective capturing of eluted protein bands is a "counter-free-flow" technique employed in each elution unit. In this technique, gravity-driven buffer solution flows against the electrophoretic force and sweeps protein molecules migrating off the bottom of the gel column into a glass capillary tube that then deposits collected fractions into a fraction collector.

We have demonstrated that this apparatus allows high-resolution separation and fast and continuous fraction collection over a broad mass range [2]. In a typical 2.5-h SDS counter-free-flow PAGE run, for example, each sample can be separated and eluted into some 48–96 fractions over the mass range of ~10–150 kDa; sample recovery rate can be 50% or higher; up to ~300 μg of protein

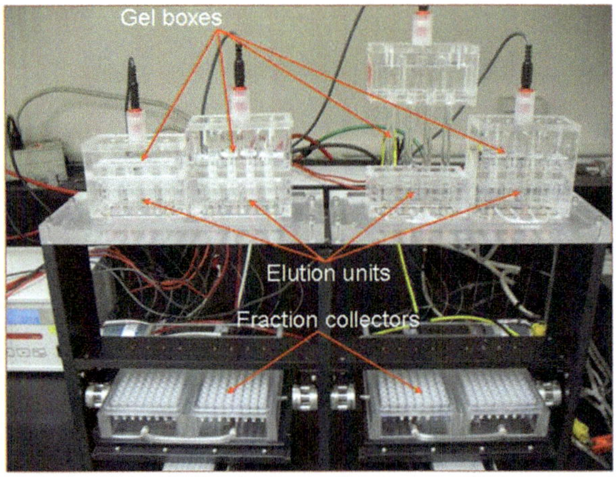

Fig. 2 Prototype 16-channel counter-free-flow electrophoresis and elution apparatus. It includes four gel boxes that each house four gel columns (top), four elution units (middle), and two motorized fraction collectors (bottom)

in 0.4 mL can be loaded onto each channel; and a purified protein band can be eluted within 400–600 μL. However, innovations can still be made to improve system performance. For example, current protocols for native PAGE gels on this system lead to lower resolution and loading capacity compared to SDS-PAGE gels.

2 Materials

All solutions were made using ultrapure water (18 MΩ/cm resistance) and analytical grade reagents. All solutions were stored at 4 °C unless otherwise specified. Gel boxes, elution units, and fraction collector were homemade.

2.1 Gel Column Preparation

1. 30% acrylamide and bis-acrylamide solution: 29:1 ratio (Bio-Rad, Hercules, CA, USA).
2. Separation gel buffer: 1.5 M Tris–HCl, pH 8.8. Weigh 18.17 g of Tris base, dissolve it in about 90 mL of ultrapure water, adjust pH with HCl, and make up to 100 mL with ultrapure water. Autoclave stock solutions and store at room temperature.
3. Stacking gel buffer: 0.5 M Tris–HCl, pH 6.8. Weigh 6.06 g of Tris base, dissolve it in about 90 mL of ultrapure water, adjust pH with HCl, and make up to 100 mL with ultrapure water. Autoclave stock solutions and store at room temperature.
4. 12% separation gel solution: 20 mL each of 30% acrylamide and bis-acrylamide, 12.5 mL of separation gel buffer and 17.5 mL of ultrapure water.

5. 2% separation gel solution: 4 mL of 30% acrylamide and bis-acrylamide solution, 15 mL of separation gel buffer and 41 mL of ultrapure water.
6. 4% stacking gel solution: 8 mL of 30% acrylamide and bis-acrylamide solution, 15 mL of stacking gel buffer and 37 mL of ultrapure water.
7. *N,N,N,N′*-Tetramethylethylenediamine (TEMED).
8. 10% Ammonium persulfate (APS): Dissolve 0.1 g of APS in 10 mL of ultrapure water. Make 100 μL/tube aliquots and store in −20 °C freezer. For each experiment, thaw and use sufficient aliquots and discard the unused solution.
9. Water-saturated butanol: Pour 25 mL of butanol and 25 mL of ultrapure water into a 50 mL centrifuge tube. Shake briefly and let it stand until layers form. Repeat three times. The top layer is water-saturated butanol.
10. Plastic film: Saran Wrap or Dura Seal.

2.2 Protein Sample Preparation

1. 6× SDS sample loading buffer: 150 mM Tris, 1.15 M glycine, 60% v/v glycerol, 18% v/v SDS, 0.06% w/v bromophenol blue. 0.91 g Tris-base, 4.32 g glycine, 9 g SDS, 5.8 g DTT, 30 mg bromophenol blue, add 15 mL ultrapure water, 30 mL glycerol, mix well, and make up to 50 mL with ultrapure water.
2. 6× stacking gel sample loading buffer: 187.5 mM Tris–HCl (pH 6.8), 60% v/v glycerol, 0.06% w/v bromophenol blue. Dissolve 30 mg bromophenol blue in 10 mL ultrapure water, add 6.25 mL stacking gel buffer, 30 mL glycerol, and make up to 50 mL with ultrapure water.
3. Cell sonication buffer: 20 mM HEPES, pH 7.4, 100 mM NaCl, 12.5 mM $MgCl_2$, 0.1 mM EDTA, 20% glycerol, 1 mM PMSF.
4. Amicon Ultra-4 Centrifugal Filter Unit with Ultracel-3 membrane (Millipore, Billerica, MA, USA).
5. Disposable PD-10 desalting columns (GE Healthcare, Piscataway, NJ, USA).

2.3 Gel Running and Protein Collection

1. 10× gel running buffer: 0.25 M Tris, 1.92 M glycine. Weigh 121.1 g of Tris base and 576 g of glycine. Dissolve in 4 L of ultrapure water and store at RT.
2. Dilute 100 mL of 10× gel running buffer to 990 mL with water, and add 10 mL of 10% SDS. Care should be taken to add SDS solution last, since it causes bubbles to form.
3. 1× native gel running buffer: 25 mM Tris, 192 mM glycine. Dilute 400 mL of 10× gel running buffer to 4 L.

4. 1× SDS gel running buffer: 25 mM Tris, 191.8 mM glycine, 0.2% SDS. Dilute 400 mL of 10× gel running buffer to 3.96 L, add 40 mL of 10% SDS, and mix well.
5. Collection plate: 96-well plate.

2.4 Slab Gel Analysis of Eluted Fractions

1. 6× native sample loading buffer: 150 mM Tris, 1.15 M glycine, 60% v/v glycerol, 0.03% w/v bromophenol blue.
2. Criterion Tris–HCl 4–15% polyacrylamide gel (Bio-Rad).
3. Criterion Tris–HCl 4–20% polyacrylamide gel (Bio-Rad).
4. SilverQuest™ Silver Staining Kit (Invitrogen, Carlsbad, CA, USA).

3 Methods (*See* Note 1)

All the steps were carried out at room temperature unless otherwise specified.

3.1 Gel Column Preparation

3.1.1 Single-Mode SDS Gel (See **Note 2**)

1. Seal the bottom of gel glass tube with plastic film (*see* **Note 3**).
2. Choose suitable gel concentration based on fractionation range (Table 1), and mix ingredients based on Table 2 (*see* **Note 4**).
3. Pour 1600 μL of separation gel solution into single-mode gel columns (*see* **Note 5**).
4. Overlay with 200 μL of water-saturated butanol (*see* **Note 6**).
5. Allow the gel to set 30–60 min at room temperature (*see* **Note 7**).
6. Remove butanol, wash gel column twice with water, and absorb the last drop of water with paper towel (*see* **Note 8**).
7. Add 500 μL 1× SDS gel running buffer on the top of gel, and SDS gel is ready for use.

3.1.2 Single-Mode Native Gel

1. Seal the gel glass tube with plastic film (*see* **Note 3**).
2. Choose suitable gel concentration based on fractionation range (Table 1), and mix ingredients based on Table 2 (*see* **Note 4**).
3. Pour 1600 μL of separation gel solution into single-mode gel columns (*see* **Note 5**).
4. Overlay with 200 μL of water-saturated butanol.
5. Allow the gel to set 30–60 min at room temperature (*see* **Note 7**).
6. Remove butanol, wash gel column twice with water, and absorb the last drop of water with paper towel (*see* **Note 8**).
7. Pour 400 μL of 4% stacking gel solution on top (*see* **Note 5**).
8. Overlay with 200 μL of water-saturated butanol.

Table 2
Separation gel recipe (*see* Note 9)

	4%	5%	6%	7%	8%	10%	12%
12% Separation gel solution (mL)	2	3	4	5	6	8	10
2% Separation gel solution (mL)	8	7	6	5	4	2	0
10% APS (μL)	30	30	30	30	30	30	30
TEMED (μL)	10	10	10	10	10	10	10

9. Allow the gel to set 30–60 min at room temperature (*see* **Note 7**).
10. Remove butanol, wash gel column twice with water, and absorb the last drop of water with paper towel (*see* **Note 8**).
11. Add 1× native gel running buffer on the top of gel (*see* **Note 10**).

3.1.3 Multimode SDS-PAGE

1. Seal the gel glass tube with plastic film (*see* **Note 3**).
2. Choose suitable gel concentration based on fractionation range (Table 1), and mix ingredients based on Table 2 (*see* **Note 4**).
3. Pour 1200 μL of separation gel solution into lower gel columns (*see* **Note 5**).
4. Overlay with 200 μL of water-saturated butanol.
5. Allow gels to set 30–60 min at room temperature (*see* **Note** 7).
6. Remove butanol, wash gel column twice with water, and absorb the last drop of water with paper towel (*see* **Note 8**).
7. Add 1× SDS gel running buffer on the top of gel (*see* **Note 10**).
8. Pour 1200 μL separation gel solution into the top gel columns (*see* **Note 11**).
9. Overlay with 200 μL of water-saturated butanol.
10. Allow gels to set 30–60 min at room temperature (*see* **Note** 7).
11. Remove butanol, wash gel column twice with water, and absorb the last drop of water with paper towel (*see* **Note 8**).
12. Add 1× SDS gel running buffer on the top of gel (*see* **Note 10**).

3.1.4 Multimode Native Gel

1. Seal the gel glass tube with plastic film (*see* **Note 3**).
2. Choose suitable gel concentration based on Table 1, and mix ingredients according to Table 2 (*see* **Note 4**).
3. Pour 800 μL of separation gel solution into lower gel columns (*see* **Note 5**).
4. Overlay with 200 μL of water-saturated butanol.
5. Allow gels to set 30–60 min at room temperature (*see* **Note** 7).

6. Remove butanol, wash gel column twice with water, and absorb the last drop of water with a paper towel (*see* **Note 8**).
7. Add 500 μL of 1× native gel running buffer on top of the gel (*see* **Note 11**).
8. Pour 800 μL of separation gel solution into top gel columns (*see* **Note 5**).
9. Overlay with 200 μL of water-saturated butanol.
10. Allow gels to set 30–60 min at room temperature (*see* **Note** 7).
11. Remove butanol, wash gel column twice with water, and absorb the last drop of water with a paper towel (*see* **Note 8**).
12. Pour 400 μL of 4% stacking gel solution on top of top gel column (*see* **Note 5**).
13. Overlay with 200 μL of water-saturated butanol.
14. Allow the gel to set 30–60 min at room temperature (*see* **Note** 7).
15. Remove butanol, wash gel column twice with water, and absorb the last drop of water with paper towel (*see* **Note 8**).
16. Add 500 μL of 1× native gel running buffer on top of the gel (*see* **Note 11**).

3.2 Protein Sample Preparation

1. For native PAGE, add one-fifth volume of 6× native sample loading buffer to a crude or partially fractionated protein sample of 0.1–20 mg/mL, vortex, and briefly spin down (*see* **Note 12**).
2. For SDS-PAGE samples, add one-fifth volume of 6× SDS sample loading buffer to a crude or partially fractionated protein sample of 0.1–20 mg/mL, vortex, briefly spin, heat at 95 °C for 10 min to denature samples, and briefly spin down (*see* **Note 12**).

3.3 Electrophoresis and Protein Collection

Single mode is fast and effective if the application targets a specific protein band. Multimode is more efficient in separating a range of different molecular weight proteins and uses fewer samples, but it requires more control and parameter optimization for operation.

3.3.1 SDS- and Native PAGE Single Mode

1. Insert gel column into conduit and put conduit into elution units (*see* Fig. 1).
2. Fill buffer tank with SDS or native running buffer for SDS-PAGE or native PAGE, respectively.
3. Adjust the position of the capillary tube to 2 mm below the tapered section of the conduit (*see* **Note 13**).
4. Load protein sample on top of gel (*see* **Note 12**).

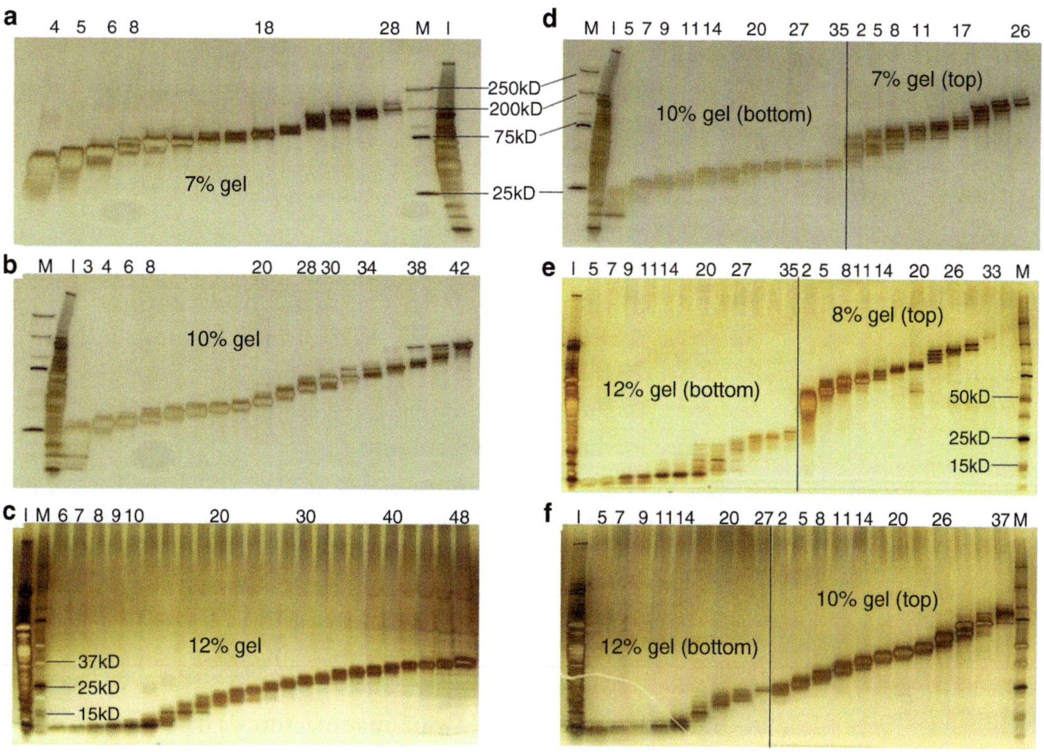

Fig. 3 Elution of SDS counter-free-flow PAGE gels. (**a–c**) Protein contents of collected fractions of a crude protein extract of *D. vulgaris* separated using the single mode and different gel concentrations. (**d–f**) Results obtained using two-gel-concentration multimode and different gel combinations. The concentrations in the gel columns used are given in each panel. The contents of each eluted fraction were visualized via slab SDS-PAGE and silver staining. The sample input lane is marked as "I." Protein markers are denoted as "M," and other lanes are denoted according to eluted fraction number

5. Set electrical field to 20–30 V/cm and the flow rate of the elution buffer to 125 mL/min.
6. Collect ~200 μL fractions in a 96-well plate (Fig. 3).

3.3.2 SDS- and Native PAGE Multimode

1. Stack multiple gel boxes together, insert gel column into conduit, and put conduit into elution units.
2. Fill buffer tank with SDS or native running buffer for SDS-PAGE or native PAGE, respectively (*see* **Note 14**).
3. Load protein sample on top of the gel.
4. Set electrical field to 20–30 V/cm and the flow rate of elution buffer to 125 mL/min.
5. After running for about 30 min, unstack the gel boxes (*see* **Note 15**).
6. Fill the gaps above the lower gel columns with running buffer and turn on the electric field for 30–60 s (*see* **Note 16**).

7. Put the upper gel segments onto another elution unit, and fill the buffer containers for each separated gel box.
8. Run upper and lower gel segments at the same time on different elution units. Set electrical field to 20–30 V/cm and flow rate to 125 mL/min.
9. Collect ~200 μL fractions in a 96-well plate.

3.4 Slab Gel Analysis of Eluted Fractions

1. Sample 12.5 μL from each eluted fraction.
2. Mix each sample with 3 μL of either 6× native sample loading buffer for native gels or 6× SDS sample loading buffer for SDS-PAGE.
3. For SDS-PAGE only, denature sample at 95 °C for 5 min before loading onto gel column.
4. For SDS gels, use Criterion Tris–HCl 4–20% precast gels and 1× SDS running buffer.
5. For native gels, use Criterion Tris–HCl 4–15% precast gel and 1× native running buffer.
6. Load samples.
7. Run electrophoresis at 200 V.
8. Silver stain the gel following manufacturer's instructions.

4 Notes

1. Our current gel column preparation method is suitable for small-scale laboratory operations. For large-scale and robust routine applications, we suggest that an industry approach must be adapted to improve gel quality and prepare gel columns in larger quantities with low cost. First, the structure of the gel box must be changed to allow quick and easy loading/unloading of individual gel columns, instead of the current batch of four. Secondly, a gel pouring manifold must be constructed so that tens or even hundreds of individual gel columns can be poured at once. With adequate QC/QA, the quality of gel columns prepared would be greatly improved. In addition, the gel column tube should be made out of plastic to make them disposable and producible in large quantities with the possibility of having new features molded onto them. These changes can be realized with a small engineering effort.

   In addition, a gel column of smaller diameter should be used if an application involves a much smaller sample volume. A smaller diameter gel column would be beneficial since better matched spot sizes between the eluting protein band and the capturing capillary would allow the use of a lower elution flow rate and/or higher electrophoresis speed. For small sample loading, alternatively, a sample loading well similar to that in a slab gel can be casted into the gel column. This alternative would eliminate the use of butanol, as the sample only migrates through the central core of the gel column during electrophoresis. Currently, the inner surface of our gel tube has not been treated (coated); therefore, there is a possibility of unwanted interactions when proteins come into contact with the wall. There was some evidence (particularly for the native gel electrophoresis) indicating that proteins were slipping through the interface between the gel and the inner glass wall, or were retained by the wall.

Another immediate benefit of using plastic tubes to house the gel columns is that they could be tapered, with a larger diameter at the upper sample loading end and a smaller one at the eluting end. This structure allows more samples to be loaded gradually by increasing the migration speed as a protein band travels into higher electric field zone. The most ideal configuration would be to have a smaller diameter at the end of the upper gel column feeding directly into the lower gel column to deposit proteins at the center of the lower gel column only.

2. Each gel box houses four gel columns; if not all the gel columns will be used, we pour gels into the unused tube to seal it.
3. Put the bottom of the glass tube on a 2″ × 2″ plastic film, and then insert the tube and film into a 1.5 mL Eppendorf tube so that the bottom of the tube is covered by plastic film, which is fixed in place by the Eppendorf tube.
4. 2% and 12% separation gel solutions are made fresh every time, and APS and TEMED are added just before pouring the gel.
5. The gel column is about 4 cm long. As a rule of thumb, a 400 μL gel is about 1 cm long.
6. The butanol layer should be about 0.5 cm high.
7. Once the gel is set, a sharp optical discontinuity at the butanol/gel interface will be visible.
8. Residual butanol can reduce resolution of the protein bands and must be carefully removed. The butanol overlay should not be left on the gel longer than 2 h.
9. SDS is not added in the gel column preparation step. For SDS-PAGE, SDS is added to the gel running buffer and will migrate into gels.
10. The gel can be stored at room temperature overnight.
11. The top and lower gel columns can be made in parallel.

12. For SDS-PAGE, the loading capacity is up to 300 μg and loading volume is 0.2–0.5 mL. For native PAGE, loading capacity is about 50 μg and loading volume is 0.2–0.5 mL.
13. If the capillary tube is set too close to the taper, the tip may become blocked by the gel if it slightly slides out or swells during electrophoresis.
14. Prior to stacking multiple gel boxes together, the space above each gel (in the hole on top of the lower gel boxes) should be filled with running buffer to prevent air bubbles from being trapped between the gel columns.
15. The time required before removing the upper gel segment needs to be determined by preliminary experiments and is typically 30 min.
16. This step is used to drive those protein molecules remaining in solution into the lower gel column.

Acknowledgments

This work was part of ENIGMA, a Scientific Focus Area Program supported by the US Department of Energy, Office of Science, and Office of Biological and Environmental Research, Genomics: GTL Foundational Science through contract DE-AC02-05CH11231 between Lawrence Berkeley National Laboratory and the US Department of Energy.

References

1. Butland G, Peregrín-Alvarez JM, Li J et al (2005) Interaction network containing conserved and essential protein complexes in Escherichia coli. Nature 433:531–537
2. Choi M, Nordmeyer R, Cornell E et al (2010) A multichannel gel electrophoresis and continuous fraction collection apparatus for high-throughput protein separation and characterization. Electrophoresis 31:440–447

Chapter 37

Cell Surface Protein Biotinylation for SDS-PAGE Analysis

Giuliano Elia

Abstract

Proteins expressed at the cell surface play important roles in physiology and represent valuable targets for new therapeutic agents. Indeed, the so-called druggable proteome consists, for about two thirds, of proteins that are either integral to or associated with the cell membrane. In spite of its importance, however, a complete characterization of the cell surface proteome has remained elusive because of the difficulty to efficiently purify these proteins from other contaminants. Methods exploiting the strong interaction between biotin and streptavidin have paved the way for the most significant advances in this field. The present chapter focuses on techniques for cell surface biotinylation with commercially available reagents and capture by avidin affinity chromatography and release of the biotinylated surface proteins for downstream analysis by electrophoretic methods.

Key words Biotin, Cell surface proteins, Electrophoresis, SDS-PAGE, Streptavidin

1 Introduction

At present, there is experimental evidence that 3541 genes in the UniProt database are involved in various disease conditions, including cancer, neurologic, systemic, and cardiovascular disease [1]. About one third of these genes represent potential drug targets, as they belong to known drug target protein classes, which include enzymes, transporters, receptors, and ion channels.

A total of 646 distinct human proteins are the target of all current FDA-approved drugs and are directly related to the mechanism of action for the drug. The distribution of cellular compartments for such targets, based on the combination of several prediction methods for transmembrane regions as well as signal peptides, indicates that almost 70% of the targets are membrane-bound or secreted [2].

Only a fraction of the cell surface proteome has been characterized so far [3, 4], as this protein category has proved difficult to study. This is due to different reasons including (1) the heterogeneity of cell surface proteins [5]; (2) hydrophobicity of most

Biji T. Kurien and R. Hal Scofield (eds.), *Electrophoretic Separation of Proteins: Methods and Protocols*,
Methods in Molecular Biology, vol. 1855, https://doi.org/10.1007/978-1-4939-8793-1_37,

integral membrane proteins, which translates into poor solubility in aqueous media; (3) relatively low abundance; and (4) relative lower frequency of tryptic cleavage sites in transmembrane hydrophobic domain of membrane proteins [6], which all have limited the applicability of the most widespread proteomic techniques (2D-PAGE, bottom-up mass spectrometry) to this category of proteins.

Protocols for purification of cell surface proteins have principally exploited differences in density of subcellular fractions by ultracentrifugation [7, 8], hydrophobicity of the proteins [two-phase partitioning [9, 10], and sequential extraction [11, 12]] or affinity for the isolation of modified or nonmodified cell surface proteins [antibody- or lectin-based approaches] [13, 14]. However, these methods have failed to produce highly pure cell surface proteomes. Other techniques including silica bead coating [15, 16], chemical capture of glycosylated cell surface proteins [17], and cell surface biotinylation [18–20] have produced better and more consistent results.

This chapter discusses cell surface biotinylation with a commercially available, cell membrane-impermeable, cleavable reagent for biotinylation of primary amino groups in side chains of lysines (sulfosuccinimidyl-2-(biotinamido)ethyl-1,3-dithiopropionate; Sulfo-NHS-SS-biotin), downstream capture and release of the biotinylated proteins by means of streptavidin (SA) affinity chromatography, and analysis of the purified protein sample by SDS polyacrylamide gel electrophoresis (SDS-PAGE) and Western blotting with SA-conjugated detection enzymes.

Sulfo-NHS-SS-biotin has a number of distinctive features that make it ideal for studying cell surface proteins. First, it is a polar molecule because of the negative charge contributed by the sulfonate group, very soluble in aqueous buffers even at low temperatures. Second, the negative charge reduces uptake of the reagent by cells, thus limiting contamination due to labeling of intracellular proteins (*see* Notes below). Third, the long spacer arm present in this compound (about 10% longer than that of the corresponding long-chain Sulfo-NHS-LC-biotin) improves binding to avidin-based reagents or supports. Finally, the disulfide bridge contained in the linker can be cleaved by reducing agents for a facilitated release of proteins after capture on avidin-based supports.

Other reagents can also be used for the derivatization of different chemical functional groups in proteins with biotin (e.g., thiol groups in cysteine residues). These compounds, as well as methods for their usage in vitro, in vivo, and ex vivo, have been extensively discussed elsewhere [21, 22].

2 Materials

2.1 Biotinylation of Cell Surface Proteins

1. Cells growing in monolayer or as a suspension in T75 tissue culture flasks in the appropriate cell culture medium.
2. Biohazard-grade, laminar flow sterile hood, equipped with vacuum flasks for aspiration and connected through an HEPA filter to a vacuum line.
3. Phosphate-buffered saline (PBS): 50 mM NaH_2PO_4, 50 mM Na_2HPO_4, and 100 mM NaCl in Milli-Q water, pH 7.4.
4. Sulfosuccinimidyl-2-(biotinamido)ethyl-1,3-dithiopropionate (EZ-Link™ Sulfo-NHS-SS-biotin, e.g., Thermo Fisher Scientific, Rockford, IL).
5. *Tris*(hydroxymethyl)aminomethane hydrochloride (Tris): Dissolve in Milli-Q water to obtain 50 mM concentration, pH 7.4.
6. Oxidized glutathione: Dissolve in PBS at 1 mM concentration.

2.2 Capture and Release of Biotinylated Proteins from Streptavidin Sepharose

1. Protein concentration measurement kit (e.g., Bio-Rad Protein Assay kit; Bio-Rad, Hercules, CA).
2. Streptavidin Sepharose™ High Performance (GE Healthcare, Little Chalfont, Buckinghamshire, UK).
3. Buffer A: 1% NP-40, 0.1% SDS in PBS.
4. Buffer B: 0.1% NP-40, 0.5 M NaCl.
5. Centrifugal filter units with microporous membrane (e.g., Ultrafree-MC 5.0 μm Durapore filter units; Millipore Co., Billerica, MA).
6. Elution solution: 6 M urea, 2 M thiourea, 30 mM D-biotin, 2% SDS in PBS, pH 12.
7. Rotating mixer.
8. Tabletop refrigerated microcentrifuge.
9. Heating block.

2.3 SDS-Polyacrylamide Gel Electrophoresis (SDS-PAGE) with Precast Gels

1. XCell SureLock Mini-Cell (Thermo Fisher Scientific).
2. NuPAGE Bis-Tris acrylamide gradient gels: 4–12%; store at 4 °C.
3. MOPS: 20× concentrated running buffer. Store at room temperature.
4. Reducing loading buffer, fivefold concentrated: 235 mM SDS, 33.6% glycerol v/v, 5% 2-mercaptoethanol v/v, 0.67% bromophenol blue w/v, 210 mM Tris–HCl, pH 6.8. Store at room temperature.
5. Amersham™ ECL™ Rainbow™ Molecular Weight markers (GE Healthcare).

2.4 Western Blotting of Biotinylated Proteins

1. XCell II Blot Module (Thermo Fisher Scientific).
2. Transfer buffer (20×): 500 mM Bicine, 500 mM Bis-Tris (free acid), 20 mM NaEDTA.
3. Methanol analytical grade.
4. Nitrocellulose membrane (e.g., Amersham™ Protran™ 0.45 NC; GE Healthcare) and blotting paper (Grade 3MMChr).
5. Tris-buffered saline with Tween (TBS-T): Prepare 10× stock with 1.37 M NaCl, 27 mM KCl, 250 mM Tris–HCl, pH 7.4, and Tween 20 1%. Dilute 100 mL with 900 mL Milli-Q water before use.
6. Blocking buffer: 5% (w/v) nonfat dry milk in TBS-T.
7. Streptavidin conjugated to horseradish peroxidase (GE Healthcare).
8. Enhanced chemiluminescent (ECL) reagents and BioMax Light film.
9. Saran Wrap.
10. X-ray film developing apparatus (e.g., Curix 60 automatic film processor; Agfa Healthcare, Mortsel, Belgium).

3 Methods

3.1 Biotinylation of Cell Surface Proteins

1. Seed cells growing as monolayers at a density of 5×10^4 cells/cm^2; cells growing in suspension should be seeded at 2–-5×10^5 cells/mL. Depending on the cell type, cells seeded as above should reach confluence within 3–5 days from seeding (*see* **Note 1**).
2. Use 1 mL of the appropriate growth medium *per* 5 squared centimeters of growing surface.
3. Replace seeding medium with the same volume of fresh medium, 24 h after seeding.
4. On the day of the experiment, fill a tray of appropriate size with crushed ice, and place a metal surface (e.g., a second, smaller stainless steel tray) in direct contact with ice.
5. Remove the medium from the flasks by aspiration (or centrifugation in case of cells growing in suspension), and wash twice with 10 mL of ice-cold PBS (*see* **Note 2**). Remove each washing carefully, operating rapidly.
6. Add 15 mL of an ice-cold solution of Sulfo-NHS-SS-biotin in PBS (250 μg/mL; 4.1×10^{-4} M) to each flask, and place them flat on the ice-cold (*see* **Note 3**) stainless steel tray (or resuspend the cell pellet in the same volume of PBS/biotinylating reagent, and place tubes on ice). Allow reaction to occur for 10 min, gently rocking the tray on an orbital shaker or manually inverting the tubes a few times.

7. Quench the reaction by adding 10 mL of Tris solution per flask/tube.
8. Monolayer cells at this point are rounded up and can normally be detached by repeatedly pipetting the liquid contained in the flask over them. In case cells were still holding firmly to the plastic surface at the end of the labeling procedure, gently scrape them off the flask surface and collect them in a centrifuge tube. In both cases, after removing cells from the flask, wash the surface once with additional 5 mL of oxidized glutathione solution (*see* **Note 4**), and add this washing to the collected cells. For cells in suspension, simply add 5 mL of oxidized glutathione solution to each tube.
9. Centrifuge cells in a refrigerated benchtop centrifuge, discard the supernatant, and wash two times with 10 mL of glutathione-containing PBS. Carefully discard the last wash, and add lysis buffer of choice to the cell pellet (*see* **Note 5**).

3.2 Capture and Release of Biotinylated Proteins from Streptavidin Sepharose

1. Measure protein concentration on a fraction of the solubilized cell pellet with one of the many available commercial kits (e.g., the Bio-Rad Protein Assay kit), according to manufacturer's instructions. Adjust concentration to a final value of 0.2 mg/mL with PBS.
2. Transfer an aliquot of 500 μL to a fresh tube, and add 72 μL of pre-washed streptavidin Sepharose resin (*see* **Note 6**). After 2 h of incubation at RT with occasional mixing by manual inversion of the tubes, transfer the slurry to an Ultrafree centrifugal filter, and spin down for 30 s at 16,100 × *g* in a tabletop centrifuge.
3. Remove or discard the filtrate, wash the resin three times with 400 μL of buffer A, followed by two additional washings with 400 μL of buffer B. Finally, quantitatively recover the washed resin from the filtering unit by means of two additional washing steps (each 50 μL) with PBS, and transfer to a fresh tube.
4. Centrifuge the resin for 30 s at 16,100 × *g* in the tabletop centrifuge, and remove the supernatant carefully, as the pellet can be very loose. Add 500 μL of the elution solution (*see* **Note 7**) to the pelleted resin, resuspend the pellet by flicking the tube gently with fingertips, and incubate for 15 min at room temperature.
5. Transfer the tube to a heating block set at 96 °C, and incubate for additional 15 min with occasional agitation. Finally, transfer quantitatively the tube content to a new Ultrafree centrifugal filter, spin down for 30 s at 16,100 × *g* in the table top centrifuge, and collect and save the filtrate.

3.3 SDS-PAGE with Precast Gels

This protocol is based on the use of precast gels, which allow a considerable gain of time and yield more reproducible results than self-casted gels, with an acceptably moderate increase of costs.

1. Remove one Bis-Tris 4–12% acrylamide gel from its package, discard the preserving buffer, and gently take out the well-forming comb and the white tape strip at the bottom of the gel cassette.
2. Carefully rinse the wells twice with Milli-Q water, paying attention not to distort the well walls; drain the liquid each time by inverting the gel upside down. Assemble the gel in the apparatus, and seal the inner chamber (*see* **Note 8**).
3. Dilute 30 mL of the 20-fold concentrated MOPS running solution with Milli-Q water to a final volume of 600 mL. Fill the inner chamber of the gel with enough buffer as to completely cover the gel wells. Pour the remaining running buffer in the outer chamber, making sure that the buffer level is such as to cover the slot in the bottom part of the gel cassette. The gel is now ready for loading.
4. Twenty-four microliters of the filtrate from the preceding section is mixed with 6 μL of fivefold concentrated loading buffer and heated at 95 °C in a heating block for 5 min.
5. After cooling and a quick spin to collect evaporated liquid, the sample is applied onto the wells of the SDS-PAGE gel. Ten microliters of the Rainbow Molecular Weight Markers are also loaded in one of the outermost wells.
6. Electrophoretic separation is allowed to proceed at 110 mA, 180 V for 1 h or until the tracking dye front reaches the bottom of the gel. Gels are then removed from the casting assembly and either processed for Western blotting or fixed for 30 min and stained with SYPRO Ruby (*see* **Note 9**).

3.4 Western Blotting for Biotinylated Proteins

1. Prepare 1 L of 1× NuPAGE transfer buffer by adding 50 mL of 20× NuPAGE transfer buffer and 200 mL methanol to 750 mL of Milli-Q water.
2. Gels obtained as described above are soaked for 10 min at RT in 1× NuPAGE transfer buffer with gentle agitation. During this time, cut one piece of nitrocellulose membrane and several pieces of 3MM filter paper in rectangles of the size of the internal dimensions of the inner core of the XCell II Blotting module (e.g., 9.5 × 5 cm). Soak the membrane, the filter paper, and the blotting pads (*see* **Note 10**) in 1× NuPAGE transfer buffer.
3. The blotting device is assembled as follows: Place two soaked blotting pads into the cathode(−) core of the blot module, followed by 4–5 sheets of soaked filter paper and by the soaked gel, flipped by 180° around its vertical axis (i.e., with the Rainbow marker position inverted with respect to the original loading sequence). Carefully place the pre-soaked nitrocellulose membrane on the gel, paying attention to remove any

trapped air bubbles. Place another 4–5 sheets of soaked filter paper on top of the membrane and again two soaked blotting pads on top of the filter paper. Close with the anode (+) cover, squeezing firmly together the blot module, and slide it into the guide rails of the lower buffer chamber.

4. Fill the blot module with 1× NuPAGE transfer buffer and the outer chamber with cold water for refrigeration, place the lid on the Blot Module unit, and start the run. The power supply is set at 30 V constant, with maximum current limited to 220 mA, and the run is allowed to proceed for 2 h.
5. Once transfer is complete, open the blotting module, carefully remove the nitrocellulose, and transfer it to a clean vessel.
6. Wash twice at RT with blocking buffer (30 min each) on a rocking platform.
7. Dilute the streptavidin-horseradish peroxidase conjugate 1:1000 with blocking buffer, and pour 20 mL of this solution onto the membrane. Allow the biotin-streptavidin binding reaction to proceed at room temperature for 45 min.
8. Remove the streptavidin-horseradish peroxidase solution, and wash the membrane three times for 10 min each with 50 mL of TBS-T and once with PBS. The remaining steps are performed in a dark room under safe light conditions.
9. Place the membrane flat on a piece of Saran wrap, and mix the ECL reagents as specified by the manufacturer's instructions (100 μL per centimeter square of nitrocellulose membrane). Pour on the membrane, and allow to react for 60 s.
10. After this time, drain away the excess of reagent, tap the membrane on clean Kimwipes, and envelope it in clean Saran wrap. The membrane is exposed to X-ray film in an X-ray cassette for variable length of time, and films are developed in a developing machine.

4 Notes

1. In order to obtain sufficient starting material for downstream purification and analysis by SDS-PAGE, it is recommended to use ~0.5–1 × 10^8 cells. Normally, cells contained in two T-75 confluent flasks (per experimental condition) should represent an adequate amount to begin with. Only actively growing cell cultures, exhibiting normal growth curves, should be used for labeling. Avoid using cells that grow sluggishly and/or cells that have been subcultured for an elevated number of doublings. The manipulations required to carry out the present protocol make it ideally suited to labeling of cells which grow as monolayers; however, it can also be applied to cells growing in

suspension. In the latter case, all washing and medium/buffer replacement steps have to be performed by centrifugation of cells at 500 × *g* for 5 min at 4 °C. Gentle centrifugation conditions are needed to prevent cell damage with consequent release of cytoplasmic proteins into the medium. Some cells in suspension have a tendency to form clumps; these should be gently but extensively disaggregated by pipetting prior to labeling, in order to ensure an even and effective access of reagents to the cell surface.

2. For cells growing as monolayers, care should be taken not to lose cells by accidentally detaching them from the plastic surface during experimental manipulations. Avoid pipetting cold buffer directly on the cells, and direct the pipette jet toward the opposite flask surface, which is best done holding the flask with the top side oriented downward. Glass Pasteur pipettes connected to a vacuum device represent a fast and efficient way to remove and discard media and washes.

3. Maintaining cells at 0 °C before and during the labeling procedure helps in reducing the uptake of biotinylating reagents into the cell. Sulfonate group-containing reagents, which are supposed to be membrane-impermeable, are however able to cross the membrane barrier and to label intracellular proteins [23]. A 4% paraformaldehyde short treatment has been shown to limit Sulfo-NHS-SS-biotin diffusion to the cytosolic compartment [23], but it is not entirely clear to which extent this impacts on further characterization of biotinylated proteins.

   Sulfo-NHS-based reagents are inherently prone to hydrolysis. Dissolve Sulfo-NHS-SS-biotin in PBS immediately before addition to cells. In experiments requiring labeling of very high number of samples, consider splitting the work among different operators to minimize artifacts due to partial hydrolysis of the reagent during the experimental time.

4. Oxidized glutathione is added to the solution to prevent untimely reduction of the disulfide bridge-containing cleavable linker in the biotinylating reagent. Oxidized glutathione should also be present during manipulations carried out to lyse cells and capture biotinylated proteins on avidin-based supports, prior to their release with reducing agents.

5. It is not possible to anticipate the precise degree of labeling that can be obtained by this protocol. Biotinylation of cell surface proteins is variable with the cell type used and also from protein to protein. The efficiency of reaction, estimated by Western blotting and avidin-horseradish peroxidase detection, using the conditions reported in this protocol is usually good; however, in case of poor outcome for a particular cell type, modification

of reaction parameters such as pH or ionic strength might help in improving the results [24].

6. Preservatives (e.g., ethanol, sodium azide) are normally included in commercial preparations of (strept)avidin-based affinity chromatography resins. Slurries have to be carefully washed before use to completely remove these preservatives. Pre-wash the desired amount of resin (1 mL of slurry binds 300 nmol of biotin) twice with wash buffer A (1% NP-40, 0.1% SDS in PBS) and twice with PBS. After each washing step, pellet the resin by short centrifugation in a tabletop microcentrifuge. As these pellets are extremely loose, particular care should be taken not to aspirate resin material while removing supernatants of each washing step.
7. D-Biotin is not soluble in water but can be dissolved in concentrated alkali. Prepare a stock 300 mM solution of D-biotin by adding 366.5 mg of the vitamin to 400 μL of 8 M NaOH, and then bring to a final volume of 5 mL with Milli-Q water.
8. Electrophoresis apparatuses can normally accommodate two gels at a time. If only one gel is needed, a plastic part of the same shape of a gel cassette (usually provided) must be used to close the opposite side of the inner core and form the inner buffer chamber.
9. SYPRO Ruby is a highly effective, though expensive, alternative to silver staining for sensitive detection of proteins in SDS-PAGE experiments. Remove the gel from the precast gel cassette, transfer it to a clean vessel, and soak it in fix solution (10% methanol, 7% acetic acid) for 30 min at RT. Stain the gel for 2 h (or overnight) with SYPRO Ruby protein stain (Invitrogen; Cat. No. S12000). Destain by repeated changes of fix solution (30 min each) on an orbital shaker, and finally rinse in Milli-Q water. After destaining, gels are imaged using a UV-transilluminator and a suitable recording device (e.g., DIANA 2 imager, Raytest, Straubenhardt, Germany). The used SYPRO Ruby solution can be recovered and reused two additional times without significant loss of activity.
10. For an efficient transfer of proteins to occur, a perfect contact between the gel and the nitrocellulose needs to be established. Air bubbles trapped at the interface between the two can generate a disruption of the buffer ion flow, and prevent proteins from migrating out of the gel or seriously distort their profile. Care should therefore be taken to thoroughly remove any macroscopically visible air bubbles. However, nitrocellulose should carefully be pre-wetted in order to also avoid air microbubble trapping in the bulk of the membrane itself. To effectively prevent microbubble formation, lay the nitrocellulose membrane in such a way as to make it float on the surface

of the transfer buffer, without submerging it or pouring liquid on top of the membrane itself. Allow the buffer to be absorbed by capillarity through the membrane pores for some minutes, until tiny droplets appear on the upper surface (the membrane "sweats"). The membrane can then be wetted with excess buffer without problems.

Another major source of air bubbles is the blotting pads, especially when they have already been used several times. In our experience, wetting blotting pads with deionized water under pressure helps dislodging also very tiny air bubbles. Eliminate excess water by gently squeezing the pads, and soak in 2–3 changes of 1× NuPAGE transfer buffer prior to assembling the blotting device.

References

1. Uhlen M, Fagerberg L, Hallstrom BM et al (2015) Proteomics. Tissue-based map of the human proteome. Science 347:1260419
2. Wishart DS, Knox C, Guo AC et al (2006) DrugBank: a comprehensive resource for in silico drug discovery and exploration. Nucleic Acids Res 34:D668–D672
3. Wu CC, Yates JR 3rd (2003) The application of mass spectrometry to membrane proteomics. Nat Biotechnol 21:262–267
4. Macher BA, Yen TY (2007) Proteins at membrane surfaces—a review of approaches. Mol Biosyst 3:705–713
5. Tan S, Tan HT, Chung MC (2008) Membrane proteins and membrane proteomics. Proteomics 8:3924–3932
6. Zheng YZ, Foster LJ (2009) Biochemical and proteomic approaches for the study of membrane microdomains. J Proteome 72:12–22
7. Huber LA, Pfaller K, Vietor I (2003) Organelle proteomics: implications for subcellular fractionation in proteomics. Circ Res 92:962–968
8. Castle JD (2003) Purification of organelles from mammalian cells. Curr Protoc Immunol. Chapter 8:Unit 8.1B
9. Schindler J, Lewandrowski U, Sickmann A et al (2006) Proteomic analysis of brain plasma membranes isolated by affinity two-phase partitioning. Mol Cell Proteomics 5:390–400
10. Schindler J, Nothwang HG (2006) Aqueous polymer two-phase systems: effective tools for plasma membrane proteomics. Proteomics 6:5409–5417
11. McCarthy FM, Cooksey AM, Burgess SC (2009) Sequential detergent extraction prior to mass spectrometry analysis. Methods Mol Biol 528:110–118
12. McCarthy FM, Burgess SC, van den Berg BH et al (2005) Differential detergent fractionation for non-electrophoretic eukaryote cell proteomics. J Proteome Res 4:316–324
13. Lawson EL, Clifton JG, Huang F et al (2006) Use of magnetic beads with immobilized monoclonal antibodies for isolation of highly pure plasma membranes. Electrophoresis 27:2747–2758
14. Ghosh D, Krokhin O, Antonovici M et al (2004) Lectin affinity as an approach to the proteomic analysis of membrane glycoproteins. J Proteome Res 3:841–850
15. Robinson JM, Ackerman WE, Tewari AK et al (2009) Isolation of highly enriched apical plasma membranes of the placental syncytiotrophoblast. Anal Biochem 387:87–94
16. Simonson AB, Schnitzer JE (2007) Vascular proteomic mapping in vivo. J Thromb Haemost 5(Suppl 1):183–187
17. Wollscheid B, Bausch-Fluck D, Henderson C et al (2009) Mass-spectrometric identification and relative quantification of N-linked cell surface glycoproteins. Nat Biotechnol 27:378–386
18. Scheurer SB, Roesli C, Neri D et al (2005) A comparison of different biotinylation reagents, tryptic digestion procedures, and mass spectrometric techniques for 2-D peptide mapping of membrane proteins. Proteomics 5:3035–3039
19. Rybak JN, Ettorre A, Kaissling B et al (2005) In vivo protein biotinylation for identification of organ-specific antigens accessible from the vasculature. Nat Methods 2:291–298
20. Scheurer SB, Rybak JN, Roesli C et al (2005) Identification and relative quantification of membrane proteins by surface biotinylation

and two-dimensional peptide mapping. Proteomics 5:2718–2728

21. Elia G (2010) Protein biotinylation. Curr Protoc Protein Sci. Chapter 3:Unit 3.6
22. Elia G (2008) Biotinylation reagents for the study of cell surface proteins. Proteomics 8:4012–4024
23. Yu M-J, Pisitkun T, Wang G et al (2006) LC-MS/MS analysis of apical and basolateral plasma membranes of rat renal collecting duct cells. Mol Cell Proteomics 5:2131–2145
24. Gottardi CJ, Dunbar LA, Caplan MJ (1995) Biotinylation and assessment of membrane polarity: caveats and methodological concerns. Am J Phys 268:F285–F295

Chapter 38

Isolation of Proteins from Polyacrylamide Gels

Stefanie Koristka, Claudia Arndt, Ralf Bergmann, and Michael Bachmann

Abstract

Minute amounts of proteins are required for immunization of mice for the development of antibodies including monoclonal antibodies. Here we describe a rapid procedure for the isolation of proteins from polyacrylamide gels after sodium dodecyl sulfate polyacrylamide gel electrophoresis in sufficient amounts for immunization of animals.

Key words Polyacrylamide gels, Isolation of proteins

1 Introduction

For the development of polyclonal or monoclonal antibodies (mAbs), enriched or better homogenously purified antigens are required. The purer the antigen, the easier to establish a polyclonal serum or mAbs. Nowadays recombinantly expressed proteins are useful starting materials for immunizations of animals. Frequently, proteins are recombinantly expressed in *E. coli* or eukaryotic expression systems (e.g., Chinese hamster ovary (CHO) cells) as His-tagged fusion proteins. Such proteins can easily be purified using an affinity chromatography step on nickel-NTA columns. Such affinity-purified proteins are usually contaminated with bacterial or other host cell proteins, including degradation or truncated expression products. The immune system is by far more sensitive than any staining procedure and will therefore mount an immune response against these contaminants. Consequently, screening of hybridoma supernatants will result in many positive clones which are not directed to the actual antigen of interest. Even worse is the situation in case of animal sera which therefore have to be further affinity purified which again requires highly purified antigens. One simple step to remove lots of these contaminations is to separate the antigen by SDS-polyacrylamide gel electrophoresis and elute the protein of interest from the polyacrylamide gel.

Biji T. Kurien and R. Hal Scofield (eds.), *Electrophoretic Separation of Proteins: Methods and Protocols*, Methods in Molecular Biology, vol. 1855, https://doi.org/10.1007/978-1-4939-8793-1_38,

Here we describe a rapid and simple method for the isolation of proteins for immunization of mice and the preparation of mAbs based on sodium dodecyl sulfate polyacrylamide gel electrophoresis (SDS-PAGE).

2 Materials

1. Mortar and pestle.
2. Plastic columns: e.g., Mobicol tubes (MoBiTec, Göttingen, Germany).
3. Phosphate-buffered saline (PBS).
4. Distilled water.
5. SDS-polyacrylamide gels (*see* Chapters 10 and 42).
6. SDS sample buffer.
7. Coomassie Brilliant Blue for staining and destaining solutions.
8. Nitrocellulose for immunoblotting.
9. Anti-his mAb and anti-mouse antibodies conjugated with alkaline phosphatase.

3 Methods

3.1 SDS-PAGE

For immunization of a mouse, about 30–50 μg of antigen is sufficient per shot. Therefore, the antigen of interest has to be separated on a preparative polyacrylamide gel as schematically shown in Fig. 1A. In dependence on the molecular weight and the epitope of the antigen, the size and percentage of acrylamide in the gel and also the procedure has to be adapted. However, adaptations are easy and not critical. Both native and denaturing gel electrophoresis conditions can be applied (*see* Chapters and). The aim of the here presented experiment was to separate the components of a protein complex consisting of a 50 kDa and a 29 kDa protein. Such isolated 50 kDa protein was used to elicit the mAb La4B6 [1].

1. For this experiment, use a 10% discontinuous SDS-polyacrylamide gel.
2. Dissolve the sample containing about 200 μg protein mixture in SDS sample buffer and denature by heating for 3 min to 95 °C.

3.2 Staining of the Gels and Identification of the Protein of Interest

1. Stain the gels with Coomassie Brilliant Blue (*see* Chapter 40 for the identification of the protein of interest).
2. Destain the gels with methanol and acetic acid (*see* Chapter 40).
3. Destain the gel in distilled water (*see* **Note 1**) for several times (at least 5× 100 mL for 1 h each) to remove methanol and acetic acid out of the gel. In case of the presented work, an unstained gel was used.

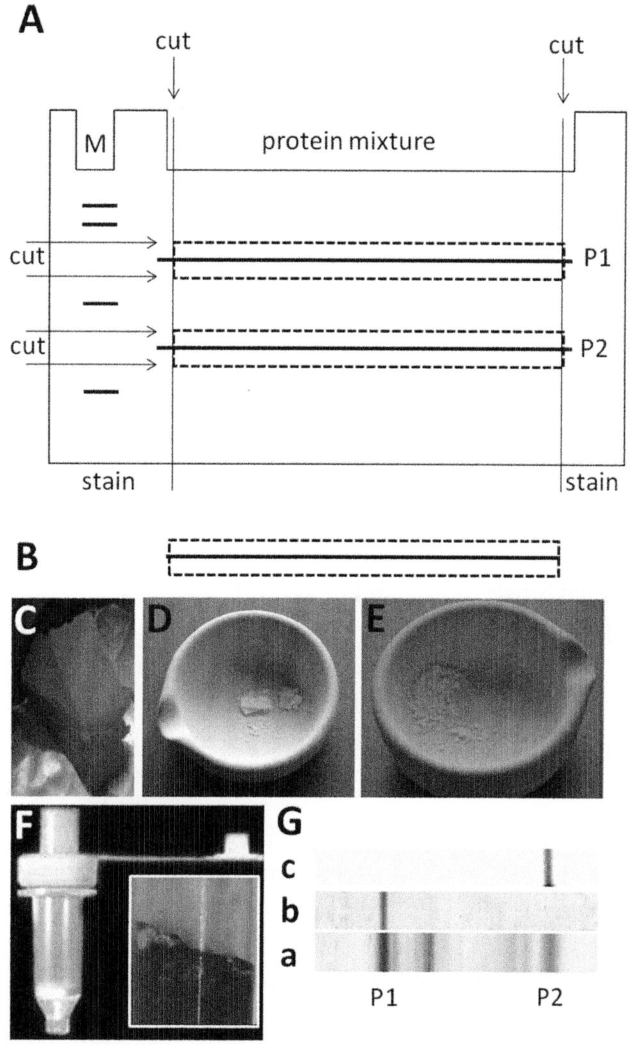

Fig. 1 Example of a protein isolation procedure. (**A**) For separation of the protein complex containing several His-tagged proteins (*see* also **G**, lane a), we prepared a preparative 10% polyacrylamide gel. (**B**) The unstained gel was cut according to the expected position based on pre-stained molecular weight markers as indicated which resulted in two gel samples containing the respective protein. (**C**) The gel samples were frozen at −80 °C overnight. (**D**) After transfer of the respective frozen gel portion (**C**) to a cooled mortar (**D**, **E**) the gel was ground to a fine powder. The powder was transferred into a suitable plastic column (e.g., Mobicol tube, **F**). (**F**, insert) Elution buffer (here for visualization, blue SDS-PAGE sample solution was used) was added dropwise leading to a re-swelling of the gel. The protein samples were then eluted by adding further elution buffer. (**G**) The purity and elution efficiency were finally checked by SDS-PAGE and immunoblotting. The two proteins P1 (50 kDa) and P2 (29 kDa) present in the starting material (lane a) were sufficiently separated and purified by the gel purification step (lanes b, c). The proteins were detected by an anti-His antibody followed by an anti-mouse antibody conjugated with alkaline phosphatase. *M* marker proteins

3.3 Cutting Out the Protein Gel Region

1. Identify the region containing the protein of interest. For this purpose, cut small lanes at either side (and if necessary also in the middle of the gel), and stain them as schematically shown in Fig. 1A. Alternatively, the positions of the proteins can be roughly estimated relative to the positions of the molecular weight markers.
2. Based on the identified positions of the proteins of interest, cut suitable gel fragments from the gel with a scalpel or razor plate (Fig. 1A, B). As we did not stain the gel before cutting, we cut relatively wide (5–7 mm) gel regions out of the gel where we expected to run the proteins of interest relative to the position of pre-stained marker proteins.
3. Stain the remaining gel with Coomassie Brilliant Blue to see that the proteins were correctly removed from the gel (data not shown).

3.4 Grinding of the Polyacrylamide Gel to a Fine Powder (See Note 2)

1. Freeze the isolated gel fragments containing the proteins overnight at −80 °C.
2. Transfer each frozen gel (Fig. 1C) into a mortar (Fig. 1C). The mortar and pestle was cooled down overnight at −20 °C (*see* **Note 2**).
3. Grind the gel to a fine powder using the pestle (Fig. 1D, E). Perform the grinding on dry ice.

3.5 Elution of the Proteins from the Gel Powder

1. Transfer the resulting gel powder (Fig. 1E) containing the protein into a suitable plastic column, e.g., Mobicol tube (MoBiTec, Göttingen, Germany; http://www.mobitec.com; (or alternatively use a small syringe) Fig. 1F).
2. Add SDS-PAGE sample buffer dropwise containing bromophenol blue (*see* **Note 3**) after transfer of the powder into the column. The powder swells, reforming a blue gel in the column (*see* Fig. 1F, insert).
3. Elute the protein by passing further SDS-PAGE sample solution (2×, 500 μL) through the column comparably to a gel filtration (*see* **Note 4**). Using Mobicol tubes, the elution procedure can be facilitated either by the pressure of a syringe or by centrifugation of the column in an Eppendorf tube using an Eppendorf centrifuge (5 min, 10,000 × *g*). The collected eluates contained the respective protein (*see* **Note 5**).
4. After gel isolation, confirm the separation of the purified proteins by a further SDS-PAGE (Fig. 1G).
5. Aliquots of samples separated on 10% SDS-polyacrylamide gels were transferred to nitrocellulose membrane. As both of the selected antigens were tagged with 6×His, the isolated antigens could be detected with a murine monoclonal antibody

directed to the His-tag and an anti-mouse antibody conjugated with alkaline phosphatase. Figure 1G, lane a shows the starting material. As shown in Fig. 1G, the gel purification procedure resulted in an efficient separation of the 50 kDa (Fig. 1G, lane b) and the 29 kDa protein (Fig. 1G, lane c).

4 Notes

1. All solutions should be prepared in water that has a resistance greater 18 M ohm and total organic content of less than five parts per billion. This standard is referred to as "water" in this text.
2. Please cool down the mortar and pestle overnight in a freezer to at least −20 °C. Handle with care, and wear safety googles! Mortar and pestle can break under these conditions especially if they were not cooled down. This can lead to severe injuries!
3. For documentation we used sample buffer solution (containing bromophenol blue) for elution of the gel. For recovery of protein useful for immunization of animals, use PBS instead. The amount of buffer requested for the formation of the gel column depends on the amount of gel powder. Add your selected buffer dropwise until a clear gel with little if any supernatant has formed in the column.
4. Usually, two to three times the gel volume in the column is sufficient for elution (depends on the molecular weight of the protein and the pore size of the gel). If necessary, the buffer volume has to be increased up to ten times of the gel volume.
5. After gel elution, the samples need to be dialyzed (e.g., against PBS) prior to immunization of animals.

References

1. Tröster H, Bartsch H, Klein R, Metzger TE, Pollak G, Semsei I, Schwemmle M, Pruijn GJ, van Venrooij WJ, Bachmann M (1995) Activation of a murine autoreactive B cell by immunization with human recombinant autoantigen La/SS-B: characterization of the autoepitope. J Autoimmun 8:825–842

Chapter 39

Continuous Elution SDS-PAGE with a Modified Standard Gel Apparatus to Separate and Isolate an Array of Proteins from Complex Mixtures

Robert G. E. Krause and J. P. Dean Goldring

Abstract

Sodium dodecyl sulfate polyacrylamide gel electrophoresis is a powerful tool to separate proteins according to their relative sizes. The technique provides information about both the size of a number of proteins and potentially the comparative amounts of each protein. To confirm the identity of proteins, proteins can be eluted from the gel and transferred electrophoretically to nitrocellulose for antibody-based detection. During electrophoresis, if the current is not stopped, proteins continue to pass down the gel and elute from the bottom of the gel. The standard electrophoresis gel apparatus can be employed with the addition of some tubing and alterations to the separating gel to collect proteins separated by size as they elute from the base of the gel as described in this chapter. Complex protein mixtures can be separated into multiple fractions containing single proteins in a few hours. Small amounts (<500 μg of protein) of protein sample can be fractionated.

Key words SDS-PAGE, Continuous elution, Protein separation

1 Introduction

Once separated by SDS-PAGE, proteins can be stained for detection and to determine the relative concentration of each protein. Recovering protein from an unstained SDS-PAGE gel for further analysis can be achieved by electrophoretic transfer to nitrocellulose and the area containing protein on the nitrocellulose cut out and the nitrocellulose dissolved with dimethyl sulfoxide [1]. Protein can be recovered from the gel by diffusing proteins from the gel, dissolving the gel, or using electro-elution devices [2, 3]. Once the current is supplied to the electrophoresis apparatus, proteins will continue to move through the gel until the current is stopped. Should the current be left on, the migrating proteins will elute from the bottom of the gel in order of their size, from the smallest to the largest protein in turn. The SDS-PAGE electrophoresis

Biji T. Kurien and R. Hal Scofield (eds.), *Electrophoretic Separation of Proteins: Methods and Protocols*, Methods in Molecular Biology, vol. 1855, https://doi.org/10.1007/978-1-4939-8793-1_39,

apparatus can be modified so that proteins passing through the gel can elute from the gel and can be recovered. Normally proteins eluting from the base of the gel would enter the lower electrophoresis buffer solution and diffuse. The separating gel has been modified in the present adaptation to form a channel to collect proteins. The channel is connected to piping to transport the eluted proteins to a fraction collector.

The first, 18% acrylamide separating gel, is poured to occupy the lower 1/4 of the volume between the glass plates. A layer of sucrose solution to form the "protein collecting" channel is added on top of the lower separating gel, and a second, upper, separating gel of appropriate acrylamide concentration is added. Once the two gels are set, there remains a liquid channel of sucrose between the upper and lower separating gels. A normal stacking gel is poured on top of the separating gel and either multiple wells or a single well is formed with the well-former.

The proteins are prepared for electrophoresis by boiling in sample buffer and loading into the well(s) in the same manner as standard electrophoresis. Once the current is applied, the proteins run through the gel similarly to an intact SDS-PAGE gel. As the bromophenol tracking dye approaches the sucrose interface, an "elution" buffer, pumped by a peristaltic pump through tubing connected to the sucrose channel, replaces the sucrose and carries the eluting proteins to a fraction collector. Proteins of different sizes are well-separated ([4–7] and Fig. 3).

The technique has been employed previously using a tube gel and typically separated protein samples containing 10–50 mg of protein [5–7]. The adaptation described here does not require a special electrophoresis apparatus and separates 500 μg or less of a complex protein mixture. A standard electrophoresis chamber is used with two additional spacers and plastic tubing connected to a peristaltic pump. We also describe a version for a large-gel apparatus for larger sample volumes which require two modified spacers (*see* below).

The eluted protein, though diluted in the elution buffer, still contains residual SDS which can be removed [8]. The percentage of the separating gel will depend on the range of molecular masses of proteins being separated. Tris-tricine gels can be used to separate small molecular mass proteins. The details required to modify the spacers, assemble the electrophoresis apparatus (Figs. 1 and 2), pour the gels, and run a continuous elution SDS-polyacrylamide electrophoresis gel are described. A result illustrating the separation of *E. coli* BL21 (DE3) proteins on a 7.5% SDS-PAGE continuous elution electrophoresis system is illustrated (Fig. 3). The role of components, explanations for steps in gel assembly, suggestions and the advantages and disadvantages of the method are included (*see* **Notes 1–12**).

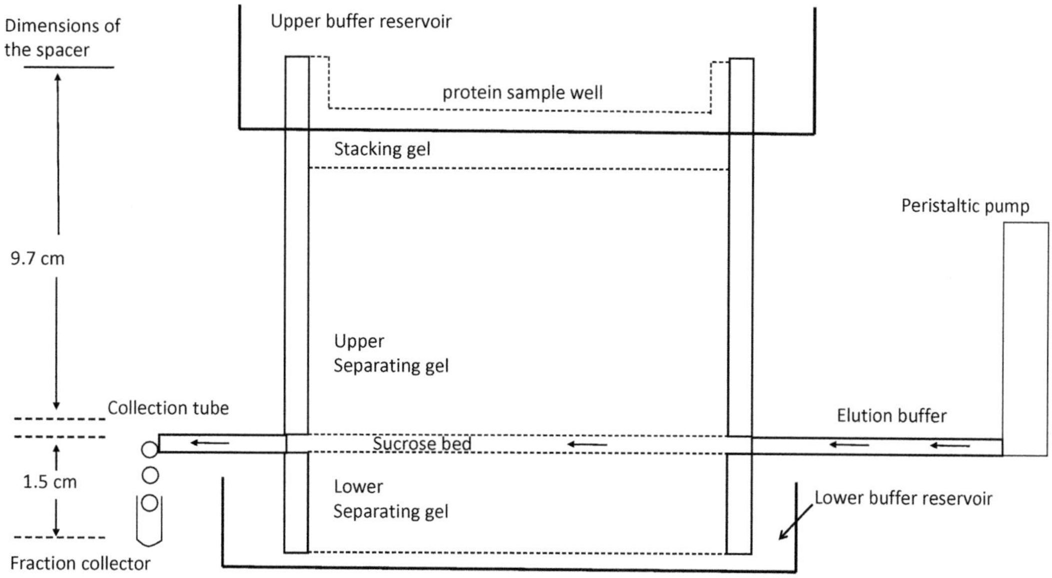

Fig. 1 Diagram of a large-gel apparatus for the fractionation of proteins by continuous elution electrophoresis

2 Materials

Gloves should be worn when handling polyacrylamide solutions. Acrylamide is a neurotoxin.

1. Large-gel: SDS-PAGE electrophoresis apparatus including casting trays and glass plates (Hoefer SE600). Use normal glass plates.
2. Large-gel: Spacers (1.5 mm thick × 16 cm in length) with a small hole (diameter will depend on the thickness of the tubing) to accommodate 1-mm-diameter plastic tubing drilled horizontally through the spacer 1/4 of the spacer length from the base of the spacer. Glue the tubing in place to seal the connection with the spacer.
3. Large-gel: Comb, with all or several of the "teeth" covered with cellotape to form a long well.
4. Mini-gel: SDS-PAGE electrophoresis apparatus including glass and alumina plates and casting trays (Hoefer Mighty Small SE260). We used a 10.6 × 10.1 cm glass plate to generate the data portrayed in Fig. 3.
5. Mini-gel: Spacers—two sets are required (1 mm thick × 1.0 cm × 10.5 cm × 2 for Hoefer and 1 mm thick × 8.8 cm × 0.9 cm Bio-Rad × 2).
6. Mini-gel Comb: Remove the second and ninth tooth of the ten-tooth comb. Cover the remaining teeth in cellotape to make a long individual well (Fig. 2f).

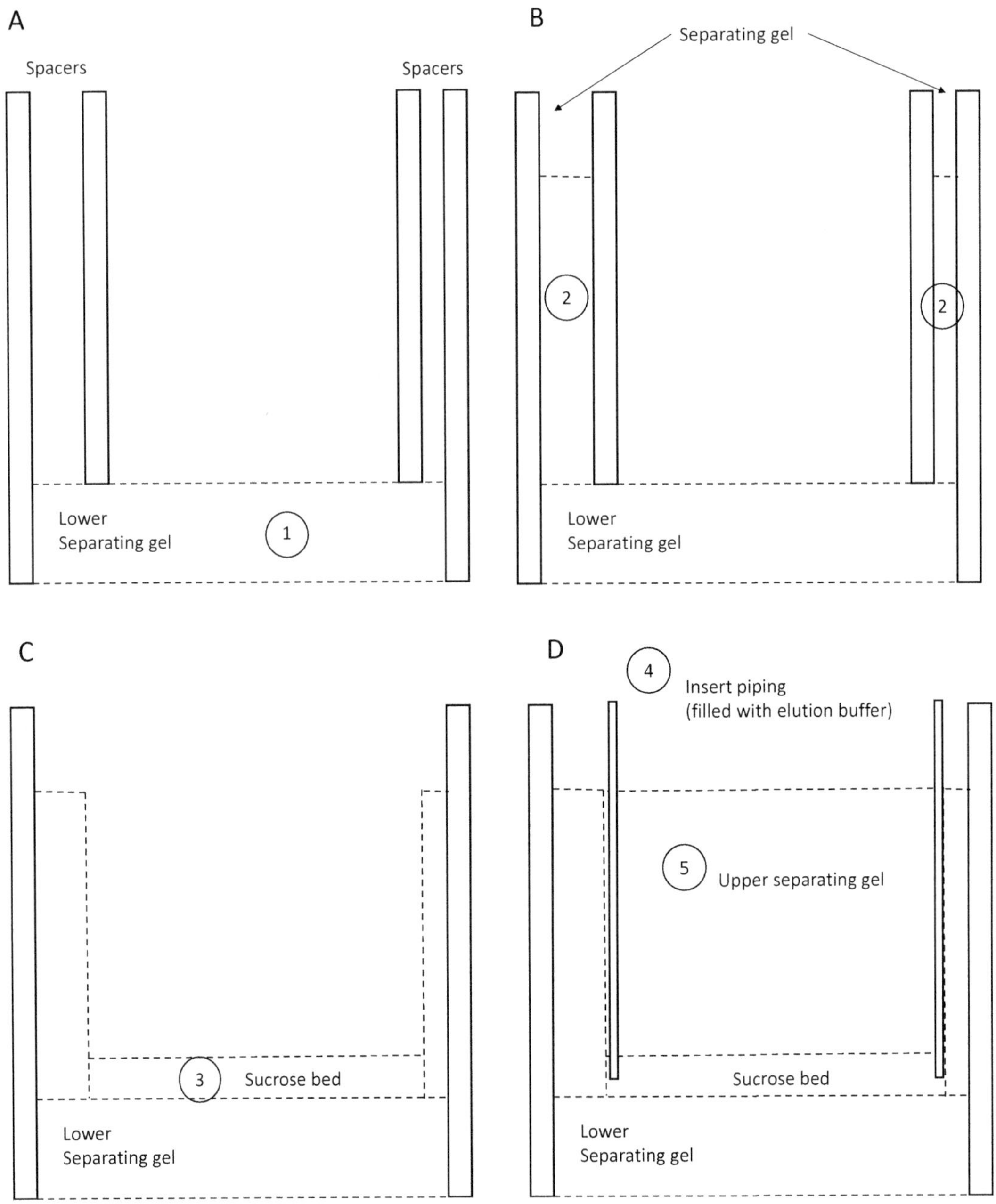

Fig. 2 Diagram of the steps for the assembly of the mini-gel apparatus for the fractionation of proteins by continuous elution electrophoresis. (**a**) Location of four spacers and (1) pouring of lower separating gel, (**b**) pouring of the outer separating gel (2), (**c**) removal of two spacers and addition of sucrose bed (3), (**d**) insertion of elution pipes containing elution buffer (4) and pouring of main separating gel (5), (**e**) pouring of stacking gel (6), (**f**) diagram of the comb, (**g**) final gel assembly illustrating connections for the elution buffer

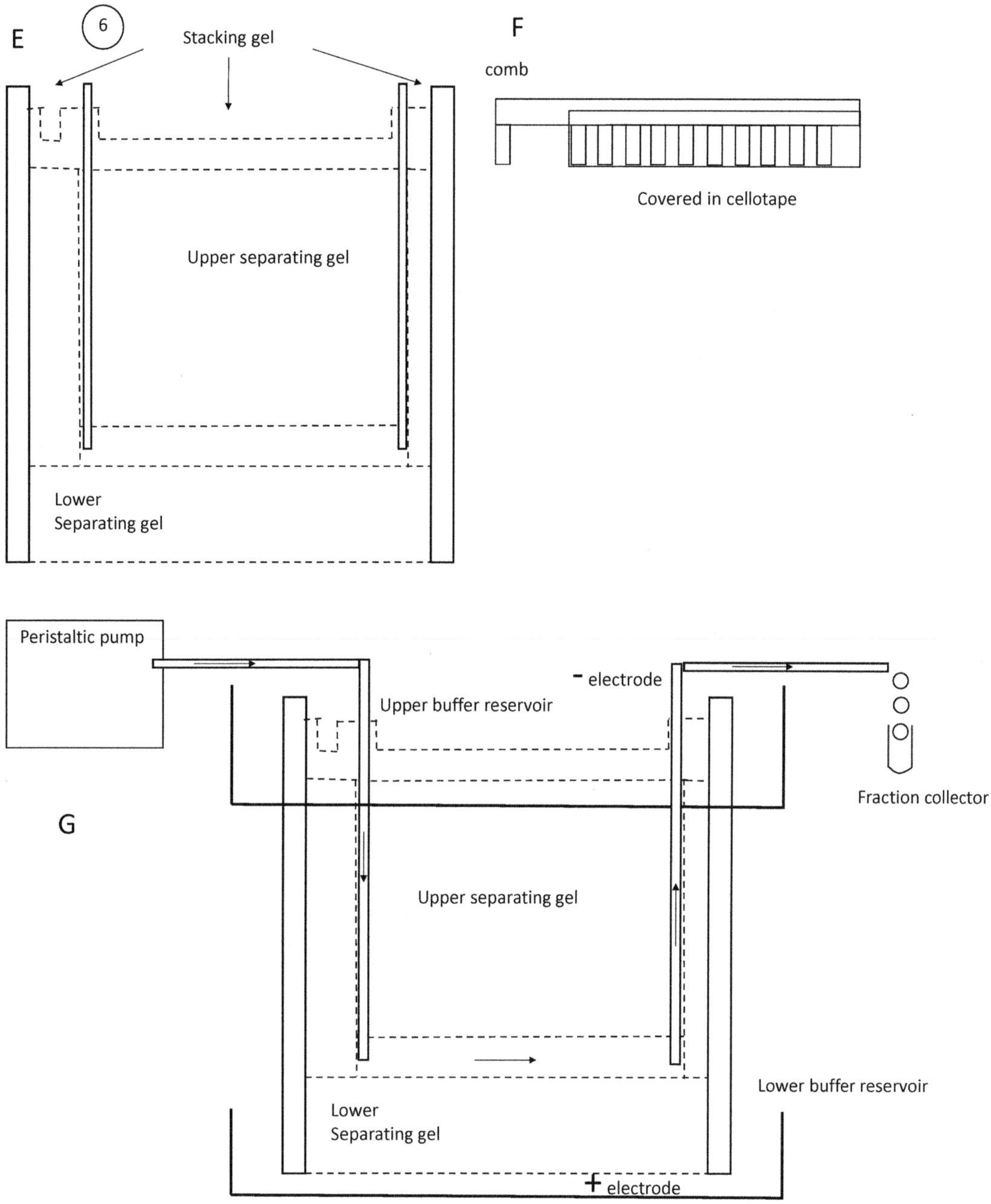

Fig. 2 (continued)

7. Acrylamide: Bis-acrylamide solution (30% (w/v) solution). Dissolve 29.2 g acrylamide and 0.8 g bis-acrylamide in 100 mL distiled H_2O, and filter through Whatman no. 1 paper. The solution should be stored in an amber and glass bottle at 4 °C.

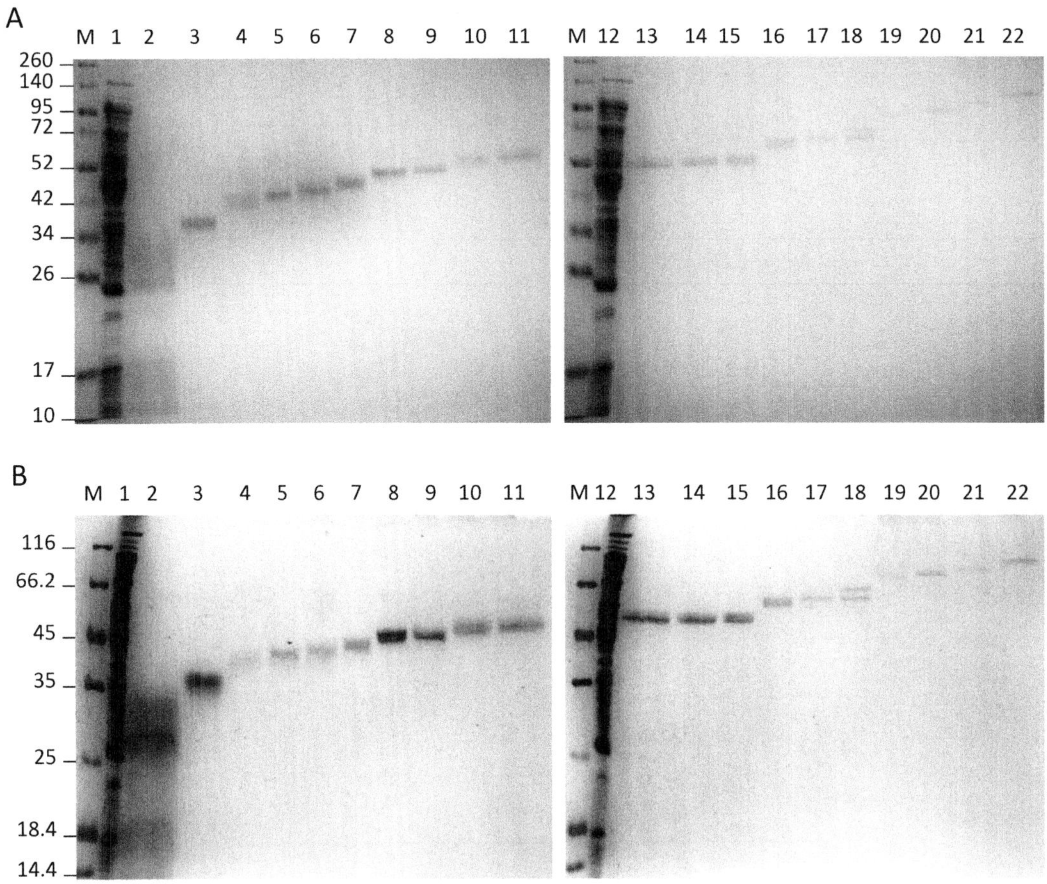

Fig. 3 Example of *E. coli* BL21 (DE3) lysate samples fractionated from a continuous elution electrophoresis experiment and run on two SDS-PAGE gels. Fractions eluted from a 7.5% SDS-PAGE gel with an absorbance at 280 nm above background, were precipitated by the KDS method and electrophoresed on a second 12.5% SDS-PAGE gel under reducing conditions. Samples were stained in (**a**), with Coomassie Brilliant Blue R250 and, in (**b**), with silver. M molecular mass markers (kDa), lanes 1 and 12 *E. coli* BL21(DE3) lysate, lanes 2–11 and 13–22 samples in order of elution from the continuous elution electrophoresis experiment. The elution samples were loaded on both gels

8. Separating gel buffer: 1.5 M Tris–HCl buffer, pH 8.8. Dissolve Tris base (18.148 g) in approximately 80 mL of distilled H_2O, and adjust the pH to 8.8 with HCl. Make the volume up to 100 mL with distiled H_2O, and filter the solution through Whatman no. 1 paper, and store at 4 °C.
9. Stacking gel buffer: 0.5 M Tris–HCl buffer, pH 6.8. Dissolve Tris base (1.5 g) in 20 mL distilled H_2O, and adjust pH to 6.8 with HCl. The volume is made up to 25 mL with distilled H_2O. The solution is filtered through Whatman no. 1 paper and stored at 4 °C.
10. 10% (w/v) SDS (lauryl/sodium dodecyl sulfate): Dissolve SDS (10 g) in 100 mL distilled H_2O.

11. 10% (w/v) ammonium persulfate: Make up ammonium persulfate (0.1 g) to 1 mL with distilled H_2O, and store at 4 °C (*see* **Note 1**).
12. Reservoir/tank/running buffer: 25 mM Tris–HCl buffer, pH 8.6, 192 mM glycine, 0.1% (w/v) SDS. Dissolve Tris base (3 g) and glycine (14.4 g) in distilled H_2O, and make up to 1 L. Just prior to use, add 0.1% (w/v) SDS.
13. 2× SDS-PAGE sample buffer: Mix 1.2 mL 0.5 M Tris–HCl, pH 6.8, 1.0 mL glycerol, 2 mL 10% (w/v) SDS (*see* Subheading 2, **item 10**), and 0.5 mL 0.1% (w/v) Bromophenol Blue, and make up to 9.5 mL for a reducing gel (*see* Subheading 3.3, **step 1**) or 10 mL for a nonreducing gel with distilled H_2O. The buffer was stored at 4 °C (*see* **Note 2**).
14. *N,N,N,N*-Tetramethylenediamine (TEMED).
15. 40% (w/v) sucrose solution.
16. Elution buffer: 0.375 M Tris–HCl, pH 8.8. Dissolve Tris base (22.7 g) in 450 mL distilled H_2O, and adjust the pH to 8.8 with HCL, and make up the volume to 500 mL.
17. Peristaltic pump and 1-mm-diameter connecting plastic/Teflon tubing.
18. Large-gel: Two small lengths (3 cm) of plastic/Teflon tubing sealed at one end to be placed over the end of the piping extending through the spacers. These act as taps preventing leakage when pouring the gels.
19. Microfuge tubes or test tubes.
20. Spectrophotometer.
21. Fraction collector.
22. Protein samples boiled in SDS-PAGE loading buffer.
23. Pre-stained molecular weight marker proteins.

3 Methods

All the steps are carried out at room temperature unless otherwise specified.

3.1 Large-Gel Casting

1. Assemble the glass plates for electrophoresis with the spacers containing the horizontal holes placed on either side of the glass plates. At this stage the tubing spanning the spacer has a small, sealed additional length of tubing as a plug.
2. Prepare an 18% (w/v) separating gel. Pour the 18% separating gel into the apparatus until the volume of liquid is just below the holes in the spacers (approximately 4 cm from the base of the spacer and glass plates). Overlay a thin layer of water or water-

saturated butanol to separate the gel from air. Allow the gel to polymerize. After polymerization, remove the water layer.

3. Place a layer of 40% (w/v) sucrose over the polymerized gel to a depth just above the holes in the spacers.
4. Prepare a separating gel with acrylamide concentration (between 4% and 15% acrylamide, or a gradient gel can be used), suitable to separate the range of sizes of proteins of particular interest. Carefully pour the separating gel on top of the sucrose bed. Add a layer of water or water-saturated butanol on top of the separating gel to separate the gel from the air, and the gel is left to polymerize. After polymerization, remove the water layer. Tissue or blotting paper can be used to remove the remaining traces of water.
5. Pour a stacking gel on top of the separating gel in the gel apparatus, and insert either a normal comb to establish lanes or a comb with teeth covered in cellotape, to make a single large well. Allow the stacking gel to polymerize.
6. Remove the comb once the stacking gel has polymerized.
7. A diagram of the large-gel apparatus for continuous elution electrophoresis is shown in Fig. 1.

3.2 Mini-Gel Casting

1. Assemble glass plates for electrophoresis with normal spacers in place (Fig. 2a).
2. Prepare an 18% (w/v) separating gel. Pour the 18% (w/v) acrylamide lower separating gel into the apparatus until the volume of liquid reaches 12 mm from the base of the glass plates (approximately 2.0 mL). The glass can be marked for reference. Allow the gel to polymerize with a layer of water or water-saturated butanol on top. After polymerization, remove the water layer.
3. Insert the second set of spacers to sit flush with the 18% separating gel. Place the left spacer 1 cm away from the existing spacer to accommodate a single lane for the molecular mass markers. Insert the second spacer 0.5 cm from the existing right side spacer (*see* Fig. 2a).
4. Prepare a separating gel with acrylamide concentration (between 4% and 15% acrylamide), suitable to separate the range of sizes of proteins of particular interest. Pour the separating gel into the two gaps between the two spacers on either side of the gel (Fig. 2b), and leave to polymerize (*see* **Note 3**).
5. Remove the inner set of spacers leaving a U-shaped gel. Layer 250 μL of a 40% (w/v) sucrose over the lower polymerized gel (Fig. 2c).
6. The pipes to carry the elution buffer are inserted between the glass plates adjacent to the polymerized separating gel (Fig. 2d).

Important: The pipes should be filled with the elution buffer before placing in the apparatus. The pipes should be placed with the lower end intruding halfway down the sucrose bed layer (Fig. 2d). Additional separating gel (at the same acrylamide concentration used above—*see* Subheadings 3.2, **step 2** and **step 4**) is carefully poured into the apparatus, on top of the sucrose bed, and around the piping. Approximately 8 mL of gel solution was used (*see* **Note 4**). A layer of water or water-saturated butanol is carefully layered on top of the separating gel. The gel is left to polymerize before the water is removed.

7. A stacking gel is introduced, and the comb inserted to generate a single lane for the molecular mass markers and a long well for the protein sample to be loaded (Fig. 2e–g).
8. Once the stacking gel has polymerized, the comb is removed.
9. A diagram showing the mini-gel apparatus for continuous elution electrophoresis is shown in Fig. 2g.

3.3 Protein Sample Preparation

1. To run a sample under reducing conditions: Add 50 μL of ß-mercaptoethanol to 450 μL sample buffer just before use to make the reducing sample buffer (under a fume hood).
2. To run a sample under nonreducing conditions: Use the nonreducing sample buffer described in Subheading 2, **item 13** above.
3. An equal volume of proteins solution is added to either the reducing or the nonreducing sample buffer and boiled for 2 min in a boiling water bath.

3.4 Electrophoresis, Protein Elution, and Collection

1. Assemble the apparatus, and fill the upper and lower reservoir tanks with SDS-PAGE reservoir/tank/running buffer (*see* Fig. 1 large-gel or Fig. 2g mini-gel).
2. Large-gel: The two "taps" over the piping in the spacer are removed, and tubes connecting the apparatus to the peristaltic pump and the fraction collector are added. The tube from the pump to the gel should be full of buffer before connection.
3. Mini-gel: The piping emerging from the top of the gel are connected via tubing to the peristaltic pump and the fraction collector. The tubing from the pump to the gel should be full of buffer before connection.
4. Load the protein sample in SDS sample buffer (with or without ß-mercaptoethanol) into the wells or the single large well at the top of the gel.
5. The apparatus is connected to an electrical supply and the electrical field set to 120–200 V and the current turned on.
6. Large and small apparatus: When the Bromophenol Blue tracking dye nears the sucrose bed, the peristaltic pump is turned on. The 0.375 M (pH 8.8) Tris–HCl elution buffer is pumped

through the channel (made by the sucrose layer) between the two gels (0.5 mL/min; *see* **Note 5**).

7. Samples eluting from the gel are collected with a fraction collector set to collect volumes of between 200 μL and 1.0 mL (*see* **Notes 6–12**).
8. The spectrophotometer is turned on 10 min before it is required in order to warm up.
9. Eluted fractions are monitored at 280 nm in the spectrophotometer to detect eluting protein.
10. Samples from each fraction can be run on normal SDS-PAGE gels and stained with Coomassie Blue, Amido Black, Direct Red 81 [9], or Silver to determine the size of proteins in each fraction and to confirm the separation of proteins in each fraction.

4 Notes

1. Ammonium persulfate when made up fresh should make a popping sound when dissolved (akin to the noise made when liquid is added to "rice crispies" cereal). Ammonium persulfate solutions can be aliquoted and frozen or stored at 4 °C. It should be noted that ammonium persulfate solutions can go off. If a gel does not polymerize, it is often due to the ammonium persulfate solution, and a fresh batch should be prepared.
2. Each component of the sample buffer has an essential function. The buffer ensures proteins are at the correct pH in the stacking gel; the SDS denatures and coats the protein sample in an overall negative charge; the glycerol ensures that proteins "sink" into the sample well; and the Bromophenol Blue allows one to see the loading of sample in the sample well and then the progress of the dye as it precedes proteins through the gel so that the progress of proteins through the gel can be tracked.
3. The left side of the mini-gel contains a well so that colored molecular mass marker proteins can be run simultaneously. The colored markers allow tracking of the progress of proteins and their molecular mass (sizes) through the gel.
4. The gel either side of the piping in the mini-gel set-up is necessary to prevent leakage of the elution buffer. When the piping was placed next to the outer spacers, it was found that the pressure in the elution system forced buffer up next to the spacers to leak from the top of the gel.
5. The volume of liquid pumped per minute by the peristaltic pump can be established in a trial run noting the time taken between the elution of the Bromophenol Blue and the largest

protein. On our apparatus we used the slowest speed possible (0.5 mL/min).

6. For a 7.5% acrylamide electrophoresis gel, the time from entry of the dye front into the separating gel until elution of the dye was 2 h, and it took a further 2.5 h for the proteins to elute from the gel (total = 4.5 h). The time increased by approximately 30 min when a 12.5% polyacrylamide gel was used.
7. For a 7.5% polyacrylamide gel 100 × 0.5 mL samples were collected and 180 for a 12.5% gel.
8. The progress of proteins can be monitored using the Bromophenol Blue tracking dye and the progress of the molecular mass markers in the lane on the left. The design of the apparatus enables the molecular mass markers to run off the gel into the electrophoresis tank.
9. Advantages of the modified apparatus to separate proteins with continuous elution: (a) the modification of the apparatus is very simple; (b) a range of different concentrations of acrylamide or a gradient gel can be used in the same apparatus; (c) Tris-tricine gels can be used for proteins of small molecular weight; (d) different lengths of separating gel can be used to separate proteins of a particular size range; (e) proteins are well-separated; (f) proteins that are separated and eluted from the gel can be used for other experiments, e.g., for the immunization of animals; (g) a continuous polyacrylamide gradient gel can be used to separate proteins over a large range of sizes; (h) eluted samples can be concentrated with precipitation by the KDS method [10]; and (i) samples containing as little as 500 μg total protein can be separated (*see* Fig. 3).
10. Disadvantages of the modified apparatus to separate proteins with continuous elution: (a) setting optimal pump volume per minute to obtain good separation requires trial and error, (b) eluted proteins are denatured, (c) different concentrations of the gel may be necessary to separate very small proteins from very large proteins, and (d) further treatment of the eluted proteins is required to remove any traces of SDS that may remain.
11. SDS can be removed with acetone precipitation [8].
12. A result showing *E. coli* proteins fractionated with the continuous elution electrophoresis method can be seen in Fig. 3.

References

1. Goldring JPD, Ravaioli L (1996) Solubilisation of protein dye complexes on nitrocellulose to quantify proteins spectrophotometrically. Anal Biochem 242:197–201
2. Goldring JPD (2015) Measuring protein concentration on nitrocellulose and after the electrophoretic transfer of protein to nitrocellulose. In: Kurien BT, Scofield RH

(eds) Detection of blotted proteins, Methods in molecular biology, vol 1314. Springer, New York, pp 19–25

3. Scofield RH (2015) Extraction of proteins from gels: a brief review. In: Kurien BT, Scofield RH (eds) Detection of blotted proteins, Methods in molecular biology, vol 1314. Springer, New York, pp 403–405
4. Kima PE, Serivastava IK, Long CA (1992) Proteins with molecular masses of 25 to 40 kilo Daltons elicit optimal protective Reponses against *Plasmodium chabaudi adami* infection. Infect Immun 60:5065–5070
5. Mulvey C, Ohlendieck K (2003) Use of continuous-elution electrophoresis as a preparative tool for blot overlay analysis. Anal Biochem 319:122–130
6. Balagopal P, Nair KS, Stirewalt WS (1994) Isolation of myosin heavy chain from small skeletal muscle samples by preparative continuous elution gel electrophoresis: application to measurement of synthesis rate in human and animal tissue. Anal Biochem 221:72–77
7. Treuheit MJ, Ataei A, Wallick ET, Kirley TL (1993) Purification of the alpha and beta subunit of Na^+, K^+-ATPase by ontinuous elution electrophoresis. Prep Biochem 23:375–387
8. Hager DA, Burgess R (1980) Elution of proteins from dodecyl sulfate-polyacrylamide gels, removal of sodium dodecyl sulfate, and renaturation of enzymatic activity: results with sigma subunit of *Escherichia coli* RNA polymerase, wheat germ DNA topoisomerase and other enzymes. Anal Biochem 109:76–86
9. Choveaux D, Krause RGE, Goldring JPD (2012) Rapid detection of proteins in polyacrylamide electrophoresis gels by Amido Black and Direct Red. In: Kurien BT, Scofield RH (eds) Protein electrophoresis: methods and protocols, Methods in molecular biology, vol 869. Springer, New York, pp 585–589
10. Goldring JPD (2015) Methods to concentrate proteins for protein isolation, proteomic and peptidomic evaluation. In: Kurien BT, Scofield RH (eds) Detection of blotted proteins, Methods in molecular biology, vol 1314. Springer, New York, pp 5–18

Chapter 40

Protein Extraction from Gels: A Brief Review

Biji T. Kurien, Rachna Aggarwal, and R. Hal Scofield

Abstract

Protein gel electrophoresis is an important procedure carried out in protein studies. Elution and recovery of proteins separated by sodium dodecyl sulfate polyacrylamide gel electrophoresis (SDS-PAGE) are often necessary for further downstream analyses. The process involves localizing the protein of interest on the gel following SDS-PAGE, eluting the protein from the gel, removing SDS from the eluted sample, and finally renaturing the protein (e.g., enzymes) for subsequent analyses. Investigators have extracted proteins from gels by a variety of techniques. These include dissolution of the gel matrix, passive diffusion, and electrophoretic elution. Proteins eluted from gels have been used successfully in a variety of downstream applications, including protein chemistry, proteolytic cleavage, determination of amino acid composition, polypeptide identification by trypsin digestion and matrix-assisted laser desorption ionization-time of flight mass spectroscopy, as antigens for antibody production, identifying a polypeptide corresponding to an enzyme activity, and other purposes. Protein yields ranging from nanogram levels to 100 μg have been obtained. Here, we review some of the methods that have been used to elute proteins from gels.

Key words Elution, SDS-PAGE, Passive diffusion, Electroelution, Renaturation

1 Introduction

In order to obtain a specific protein of interest from a gel, it has to be extracted out of the gel matrix that imprisons them. For this, one has to just simply reverse the way these molecules entered the gel. But for all practical purposes this process is not easy, especially when working with large protein complexes [1]. Omitting the gel matrix and performing preparative electrophoresis in solution is an alternative to the arduous process of recovering proteins from gels [2]. However, free-flow electrophoresis is even more challenging with respect to the technical requirements compared to the traditional method of gel electrophoresis followed by protein elution.

Proteins recovered from gels have been used in protein chemistry, proteolytic cleavage, determination of amino acid composition, and polypeptide identification by trypsin digestion and matrix-assisted laser desorption ionization-time of flight mass

Biji T. Kurien and R. Hal Scofield (eds.), *Electrophoretic Separation of Proteins: Methods and Protocols*,
Methods in Molecular Biology, vol. 1855, https://doi.org/10.1007/978-1-4939-8793-1_40,

spectroscopy, as antigens for antibody production, identifying a polypeptide corresponding to an enzyme activity, and other purposes. Protein yields ranging from nanogram levels to 100 μg have been obtained. Denatured enzymes obtained in this manner are often amenable to renaturation and regain enzymatic activity [3].

The procedure basically involves (a) gel electrophoresis (b) localizing the protein of interest on the gel (c) protein elution from the gel (d) SDS removal from the eluted sample and (e) protein renaturation for subsequent analysis [3].

In this chapter we report the various methods that have been employed to elute proteins from gels.

2 Elution Procedures

2.1 Diffusion-Mediated Elution of Proteins from Gels

This method, known as elution by diffusion [4] or passive elution, is the simplest method to release protein molecules from gel matrix and can be accomplished by adding water (or buffer) to a piece of gel containing the protein of interest. Trapped protein molecules, generally proteins below 60 kDa [5], diffuse out of the gel matrix into the surrounding medium if one waits sufficiently long and can be separated from the gel by centrifugation or filtration. Contaminants such as acrylamide monomers or remainders of the polymerization initiators which also diffuse out of the matrix may necessitate additional purification procedures (such as ion exchange chromatography). In addition, this method does not work well with protein complexes in acrylamide gels [6], unless harsh detergents like sodium dodecyl sulfate (SDS) are added to the elution process [7]. Complete elution by passive diffusion has been achieved by crushing a gel slice with a small Teflon pestle in an elution buffer containing 0.1% SDS and incubating the crushed gel fragments for 4 h (for a 36 kDa protein) and 16–24 h (for a 150 kDa protein) on a rotator [3].

2.1.1 SDS Removal and Concentration

SDS removal following elution has been effectively achieved by acetone precipitation. This procedure also concentrates the protein [3].

The use of acetone precipitation to remove SDS and concentrate protein after elution from gel slices has been proposed. For this, a dry ice-ethanol bath is first set up. Cold acetone (−20 °C) is added to the eluted protein sample. The protein in the sample is then permitted to precipitate for 30 min in the dry ice-ethanol bath. Following this, the sample is thawed and centrifuged at maximum speed in a microcentrifuge for a maximum of 5 min for spinning down the precipitated protein. In a test sample of 0.16 μg radioactive RNA polymerase, the authors obtained a recovery of 70% with no carrier BSA in the elution buffer. However, with the use of 15 and 100 μg/mL of BSA in the elution buffer, the authors obtained a yield of 90% and 99%, respectively. They also found that 99.9% of SDS in the eluted protein sample was removed by using the acetone precipitation method. If desired, the authors suggest that the remaining traces may be eliminated by another round of centrifugation after a quick wash of the pellet with the use of 1 mL ice-cold (80%) acetone [3].

To renature the activity of an enzyme after acetone precipitation, the authors carried out the following protocol:

Prepare a 50 mM Tris (pH 7.9) dilution buffer containing 20% glycerol, 0.1 mM EDTA, 1 mM DTT, 0.15 M NaCl, 25–100 μg/mL of BSA (optional), and 0.1% SDS. Prepare 6 M guanidine hydrochloride (GuHCl) in the prepared dilution buffer. Add 20 μL of 6 M GuHCl to solubilize and denature the pellet, a process that takes about 20 min at room temperature. After this, quickly dilute the solubilized protein 50-fold with addition of dilution buffer, and allow to renature at room temperature for 1–12 h [3].

2.2 Protein Recovery by Dissolving Gels

The easiest way to elute proteins from a gel matrix would be to dissolve this matrix. However, the ability of acrylamide to form stable and chemically inert matrix becomes a disadvantage in this context. Very harsh conditions are required to recover proteins from acrylamide gels containing the standard cross-linker *N,N*-′-methylene bisacrylamide. Typically a gel slice is incubated in 30% hydrogen peroxide at 50 °C for this purpose. This results in irreversible damage to all proteins. Other cross-linkers, such as *N, N*′-diallyltartardiamide (gel slice that has this cross-linker needs to be dissolved with 2% periodic acid), *N,N*′-(1,2-dihydroxyethylene) bisacrylamide (protein recovery requires harsh conditions), and ethylene diacrylate (gel slice needs to be dissolved with 1 M sodium hydroxide), have been used. However, these methods have serious problems as well (as noted above). Cleavable disulfide cross-linkers have been used that require milder dissolving conditions. However, the cleavable bond is unstable under alkaline conditions [1].

2.3 Electroelution Devices

Several methods have been described to elute proteins out of gels by the use of an electric field (*reviewed in* [6]). The simplest method requires no special device, except for a dialysis tubing and a flatbed electrophoresis chamber. The proteins are trapped by the large surface of the tubing as they migrate electrophoretically out of the gel. The nonspecific adsorption by the dialysis tube may be problematic.

Other electroelution methods include devices like the vertical-type eluter, horizontal-type electroeluter, and bridge-type electroeluters and methods such as reversed electrophoresis using a discontinuous conductivity gradient, elution by employing a steady-state stacking buffer system, and continuous elution devices. These methods have been reviewed nicely by Seelert and Krause [1]. We discuss below a simple column electroelution method described by Li et al. [8].

2.3.1 A Simple Monolithic Column Electroelution Method for Recovering Proteins from Gel

As per authors Li et al. [8], the four most coveted methods used for protein extraction from gel slices are based upon the precepts of protein diffusion extraction, namely, the physical extrusion of protein, gel dissolution using chaotropic agents, and of course electroelution based on electrophoresis—all with myriad pros and cons with respect to complexity of procedure, recovery, and time. As a variation to the traditional setup, the authors designed a unique novel monolithic column electroelution (MCE) device to demonstrate a more efficient method for protein extraction from protein gel slices. The authors then used the aforementioned common electroelution methods and drew comparisons with their novel device by use of "standard proteins," namely hemoglobin, myoglobin, as well as bovine serum albumin. Interestingly their results evinced that the MCE-mediated experiments were basic and more time effective. It took between 15 and 30 min with a yield of approximately 82%. Further, they remark that in comparison with the approximately 240 min it takes in typical devices, the MCE was able to provide a 70% protein recovery in 15 min. They further detail that the recovery of proteins which is usually expected to be 41–50% using standard methods in standard machines paled in comparison with the 83% protein recovery they obtained with the MCE. They advise the use of this novel device for traditional protein extraction requirements in proteomics as well as in DNA studies.

As described in this chapter, several interesting methods have been employed to elute proteins out of gels. An interesting method to elute proteins out of gels has been described in detail in this book (*see* Chapter 38). The method is simple, convenient, and easy to perform.

References

1. Seelert H, Krause F (2008) Preparative isolation of protein complexes and other bioparticles by elution from polyacrylamide gels. Electrophoresis 29:2617–2636 Review
2. Ressler N (1958) Application of direct photographic photometry to preparative electrophoresis. Nature 182:463–464
3. Burgess RR (2009) Elution of proteins from gels. Methods Enzymol 463:565–572 Review
4. Lewis UJ, Clark MO (1963) Preparative methods for disk electrophoresis with special reference to the isolation of pituitary hormones. Anal Biochem 6:303–315
5. Jacobs E, Clad A (1986) Electroelution of fixed and stained membrane proteins from preparative sodium dodecyl sulfate-polyacrylamide gels into a membrane trap. Anal Biochem 154:583–589
6. Shoji M, Kato M, Hashizume S (1995) Electrophoretic recovery of proteins from polyacrylamide gel. J Chromatogr A 698:145–162
7. May LT, Santhanam U, Sehgal PB (1991) On the multimeric nature of natural human interleukin-6. J Biol Chem 266:9950–9955
8. Li GQ, Shao J, Guo CG, Dong JY, Fan LY, Cao CX (2012) A simple monolithic column electroelution for protein recovery from gel electrophoresis. Anal Biochem 430:24–31

Chapter 41

Gel Absorption-Based Sample Preparation Method for Shotgun Analysis of Membrane Proteome

Xianchun Wang and Songping Liang

Abstract

Membrane proteins solubilized in a starting buffer containing high concentration of SDS are directly entrapped and immobilized into gel matrix when the membrane protein solution is absorbed by the vacuum-dried polyacrylamide gel. After the detergent and other salts are removed by washing, the proteins are subjected to in-gel digestion, and the tryptic peptides are extracted and analyzed by CapLC-MS/MS. The newly developed method not only avoids protein loss and the adverse protein modifications during gel-embedment but also improves the subsequent in-gel digestion and the recovery of tryptic peptides, particularly hydrophobic peptides. Thus, this method facilitates the identification of membrane proteins, especially integral membrane proteins.

Key words Gel absorption, In-gel digestion, Membrane proteome, Mass spectrometry, Shotgun

1 Introduction

Membrane proteins as critical components of cellular structure and function play central roles such as signal transduction, cellular homeostasis, cell–cell recognition, and host defense [1, 2]. A comprehensive analysis of membrane proteins would facilitate our understanding of their roles in cell life. However, the analysis of membrane proteins presents a special challenge due to their hydrophobic nature and low abundance, which seriously complicates their solubilization, separation, and identification [3, 4].

To enhance the solubility of membrane proteins, researchers have commonly used some organic solvents and detergents in their extraction buffer [3–11]. Although these additives can improve the solubilization and extraction of membrane proteins, most of them reduce enzyme activity, interfere with high-performance liquid chromatography (HPLC) separation, and affect mass spectrometric analysis [8, 11, 12]. On the other hand, although the conventional two-dimensional electrophoresis (2DE) has had some successes in

Biji T. Kurien and R. Hal Scofield (eds.), *Electrophoretic Separation of Proteins: Methods and Protocols*, Methods in Molecular Biology, vol. 1855, https://doi.org/10.1007/978-1-4939-8793-1_41,

the analysis of soluble proteins, this technique is unsuitable for the separation of membrane proteins, mainly due to protein aggregation during the isoelectric focusing (IEF) step [13]. In view of this drawback, one-dimensional electrophoresis (SDS-PAGE) combined with mass spectrometry (1D-PAGE-MS) has been widely used for analysis of membrane proteins. In this protocol, a certain concentration of detergents can be used to enhance the solubility of membrane proteins during sample preparation and SDS-PAGE separation, because the detergents, as well as other interfering substances, can be removed by washing, thus giving a clean environment for subsequent in-gel digestion and MS/MS analysis. However, problems such as protein insolubility still remain [14, 15].

For high-throughput analysis of membrane proteome, an alternative protocol named "tube-gel digestion" has been recently developed and applied [14–18]. In this protocol, detergent-containing protein lysate was mixed with acrylamide solution, and the proteins were incorporated into the polyacrylamide gel when the gel solution polymerized. Detergents were removed by washing prior to protein digestion and would not interfere with the subsequent analyses. However, this method unavoidably led to significant protein loss, because there was usually a part of the protein solution excluded from the gel during gel polymerization. In addition, proteins incorporated into gel in this method have a higher risk of being chemically modified by reagents such as acrylamide [19, 20]. Such limitations of the protocol will reduce the efficiency and confidence of protein identification and quantification. Therefore, its applicability to membrane proteome is limited, and there still exists a substantial need for more applicable methods to be explored and optimized.

Here we present a new method for gel absorption-based sample preparation. In the method, the protein sample prepared using high concentration of sodium dodecyl sulfate (SDS) is directly and completely absorbed by vacuum-dried polyacrylamide gel, and then the interfering substances including SDS and some other salts are removed by in-gel washing steps while retaining the denatured proteins in the gel, thus offering a clean environment amenable to downstream buffer exchange, proteolytic digestion, digest recovery, etc. The newly developed protocol not only can more efficiently avoid protein loss and adverse chemical modifications but also improve the protein digestion and tryptic peptide recovery, thereby significantly facilitating the identification of membrane proteins, especially hydrophobic integral membrane proteins. In addition, it is worth mentioning that the volume of the used vacuum-dried gel can be conveniently adjusted according to the volume and quantity of a sample. Therefore, the method can be used for the preparation of protein samples varying in protein molecule size, sample volume, kind, and concentration of detergents and salts.

2 Materials

Prepare all solutions using ultrapure water obtained from a Milli-Q Plus purification system (Millipore, Bedford, MA) and reagents of the highest grade available. Prepare and store all reagents at room temperature unless indicated otherwise. Diligently follow all waste disposal regulations when disposing waste materials.

1. Thirty percent acrylamide/Bis solution (29:1 acrylamide/Bis): Weigh 29 g of acrylamide monomer and 1 g Bis (cross-linker), and transfer to a 100 mL graduated cylinder containing about 40 mL of water. Use a glass rod to mix the solution until the reagents are completely solubilized. Make up to 100 mL with water, and filter the solution through a filter paper. Store at 4 °C in a bottle (*see* **Note 1**).
2. Ammonium persulfate (AP): 10% solution in water (*see* **Note 2**).
3. *N*,*N*,*N*′,*N*′-tetramethylenediamine (TEMED): 10% solution in water. Store at 4 °C.
4. Ammonium bicarbonate (NH_4HCO_3): 25 mM solution in water.
5. Sodium dodecyl sulfate (SDS) lysis buffer: 4% SDS in 25 mM NH_4HCO_3 (*see* **Note 3**).
6. Dithiothreitol (DTT): 50 mM DTT in 25 mM NH_4HCO_3 (*see* **Note 4**).
7. Iodoacetamide (IAA): 40 mM IAA in 25 mM NH_4HCO_3 (*see* **Note 5**).
8. Sucrose solutions: Prepared in weight/weight (W/W) percent concentration (*see* **Note 6**).
9. Homogenization buffer: 50 mM 4-(2-hydroxyethyl)-1-piperazineethanesulfonic acid (HEPES), 1 mM $CaCl_2$, and 0.1 mM phenylmethylsulfonyl fluoride (PMSF), pH 7.4.
10. Automated Agilent 1200 LC system coupled with an ion trap mass spectrometer (Bruker Daltonics, Bremen, Germany).
11. C18 precolumn (500 μm i.d., 3.5 cm long, Bruker).
12. C18 reverse-phase capillary column (PepMap, 180 μm i.d., 15 cm long, LC Packings).
13. Data Analysis™ 3.4 software (Bruker Daltonics).
14. Chemstation B01 (Agilent Technologies, Waldbronn, Germany).
15. EsquireControl™ 6.0 software (Bruker Daltonics).
16. Mascot software 2.0 (Matrix Science, London, UK).

3 Methods

Carry out all the procedures at room temperature unless otherwise specified. Obtain appropriate approval for carrying out animal experiments from the Institutional Animal Care and Use Committee.

3.1 Enrichment of Rat Liver Plasma Membrane

1. Sacrifice the rats, excise the liver, and remove liver blood. Cut the liver into pieces, and homogenize them in ice-cold homogenization buffer.
2. Centrifuge the homogenate at 1500 × *g* for 10 min at 4 °C. Recover the resulting supernatant, and centrifuge it again at 2400 × *g* (Beckman, SW-28 rotor) for 30 min at 4 °C. Recover the pellet.
3. Mix the pellets with 44% sucrose, over the top of which layer 42.3% sucrose carefully. After centrifugation at 90,000 × *g* for 2.5 h at 4 °C, the crude plasma membranes at the top of 42.3% sucrose were collected.
4. For further enrichment, mix the obtained membrane fraction with 50% sucrose, and add 30% sucrose to the mixture until the final concentration of sucrose solution was 44%, over which layer 42.8, 42.3, 41.8, 41.0, 39.0, and 37.0% sucrose sequentially.
5. After centrifugation at 100,000 × *g* for 6 h at 4 °C, collect the plasma membranes floating in the upper part of the sucrose solution column, wash the sample three times with homogenization buffer (*see* **Note 7**), and then store it at −80 °C until further analysis.

3.2 Protein Extraction and Their Reduction and Alkylation

1. Extract the proteins from plasma membranes with 4% SDS/25 mM NH_4HCO_3 (*see* **Note 8**).
2. Reduce the proteins with 50 mM DTT at 50 °C for 1 h, and alkylate them with 40 mM IAA at room temperature in the dark for 45 min.

3.3 Incorporation of Membrane Proteins into Polyacrylamide Gel and In-Gel Digestion

1. For incorporation of the proteins, an 80 μL polyacrylamide gel was firstly prepared (*see* **Note 9**). Mix 26.7 μL of acrylamide/bisacrylamide (30%, w/v, 29:1) solution, 48.5 μL of 25 mM NH_4HCO_3, 4 μL of 1% AP, and 0.8 μL of 10% TEMED in an Eppendorf tube.
2. After polymerization, wash the gel thoroughly with Milli-Q water (*see* **Note 10**), followed by dehydration in 100% acetonitrile (ACN) and lyophilization in a SpeedVac.
3. Add 20 μL of the membrane protein solution (100 μg of protein) into the tube to make the proteins absorbed by the dried gel at 4 °C.

4. Cut the gel lumps into small pieces, and wash thoroughly with 25 mM NH_4HCO_3 and 50% ACN/25 mM NH_4HCO_3 sequentially. Dehydrate the gel pieces in 100% ACN, and then dry them in a SpeedVac.
5. Add trypsin (1:100 w/w enzyme-to-protein ratio) prepared in 25 mM NH_4HCO_3 (*see* **Note 11**) to the gel pieces. Incubate the gels overnight at 37 °C.
6. Extract the tryptic peptides from the gel pieces twice, each by adding 40 μL 67% ACN containing 5% formic acid with sonication. Combine the supernatants, and then concentrate the sample solution in a SpeedVac. Store it at −20 °C until CapLC-MS/MS analysis.

3.4 CapLC-MS/MS Analysis

1. Analyze the tryptic digests prepared as above by an automated Agilent 1200 LC system coupled with an ion trap mass spectrometry, equipped with an autosampler and a C18 reverse-phase capillary column (PepMap, 180 μm i.d., 15 cm long, LC Packings), using the following eluting solvents: solvent A (98% H_2O, 1.9% ACN, and 0.1% formic acid) and solvent B (95% ACN, 4.9% H_2O, and 0.1% formic acid).
2. Before separation on the capillary column, pre-concentrate the sample on a C18 precolumn (500 μm i.d., 3.5 cm long, Bruker) that is connected in the front of the capillary column using solvent A at a flow rate of 20 μL/min (*see* **Note 12**).
3. When the sample is eluted from the precolumn and further separated on the C18 PepMap capillary column, set the flow rate at 3 μL/min and the column temperature at 25 °C. Use a gradient from 5% to 31% B in 137 min and then 31% to 55% B in 10 min, followed by 80% B for 10 min, and then by 5% B for 10 min (*see* **Note 13**).
4. Online direct the peptides eluted from the capillary column into the ion source of the mass spectrometry. Control the LC-MS system using Chemstation B01 and EsquireControl™ 6.0 software. Set the main instrumental parameters as follows: nebulizer pressure, 10 psi; drying gas flow rate, 5 L/min; drying gas temperature, 300 °C; capillary voltage, 4000 V; and the full MS scan mode, standard enhanced (m/z 350–1600 Da). Select the seven most abundant ions (*see* **Note 14**) in each MS scan for collision-induced dissociation (CID) analysis, using data-dependent MS/MS mode over the m/z range of 100–2000 Da.

3.5 Data Processing and Bioinformatics Analysis

1. Process the obtained raw mass spectrometry data, and create Mascot-compatible mgf files using Data Analysis™ 3.4 software with the following main parameters: compound (autoMS) threshold, 10,000; number of compounds, 300; and retention time windows, 1.0 min.

2. For protein identification, perform database searches using Mascot software 2.0 against the international protein index (IPI) rat combined forward/reverse database (vision 3.56, www.ebi.ac.uk/IPI, 79768 protein sequences). Set the search parameters as follows: enzyme, trypsin; allowance of up to one missed cleavage; peptide mass tolerance, 1.2 Da, and MS/MS mass tolerance, 0.6 Da; fixed modification, carbamidomethylation (C); and variable modification, oxidation (at Met) (*see* **Note 15**).
3. Identify proteins generally on the basis of two or more peptides whose ion scores exceed the threshold, $P < 0.05$, which indicates identification at the 95% confidence level. If a protein is identified by a single peptide, inspect manually its spectrum systematically to confirm or cancel the Mascot suggestion. For a protein to be confirmed, its spectrum should be checked and interpreted in detail (*see* **Note 16**).
4. Retrieve the theoretical molecular weight (MW) and isoelectric point (p*I*) of the identified proteins from Mascot output files.
5. Carry out the mapping of putative transmembrane domains (TMDs) for identified proteins using the TMHMM 2.0 program based on transmembrane hidden Markov model (http://www.cbs.dtu.dk/services/TMHMM) by submitting the FASTA files [21].
6. Calculate the grand average of hydropathy (GRAVY) values for identified proteins using the ProtParam program, available at http://tw.expasy.org/tools/protparam.html.
7. Predict the information on the subcellular location and function of identified proteins by using gene ontology (GO) component and function terms, respectively. The text-based annotation files are available for download from GO database ftp site at ftp://ftp.geneontology.org/pub/go [22].

   By using this method, a total of 190 nonredundant membrane proteins were identified from a rat liver membrane-enriched fraction in triplicate experiments. The identified membrane proteins were distributed in the ranges of MW 6.5–538 kDa, pI 4.37–10.72, and GRAVY values −0.86 to +0.80. One hundred twenty-five (65.8%) of the 190 identified membrane proteins were transmembrane proteins with 1–15 transmembrane domains [23].

4 Notes

1. The acrylamide solution can be stored at 4 °C for 1 month. Acrylamide monomer is a neurotoxin. Wear a mask when weighing acrylamide. Wear gloves even while handling polymerized gels.
2. It is best to prepare fresh AP solution each time.

3. SDS precipitates at 4 °C. The SDS solution needs to be warmed prior to use.
4. It is best to prepare fresh solution each time.
5. The concentration of sucrose solution is expressed in percentage by weight.
6. Centrifuge the sample solution at 100,000 × *g* for 30 min at 4 °C to retrieve the plasma membranes each time.
7. About 2% final concentration.
8. The volume of vacuum-dried gel is adjusted according to the amount of a protein sample to be analyzed, guaranteeing that all the protein solution is completely absorbed by the gel.
9. Thoroughly washing guarantees that some residual reagents such as acrylamide in polymerized gel do not react with amino acid residues in proteins, such as His, Lys, Tyr, Cys, or the N-terminus [19].
10. It is helpful to add a certain concentration of organic solvent or enzyme-friendly detergent such as 10% ACN or 0.1% sodium deoxycholate (SDS) [18].
11. The residual detergents and other interfering substances, if any, in the sample are removed prior to mass spectrometric analysis.
12. The eluting gradient can be adjusted according to the type of samples.
13. The number of the ions chosen can be set at 5 or so.
14. The search parameters can be modified as desired.
15. Protein identification and characterization should adhere to the Paris guidelines (http://www.mcponline.org/misc/ParisReport_Final.dtl).

Acknowledgments

This work was supported by grants from the Special Program for Key Basic Research of the Ministry of Science (2004CCA00300), National High Technology Research and Development Program or "863 Program" of China (2006AA02Z141), and National Basic Research Program or "973 Program" of China (2010CB529800, 2007CB914203).

References

1. Josic D, Clifton JG (2007) Mammalian plasma membrane proteomics. Proteomics 7:3010–3029
2. Speers AE, Wu CC (2007) Proteomics of integral membrane proteins–theory and application. Chem Rev 107:3687–3714
3. Blonder J, Goshe MB, Moore RJ et al (2002) Enrichment of integral membrane proteins for proteomic analysis using liquid chromatography-tandem mass spectrometry. J Proteome Res 1:351–360

4. Lin Y, Chen Y, Yang X et al (2009) Proteome analysis of a single zebrafish embryo using three different digestion strategies coupled with liquid chromatography–tandem mass spectrometry. Anal Biochem 394:177–185
5. Ruth MC, Old WM, Emrick MA et al (2006) Analysis of membrane proteins from human chronic myelogenous leukemia cells: comparison of extraction methods for multidimensional LC-MS/MS. J Proteome Res 5:709–719
6. Quach TT, Li N, Richards DP et al (2003) Development and applications of in-gel CNBr/tryptic digestion combined with mass spectrometry for the analysis of membrane proteins. J Proteome Res 2:543–552
7. Mitra SK, Gantt JA, Ruby JF et al (2007) Membrane proteomic analysis of *Arabidopsis thaliana* using alternative solubilization techniques. J Proteome Res 6:1933–1950
8. Zhou J, Lin Y, Deng X et al (2008) Development and application of a two-phase, on-membrane digestion method in the analysis of membrane proteome. J Proteome Res 7:1778–1783
9. McCarthy FM, Burgess SC, van den Berg BHJ et al (2005) Differential detergent fractionation for non-electrophoretic eukaryote cell proteomics. J Proteome Res 4:316–324
10. Nagaraj N, Lu A, Mann M et al (2008) Detergent-based but gel-free method allows identification of several hundred membrane proteins in single LC-MS runs. J Proteome Res 7:5028–5032
11. Zhang N, Chen R, Young N et al (2007) Comparison of SDS- and methanol-assisted protein solubilization and digestion methods for *Escherichia coli* membrane proteome analysis by 2-D LC-MS/MS. Proteomics 7:484–493
12. Zhang N, Li L (2004) Effects of common surfactants on protein digestion and matrix-assisted laser desorption/ionization mass spectrometric analysis of the digested peptides using two-layer sample preparation. Rapid Commun Mass Spectrom 18:889–896
13. Santoni V, Molloy M, Rabilloud T (2000) Membrane proteins and proteomics: un amour impossible. Electrophoresis 21:1054–1070
14. Lu X, Zhu H (2005) Tube-gel digestion: a novel proteomic approach for high throughput analysis of membrane proteins. Mol Cell Proteomics 4:1948–1958
15. Cao R, He Q, Zhou J et al (2008) High-throughput analysis of rat liver plasma membrane proteome by a nonelectrophoretic in-gel tryptic digestion coupled with mass spectrometry identification. J Proteome Res 7:535–545
16. Han CL, Chien CW, Chen WC et al (2008) A multiplexed quantitative strategy for membrane proteomics: opportunities for mining therapeutic targets for autosomal-dominant polycystic kidney disease. Mol Cell Proteomics 7:1983–1997
17. Yu H, Wakim B, Li M et al (2007) Quantifying raft proteins in neonatal mouse brain by "tube-gel" protein digestion label-free shotgun proteomics. Proteome Sci 5:17
18. Zhou J, Huang S, Bi D et al (2009) Analysis of integral membrane proteins by heat gel-embedment combined with improved in-gel digestion system. Electrophoresis 30:4109–4117
19. Righetti PG, Gelfi C, Bosisio AB (1981) Polymerization kinetics of polyacrylamide gels. III. Effect of catalysts. Electrophoresis 2:291–295
20. Liu T, Martin AM, Sinai AP et al (2008) Three-layer sandwich gel electrophoresis: a method of salt removal and protein concentration in proteome analysis. J Proteome Res 7:4256–4265
21. Krogh A, Larsson B, von Heijne G et al (2001) Predicting transmembrane protein topology with a hidden markov model: application to complete genomes. J Mol Biol 305:567–580
22. Ashburner M, Ball CA, Blake JA et al (2000) Gene ontology: tool for the unification of biology. The gene ontology consortium. Nat Genet 25:25–29
23. Zhou J, Xiong J, Li J et al (2010) Gel absorption-based sample preparation for the analysis of membrane proteome by mass spectrometry. Anal Biochem 404(2):204–210

Chapter 42

Ultrarapid Sodium Dodecyl Sulfate Polyacrylamide Mini-Gel Electrophoresis

Rohit Thomas and Biji T. Kurien

Abstract

We describe here an ultrafast method for electrophoresing proteins on SDS-PAGE. Previously we reported a method to complete SDS-PAGE and immunoblotting in an hour, including electrophoresing proteins at 70°C in 10 min. Here we show that we can electrophorese molecular weight standards and bovine serum albumin on a 4–20% gradient gel in well under 10 min using heated (44 °C) Laemmli running buffer and high voltage.

Key words SDS-PAGE, Molecular weight standards, Heat, Gradient gel, Precast gel

1 Introduction

Polyacrylamide gel electrophoresis involving sodium dodecyl sulfate and β-mercaptoethanol separates proteins on the basis of size (SDS-PAGE) [1, 2]. Proteins electrophoresed on SDS-PAGE can be visualized with protein stains. Electrophoresis of proteins on a mini-gel takes at least 2 h to complete to ensure ideal separation without "smiling" artifacts. However, Haeberle showed that it was possible to electrophorese proteins in 5 min using a special gel and a special buffer that was heated to 70 °C [3].

Heat has been used to rapidly transfer proteins from gels to membrane [4–6]. The hypothesis behind this quick protein transfer appears to be that the permeability of the gel is increased by heat, allowing the proteins to transfer easily from the polyacrylamide gel matrix to the membrane [4]. We had also shown previously that the process of SDS-PAGE to immunoblotting can be achieved in 1 h [7], including using heated running buffer to electrophorese proteins in 10 min.

Biji T. Kurien and R. Hal Scofield (eds.), *Electrophoretic Separation of Proteins: Methods and Protocols*, Methods in Molecular Biology, vol. 1855, https://doi.org/10.1007/978-1-4939-8793-1_42,

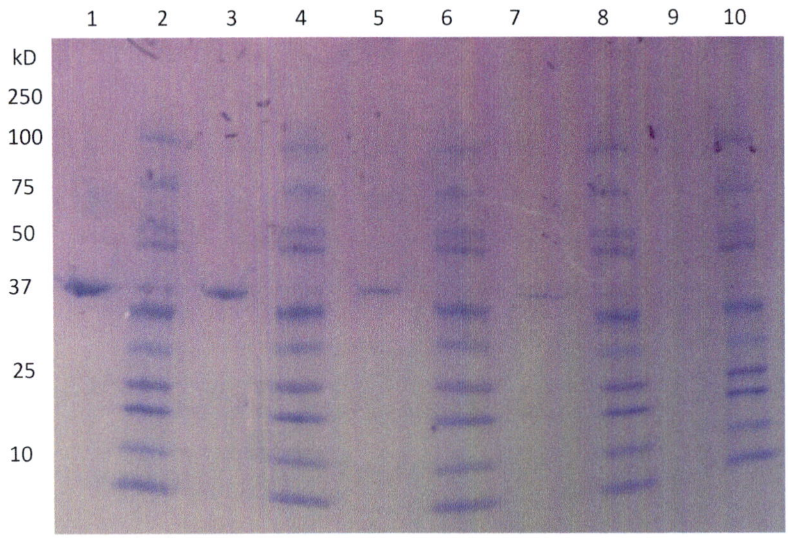

Fig. 1 Ultrafast SDS-PAGE electrophoresis: Pre-stained high molecular weight proteins and increasing amounts of bovine serum albumin were electrophoresed on a 4–20% gradient gel in <6 min and stained with Coomassie Brilliant Blue. Lanes 2, 4, 6, 8, and 10 show pre-stained high molecular weight protein standards. Lanes 1, 3, 5, 7, and 9 show decreasing amounts of BSA (8, 4, 2, 1, and 0.5 μg, respectively)

Here we show that we can electrophorese proteins (Fig. 1) in less than 6 min, a process normally carried out in a period of 2–3 h, using SDS running buffer heated to 50 °C and high voltage (250 V).

2 Materials

Prepare all solutions using ultrapure water (prepared by purifying deionized water, to attain a sensitivity of 18 MΩ cm at 25 °C) and analytical grade reagents. Prepare and store all reagents at room temperature (unless indicated otherwise). Diligently follow all waste disposal regulations when disposing waste materials. We do not add sodium azide to our reagents.

1. Ten percent regular gels or 4–20% gradient SDS-PAGE precast gels (10-well).
2. SDS lysis buffer (5×): 0.3 M Tris–HCl (pH 6.8), 10% SDS, 25% β-mercaptoethanol, 0.1% bromophenol blue, and 45% glycerol. Leave one aliquot at 4 °C for current use, and store the remaining aliquots at −20 °C (*see* **Note 1**).
3. SDS-PAGE running buffer: 10× running buffer (*see* **Note 2**).
4. BenchMark pre-stained molecular weight standards.
5. FB300 power supply.

6. Corning PC-351 magnetic stirrer.
7. Coomassie Brilliant Blue (0.05%) in 25% methanol and 10% acetic acid.

3 Methods

All procedures are carried out at room temperature unless otherwise specified.

3.1 SDS-PAGE

1. Set up commercial (10% or 4–20%) gradient gels in electrophoresis chamber for electrophoresis.
2. Heat running buffer to about 50 °C (*see* **Notes 3** and **4**).
3. Add molecular weight marker to the wells prior to adding the buffer.
4. Prepare bovine serum albumin sample and add to wells similarly.
5. Fill up the wells with the hot buffer using a pipette.
6. Transfer the heated buffer very carefully into the electrophoresis chamber without disturbing the samples in the wells (*see* **Notes 5** and **6**). The buffer reaches to about 44 °C prior to the start of the gel (*see* **Note** 7).
7. Add couple of drops of 0.1% bromophenol blue to the running buffer, and mix gently.
8. Electrophorese the sample at top settings of the FB power supply (~250 V).
9. The samples can be immediately seen to start migrating from the wells into the gel. The sample dye reaches the bottom of the gradient gel in about 6 to 10 min (*see* **Note 8**).
10. Stain the gel with Coomassie Brilliant Blue.

4 Notes

1. SDS precipitates at 4 °C. Therefore, the lysis buffer needs to be warmed prior to use.
2. Simple method of preparing running buffer (0.025 M Tris, pH 8.3, 0.192 M glycine, 0.1% SDS): Prepare 10× native buffer (0.25 M Tris, 1.92 M glycine). Weigh 30.3 g Tris and 144 g glycine, mix, and make it to 1 L with water. Dilute 100 mL of 10× native buffer to 990 mL with water, and add 10 mL of 10% SDS. Care should be taken to add SDS solution last, since it makes bubbles. Here we have used Bio-Rad's 10× running buffer, diluted to 1× with water.

3. If using 1 L of buffer, use at least a 2 L glass beaker to heat. This is because the SDS in the buffer creates bubbles. We used Bio-Rad's running buffer with SDS already in it. If making own running buffer, add SDS after heating.
4. Cover beaker with Saran wrap or some kind of cling wrap. Use extreme caution when heating buffer.
5. Exercise caution in handling the hot contents. Use thermo-gloves to hold the hot beaker and pour carefully.
6. Transfer hot buffer directly into outer electrophoresis chamber. However, when transferring into inner chamber, the hot buffer comes directly in contact with the antigens in the wells. Therefore, transfer the buffer gently into chamber along the sides of a 10 mL pipette so that the antigens in the wells are not disturbed significantly. Alternatively, the molecular weight marker and other antigens can be added as done normally, after filling up the electrophoresis chamber with buffer.
7. The buffer reaches to about 60 °C in the inner chamber at the conclusion of the run.
8. Using the Haeberle gel electrophoresis system [3], we were able to electrophorese proteins in about 5 min (data not shown).

References

1. Laemmli UK (1970) Cleavage of structural proteins during assembly of the head of bacteriophage T4. Nature 227:680–685
2. Srinivas PR (2012) Introduction to protein electrophoresis. Methods Mol Biol 869:23–28
3. Haeberle JR (1997) High-temperature sodium dodecyl sulfate polyacrylamide gel electrophoresis. BioTechniques 23:638–640
4. Kurien BT, Scofield RH (1997) Multiple immunoblots after non-electrophoretic bidirectional transfer of a single SDS-PAGE gel with multiple antigens. J Immunol Methods 205:91–94
5. Kurien BT, Scofield RH (2002) Heat-mediated, ultra-rapid electrophoretic transfer of high and low molecular weight proteins to nitrocellulose membranes. J Immunol Methods 266:127–133
6. Kurien BT, Scofield RH (2000) Ultrarapid electrophoretic transfer of high and low molecular weight proteins using heat. Methods Mol Biol 536:181–190
7. Kurien BT, Danda D, Bachmann MP, Scofield RH (2015) SDS-PAGE to Immunoblot in one hour. Methods Mol Biol 1312:449–454

Chapter 43

A Brief Review of Other Notable Electrophoretic Methods

James D. Fesmire

Abstract

Electrophoretic methodologies for qualitative and preparative purposes are commonly used in biological research and have been well established as an integral analytical tool for a long time in most research laboratories. Listed here are some of the more specialized innovations that have been developed in recent times for special purposes of study. These include micropreparative isoelectric focusing in liquid suspension, accelerated protease digestion by SDS-PAGE, two-dimensional SDS-PAGE for membrane protein resolution, carbon nanotube-modified page for resolution of complement C3, electrophoretic resolution of ultra-acidic proteomes in acidic media, and two-dimensional immunoelectrophoresis of pre-beta/alpha lipoprotein A-I in agarose. All these methods are briefly reviewed in this chapter.

Key words Micropreparative IEF, Controllable pore expansion, Crosslinker cleavage, Doubled SDS-PAGE, Carbon nanotube-assisted PAGE

1 Micropreparative Isoelectric Focusing Protein Separation in a Suspended Drop

This methodology [1] offers the resolving power of isoelectric focusing combined with the potential advantage of analysis of samples in very low concentrations (0.6 μg) contained in volumes as low as 12 μL.

Electrophoresis is performed by suspending the microvolume of sample between the anode and cathode in open air. The electrophoresis procedure is completed when the buffer suspension is evaporated to the extent that it can no longer maintain adhesion connecting both electrodes and is therefore separated. The proteins are recovered from each electrode separately according to their migrational positioning.

The mild electrophoretic conditions of 5 V or less for the procedure generally protect the sample integrity, therefore permitting recoveries of the intact, separated proteins of 50% or more.

James D. Fesmire was deceased at the time of publication.

Biji T. Kurien and R. Hal Scofield (eds.), *Electrophoretic Separation of Proteins: Methods and Protocols*, Methods in Molecular Biology, vol. 1855, https://doi.org/10.1007/978-1-4939-8793-1_43,

The procedure is comparably fast requiring 1 h or less running time, depending on migrational rates of the sample proteins in question.

The method is relatively simple requiring little or no sample preparation for complex equipment. Samples may be applied in an intact or denatured state for resolution.

2 Accelerated Protease Digestion and MS Recovery by Modified SDS PAGE Media

The authors [2] present a novel and effective innovation to the study of proteomic research.

The procedure consists of adjusting the pore size of the bis-acrylamide PAGE gel to permit increased access to the peptides of study and thereby increase the peptide numbers as much as 15–45% as determined by mass spectroscopy.

This increase is achieved by the addition of the alkaline hydrolysis reagent, ethylene glycol diacrylate (EDA), with the bis-acrylamide which cleaves the cross-linkages permitting greater access of the proteases to the peptides of interest.

Acrylamide pore size is controlled by means of the proportions of the additions in accordance with the requirements of the peptides in question.

3 Two-Dimensional Bicine-Doubled SDS-PAGE Method for Enhanced Membrane Protein Resolution

This method [3] consists of a two-dimensional bis-tricine (bicine) acrylamide media in both dimensions using a modified bicine buffer as an effective approach to resolve the difficulties of insolubility in membrane protein chemistry.

To further enhance resolution, acrylamide was increased from 10% to a 12% gradient in bicine buffer from the first to the second dimension.

Large format one-dimensional gels were run at 24 mA/gel for approximately 7 h. Gel thickness ranged from 0.75 mm for small gels to 1.0 mm in thickness for the large format gels.

The authors demonstrated a 151% increase in spot count number applying this method in a bicine buffer system to membrane proteins from the anaerobic bacterium *Clostridium thermocellum*.

4 Carbon Nanotube Modified Polyacrylamide Gel Electrophoretic Method for Enhanced Resolution of Complement C3 Proteins

A modified form of carbon nanotube technology combined with polyacrylamide gel electrophoresis was developed in this study [4] for the purpose of separation and identification of proteins in small concentration related to liver dysfunction.

Due to the inherently high retention properties owing to the structure and composition of the single-walled, water-soluble, carbon nanotube, a carboxylated form in the presence of Triton X-100 was developed for use in combination with acrylamide separating gels in the polyacrylamide gel electrophoretic methodology to be applied toward resolving the protein complement C3 and haptoglobin as may be diagnostic in hepatic cirrhosis patients.

Theory suggests that the adsorption kinetics is improved by means of the addition of carboxylated nanotubes with the proteins of study.

This combined with the driving electrophoretic forces results in a much greater resolving power.

Therefore, this novel modification may be of great advantage for use as a discerning diagnostic tool for a variety of disorders as compared with normal responses in human serum.

5 Electrophoretic Resolution of Ultra-Acidic Proteomes in Acidic Acrylamide Media

This innovative methodology was successfully developed for the purpose of resolving and identifying proteomes in hyperphosphorylated forms which characteristically result in developing an isoelectric point of 2.5 or less [5].

This acidic nature categorically presents a restrictive obstacle for study using traditional two-dimensional electrophoretic systems. One protein of interest of this type is the tau (t) protein, which has been identified as being related to the neurodegenerative Alzheimer's disease.

Authors of the method provide a suitable modification here to circumvent the problem by devising a compatible acidic gel system of low pH (2.5–5.0).

The method consists of first dimension being comprised of an anodic acidic gel with up to 5 mg per lane. The lanes were cut, transferred, and separated by SDS-PAGE as the second dimension.

Proteins are visualized using Coomassie G-250 as stain.

6 Two Dimensional Electrophoresis

This modified immunological technique, originally developed by Laurel [6], provides a powerfully discerning immunological tool for the purpose of examination of antibody specificity and/or homogeneity of antigens.

The method presented here is a modified form presented by Weeke [7] in that the 4 × 4 inch gels are poured separately.

First, 1.5 mL of 2% indubiose (agarose) without antibody is poured into a 4 × 4 inch glass frame separated by a 1.5-mm-thick u-frame and 0.004″ clear Cronar hydrophilic photography film, to which the gel will bond permanently.

This casting is held together with binder clips which also provide a vertical support stand during the curing process.

As soon as the bottom gel is set, but still warmest, the second layer containing the antibody in 12.5 mL of agarose is poured into the casting. Pouring both layers while hot, but separately, creates a more consistent matrix at the boundary which permits the protein to smoothly migrate into the gel containing the antisera.

The casting is stood at room temp until cured sufficiently for use or can be sealed and stored overnight at 4 degrees centigrade with no appreciable diffusion.

Samples may be applied to wells or troughs for the first dimension as determined by the assay requirements in 2% agarose, with veronal buffer, pH 8.6, and 80–100 V DC measured at 4.5 inches across the gel, using cotton wicks producing 50–60 mA; the gel resolves the protein bands in approximately 45 min to 1 h for human high-density lipoproteins.

The plate is rotated 90 degrees and electrophoresed at the same rate for 3–4 h. Upon completion of the electrophoresis, the gel is washed in saline overnight, or 2 h minimum with stirring.

The washed plate is blotted and dried with filter paper and then stained as desired with Coomassie blue protein stain.

Areas measured under the individual completed peaks are directly proportional to the concentration of the protein in question in one- and two-dimensional assays, making this methodology a powerful analytical tool.

The two-dimensional pattern shown in Fig. 1 is presented to represent expressions of the lipoprotein A-I as separate peaks represented by the pre-beta A-I as the slower peak and the alpha A-I as the faster moving, larger peak. The migration is from left to right.

This gel also contains antisera to the lipoprotein A-II and is shown as the precipitin line closely associated with the A-I in the normal sample but quite slower in migration in the hypertriglyceridemic sample, with the pre-beta A-II peak slower than pre-beta A-I.

Overall, the hypertriglyceridemic sample shows the A-I distribution to be greatly in favor of the pre-beta position and therefore much slower in migration of the A-I peaks.

These resulting areas of the lipoprotein A-I peaks are interpreted and presented here to be a reflection of the effects of the enzyme LCAT (lecithin-cholesterol acetyltransferase) activity on the reverse cholesterol transport system in accordance with [8].

The status of metabolic HDL cholesterol transport efficiency is expressed as a function of A-I clearance to the higher-density region (alpha) in HDL. The distribution of the measured areas is presented as reflective of the status of risk for HDL particle clearance.

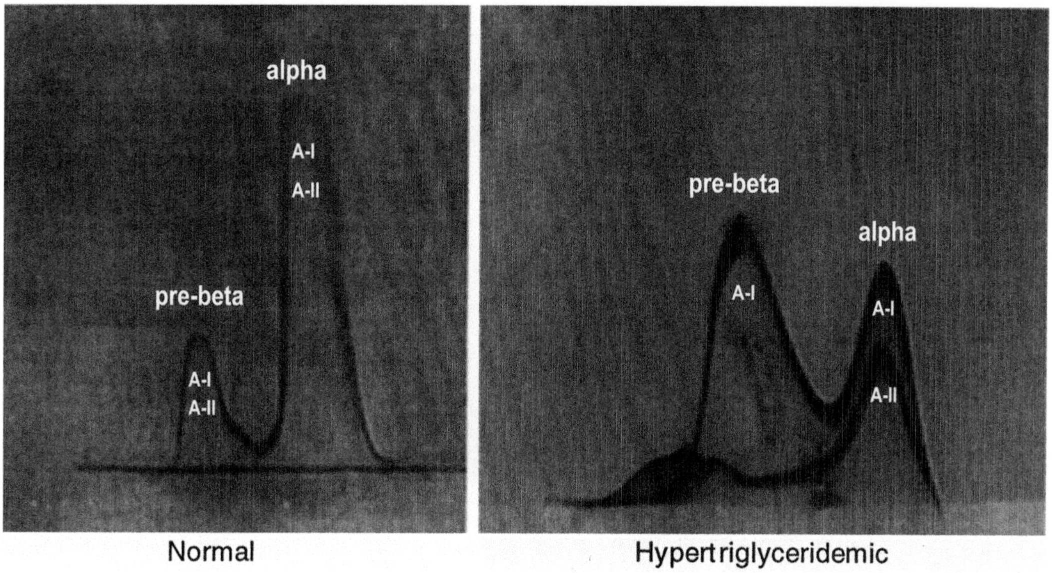

Fig. 1 Second dimension agarose patterns of apolipoprotein A-I and apolipoprotein A-II expressions of normal and hyperlipemic sera. Immunological continuity is shown to represent concentrations of particles positioned within the pre-beta and alpha regions for both the A-I and A-II lipoproteins, both of which migrate in advance of the beta position. Areas in favor of the alpha position peaks for A-I and A-II reflect lipid transport efficiency

Pre-beta A-I is shown to be a positive risk factor in this assay significantly correlating with LDL cholesterol and total serum cholesterol.

$$\text{The distribution ratio} = \frac{\text{Alpha A} - \text{I mm}^2}{\text{Pre} - \text{betaA} - \text{I mm}^2} = (\text{AIDR})$$

A favorable percent clearance to the alpha region (AIDR) = 85% or higher.

References

1. Egatz-Gomez A, Thormann W (2011) Micro-preparative isoelectric focusing protein separation in a suspended drop. Electrophoresis 32:1433–1437
2. Bornemann S, Rietschel B, Baltruschat S et al (2010) A novel polyacrylamide gel system for proteomic use offering controllable pore expansion by crosslinker cleavage. Electrophoresis 31:585–592
3. Williams TI, Combs JC, Thakur AP et al (2006) A novel Bicine running buffer system for doubled sodium dodecyl sulfate—polyacrylamide gel electrophoresis of membrane proteins. Electrophoresis 27:2984–2995
4. Jiang F, Wang Y, Hu X et al (2010) Carbon nanotubes-assisted polyacrylamide gel electrophoresis for enhanced separation of human serum proteins and application in liverish diagnosis. J Sep Sci 33:3393–3399
5. Hempel K, Rosen R, Becher D et al (2009) Analysis of ultra acidic proteins by the use of anodic acidic gels. Anal Biochem 385:208–214
6. Laurel CB (1965) Antigen-antibody crossed electrophoresis. Anal Biochem 10:358, 1965
7. Weeke B (1973) Crossed Immunoelectrophoresis. In: Axelsen NH, Krell J, Weeke B (eds) Quantitative immunoelectrophoresis. Universitetsforlaget, Oslo
8. Fielding PE, Miida T, Fielding CJ (1991) Metabolism of low-density lipoprotein free cholesterol by human plasma lecithin-cholesterol acyltransferase. Biochemistry 30:8551–8557

Chapter 44

Single-Cell High-Resolution Detection and Quantification of Protein Isoforms Differing by a Single Charge Unit

Kristi A. Koelsch

Abstract

Isoelectric focusing (IEF) is an electrophoretic technique that enables the separation of proteins based on their isoelectric points. Until recently, this valuable method was not feasible for single-cell applications, which are necessary to interrogate heterogeneous cell populations. Herein we highlight a recently published method enabling the analysis of single-cell proteomics, which utilizes microfluidics coupled with IEF, photocapture, and immunoprobing of the protein in the same micro-gel, which can be stripped and reprobed multiple times.

Key words Single-cell, Isoelectric focusing, IEF, scIEF, Microfluidic separation, Isoelectric point

1 Introduction

Cells within populations once believed to be functionally distinct have since been shown to have wide variances in phenotypes [1]. The heterogeneity of cellular populations involved in various physiological processes requires higher-resolution analyses that will provide information at the single-cell level. Genomic and transcriptomic research with recent technologies have bridged this gap making single-cell analyses possible [2, 3]. However, single-cell proteomic studies have lagged due to the limited availability and specificity of antibodies, which can also have off-target binding causing artifacts that may hinder the resolution of small differences between single cells [4]. The problems associated with antibodies are complicated further in functional proteomic studies involving protein variants (proteoforms) arising from posttranslational alternate splicing and chemical modification events [5]. Generating antibodies that are specific for discrete isoforms have been problematic and limiting.

In this chapter we highlight a method recently reported by Tentori et al. [6] for the detection of protein isoforms in individual

Biji T. Kurien and R. Hal Scofield (eds.), *Electrophoretic Separation of Proteins: Methods and Protocols*, Methods in Molecular Biology, vol. 1855, https://doi.org/10.1007/978-1-4939-8793-1_44,

cells that differ by only one charge unit by utilizing a method of high-resolution single-cell isoelectric focusing (IEF) and high-selectivity immunoprobing. The method employs microfluidic integration of all steps to minimize diffusion-based dilution of the lysates from single cells. It utilizes a custom-cast IEF cassette that is comprised of two parts: The bottom layer is a polyacrylamide gel-coated standard glass microscope slide with customizable molded microwells for the passive capture of single cells, providing the framework for immobilizing the focused proteins for subsequent blotting. The top layer "lid" consists of a cast gel with chemically distinct regions. The center focusing region is comprised of lysis reagents with carrier ampholytes, and on either end, there are segments of catholyte and anolyte immobilines, respectively. When the two layers are assembled, it enables the diffusion of the lysis reagent and pH gradients from the top layer to the bottom layer without significant dilution of the lysates from the single cell. Following IEF, the proteins are UV photocaptured and immunoprobed within the same bottom layer gel. In this protocol single cells expressing tGFP protein are analyzed as a proof of concept. The method can be extended to any protein of interest in any cell type with adjustments to the experiment parameters.

2 Materials

For all reagents, use those that meet or exceed the standards set by the American Chemical Society (ACS) for purity unless otherwise noted.

2.1 Reagents

1. 3-(Trimethoxysilyl)propyl methacrylate.
2. Acrylamide/bis-acrylamide solution (29:1, electrophoresis grade).
3. BPMAC monomer (*N*-[3-[(3-benzoylphenyl)formamido]propyl]methacrylamide) (custom synthesis by PharmAgra Laboratories, Brevard, NC) [7].
4. Gel Slick™ (Lonza, Rockland, ME).
5. CHAPS (3-[(3-cholamidopropyl)dimethylammonio]-1-propanesulfonate hydrate), electrophoresis grade, ≥98%.
6. GE Healthcare Polybuffer® 74, PB74
7. GE Healthcare Polybuffer® 96, PB96.
8. Fluorescent IEF markers: pI 4.5 solution (1 mg/mL in 5 mM HCL), pI 5.5 solution (3 mg/mL in 15 mM HCL), and pI 6.8 (1 mg/mL in 5 mM HCL)(all from Sigma Chemical Co., St. Louis, MO).

9. Immobilines: acrylamido buffer pK_a 3.6 (~0.2 M in H_2O) and acrylamido buffer pK_a 9.3 (~0.2 M in 1-propanol) (both from Sigma).
10. VA-086 UV photoinitiator (2,2′-azobis[2-methyl-*N*-(2-hydroxyethyl)propionamide, Wako Chemicals, Richmond, VA).
11. 10× PBS (phosphate-buffered saline, pH 7.4, molecular biology grade).
12. TBST (Tris-buffered saline, 0.1% Tween 20).
13. Tris–HCL, pH 6.8.
14. Dulbecco's modified Eagle's medium (DMEM), high glucose.
15. Sodium pyruvate solution (100 mM).
16. MEM nonessential amino acid solution (100×).
17. Penicillin–Streptomycin solution (100×; penicillin 10,000 U/mL, streptomycin 10,000 μg/mL).
18. Fetal bovine serum, heat inactivated.
19. Trypsin-EDTA (0.05%).

2.2 Antibodies

2.2.1 Primary Antibodies

1. Rabbit anti-TurboGFP (tGFP) (1:30, Pierce Antibody Products, Rockford, IL).
2. Mouse anti-lamin A/C (1:5, mab 636, Pierce Antibody Products).
3. Rabbit anti-β-tubulin (1:10, mab 6046, Abcam, Cambridge, UK).

2.2.2 Secondary Antibodies

1. Alexa Fluor 555-labeled donkey anti-mouse IgG (1:30, Life Technologies, Grand Island, NY).
2. Alexa Fluor 647-labeled donkey anti-rabbit IgG (1:30, Life Technologies).

2.3 Cell Lines and Cell Culture

1. U-251-tGFP cells. Human glioblastoma astrocytoma cell line, U-251 MG (Sigma, stably transduced with tGFP [7]) (*see* **Note 1**).
2. Culture cells in high-glucose DMEM supplemented with 1 mM sodium pyruvate, 1× MEM nonessential amino acids, 1% penicillin and streptomycin, and 10% fetal bovine serum in a humidified incubator at 37 °C with 5% CO_2.
3. Prior to experiment, trypsinize cells to remove from culture surface, and wash carefully with 1× PBS to remove residual media proteins.
4. Count cells and resuspend at 1×10^6 cells/mL in 1× PBS and keep on ice.

2.4 Equipment

1. Custom 3D-printed acrylonitrile butadiene styrene (ABS) electrophoresis chamber with graphite bar electrodes embedded in the floor of the chamber.
2. Bright-field microscope.
3. pH meter.
4. High-voltage power supply.
5. UV light source.
6. Flexible fiber-optic light guide (for delivery of UV to gel).
7. UV meter Hamamatsu C6080-365.
8. Inverted fluorescence microscope.

3 Methods

This protocol requires custom fabricated components. See detailed instructions for the fabrication of the SU-8 2025 photoresist silicon wafer molds required for casting microwells into the bottom slide gel layer, the top and bottom layer component casting and assembly [7–9], and the custom 3D-printed electrophoresis chamber [6] for IEF separation.

3.1 Fabrication of IEF Cassette

Important! 3-(Trimethoxysilyl)propyl methacrylate, acetic acid, and methanol are flammable, and methanol is also toxic. For safety, perform **steps 4–7** inside a chemical fume hood with sash lowered using proper personal protective equipment such as fire-resistant lab coat and nitrile gloves.

3.1.1 Silane-Coated Microscope Slide Preparation

Microscope slides must be silanized to form a surface monolayer of methacrylate functional groups for proper adherence of bottom gel layer.

1. Prepare 400 mL silane solution by adding 80 mL of 3-(trimethoxysilyl)propyl methacrylate and 120 mL acetic acid to 200 mL of ddH_2O in a glass flask. Degas solution in a sonicator for 30 min (*see* **Note 2**).
2. To later identify the silanated side of the slide, make an etch mark on the same corner of each slide with a glass etching tool. Situate the markings in such a way that the silanated side will be face down when the markings are in the upper right-hand corner. Slides can be used whole or cut to desired size by scoring with a pencil style cutter using a ruler as a guide.
3. Place marked slides back to back into a slide rack so that the side to be silanated faces the solution. Place the loaded slide rack into a glass slide-staining dish.

4. Pour the silane solution over the slides to completely cover them. Ensure that any bubbles do not remain between slides and glass staining dish. Let set at room temperature for 30 min for reaction to occur.
5. Remove slide rack from silane solution, and transfer to a staining dish containing methanol. Gently rock the rack within the dish to ensure complete rinsing of silane solution from slides. Repeat with fresh methanol.
6. Remove slide rack from methanol, and transfer to a staining dish containing ddH_2O. Gently rock the rack within the dish to remove the methanol.
7. Repeat **steps 5** and **6** two more times. It is not necessary to replace with fresh methanol or ddH_2O for these steps.
8. Place slides in a dish with fresh ddH_2O, and remove one at a time, quickly drying them with a nitrogen gun.
9. Store slides in a slide storage box at room temperature until used for up to 2 months.

3.1.2 Bottom Gel Layer Construction

The bottom layer contains the microwells for capturing single cells and will serve as the scaffolding for the immunoprobing of the BPMAC-mediated photoimmobilized proteins (*see* **Note 3**).

1. Fabricate a SU-8 wafer mold with a central row of microposts (32 μm diameter × 40 μm high, spaced 500 μm apart) and two parallel flanking rails spanning the length of the mold (1 mm thick × 40 μm high, spaced 24 mm apart) which serve to mold the microwells in the gel and to align the mold to the bottom slide (*see* refs [7, 8]).
2. Place SU-8 mold faceup onto lab bench top, and secure with tape on the edges to immobilize.
3. Place prepared slide silanized side-down onto SU-8 mold, manually aligning to the rails for correct positioning of microposts and subsequent microwells in the gel.
4. Prepare a polyacrylamide gel precursor solution of 6% T (w/v total acrylamides) and 3.3%C (w/w of the cross-linker *N,N*-methylenebis-acrylamide) containing 5 mM BPMAC as described in Tentori et al. [6].
5. Degas the precursor gel by sonication, and vacuum to remove any air bubbles. Add initiators, APS (final concentration 0.08% (vol/vol)) and TEMED (final concentration 0.08% (vol/vol)), to the gel precursor, and mix well with a pipette tip without forming bubbles (*see* **Note 4**).
6. Using a 200 μL pipette, gently inject the gel solution in the gap between the SU-8 mold and glass slide allowing it to completely fill the void between by capillary action.

7. Immediately, before polymerization, gently press down on glass slide to ensure microposts contact the glass slide and to remove excess gel. Allow to sit undisturbed until polymerized, approximately 15 min.
8. To remove slide from SU-8 mold following polymerization, re-wet the edges of the slide with 2 mL 1× PBS, and then gently slide a razor blade under the edge of the slide, and raise the slide up and off of the mold.
9. Rinse SU-8 mold to remove residual precursor solution, air-dry, and store for reuse.
10. Visually inspect the bottom cast gel for integrity of microwells using a bright-field microscope.
11. The bottom gel can be used immediately or stored faceup in 1× PBS at 4 °C for later use (for up to 1 week).

3.1.3 Top Layer Construction

The top layer serves a dual purpose. It contains the cell lysis and IEF reagents that will diffuse into the bottom layer, and it is a template for the pH and electric potential gradients required for IEF, thus eliminating the need for excess fluid and maximizing protein concentrations. Detailed instructions and images for the construction of the top layer gel are described by Tentori et al. [6, 10].

1. Thoroughly clean and dry two glass slides. To make slides hydrophobic, coat with Gel Slick™ as per manufacturer's recommendations.
2. To enable the casting of the three continuous but chemically heterogeneous gel sections (basic, focusing, and acidic) of the lid, place a 500-μm-thick × 9-mm-wide spacer (C.B.S. Scientific, Del Mar, CA) between the two hydrophobic slides in the center focusing region. The width of the focusing region is determined by the width of the spacer and can be optimized for your experiment. This will later contain the focusing region gel and will be flanked on either side by basic and acidic gel regions.
3. Prepare the acidic (anolyte) region precursor gel solution using a gel composition of 15%T (3.3%C), with immobilines, 13.6 mM pK_a 3.6 and 6.4 mM pK_a 9.3, and initiator VA-086 0.2% (w/v).
4. Prepare the basic (catholyte) region precursor gel solution using a gel composition of 15%T (3.3%C) with immobilines, 14.4 mM pK_a 9.3 and 5.6 mM pK_a 3.6, and initiator VA-086 0.2% (w/v).
5. Prepare focusing region precursor gel solution using a gel composition of 15%T (3.3%C), carrier ampholytes PB74 1:10 (w/v) and PB96 1:10 (w/v), native detergents CHAPS 3.6% (w/v), Triton x-100 1% (v/v), digitonin 0.125% (w/v), and

initiator VA-086 0.2% (w/v). For denaturing conditions, include chaotropes, 7 M urea and 2 M thiourea (*see* **Note 5**).

6. To polymerize the basic boundary region, pipette the basic (catholyte) gel precursor into the basic region space between the hydrophobic slides. Mask the regions not to be polymerized, and photopolymerize the section using a UV light source for 4-min exposure time at 20 mW/cm^2 power with a 390 nm longpass UV filter.
7. Wash out any unpolymerized basic gel precursor, and then apply the acidic gel precursor to the other side. Photopolymerize as in **step 6**.
8. Wash out any unpolymerized acidic gel precursor, remove the center spacer, and apply the focusing region precursor gel solution. Repeat photopolymerization step as in **steps 6** and 7.
9. Remove glass slides, releasing the polymerized lid gel layer for assembly into the IEF cassette.

3.2 Assembling Cassette, Loading, and IEF Focusing

1. Remove previously prepared cell suspension from ice. Dehydrate the bottom layer with a stream of nitrogen, and then pipette cell suspension directly over the microwells on the bottom layer. Incubate at 4 °C for 10 min to allow cells to settle by gravity into the microwells.
2. Rinse slides 2× with a buffer that contains carrier ampholytes but no detergents.
3. Place bottom layer with cells loaded into the custom-printed 3D IEF chamber between the electrodes.
4. Position lid layer over bottom slide so that the focusing region of the lid layer is aligned over the microwells on the bottom layer with the lid catholyte and anolyte gel regions resting on the chamber's graphite cathode and anode, respectively.
5. Upon contact between the lid and bottom gel layers, there will be diffusion of the pH patterning and nonionic lysis reagents from the lid layer into the bottom gel matrix causing rapid cell lysis (between 30 and 60 s) and transfer of the pH gradient.
6. Place a clean glass slide on top of the lid to prevent excess evaporation during IEF process.
7. Using a high-voltage power supply, apply an electrical field with potential difference of 600 V for 5.5–6.0 min (*see* **Note 6**).
8. Gel-bound proteins are then covalently bound in gel by BPMAC-mediated photocapture using a light source with UV filter (300–380 nm bandpass) held at 5 cm above the slide for 45 s with a power of 400 mW/cm^2 at the slide surface (320–400 nm UV meter).

3.3 Immunoprobing

1. Place the bottom layer (with gel-bound proteins) face down and suspended above a glass surface with strips of 80-μm-thick Kapton tape spacers.
2. Pipette 40 μL of TBST solution containing the primary antibody into the space between the bottom layer and glass surface. Incubate for 2 h at room temperature.
3. Remove the bottom layer from the spacers/glass, and wash with TBST for 30 min while on a slow rocker. Repeat using fresh TBST.
4. Using the same method, incubate bottom layer with secondary antibodies for 1 h. Wash in TBST 2× as done in **step 3**.
5. A final wash is performed in deionized water for 1 min.
6. Carefully dry the gel with a stream of nitrogen, and then image (see below).

3.4 Stripping and Reprobing

To probe multiple targets using primary antibodies raised in the same host, the bottom gel may be stripped and reprobed.

1. Incubate the bottom layer in harsh stripping buffer (62.5 mM Tris–HCL, pH 6.8 2.5% (w/v) SDS, and 1% (v/v) β-mercaptoethanol) at 50 °C overnight.
2. Wash stripped slides in TBST for 20 min, then gently dry with nitrogen stream, and image to confirm complete removal of previous antibodies.
3. Rehydrate slides in TBST for 20 min, and then reprobe with desired antibodies.

3.5 Imaging and Analysis

Detailed imaging and analysis methods are described by Tentori et al. [6].

4 Notes

1. The cell line, U-251 MG, was formerly known as U-373 MG and distributed by The American Type Culture Collection (ATCC).
2. Silane solution must be prepared immediately prior to use.
3. The components of the gel precursor solution are toxic and skin irritant. Use personal protective wear while preparing.
4. Prepare APS stock fresh immediately before use.
5. Since the carrier ampholytes and lysis reagents are not immobilized in the focusing region, they will diffuse into the catholyte and anolyte regions. Therefore, the focusing region gel solution should not be applied and polymerized until just prior to use.
6. The voltage application time will need to be determined for any given experiment.

References

1. Altschuler SJ, Wu LF (2010) Cellular heterogeneity: do differences make a difference? Cell 141(4):559–563
2. Dalerba P, Kalisky T, Sahoo D, Rajendran PS, Rothenberg ME, Leyrat AA et al (2011) Single-cell dissection of transcriptional heterogeneity in human colon tumors. Nat Biotechnol 29(12):1120–1127
3. Van Loo P, Voet T (2014) Single cell analysis of cancer genomes. Curr Opin Genet Dev 24:82–91
4. Stadler C, Rexhepaj E, Singan VR, Murphy RF, Pepperkok R, Uhlen M et al (2013) Immunofluorescence and fluorescent-protein tagging show high correlation for protein localization in mammalian cells. Nat Methods 10 (4):315–323
5. Egelhofer TA, Minoda A, Klugman S, Lee K, Kolasinska-Zwierz P, Alekseyenko AA et al (2011) An assessment of histone-modification antibody quality. Nat Struct Mol Biol 18 (1):91–93
6. Tentori AM, Yamauchi KA, Herr AE (2016) Detection of isoforms differing by a single charge unit in individual cells. Angew Chem Int Ed Engl 55(40):12431–12435
7. Kang CC, Lin JM, Xu Z, Kumar S, Herr AE (2014) Single-cell Western blotting after whole-cell imaging to assess cancer chemotherapeutic response. Anal Chem 86 (20):10429–10436
8. Hughes AJ, Spelke DP, Xu Z, Kang CC, Schaffer DV, Herr AE (2014) Single-cell western blotting. Nat Methods 11(7):749–755
9. Kang CC, Yamauchi KA, Vlassakis J, Sinkala E, Duncombe TA, Herr AE (2016) Single cell-resolution western blotting. Nat Protoc 11 (8):1508–1530
10. Tentori AM, Hughes AJ, Herr AE (2013) Microchamber integration unifies distinct separation modes for two-dimensional electrophoresis. Anal Chem 85(9):4538–4545

Chapter 45

Artifacts and Common Errors in Protein Gel Electrophoresis

Biji T. Kurien and R. Hal Scofield

Abstract

In spite of taking precautions, some common mistakes creep into well-planned gel electrophoresis experiments. This occurs commonly in relation to calculating the cross-linking factor of a gel, polymerization temperature and time for a polyacrylamide gel, inducing aggregates in samples for electrophoresis, titrating the running buffer in electrophoresis, proper sample preparation, amount of protein to be loaded on a gel, sample buffer-to-protein ratios, incompletely removing phosphate-buffered saline from cells prior to cell lysis, and over-focusing of IPG strip in two-dimensional gel electrophoresis. In addition, subtle artifacts can have significant deleterious effects on carefully planned and executed experiments. Proteases that act at room temperature upon proteins in the sample buffer prior to heating, cleavage of the Asp-Pro bond upon prolonged heating of proteins at high temperatures, contamination of sample or sample buffer with keratin, leaching of chemicals from disposable plastic ware, and contamination of urea with ammonium cyanate are some of the common reasons for artifacts in gel electrophoresis. Taking proper heed to all these factors can greatly help generate good experimental results.

Key words Artifacts, Common mistakes in electrophoresis, Aggregates, Keratin, Asp-Pro bond, Proteases, Ammonium cyanate, Cross-linking factor

1 Introduction

Protein repertoire of a cell in health and disease has been studied well with polyacrylamide gel electrophoresis (PAGE). Westermeier [1] and Burgess [2] have described mistakes made commonly in electrophoresis and important but little known artifacts in protein biochemistry. There are plenty of erroneous protocols that are available, in spite of widespread information available via text books, instruction manuals, and online tutorials [1]. Sometimes subtle and obscure artifacts can significantly affect the outcomes of carefully performed experiments [2]. This chapter reviews important artifacts and commonly made mistakes that can derail an otherwise perfectly executed experiment.

Biji T. Kurien and R. Hal Scofield (eds.), *Electrophoretic Separation of Proteins: Methods and Protocols*, Methods in Molecular Biology, vol. 1855, https://doi.org/10.1007/978-1-4939-8793-1_45,

2 Artifacts Seen in Gel Electrophoresis

2.1 Sample Preparation for SDS-PAGE

There are three different ways in which one may find contaminants in a purified or nearly purified protein sample by sodium dodecyl sulfate (SDS)-PAGE.

2.1.1 Proteases

Proteases bring about proteolytic artifacts. The occurrence of a diffuse smear in the lower portion of the gel and also a general lack of good resolution of protein bands may be a consequence of the presence of proteases. One may also observe the loss of high-molecular-weight protein bands, in certain cases, and the simultaneous appearance of lower-molecular-weight protein bands. In addition, the presence of proteases can lead to loss in biological activity or significant modification of the isoelectric point of proteins [3].

Samples containing proteins are normally added to sample buffer, containing SDS, β-mercaptoethanol or dithiothreitol, and sucrose or glycerol, and heated at 95–100 °C for 5 min. Heating enables better denaturation and reduction of the proteases and thus bring about its inactivation [4]. Gradually, it was recognized that the heating had to be carried out as soon as the samples were diluted in the sample buffer. This is because SDS, even though it easily unfolds most proteins, does not unfold some proteases. Therefore, the proteases will have time to digest the proteins of interest in the sample buffer if the heating step is not carried out immediately. One can design an experiment to find out if proteases are causing multiple bands to appear in SDS-PAGE analysis of a purified protein or a nearly pure protein. One way would be adding the protein of interest to two portions of the sample buffer. Mix well and heat one portion immediately. Leave the other at room temperature for 2–4 h and then heat. Analyze both samples on SDS-PAGE, and check for degradation of the protein in the sample not heated immediately. As little as 1 pg protease in a protein sample has been shown to bring about major degradation, if the sample is added to the lysis buffer and not heated immediately [2].

2.1.2 Cleavage of Asp-Pro Bond at 100 °C

The bond between aspartic acid and proline is the most susceptible bond for cleavage by heat or acidic conditions [5]. The bond is cleaved by heating at high temperatures for too long. Heating at 75 °C for 5 min instead of 95–100 °C for 5 min has been found to avoid D-P bond cleavage, as well as completely inactivating proteases [2]. However, it has to be also noted that several proteins have been found to be stable at 100 °C for several hours [6].

2.1.3 Keratin Can Contaminate SDS Lysis Buffer or Sample

The protein keratin migrates as a heterogeneous cluster of contaminating bands around 55–65 kDa on reducing SDS gels (with β-mercaptoethanol) and near the top of the gel on nonreducing

SDS gels (without β-mercaptoethanol). Keratin occurs in the skin, dander, etc. Commonly the lysis buffer itself gets contaminated with skin contact or by a flake of dandruff. This is normally a minor contaminant and can be visualized mainly in silver-stained gels and occasionally on Coomassie blue-stained gels. Contamination of lysis buffer with keratin can be ruled out by running sample buffer alone without the addition of proteins to a lane in the gel. The buffer should be remade if keratin bands are observed in this lane. Aliquoting lysis buffer and storing them at −80 °C are practiced by many laboratories. An aliquot can be thawed and used within a day or 2. Keratin has been observed occasionally in western blots as a result of contamination of antigen used to prepare a polyclonal antibody and reaction with keratin transferred from SDS gel to membrane [2].

2.2 Leaching of Chemicals from Plastic Ware

Some chemicals leach out from common laboratory plasticware (disposable) into standard aqueous buffers. In certain cases, this can have significant effects on experimental results. Chemicals like oleamide are employed as lubricating agents in the molding process. Other chemicals, such as certain cationic biocides, are used to help prevent bacterial colonization of the plastic surface. Most of this material can be removed by washing the plastic ware in water or even better with methanol or DMSO [2, 7].

2.3 Contamination of Urea with Cyanate

Urea is used widely as a denaturing agent for proteins. However, urea solutions contain significant amounts of ammonium cyanate. Ammonium cyanate is in chemical equilibrium with urea [$(H_2N)_2C{=}O$ in equilibrium with NH_4^+ + NCO^-)]. Isocyanic acid (H–N=C=O) has been shown to react with lysine's ε-amino group and the amino terminus as well as with arginine and cysteine to a lesser extent to form a carbamylated protein [8]. Carbamylation can interfere with enzyme function in some cases, alter charge, and block certain protease cleavage reactions. In addition, as measured by mass spectrometry, it can add 43 Da per carbamylation event to the mass of the protein. To remove these contaminant ions (to diminish or prevent carbamylation), one can treat a urea solution with a mixed bed resin (Bio-Rad AG 501-X8). However, since this is a chemical equilibrium, the ammonium cyanate level builds up again to levels in the 0.5–3 mM range in an 8 M urea solution, with the potential to reach up to 20 mM [9]. Chemical scavengers like ethylenediamine, glycylglycine, or glycinamide (in the 5–25 mM range) hve been shown to reduce cyanate to less than 0.1 mM in 8 M urea, Tris–HCl (pH 8) [9]. Burgess [2] suggests that the best general practice, if urea is to be used, is to replace some of the NaCl in the buffer with some ammonium salt (like 25–50 mM ammonium chloride) to push the equilibrium back by the common ion effect toward less cyanate. Lower temperature and acidic conditions have been shown to slow the reaction of isocyanic acid with amino groups in proteins. This effect can be also minimized by restricting the time of protein exposure to the urea solution to the shortest time possible [2].

3 Commonly Made Mistakes in Electrophoresis

3.1 Sample Preparation

Errors commonly made [10] during sample preparation include employing an incorrect protein-to-sample buffer ratio, failure to remove insoluble material, and overloading and under-loading of protein. In order to prevent inadequate sample buffer-to-protein ratios, overloading, and under-loading of samples, the protein concentration of the sample should be determined using a standard protein assay.

Poorly resolved, distorted bands in the overloaded lane and distorted protein bands in adjacent lanes result from loading too much protein. On the contrary, under-loading leads to lack of detection of minor protein bands, and even makes major bands too faint for photographic reproduction. One should load purified protein in the 0.5–4.0 μg range (depending on well size and gel thickness) and from 40–60 μg for crude samples if a Coomassie blue stain is to be used for gel staining. Less protein per sample is required when silver staining method is employed, since it is about 100-fold more sensitive than Coomassie staining [10]. Enough sample buffer should be used in order to maintain an excess of SDS. Most proteins bind SDS in a constant mass ratio of 1.4 μg SDS per 1.0 μg protein. However, Hames [11] recommends a ratio of 3:1.

Proteins like histones and membrane proteins may need addition of 6–8 M urea or a nonionic detergent like Triton X-100 since they may not completely solubilize by heating in SDS sample buffer alone [12, 13]. The insoluble material should be removed by a 2-min centrifugation (at 17,000 × *g*) following heat treatment in SDS lysis buffer. Streaking within the gel will happen if precipitated insoluble material is not removed. If the supernatant is not loaded immediately, it can be stored at 4 °C overnight or frozen at −20 °C for longer periods. However, the stored samples need to be warmed briefly at 37 °C to redissolve the SDS and centrifuge once again to remove insoluble material before loading [10].

Pre-treatment methods such as lyophilization, spin concentration, dialysis against concentrated polyethylene glycol (PEG), and excess solvent absorption by exposing dialysis bag with sample to dry PEG, Aquacide or Sephadex® (gel filtration media), can be employed to concentrate samples that are too dilute for analysis. Prior to addition of SDS sample buffer, samples concentrated by the above said methods may be dialyzed against 50 mM Tris–HCl, pH 6.8, to get rid of low-molecular-weight contaminants. Trichloroacetic acid or acetone can be used to precipitate and concentrate proteins from dilute samples and acidic samples as well as get rid of contaminants such as potassium, guanidine hydrochloride, or ionic detergents from samples [1, 14].

Extremely viscous samples such as crude cell extracts owe their viscosity to the high concentration of unsheared nucleic acids. Treatment of samples with Benzonase® nuclease (recombinant endonuclease) prior to addition of sample buffer can help eliminate the viscosity. This endonuclease lacks proteolytic activity and completely degrades all forms of DNA and RNA. Vigorous vortexing of the heated sample or physical shearing of the nucleic acids through sonication can also help reduce viscosity [10].

3.2 Miscalculating Cross-Linking Factor of a Polyacrylamide Gel

Two factors control the pore size of a polyacrylamide gel; (a) the total concentration of acrylamide T and (b) the degree of cross-linking C:

$$T = \frac{(a + b) \times 100}{V} [\%], \quad C = \frac{b \times 100}{a + b} [\%]$$

a = mass of acrylamide in g.
b = mass of methylenebisacrylamide in g.
V = volume in mL.

Sometimes, one mistakenly assumes that the given total concentration of acrylamide T is the percentage of acrylamide per volume and C (the cross-linking factor) is the percentage of methylenebisacrylamide per volume. This will lead to too much amounts of cross-linker in the gel. This will result in gels that are opaque, brittle, and highly hydrophobic. The correct method is to prepare the solutions as per the equation noted above. Alternatively, it would be better to use commercially available acrylamide/methylenebisacrylamide stock solutions that are ready to use [1].

3.3 Temperature and Time of Polymerization for a Polyacrylamide Gel

It takes about 30 min to 1 h for the polymerization of acrylamide and methylenebisacrylamide to occur. The matrix formation is unfinished at this time point, however. Polymerization continues, in a silent manner for several hours for the complete matrix formation. In general, complete polymerization is brought about efficiently, only when it is carried out at room temperature [1]. However, Haeberle [15] has described a high-temperature SDS-PAGE (running gels at 70 °C) that can be run in as little as 5 min for a mini gel. An added advantage of this method is the greatly accelerated rate of polyacrylamide cross-linking. According to this paper, the gel can be transferred to the 70 °C buffer as soon as it becomes rigid enough to be removed from the casting stand, and by the time the samples are loaded, the cross-linking is almost complete [15]. Kurien et al. also showed that it is possible to electrophorese proteins using a pre-cast gel in 10 min using heated standard running buffer [16].

Polyacrylamide gels are polymerized in the refrigerator or cold room in some laboratories, or the gels are used one or a few hours following the initiation of polymerization. Using incompletely polymerized gels can lead to disturbances during the protein separation, especially on a native gel. In addition, high background noise will be a problem in downstream mass spectrometry analysis as a consequence of the incompletely polymerized acrylamide mono- and oligomers. The correct way is to allow the gel to polymerize overnight at room temperature. The gel can be stored in a refrigerator, if necessary, afterward [1].

3.4 Protein Aggregates in SDS Samples

After boiling, the sample is often directly loaded on the SDS gel after it has cooled down. Consequently, the reductant often becomes partly oxidized; a part of the cysteines remain unprotected and result in back-folding and the creation of inter-polypeptide aggregates. Back-folding leads to blurred zones and sometimes the formation of double bands. While some aggregates precipitate in the sample well (too big to enter the gel), some induce artifactual zones in the high-molecular-weight area. Two or three horizontal lines across the entire gel in the molecular weight range between 40 and 60 kDa are formed by an excess of reductant. The right procedure is to allow the sample to cool to approximately 60 °C (after boiling) and then add iodoacetamide [10 μL of a 20% (w/v) aqueous iodoacetamide solution to 100 μL sample] and incubate the sample for 30 min at room temperature. Sharper bands are obtained with this procedure, and artifacts (double bands, additional high molecular weight bands, precipitate in the sample well, and lines across the gel) are abolished [1].

3.5 Titrating Running Buffer in SDS PAGE

The discontinuous buffer system based on Läemmli's method [17] consists of the stacking and resolving gel buffers that are composed of Tris and chloride buffers with defined pH values (with or without SDS). The running buffer has SDS, Tris, and glycine. The discontinuity between the chloride (acting as the leading ion) and the glycine (acting as the trailing ion) permits a controlled slow protein entry into the polyacrylamide gel and the stacking effect, resulting in very sharp and well-resolved zones.

However, some protocols call for adjusting the pH of the Tris-glycine buffer to a pH of 8.3 or 8.4. Doing so brings about a high load of chloride in the upper buffer chamber, and this causes the following effects. (a) The protein separation will take much longer than expected, with some zones remaining poorly resolved. (b) Since chloride has a very high electrophoretic mobility compared to all other ions, only chloride ions will migrate toward the anode until there is no chloride remaining in the upper buffer chamber. Only after the depletion of chloride in the upper chamber will the protein ions start to migrate. Therefore the electrophoretic run will take several hours to overnight in a large format chamber.

(c) Finally, the stacking will not be effective as in a perfect discontinuous buffer. The right procedure is to use only Tris-base, glycine, and SDS for the running buffer. The buffer should not be titrated even when pH is above 9 [1].

3.6 Over-Focusing of IPG Strips in 2-DE Should Be Avoided

Two-dimensional electrophoresis (2-DE) gels ideally show well-defined pattern of spots corresponding to all the individual proteins. However, the spots are sometimes blurred in certain areas. Vertical or horizontal streaks are seen instead of round spots. One reason for this is over-focusing. Of late, isoelectric focusing (the first dimension) mostly involves the use of polyacrylamide strips containing immobilized pH gradients. Since the pH gradient is fixed to the polyacrylamide matrix, it cannot go away owing to long exposure to an electric field. However, other disturbances in 2-DE maps can be brought about by over-focusing [1].

Some proteins are unstable, when they are at their isoelectric point for a certain time. It has been seen that some basic proteins are especially vulnerable to hydrolysis at their isoelectric point pH. This can bring about horizontal streaking from certain protein spots. Consequently, the focusing time applied to basic IPG strips should be limited to the minimum [1].

The addition of thiourea to urea, for protein extraction and IPG rehydration [18] results in increased solubility of hydrophobic proteins. As a result, there is a marked increase in protein spots in the 2-DE pattern compared with using urea alone. However, a strong vertical streak develops around pH 5.5 area in some cases. The protein spots along the acidic side of this streak appear more blurred than the spot pattern in the remaining areas of the gel. This phenomenon is not observed with all batches of thiourea. However, the phenomenon correlates also with the volt-hour load placed on an IPG strip. The amount of volt hours placed on an IPG strip should be reduced if such an effect is observed [1].

3.7 Remove PBS Completely from Cells Prior to Cell Lysis

Cells derived from cell culture have to be washed efficiently with an iso-osmotic solution, such as phosphate-buffered saline (PBS), prior to cell lysis to obtain antigens for high-resolution two-dimensional electrophoresis. Sometimes, the PBS solution is not removed completely from the cells. Consequently, the salt contamination of the sample causes many and long horizontal streaks in the 2-DE electrophoresis pattern.

To correct this, the PBS must be removed from the cells with a pipette completely after the final wash. Care must be taken to see that not even a tiny droplet of PBS is left behind. To ensure this, it would be good to have one to three additional washes with 250 mM sucrose and 10 mM Tris–HCl (pH 7.0) [1].

References

1. Westermeier R (2007) Frequently made mistakes in electrophoresis. Practical proteomics 1/2007. Wiley-VCH Verlag GmbH & Co. KGaA, Weinheim http://www.practical.proteomics-journal.com
2. Burgess RR (2009) Important but little known (or forgotten) artifacts in protein biochemistry. Methods Enzymol 463:813–820 Review
3. Weidner M-F, Grenier D, Mayrand D (1996) Proteolytic artifacts in SDS-PAGE analysis of selected periodontal pathogens. Oral Microbiol Immunol 11:103–108
4. Pringle JR (1975) Methods for avoiding proteolytic artefacts in studies of enzymes and other proteins from yeasts. Methods Cell Biol 12:149–184
5. Volkin DB, Mach H, Middaugh CR (1995) Degradative covalent reactions important to protein stability. In: Shirley BA (ed) Protein stability and folding: theory and practice. Methods Mol. Biol, vol 40. Humana Press, New York, pp 35–63
6. Deutsch DG (1976) Effect of prolonged 100 _C heat treatment in SDS upon peptide bond cleavage. Anal Biochem 71:300–303
7. McDonald G, Hudson A, Dunn S et al (2008) Bioactive contaminants leach from disposable laboratory plasticware. Science 322:917
8. Stark GR (1965) Reactions of cyanate with functional groups of proteins: reaction with amino and carboxyl groups. Biochemistry 4:1030–1036
9. Lin M-F, Williams C, Murray MV et al (2004) Ion chromatographic quantification of cyanate in urea solutions: estimation of the efficiency of cyanate scavengers for use in recombinant protein manufacturing. J Chromatogr B 803:353–362
10. Grabski A, Burgess RR (2001) Preparation of protein samples for SDS-polyacrylamide gel electrophoresis: procedures and tips. Innovations 13:10–12
11. Hames BD (1990) In: Hames BD, Rickwood D (eds) Gel electrophoresis of proteins. Oxford University Press, New York, pp 1–147
12. Franklin SG, Zweidler A (1977) Non-allelic variants of histones 2a, 2b and 3 in mammals. Nature 266:273–275
13. Siu CH, Lerner RA, Loomis WF (1977) Rapid accumulation and disappearance of plasma membrane proteins during development of wild-type and mutant strains of Dictyostelium discoideum. J Mol Biol 116:469–488
14. Marshak DR, Kadonaga JT, Burgess RR et al (1996) Strategies for protein purification and characterization. Cold Spring Harbor Laboratory Press, New York
15. Haeberle JR (1997) High-temperature sodium dodecyl sulfate polyacrylamide gel electrophoresis. BioTechniques 23:638–640
16. Kurien BT, Danda D, Bachmann MP, Scofield RH (2015) SDS-PAGE to immunoblot in one hour. Methods Mol Biol 1312:449–454
17. Läemmli UK (1970) Cleavage of structural proteins during the assembly of the head of bacteriophage T4. Nature 227:680–685
18. Rabilloud T (1998) Use of thiourea to increase the solubility of membrane proteins in two-dimensional electrophoresis. Electrophoresis 19:758–760

INDEX

Biji T. Kurien and R. Hal Scofield (eds.), *Electrophoretic Separation of Proteins: Methods and Protocols*, Methods in Molecular Biology, vol. 1855, https://doi.org/10.1007/978-1-4939-8793-1,

R

S

T

U

V

Z

CPI Antony Rowe
Eastbourne, UK
March 19, 2019